MECHANICAL VIBRATIONS

THEORY AND PRACTICE

MECHANICAL VIBRATIONS

THEORY AND PRACTICE

SHRIKANT BHAVE

Formerly Director (Academic), Sardar Vallabhbhai Patel Institute of Technology (SVIT, Vasad, Gujarat)

Pearson

ISBN 978-81-317-3248-9

First Impression, 2010

Published by Pearson India Education Services Pvt. Ltd, CIN: U72200TN2005PTC057128

Head Office: 1st Floor, Berger Tower, Plot No. C-001A/2, Sector 16B, Noida – 201 301,Uttar Pradesh, India.

Registered Office: 7th Floor, SDB2, ODC 7, 8 & 9, Survey No.01 ELCOT IT/ ITES SEZ, Sholinganallur, Chennai – 600119, Tamil Nadu, India. Website: in.pearson.com, Email: companysecretary.india@pearson.com

Digitally printed in India by Trinity Academy for Corporate Training Ltd, New Delhi in the year of 2023.

DEDICATION

I dedicate this book to
my beloved mother Sarla Bhave
without whom nothing
would be much worth doing

CONTENTS

Chapter 3 Two degrees-of-freedom systems 76

Chapter 4 Multi degrees-of-freedom systems 114

Chapter 5 Torsional vibrations 183

Chapter 10 Miscellaneous topics in vibration analysis and introduction to noise analysis 323

FOREWORD

The subject of theory and practice of vibration analysis has become one of the most important subjects that have to be very clearly understood by students graduating from universities. This is because when they enter the professional field they are required to resolve the vibration and vibration-related failure problems encountered in the field. Additionally, understanding the subject of vibration analysis assumes greater importance while designing the equipments and the structures such that these do not fail while operating in the field. The practicing engineers are also required to have a fundamental knowledge of the subject of analysis and diagnosis of practical vibration problems. It is very heartening to see that this book on theory and practice of vibration analysis caters to the needs of both students as well as professional engineers in a very lucid manner.

Traditionally, in earlier days, some fundamental aspects of vibration analysis used to be covered in the subject of theory of machines for electrical/mechanical engineering students and in the subject of theory of structures for civil engineering students. However, this scenario has changed significantly and the students are required to learn this subject in much more detail. Consequently, most of the universities in India as well as abroad have introduced the subject of vibration in their course curriculum. Further, the practicing engineers also require the concepts of vibration analysis in solving practical vibration and vibration-related failure problems.

Most of the available books in the market deal with the subject of theory and the practice of vibration analysis do not fully meet the requirements of the industry. Many of these deal extensively with mathematical theory of vibration analysis with practically no or very little application to practical problems; while some of them mainly serve the purpose of only solving numerical problems. Thus, the main purpose of introducing the subject of vibration analysis at graduate level gets lost. From this point of view, the present book on the theory and practice of vibration analysis authored by Dr Shrikant Bhave is extremely relevant and useful to graduating students as well as practicing engineers.

Dr Bhave is an internationally well-known expert on the subject of vibration analysis and has nearly four decades of experience in teaching. During his tenure at Bharat Heavy Electricals (BHEL, 1976–96), he has solved a large number of extremely complex problems on vibration and vibration-related failures encountered on large power plants and industrial equipments in India as well as abroad. Also, during his tenure at BHEL he guided several stress and vibration analysis-related MTech./PhD theses at IIT/IISC. He also trained graduate engineer trainees of BHEL on this subject. His book on theory and practice of vibration analysis reflects his vast experience in the field while teaching at H.B.T.I. (Kanpur 1966–76), working at BHEL (1976–96), L&T (1996–2002), and as Director—Academic at SVIT (Vasad 2002–08). He has also been working as a freelance consultant since 2003.

The book presents the basic theory of vibration analysis in an extremely lucid manner ensuring that an overdose of mathematics is not given to readers. Several practical applications of the mathematical theory of vibrations have been discussed which will make a pleasant reading and the basic concepts of vibration analysis will be understood better. Several solved problems of vibration analysis pertaining to practical situations will enable readers to know the manner in which theory can be applied to practical problems.

Most of the books on the subject of theory and practice of vibration analysis do not deal with diagnosis of vibration and vibration-related failure problems. This book is a very welcome exception in this regard. The chapter on diagnosis makes this book a very valuable reference book for the practicing engineers as it deals with several case studies.

The book also deals with fundamental concepts behind finite element analysis. This is very important since it is necessary to know the concepts before using commercial FEM software.

Finding natural frequencies and the mode shapes (modal analysis) is one of the most important activity required to be performed by vibration engineers. The book deals with theoretical aspects of modal analysis in a way which students can easily understand. However, in practice modal analysis is required to be carried out at work sites. The book presents a dedicated chapter on the fundamental concepts of experimental modal analysis. This also is a unique feature of this book which, in general, is not found in most textbooks on theory and practice of vibration analysis.

The presentation is good and the language is simple which average students are bound to understand. I would like to recommend this book as a textbook for graduate students as well as a reference book for practicing engineers.

Dr. A. L. Chandraker
Executive Director (Retd)
Corporate Research and Development
Bharat Heavy Electricals,
Hyderabad, India.

PREFACE

This book serves as an introduction to the subject of vibration analysis at the undergraduate level and also to a certain extent at the postgraduate level. Additionally, it is highly useful to professionals dealing with vibration analysis for troubleshooting of vibration problems occurring in the field. The present day industry as well as power sector/oil and gas sectors, etc. do expect that the engineers they employ should have a certain level of practical knowledge of vibration analysis. Therefore, it is strongly felt that even at the undergraduate level students should get some exposure to the problems of practical vibration before they face the practical scenario.

Considering these points, this book deals with the subject of theory and practice of vibration analysis according to the syllabi of most of the engineering institutions. Additionally, it will also provide inputs of basic vibration analysis required for designing as well as solving the practical vibration/vibration-related failure problems. As far as possible, the language is simple and, more importantly, it pertains to cases occurring in the real world.

FEATURES

The book begins with a single degree-of-freedom vibration system with detailed explanations of various sub-systems such as mass, spring and damper and on how we evaluate these for practical problems. The fundamental understanding of resonance and near-resonance conditions have been presented in a concise manner. We also explain how complex engineering structures can be simulated as single degree-of-freedom system as the first level approximation.

This is followed by a gradual introduction of the concepts of two degrees and multi degrees-of-freedom systems. Several solved examples aiming at clarifying the concepts of mathematical analysis are presented in the book. The emphasis is laid on how complex engineering systems can be expressed as lumped multi-mass, springs and damper systems.

Special emphasis has been given to the analysis of bending and torsional vibrations since a large number of engineering structures and machines consist of beams and shafts. The important concept of whirling of rotors has been very well presented. One of the most popular methods of resolving shaft vibration problems, single- and two-plane balancing, has also been discussed in detail.

The book also explains the basic methodology of determining the natural frequencies and the mode shapes using experimental modal analysis techniques. The fundamental concepts in finite element method for vibration analysis have also been discussed. Important features of MATLAB and the use of simple commands for solving polynomials, system of equations, matrix operations such as inversion and for determining the Eigen values and Eigen vectors have been discussed. Instead of showing solved numericals using MATLAB or C++ and FORTRAN programs which users generally do not view with great interest, some of the important features of this computational MATLAB tool is explained through a few solved examples.

One of the most important features of this book is the detailed presentation on vibration instrumentation and methods of diagnosis of practical vibration problems. Several case studies have been presented to explain the methodology of diagnosing the vibration problems.

This book is organized into nine chapters.

In **Chapter 1,** the overall scope of the book and the essential reason to study the vibration analysis as a separate subject instead of studying it as a portion of theory of machines has been discussed. The practical aspects of vibration analysis have also been discussed.

In **Chapter 2,** we discuss in detail the single degree-of-freedom vibration system. Initially, we describe the methodology of arriving at equation of motion for rigid body oscillations followed by a discussion on individual elements of single degree-of-freedom system, namely mass, spring and damper. We also discuss the methodology of arriving at the parameters of mass, spring and damper for practical cases. In this chapter, we also discuss various damping mechanisms such as viscous damping, Coulomb damping and hysteretic damping. We discuss the methodology of arriving at system equation for free undamped and free damped vibration. We define the important types of damping such as over-damping, critical damping and under-damping. We show that only the under-damped system is capable of vibrations. We also deal with logarithmic decrement as a powerful tool for evaluating the extent of damping in the single degree-of-freedom system. We then proceed to the case of forced vibrations of undamped as well as damped systems. We also explain resonance condition for both undamped and damped systems. One of the important aspects of rate of increase in time-dependent vibration levels in forced undamped vibrations at resonance condition is discussed. We also discuss the charts showing variation of vibration levels and phase angles at various frequency ratios of the disturbing force and the natural frequency of the system. One important diagnostic rule for identifying the natural frequency/critical speed of the rotor system is by noting the speed at which there is a phase angle shift of 90° for lightly damped system has been brought out in a very scientific manner.

A brief discussion on assessing the response of the system with Coulomb damping and hysteretic damping has been presented. We also discuss the issues related to transmissibility of vibrations displacement and force transmissibility.

Chapter 3 deals with two degrees-of-freedom system. The concept of generalized coordinates has been explained. We deal with free vibrations of undamped as well as damped systems. In this chapter, we explain how natural frequencies and mode shapes are determined using equations of motion. We also deal with the use of Lagrange's equation to find the natural frequency. Principle behind orthogonality of modes has been discussed in this chapter. We also deal with semi-definite systems. The forced vibration response of two degrees-of-freedom systems has also been discussed.

Chapter 4 deals with multi degrees-of-freedom systems. After introducing the scope of this chapter, we discuss about modelling of continuous system as a lumped system of masses, springs and dampers. We explain the method of writing the governing equation of motion for multi degrees-of-freedom system using Newton's second law of motion, using influence coefficients pertaining to stiffness influence coefficients, flexibility influence coefficients and inertia influence coefficients. We also discuss issues pertaining to generalized coordinates which are extensively used in the analysis of multi degrees-of-freedom systems. We then discuss the important aspects of energies in vibration systems which form the basis for the use of Lagrange's equation for finding the natural frequencies. The use of Lagrange's equation for determining the natural frequencies of complex system has been demonstrated through several solved examples.

This is followed by discussion on Eigen value problem, principle behind the orthogonality of modes (which has also been discussed to a certain extent in the previous chapter).

These form the basis of an important subject of modal analysis which a vibration engineer is expected to know well. This is followed by the discussion on various methods of finding natural frequencies and the mode shapes. These methods include method of matrices and matrix iterations, Holzer's method, Stodola's method, Dunkerely's method, Raleigh's method and mechanical impedance method.

Chapter 5 deals with torsional vibrations. We explain that unlike linear and transverse vibrations, torsional vibrations cannot be measured easily. This is a serious issue since all the rotating machines are subject to torsional oscillations due to alternating torques apart from transverse vibrations. Several failures have been reported in literature on the failures of rotating machinery such as electric motors, fans, compressors, turbines, reciprocating compressor, etc. due to torsional vibration. Thus, one has to depend on the accuracy of theoretical analysis of torsional vibrations. With this background, we start with single degree-of-freedom system followed by two degrees-of-freedom systems. We also explain how gear wheels can be treated in torsional analysis. The analysis is further extended to multi degrees-of-freedom systems and explains how a complex system comprising of multicylinder turbine coupled to a generator and exciter can be treated as multi degrees-of-freedom system comprising several shafts and discs. The Holzer's method which has been described in the previous chapter has been used for finding torsional natural frequencies and the mode shapes. We have also discussed forced torsional vibrations in this chapter.

Chapter 6 deals with transverse vibrations of beams and rotors supported on bearings. We derive the fundamental equation for the analysis of beam vibrations followed by the analysis of free vibrations for various end conditions. We also discuss the concept of normal functions and their property of orthogonality. We then proceed to discuss the forced vibration of beams for various support conditions. We explain through several solved numerical problems the concept behind hammer test/bang test for determining the natural frequencies of any complex system—an important step in experimental modal analysis. In the cases discussed earlier, the solutions were in the form of infinite series with which engineers do not feel comfortable. We describe various methods to analyse the vibration of the beam due to distributed load, discrete loads, etc. and show that these methods provide solutions of acceptable accuracy. The important methods discussed are Raleigh's method and Raleigh–Ritz method. A detailed discussion on whirling of rotating shafts and critical speeds has been presented. We evaluate the response of whirling rotors to forces such as unbalances in the rotor system. We then proceed to clarify the concepts behind the balancing of rotors. We consider both single-plane and two-plane balancing of rotors. We also describe different types on unbalances such as static, quasi-static and dynamic unbalance that can be present in the rotor. Quite a good number of machines operating in fields such as car engines, large high pressure compressors, etc. are reciprocating machines. The requirement of balancing the primary unbalance force/primary unbalance couple and secondary force/secondary couple in the machines is explained. The method of balancing the machine by appropriate crank arrangement has been discussed.

Chapter 7 deals with vibration diagnosis and control. We describe the nature of the problem experienced by the vibration engineer in resolving the practical vibration problems. In practical situations, many times it is required to collect and analyse the vibration data obtained from the machine and then find the reasons for the same. Having found the reasons, the next step is to eliminate them or at least reduce their intensity. We discuss various perturbation forces which cause vibrations in the rotating machines.

In this chapter, we discuss the basic concepts behind the vibration spectrum analysis. We then turn our attention towards sensing and measurements. We explain the relative merits of making vibration measurements in displacement, velocity and acceleration modes. We also show that making vibration acceleration measurements in high frequency domain gives sufficient clues about the impending failures. We also describe how a tunable filter added to vibration metre serves the purpose of a versatile vibration analyser. We also discuss the characteristics of various filters used in the vibration analyser. We then discuss the basic philosophy used in the design of vibration pickups. We describe various versions of accelerometer pickups and also how they should be mounted on the vibrating body to obtain reliable data. We then proceed to describe proximity probes meant for the measurement of shaft relative vibrations. We show that with two mutually perpendicularly mounted shaft probes, it is possible to obtain information of journal centre locus and shaft vibration orbits for various operating conditions of the machine. This information is highly useful in diagnosis of complex vibration problems. We also discuss in brief the vibration norms and standards.

We then proceed to various techniques used in the vibration analysis. We describe the importance of making measurement of the phase of vibration. We then describe the basic principles behind the fast Fourier transform (FFT) analysis and describe how they work. We also briefly discuss the principles behind remote sensing of vibrations which is used in analysing complex blade failure problems (which are rare) in turbo machinery.

A detailed discussion on the subject of vibration data reduction has been presented. We discuss conventional spectrum analysis, water fall diagram, Bode plots for identifying the critical speeds of the assembled rotor-bearing system, shaft orbit measurements, analysis of time waveform of vibration mode shape analysis, vibration scan, etc. Several case studies carried out by the author have been presented to explain the various techniques of the described data reduction.

Towards the end of the chapter, we discuss various diagnostic rules to identify malfunctions such as unbalance, misalignment, bent rotor, defects in the flow path of turbo machinery, mechanical rubs, etc. These have been supported with some case studies by the author during his professional practice on a wide range of turbo machinery.

Chapter 8 deals with the fundamental principles of finite element method for vibration analysis. This is required as many commercial grade vibration analysis softwares are available but unless the fundamental principles behind FEM are clear, their utility for solving the practical vibration problems could be quite questionable in spite of the fact the commercial softwares do provide the necessary user manuals.

The FEM involves complex computations. These computations can be easily done using MATLAB software. A brief description of MATLAB is given along with the type of commands needed for solution of polynomials, solutions of system of equations, various matrix operations including matrix inversion and solution of Eigen value problems along with typical solved problems.

Chapter 9 deals with the fundamentals of experimental modal analysis techniques for determining the natural frequencies and the associated mode shapes of the vibrating body/structure.

Vibration engineers working in the field are required to carry out diagnosis and solution of vibration problems and the associated failures. It is from this point of view that we have included chapters on diagnosis, FEM and also on modal analysis. This would greatly benefit the graduating engineers as well as engineers working in the area of vibration analysis which also happens to be one of the most important techniques in condition monitoring of the equipments working at sites.

SHRIKANT BHAVE

1 Fundamentals of Vibration Analysis

1.1 INTRODUCTION

Any motion that repeats itself after an interval of time is called vibration or oscillation. The common day-to-day examples of vibrations are swinging pendulum of the clock and motion of plucked string of a musical instrument. The theory of vibrations deals with the study of oscillatory motions of bodies/components of body and the forces associated with them. The vibrations experienced by the components/structures usually result in their stretching/unstretching or twisting/untwisting or both. As a result, they experience alternating stresses and thereby stress-induced fatigue. Depending on the magnitude and the number of cycles of alternating stresses, the component may fail earlier or later.

There are several reasons for studying the subject of theory and practice of vibration analysis. In large number of cases, the vibrations generated by the machinery/structures cause physical discomfort and in many cases, long-term exposure of components of these machinery/structures to vibrations causes mechanical failures. For example, the vibrations caused by the blades of steam turbines, gas turbines etc., which are typically high-speed machines, may lead to their failure. The consequential damage to the turbine could be catastrophic and extremely difficult and expensive to repair. We had investigated the catastrophic failure of a 236 MW nuclear steam turbine power plant. In this plant, the failure of one blade in the low-pressure turbine due to its vibration caused total destruction of the steam turbine generator unit. The civil structures under certain conditions such as earthquake may fail because of vibrations during the earthquake. These failures are colossal in a sense that they not only cause damage to the structure but may also take a heavy toll on human life. The recent failure of a mega hydro power project comprising 10 units of 600 MW each at Sayano-Shushenskya in Russia resulted in the death of 76 people. The estimated cost of the damage is of the order of $310 million and the loss of production of aluminum is of the order of 500,000 tons because of non-availability of electrical power. The failure of blade(s) in the gas turbine engine of the aircraft, caused by blade vibration, not only damages the engine but also leads to the loss of the aircraft along with those on board.

On the other hand, vibrations are used effectively and are a great advantage in vibratory screens, washing machines etc. Music is a by-product of vibrations alone. The basic life-force in all living beings is vibratory in nature. An equipment called Vibratory Stress Relief (VSR) makes use of vibrations to release the residual stresses in castings, forgings, and welded structures. Another example of beneficial effects of oscillations is the technique of Enhanced External Counter Pulsations (EECP) given to the arteries of the legs to remove heart blockages. Pulsed electromagnetic-wave therapy, which has now become popular in regeneration and repair of tissues of human knees also derives its benefits from vibrations of tissue at the cellular level. However, such beneficiary applications of vibrations are very few. Basically, engineers would like to know why a structure/machine vibrates and also the methods to eliminate these vibrations or at least reduce their intensity so that vibration-related failures do not occur during the designated life-time of the structure/machine.

The present book deals with vibration analysis, diagnostics, and control considering the potential damage associated with vibrations. The analysis is not restricted to mathematical analysis only but also takes into account the experimental analysis. This is because of certain limitations of both analytical as well as experimental methods in the analysis of complex vibration problems.

1.1.1 Definition of Vibration of Different Forms

If one observes the oscillatory motion of a pendulum, it is very easy to note that at the extreme of swings, where the pendulum momentarily stops and starts reverse motion, the velocity is zero. At this point, the kinetic energy is zero but the potential energy is largest. Thus, there is a transfer of the pendulum's potential energy to kinetic energy and kinetic energy to potential energy alternately. Similar is the case with the vibratory system consisting of a spring and a mass. The spring stores the potential energy, while the mass (or inertia) stores kinetic energy. The spring could be an elastic element such as beam, rotor, etc. If we observe the vibration of spring–mass system by disturbing it from its stable position, the vibrations over a certain period of time will diminish. Thus, the system has a damper by means of which the energy is gradually lost. The damper can be a dashpot, friction device, surrounding medium or even the vibrating component that undergoes alternate deformations. It is thus obvious that in order to sustain the vibration, a disturbing force must be applied. When such sustained vibrations are ensured by a disturbing or perturbation force, the vibrations are called forced vibrations.

Thus, we have two types of vibrations, namely, forced vibrations and free vibrations. In case the system has significant damping, the vibration can be called damped forced vibration or damped-free vibration. In case damping is zero (which never happens in reality) or is very negligible, the vibration can be considered as undamped forced vibration or undamped free vibration depending upon whether or not there is a continuous disturbing force.

Apart from this, the vibratory system may be such that the vibratory motion is rigid-body oscillation, translational vibration, bending vibration or torsional vibration or their combinations as shown in Fig. 1.1.

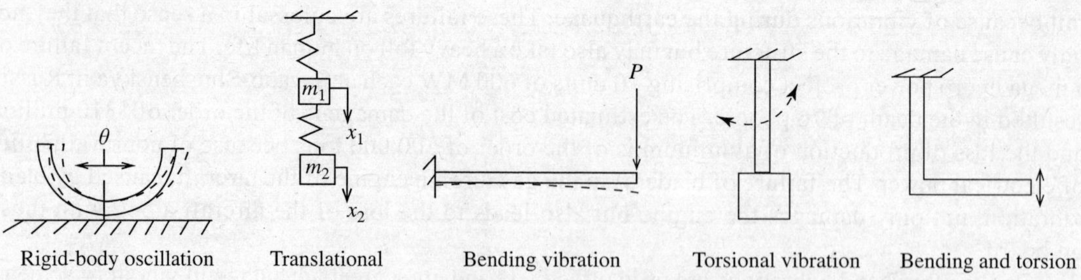

Figure 1.1 Various Types of Vibrations

In this book, all the different types of vibrations will be dealt within the various forthcoming chapters.

As mentioned earlier, the simplest kind of vibration system may consist of a single spring, a damper, and the mass or it may be just a swinging pendulum. As we shall explain later, such a system has only one parameter to be evaluated, i.e. the time-dependent (single) displacement of the mass from its position of equilibrium. Such systems are called single degree-of-freedom systems (SDOF). The analyses of such systems are relatively simple. However, in real-life situations, we deal with very complex systems having multiple degrees-of-freedom. For example, we may have two, three or many more degrees-of-freedom. Figure 1.2 below shows typical two degrees-of-freedom systems.

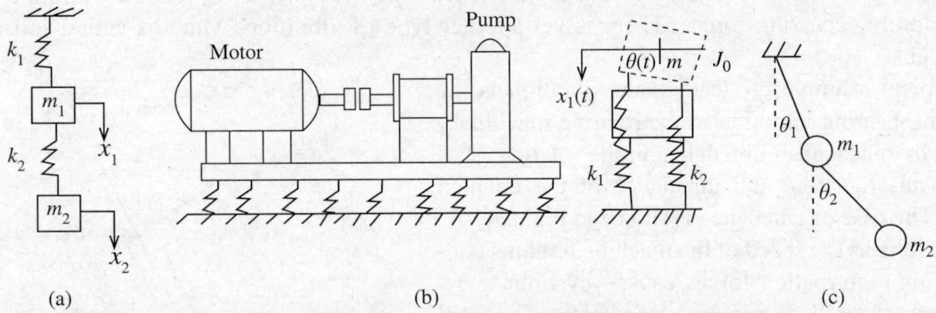

Figure 1.2 Two-Degrees-of-Freedom Systems

Figure 1.2(b) shows a motor-pump assembly supported on spring system. This can be idealized as a bar of mass m and mass moment of inertia J_0 supported on two springs k_1 and k_2. The displacement of the system at any time can be specified by a linear coordinate $x(t)$, indicating the vertical displacement of the centre of gravity (CG) of the mass and an angular coordinate $\theta(t)$, denoting the rotation of the mass m, about its CG. Instead of $x(t)$ and $\theta(t)$, we can also use $x_1(t)$ and $x_2(t)$ as independent coordinates to specify the motion of the system. Thus, the system has two degrees-of-freedom. It is very important to note that in this case, the mass m is not treated as a point mass as shown in Fig. 1.2(a), but as a rigid-body having two possible types of motion. Yet another example of two degrees-of-freedom system is shown in Fig. 1.3.

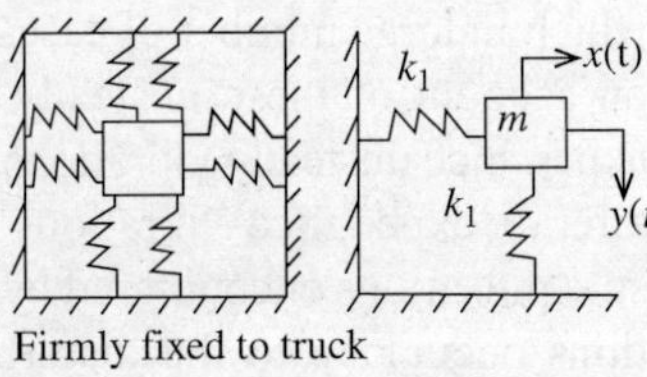

Figure 1.3 Two Degrees-of-Freedom System

Figure 1.3 shows the method of packaging a consignment of mass m, usually done while transporting or shipping. The mass m is confined to motions in x and y directions. The system can then be modelled as a mass supported by springs along the x and y directions. This is the case where we have one mass (point mass) and two degrees-of-freedom because of motions in the x, y directions. In fact, the strict definition of number of degrees-of-freedom is

$$n = \text{Number of masses} \times \text{Number of possible types of motion of each mass}$$

Similar examples can be given for three or more degrees-of-freedom systems. However, when we consider vibrations of continuous structures or components, it is quite difficult to model them in terms of an assemblage of springs, masses, and dampers. Consider for example, a blade mounted on the wheel of steam turbine/gas turbine rotor. This blade behaving as a beam can have bending as well as twisting (torsional) vibrations. This blade cannot be idealized as a simple single degree-of-freedom system since the displacement on each point on the blade will be unique and as such there are an infinite number of points on the blade whose displacements during vibration are required to be evaluated.

Components like blades or shafts are regarded as infinite degrees-of-freedom systems. However, their analysis becomes highly complex. Fortunately, there are several methods of idealizing such components/structures experiencing vibration into some finite degrees-of-freedom systems, thereby making their analysis relatively simpler. We shall discuss these in the forthcoming chapters.

1.1.2 Classifications of Vibrations Experienced in Actual Practice

The vibrations in a body are caused by the application of forces. These forces are called the perturbation forces. Depending on the nature of perturbation/disturbing force, the vibrations can be classified as (1) steady-state vibrations when the perturbation forces do not change with time, (2) unsteady-state vibrations when the forces causing vibrations change with time or (3) random vibration caused by randomly occurring forces. There is yet another type of vibration, which is called self-excited vibration.

Typical example of steady-state vibration is the shaft-and-bearing vibration in the rotating machinery caused by mechanical imbalance in the rotating parts or say misalignment and unbalance of the coupled rotors. The case of unbalance is shown in Fig. 1.4.

As long as the speed of the machine remains constant, this perturbation forces caused by imbalance/misalignment will remain in a steady-state, i.e. will not change with time. On the other hand, the blades of a turbomachinery experiences pressure pulsations, which in large number of cases change with time, and on account of these unsteady forces the blades may experience unsteady vibrations. Similarly the wings of aircraft experience vibrations due to unsteady pressure pulsations encountered by them. When the vibrations occur due to the disturbing force, which only randomly appears, they are termed as random vibrations, for example, earthquake-induced vibrations. A galloping transmission line is an example of a self-excited vibration.

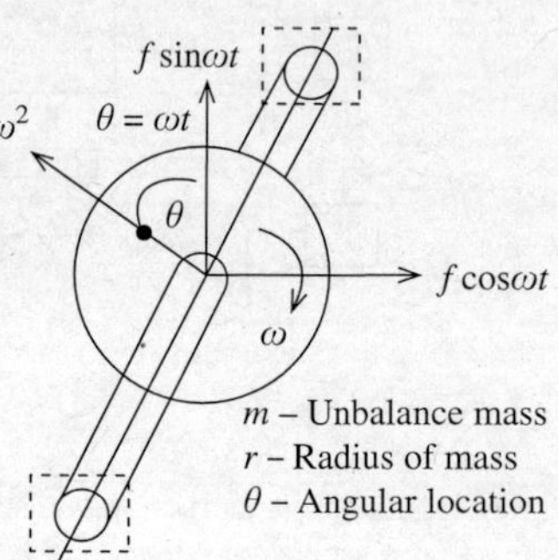

Unbalance mass produces alternating forces in horizontal and axial direction.

Figure 1.4 Unbalance of a Rotor

The nature of response of a vibratory system greatly depends upon the nature of perturbation/excitation forces and also the structural characteristics of the vibrating body. A study of the vibration response is of great value in understanding the perturbation forces as well as the structural characteristics. The waveform of the vibration response contains certain vital information. We shall now discuss these in the following section.

1.1.3 Parameters of Vibration Waveform

The simplest form of the vibration experienced by a body/component or a structure is a waveform as shown in Fig. 1.5.

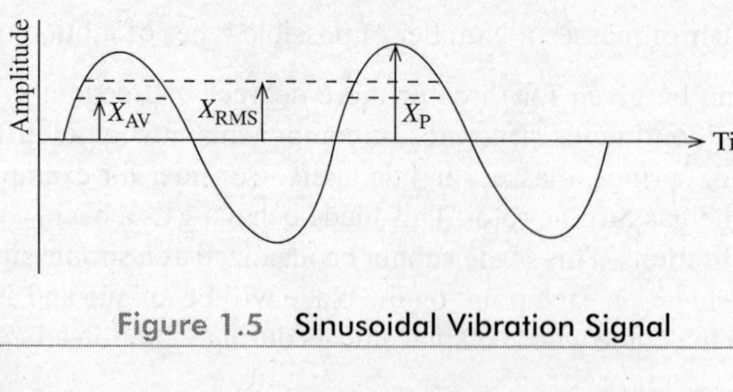

Figure 1.5 Sinusoidal Vibration Signal

This motion is an oscillatory motion of a body/component about a reference position such that the motion repeats exactly after a certain time. The simplest of this form of vibration is the harmonic motion as shown in Fig. 1.5

$$X = X_P \sin \omega t$$
$$\text{or} \quad \dot{X} = dx/dt = X_P\, \omega \cos \omega t \tag{1.1}$$
$$\text{or} \quad \ddot{X} = d^2x/dt^2 = -\omega^2 X_P \sin \omega t$$

where X corresponds to displacement, $\dot{X}$ to velocity, and $\ddot{X}$ to acceleration of vibration, respectively. It is apparent that the form and period of vibration remain the same regardless of whether it is specified in the displacement, velocity or acceleration mode. We shall revert to this issue in much more details in the chapter *Vibration Diagnosis and Control* (Chapter 7). The various parameters of the waveform are shown in Fig. 1.5.

$$f = \text{Frequency} = 1/T \tag{1.2}$$
$$\omega = \text{Circular frequency} = 2\Pi f \tag{1.3}$$
$$X = X_P \sin \omega t \tag{1.4}$$
$$X_{AV} = 1/T \int_0^T |Xdt| \tag{1.5}$$
$$X_{RMS} = \sqrt{1/T \int_0^T x^2\, dt} \tag{1.6}$$

$$\frac{X_{RMS}}{X_{AV}} = \text{Form factor } f_f \approx 1.11 \ (\approx 1 \text{ dB for Fig. 1.5})$$

$$\frac{X_P}{X_{RMS}} = \text{Crest factor } f_c = 1.414 \ (\approx 3 \text{ dB for Fig. 1.5})$$

Thus, parameters frequency (f or ω), X_P (peak amplitude), X_{AV}, X_{RMS}, form factor f_f, crest factor f_c specify the vibrations (sinusoidal in Fig. 1.5) completely.

Most of the vibrations experienced in daily life are not as simple as shown in Fig. 1.5, even though, many of them may be periodic. Typical non-harmonic periodic motions are shown in Fig. 1.6.

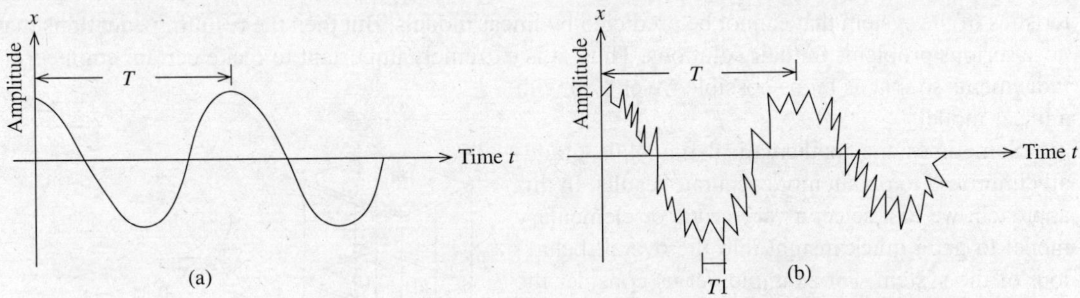

Figure 1.6 Non-harmonic Periodic Motions

For the periodic motions shown in Fig. 1.6, we can obtain the information regarding peak amplitude, average and absolute RMS, form factor, and crest factor. As will be shown later in Chapter 7, *Vibration Diagnosis and Control*, this information is only academic interest and it does throw enough light upon the causes of vibration and hence is of very little use. However, it serves the purpose of some details in vibration analysis as will be evident in the forthcoming chapters.

1.1.4 Why Do We Need Mathematical Modelling?

The vibration response waveforms can be analyzed in several ways, which shall be discussed later in this book. This analysis helps in identifying the perturbation forces that cause the vibrations. In some cases, these perturbation forces can be eliminated or at least reduced to a certain extent. However, in many cases it becomes extremely necessary to change the structural characteristics. For this purpose, we need to have a tool, which enables us to carry out a parametric study that may lead to some design modifications. Such a tool can be devised using the mathematical modelling of vibrating system.

1.2 PRACTICAL ASPECTS OF VIBRATION ANALYSIS

A vibratory system is a dynamic system for which inputs in the form of excitation forces and the outputs in the form of responses (vibration displacements, velocity or acceleration of vibrations) are time-dependent. The vibration response, in addition to the nature and magnitude of excitation force/perturbation force also depends upon the way the perturbation forces act. In other words, it means that the response is also dependent upon the initial conditions of the vibratory system. Because of this, practical vibrating systems are extremely complex from the point of view of their analysis. The mathematical analysis of practical vibration systems, therefore, requires certain simplifications which result in arriving at a reasonably well-solvable system of equations that relate vibration response to structural characteristics (mass, stiffness, and damping) and perturbation forces. The vibration analysis of a system usually involves mathematical modelling, derivation of the governing equations, solution of the equations, and finally the interpretation. We shall now briefly discuss the step-wise procedure that can be adopted for such analysis.

1.2.1 Step 1: Mathematical Modelling

The main intention of mathematical modelling is to represent all the important features of a vibrating system in such a way that we are able to derive the mathematical equations governing the behaviour of the system. The mathematical model should include enough details in order to describe the system in terms of equations without making them too complex. The model can be linear or non-linear, depending upon the behaviour of the components of a system. Linear models permit quick solutions and are very simple to handle. The non-linear models are sometimes required since they reveal certain characteristics of the system that cannot be predicted by linear models. But then the resulting equations may pose serious problems in their solutions. Thus, it is extremely important to make certain engineering judgements so that as far as possible we end up with a linear model.

Sometimes the mathematical model is gradually improved to obtain more accurate results. In this approach, we first select a very crude or elementary model to get a quick insight into the overall behaviour of the system. For example, let us consider the vibration analysis of a three-storey building (Fig. 1.7) when subjected to seismic excitation.

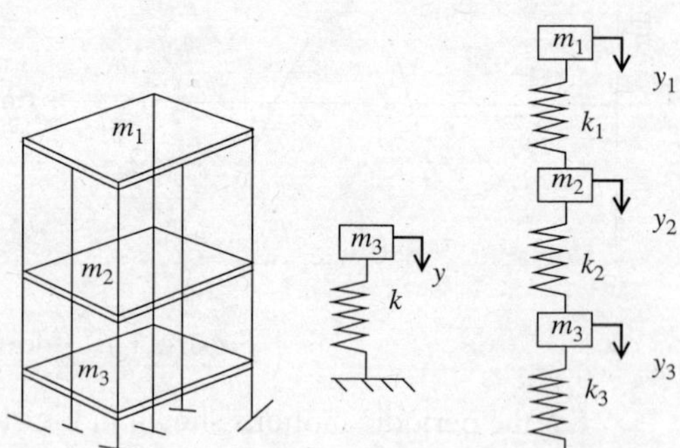

Figure 1.7 Modelling of a Three-Storey Building

As a first-level approximation, let us assume that this system can be represented as a single degree-of-freedom system consisting of mass m $(= m_1 + m_2 + m_3$ = sum of masses of all floors) supported on a single spring of stiffness k $(= k_1 + k_2 + k_3 =$ algebraic sum of stifnesses of all columns which are assumed to behave as simple springs). We can then put ground acceleration (due to seismic activity) as the disturbing force and evaluate the response. In the next step (approximation), we shall consider the building as a three-mass and a three-spring system (three degrees-of-freedom system) and evaluate the response. In this model we considered the floors only as masses. In the next improved model, we can consider the elastic behaviour of the floors as the slabs. In this fashion we can gradually enlarge the scope of modelling. We can then consider the effect of properties of the soil and foundation system. If the first-level modelling shows that it is necessary to change the design of the columns, we incorporate that change first and then proceed further to the next level of approximation.

1.2.2 Step 2: Derivation of Governing Equations

Once the mathematical model is available, we use the principles of dynamics to derive the equations of vibratory motion of a body subjected to a prescribed perturbation (excitation force). The equations of motion can be derived easily by drawing the free body diagrams for all the masses involved. The free body diagram of a mass can be obtained by isolating the mass and indicating all externally applied forces, reactive forces, and inertia forces. The equations of motion of a vibrating system are usually in the form of a set of ordinary differential equations for a discrete system (consisting of lumped masses, springs, and dampers) and partial differential equations for a continuous system. The equations may be linear or non-linear depending upon the behaviour of the components of a system. Several other approaches are available to derive the equations, which shall be discussed later in this book.

1.2.3 Step 3: Solution of Governing Equations

The equations of motion are solved to find the response of the vibrating system. Depending upon the nature of the problem, we can use one of the following techniques for finding the solution: 1) the standard method of solving the differential equations; 2) the Laplace transform method; and 3) the numerical method. In case the governing equations are non-linear, it is very difficult to obtain close-form solution. In case the governing equations are partial differential equations, the solution becomes extremely complex. In such a case, numerical methods involving computers are required to be used.

1.2.4 Step 4: Interpretation of Results

The solution of the governing equations gives the displacements, velocities, and accelerations of the various masses of the system. The results must be interpreted with a clear view of the purpose of the analysis determining whether it is for design validation or for solving a vibration problem being experienced.

The analytical methods have their own limitations. It is therefore necessary to make the practical vibration measurements under certain circumstances. The measurements involve not only the measurement of the response of the system to actual perturbation forces being experienced but also for determining the inherent structural properties, such as mass distribution, stiffness distribution (springs in discrete systems), and the damping. The measured data can then be used to make discrete system models, which can be used for parametric studies that can then become a powerful tool for design as well as for solving the existing vibration problems. It is for this reason the author has discussed experimental methods to a considerable extent in this book.

Solving a vibration problem or designing a equipment such that it will not fail prematurely requires inputs from both analytical as well as experimental methods.

CONCLUSION

We have discussed the point that vibrations of machines, their components, structures, etc. can cause discomfort not only to the operators or people living in the vicinity but can also leads to catastrophic failures. It is, therefore, extremely important to study the vibration characteristics of the components/structures and evaluate their response to various types of perturbation/disturbing forces known. A large part of the book is devoted to this aspect for which detailed mathematical analyses are required. However, we must know the nature of the perturbation forces whose determination is also an important issue in the theory and practice of vibration analysis. Sadly, this aspect of vibration analysis is not dealt in detail by majority of the textbooks on this subject.

As will be seen in forthcoming chapters, it may not be always be possible to eliminate the disturbing forces totally; however, in many cases, it is possible to minimize them as much as possible. In the extreme case, when the disturbing forces cannot be eliminated or reduced, it becomes necessary to minimize the response of the system, by appropriately changing the vibration characteristics of these components/structures.

Practical vibration problems are solved by the iterative procedures discussed here. In order to be able to do so, it is necessary to know the vibration characteristics of these components/structures. Hence, a significant portion of this book will deal with the aspect of response of a system to a given disturbing force. As mentioned earlier, this requires a great deal of mathematical analysis of vibration phenomenon. The methodology of identifying the perturbation forces and the diagnostics and methods of solving practical vibration problems are discussed in Chapter 7, *Vibration Diagnosis and Control*.

In this chapter, we have dealt with some fundamental concepts of vibrations. We also discussed about various vibrations and their classifications. We also emphasized the need for synergy between analytical methods and practical vibration measurement/analysis.

2

Single Degree-of-Freedom Vibration Systems

In Chapter 1, we discussed a few of the fundamentals of vibration theory. We discussed various types of vibrations and the important parameters of vibration waveforms. The practical aspects of analysis and the solution of vibrations were also discussed. We shall now consider the important aspects of vibration analysis. To begin with, we deal with a simplest kind of vibration systems. These systems are called a single degree-of-freedom vibration system.

2.1 DEFINITION OF DEGREES-OF-FREEDOM

As mentioned in the previous chapter, vibration is nothing but exchange of potential energy and kinetic energy alternately. The vibration system is excited by the disturbing force/perturbation force. The effect is to move the components of the system from its equilibrium position by the application of a force where system provides a restoring force. The system can also have an energy dissipation device such as a damper. The motion of the vibratory system can be one-dimensional, two-dimensional or three-dimensional. The degree-of-freedom is defined as the number of coordinates required to specify the vibration. For example, Fig. 2.1 shows various vibration systems whose degrees-of-freedom are decided by the number of coordinates/displacement parameters.

Figure 2.1(a) shows a system comprising a single mass, a spring, and a damper. The displacement (response) x specifying the vibration is a single parameter; hence it is termed as a single degree of vibration system. Figure 2.1(c) also is a single degree-of-freedom system since only one coordinate θ suffices to describe the vibratory motion. Figure 2.1(b) is a two degrees-of-freedom system since two displacement parameters x_1 and x_2 are required to describe the vibration behaviour. Similarly, since x_1, y_1 and x_2, y_2 are required to specify the vibration of system in Fig. 2.1(d); the system has four degrees-of-freedom. It is very easy to understand that the beam shown in Fig. 2.1(e) has infinite degrees-of-freedom since there are an infinite number of points on the dynamic deflection curve.

It is very important to note in all the cases shown in Fig. 2.1, barring the case shown in Fig. 2.1(c), that the spring (discrete springs or elastic members) provides the restoring force, while the damper provides the energy dissipation. Since the spring element provides the restoring force by way of stretching/unstretching, they get stressed (hence can fail depending upon the magnitude of the stress),

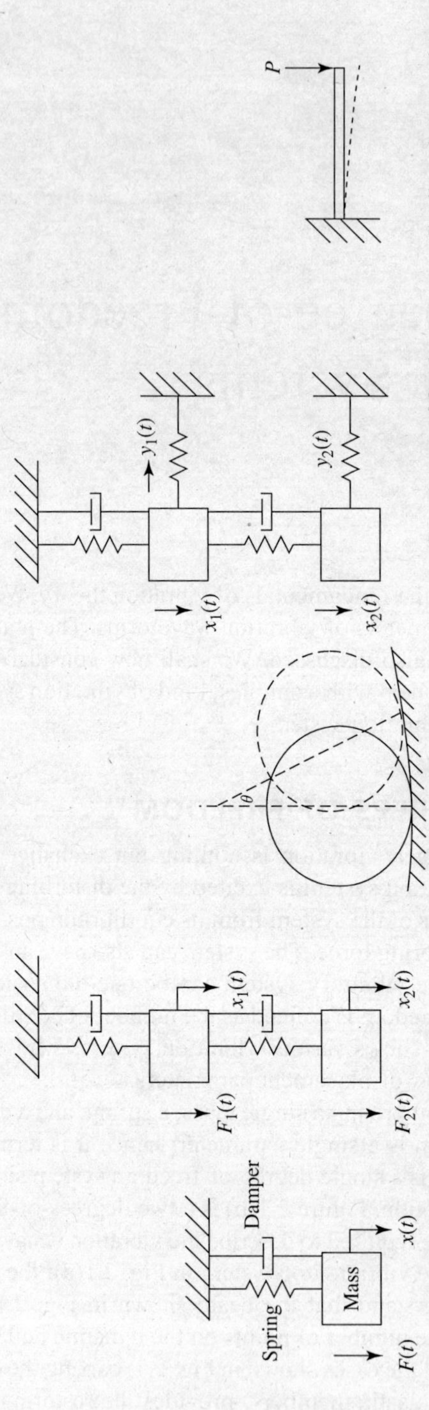

(a) Single degree-of-freedom translational vibration system Parameter x

(b) Two degrees-of-freedom translational vibration system Parameters $x_1(t)$, $x_2(t)$

(c) Cylinder rolling without slipping on a circular shape. Single degree-of-freedom system Parameter θ

(d) Four degrees-of-freedom translational vibration system. Parameters x_1, y_1 and x_2, y_2

(e) Cantilever beam continuous system. Bending vibrations. Infinite degrees-of-freedom

Figure 2.1 Degrees-of-Freedom

while the damper also provides vibrations opposing force which is merely an energy dissipation device.

The system shown in Fig. 2.1(c) belongs to the category of rigid-body oscillations. There are several cases of free oscillations of this type where the restoring force (when the body is disturbed from its equilibrium position) is not provided by physical spring or by other mechanisms. We shall now deal with this category of free vibrations.

2.2 SINGLE DEGREE-OF-FREEDOM SYSTEM

We first consider systems such as that shown in Fig. 2.1(c). We call these vibrations as rigid-body oscillations.

2.2.1 Rigid-Body Oscillations

Rigid-body oscillations are free vibrations of mechanical systems where there is no physical spring but there are other mechanisms that try to restore the body to its equilibrium position. For example, the shell shown in Fig. 2.2 when disturbed from its equilibrium position will oscillate about its equilibrium.

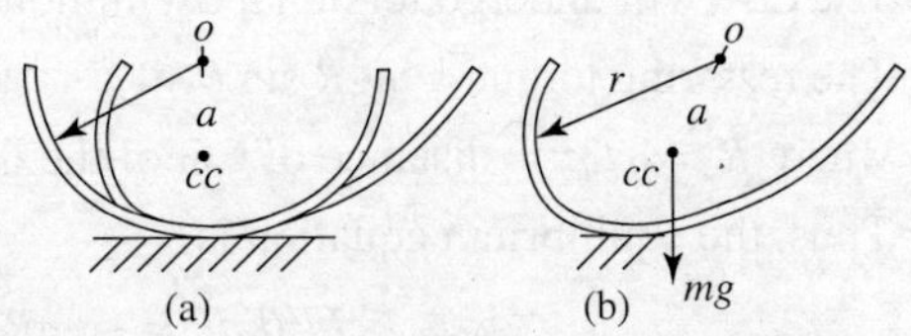

Figure 2.2 Oscillations of Shell

It is obvious that the restoring force is provided by the body weight itself. For analyzing such problems the following methods are very useful.

1. Energy method: For a conservative system (where there are no losses of energy), the total energy of the system is unchanged at all times. The total energy comprises of potential energy (positional energy) and kinetic energy (due to velocity of a vibratory motion) and the sum of these two energies is constant. This means

$$KE + PE = \text{Constant or } \frac{d}{dt}(KE + PE) = 0$$

This is the equation of motion of a system under consideration. This equation enables us to find the frequency of natural (free) oscillations or natural frequencies of vibrations. This method is also called the Rayleigh–Ritz method.

2. Rayleigh method: In this method also, we consider the conservative system which means that there are no losses of energy on account of frictional effects. We know that for such systems the total kinetic energy of the system is zero at maximum displacement but reaches maximum at the static equilibrium. Thus,

$$(KE)_{\max} = (PE)_{\max} = \text{Total energy of the system}$$

This is known as Rayleigh's method. The resulting equation will readily yield the natural frequency of the system.

Problem 2.1

Let us apply Rayleigh's method to find the natural frequency of a semi-circular shell of mass m and radius r, which rolls from side-to-side without slipping (Fig. 2.2).

Solution:

Let ω_n (rad/s) be the frequency of oscillations.

Let I_A = polar moment of inertia (mass moment of inertia) about A, since oscillations take place around point A.

$$I_A = I_{CG} + m(r-a)^2 = I_0 - ma^2 + m(r-a)^2. \text{ But } I_0 = mr^2$$
$$\therefore I_A = 2mr(r-a)$$

According to Rayleigh's theorem,

$$KE_{max} = PE_{max}$$
$$\therefore mr(r-a)\omega_n^2 = mg(1-\cos\theta) \quad \text{or}$$
$$\omega_n = \{ga(1-\cos\theta)/r(r-a)\}^{1/2}$$

Problem 2.2

A semi-circular disc of radius r and mass m is pivoted freely about its centre as shown in Fig. 2.3. Find the natural frequency.

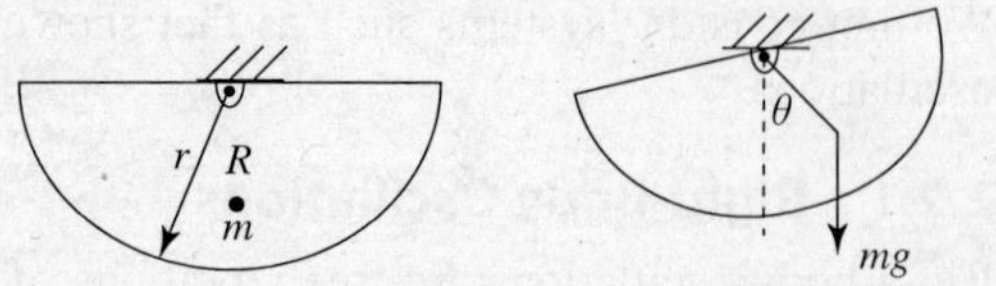

Figure 2.3 Oscillations of Semi-circular Disc

Solution:

The disc will undergo torsional oscillations.

The restoring torque $= mgR\sin\theta$

where $R = 4r/3\pi$ = distance of CG of the disc.

Thus, the equilibrium equation is

$$Jd^2\theta/dt^2 = -mgR\sin\theta \quad \text{or} \quad J\ddot{\theta} + mgR\sin\theta = 0$$

The polar moment of inertia $J = \frac{1}{2}(mr^2)$

Therefore, $\quad \frac{1}{2}mr^2\ddot{\theta} + mgR\sin\theta = 0$; for small θ, $\sin\theta \sim \theta$

Therefore, $\quad \frac{1}{2}r^2\ddot{\theta} + gR\theta = 0.$

Since $R = 4r/3\pi$, we get

$$\ddot{\theta} + (8g/3r\pi) = 0$$

Solution of above equation is $\omega_n = \sqrt{8g/3r\pi}$.

Problem 2.3

A uniform stiff rod is restrained from vertical movement by both linear and torsional springs as shown in Fig. 2.4. Find the natural frequency of the system. Torsional spring has stiffness K and the linear springs have stiffnesses k_1 and k_2. Assume $k_1 = k_2 = k$.

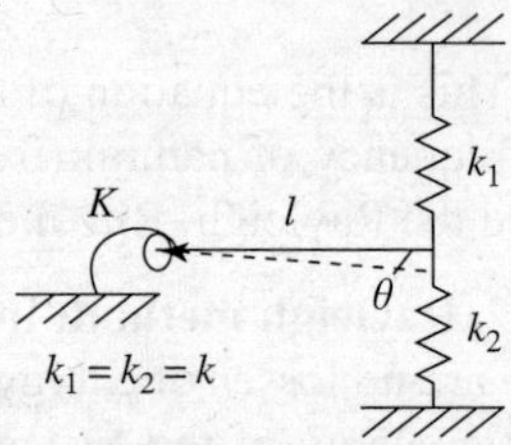

Figure 2.4 Rod with Linear and Torsional Springs

Solution:

The torsional mode is prevented by the spring having stiffness K.

Assuming that angle θ is small, the restoring torque is given by

$$K\theta + 2k_u l^2 \sin\theta$$

where $K\theta$ is torsional spring torque and $k_u l\sin\theta$ is the force in each of the linear springs k_u, and arm length for each of these forces is l.

The inertia torque is $J\ddot{\theta}$, where $J = ml^2/3$. Assuming θ to be small, $\sin\theta \simeq \theta$.

Thus, the equation of motion is

$$(ml^2/3)\ddot{\theta} + K\theta + 2k_u l^2\theta = 0.$$

Solution of this equation is given by

$$\omega_n = \sqrt{(3K + 6k_u l^2)/ml^2}$$

Problem 2.4

Figure 2.5 shows a system where both translational and rotational oscillations occur. Find the natural frequency.

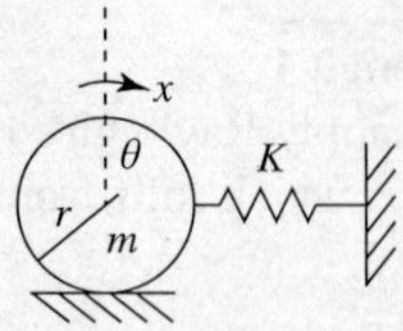

Figure 2.5 Oscillations of Disc

Solution:

The translational motion x is associated with rotational oscillation of the disc.

Translational KE is $\frac{1}{2} m\,(\dot{x})^2$

Rotational KE is $\frac{1}{2} J(\dot{\theta})^2$, where $J = \frac{1}{2} mr^2$

For small θ, $x = r\theta$ and $\dot{x} = r\dot{\theta}$

Total $KE = \frac{1}{2} m\,(\dot{x})^2 + \frac{1}{2} mr^2\,(\dot{\theta})^2 = \frac{1}{2} mr^2\,(\dot{\theta})^2 + \frac{1}{4} mr^2\,(\dot{\theta})^2 = \frac{3}{4} m\,(\dot{x})^2$

The displacement x will cause spring to store energy as potential energy $\frac{1}{2} Kx.x = \frac{1}{2} Kx^2$

Total energy $= \frac{3}{4} m\,(\dot{x})^2 + \frac{1}{2} Kx^2$

According to Rayleigh's method, the system equation is given by

$$d(KE + PE)/dt = 0 \quad \text{or}$$

$$\tfrac{3}{2} m\,\ddot{x} + Kx = 0$$

Solution of the above equation is given by

$$\omega_n = \sqrt{(2K/3m)}$$

Problem 2.5

For the system shown in Fig. 2.6, find the frequency equation.

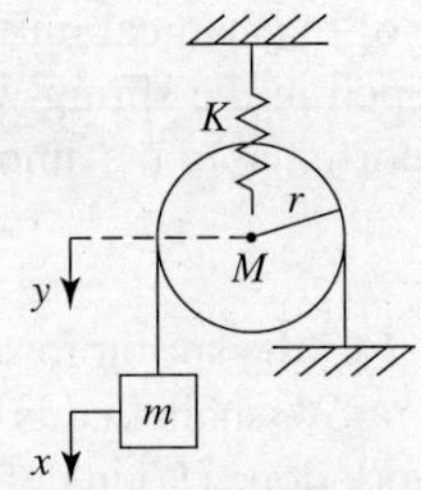

Figure 2.6 Oscillation of Pulley–Mass System

Solution:

KE of the system = KE of mass + KE of pulley.

The mass travels a distance x and the pulley travels a linear distance y and also rotates through an angle θ. Therefore, the total KE is

$$KE = \tfrac{1}{2} m(\dot{x})^2 + \{\tfrac{1}{2} M(\dot{y})^2 + \tfrac{1}{2} J(\dot{\theta})^2\},$$

where the second term in the bracket corresponds to rotational energy.

For small x and θ, it can be seen that

$y = x/2$ and $\theta = x/2r$, where r, is the radius of the pulley.

The PE of the system = Elastic energy of spring

$$= \tfrac{1}{2} Ky^2 = \tfrac{1}{2} K(\tfrac{1}{2} x)^2 = \tfrac{1}{8} Kx^2$$

$d(KE + PE)/dt = 0$ gives the frequency equation.

Problem 2.6

Figure 2.7 shows a cylinder of mass m and radius r rolling without slipping on a circular surface of radius R (this is a situation in a roller bearing often used in rotating machinery). Derive the equation of motion of the roller and the natural frequency.

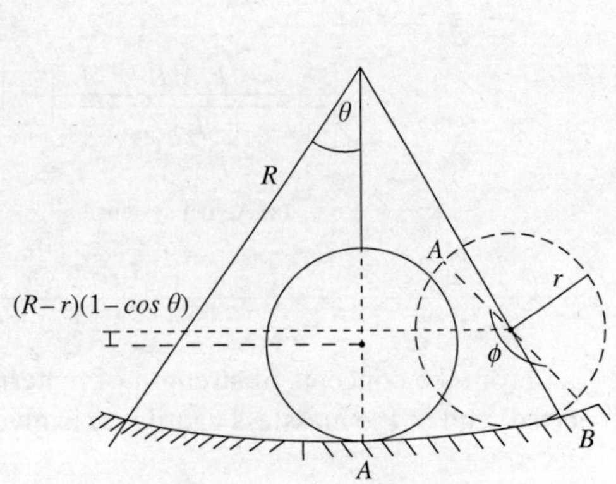

Figure 2.7 Cylinder Rolling without Shipping

Solution:

The total energy of the system consists of rotational KE, translational KE, and the PE.

One can see that the translational displacement is $(R - r)\theta$ and the translational velocity is $(R - r)\dot{\theta}$.

Thus, translational $KE = \frac{1}{2}m\{(R - r)\,\dot{\theta}\}^2$

Rotational $KE = \frac{1}{2}J(\dot{\Phi} - \dot{\theta})^2$ and $J = \frac{1}{2}mr^2$

($AB = R\theta = r\Phi$ as the roller moves without slipping. Thus, $\Phi = R\theta/r$)

$$PE = mg(R - r)(1 - \cos\theta)$$

$d(KE + PE)/dt = 0$ will give the frequency equation.

The readers may verify that the natural frequency is $\omega_n = \sqrt{2g/3(R - r)}$.

The examples discussed here are a few of the many engineering situations where vibrations can occur without the presence of a physical spring, for example, rolling and pitching of ships, vibrations-induced failures (called false brinelling) of anti-friction bearings of the machinery during their transportation, sloshing of fluid-filled containers during transport, etc.

2.2.2 Spring–Mass–Damper Systems

(1) Spring element: Normally for the ease of analysis, we assume a linear spring. A linear spring is a type of mechanical link that is generally assumed to have negligible mass and damping. A force is developed in the spring whenever there is a relative motion between the two ends. The spring force is proportional to the amount of deformation and is given by

$$F = kx \tag{2.1}$$

where F is the spring force, x is the deformation and k = force/displacement is the stiffness or spring constant. We shall, in this book, always assume that the spring force versus deformation curve is a line. The work done (U) in deforming a spring is equal to

$$U = \tfrac{1}{2}kx^2. \tag{2.2}$$

Here, U is stored as strain energy or potential energy in the spring. Elastic elements like beams (usually considered massless) also behave as springs. For example, consider a cantilever beam (massless as first assumption) with an end-mass as shown in Fig. 2.8(a).

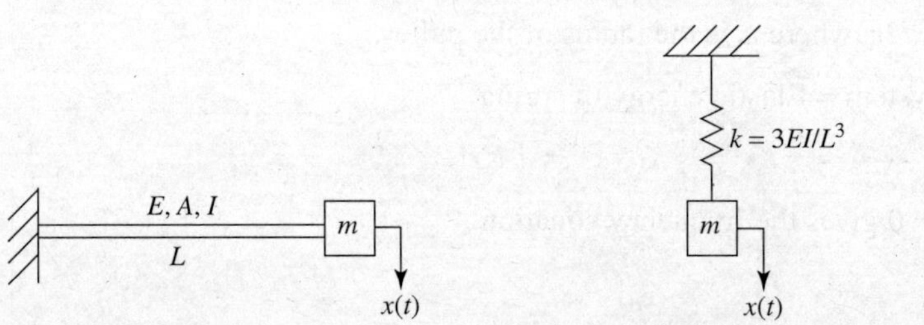

(a) Actual system (b) Equivalent SDOF system of beam

Figure 2.8 Cantilever with End-Mass

From the concepts of strength of materials, we know that the static deflection at the free (unsupported) end of the massless cantilever is given by

$$\delta_{st} = WL^3/3EI \tag{2.3}$$

where $W = mg$ is the weight of the mass m, E is the Young's Modulus and I is the area moment of inertia. Thus,

$$k = W/\delta_{st} = 3EI/L^3 \tag{2.4}$$

The cantilever shown in Fig. 2.8(a) is thus modelled as spring–mass system as shown in Fig. 2.8(b).

In many cases it is convenient to represent a continuous system as combinations of masses and springs. For example, a crude model of a three-storey building shown in Fig. 2.9(a) is shown as spring–mass system in Fig. 2.9(b). It is therefore necessary to know the stiffness of equivalent springs when the springs are in parallel or series.

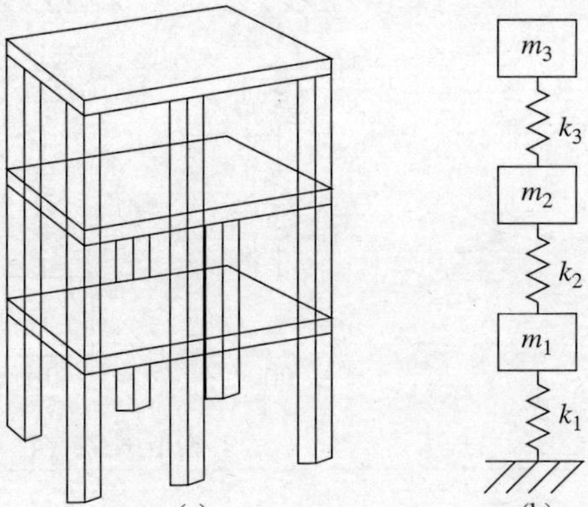

(a) Three-storey building (b) Lumped spring–mass model

Figure 2.9 Idealization of a Building

For springs in parallel as shown in Fig. 2.10, δ_{st}, the equivalent spring stiffness is calculated by the following method.

The two springs of stiffness k_1 and k_2 support the load W. Let δ_{st} be the static deflection of the weight W. Then by considering the equilibrium of the free body [Fig. 2.10(c)].

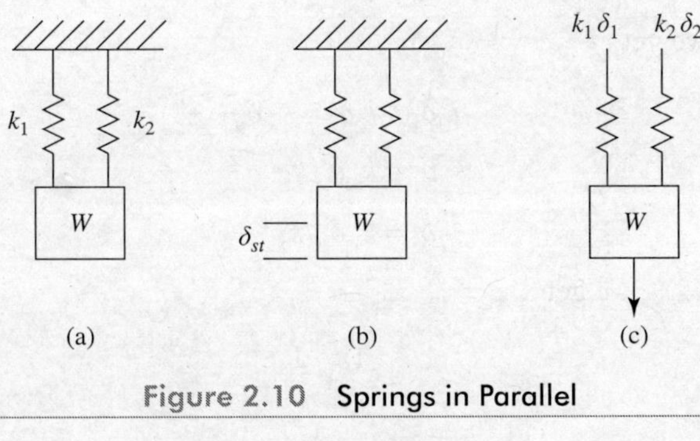

Figure 2.10 Springs in Parallel

$$W = k_1\delta_{st} + k_2\delta_{st} \tag{2.5}$$

If k_{eq} denotes the equivalent spring constant for the combination of two springs

$$W = k_{eq}\,\delta_{st} \tag{2.6}$$

From Equations 2.5 and 2.6 we find

$$k_{eq} = k_1 + k_2$$

In general, if we have n springs with spring constants $k_1, k_2, \ldots, k_n$ in parallel, then the equivalent spring constants

$$k_{eq} = k_1 + k_2 + \cdots + k_n \tag{2.7}$$

For springs in series as shown in Fig. 2.11, the equivalent spring stiffness is found as follows.

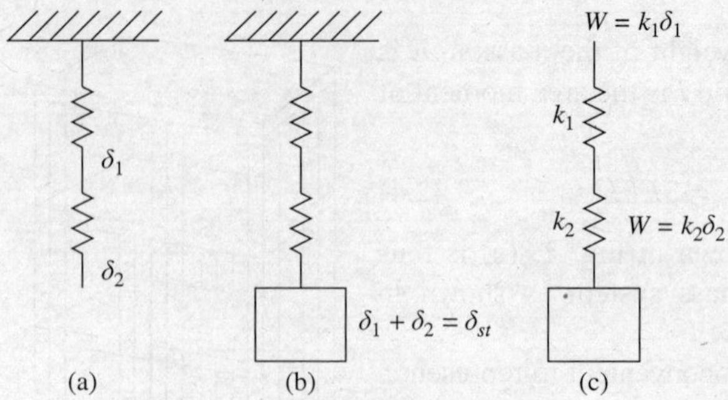

Figure 2.11 Springs in Series

Under the action of W, springs 1 and 2 undergo elongations δ_1 and δ_2 such that

$$\delta_1 + \delta_2 = \delta_{st} \tag{2.8}$$

Since both springs are subjected to the same force W, the analysis of equilibrium of free body shown in Fig. 2.11(c) shows

$$W = k_1\delta_1 = k_2\delta_2 \tag{2.9}$$

If k_{eq} denotes stiffness of equivalent spring,

$$W = k_{eq}\delta_{st} \tag{2.10}$$

Equating 2.9 and 2.10, we get

$$k_1\delta_1 = k_2\delta_2 = k_{eq}\delta_{st}$$

or

$$\delta_1 = k_{eq}\delta_{st}/k_1$$

and

$$\delta_2 = k_{eq}\delta_{st}/k_2 \tag{2.11}$$

Using these in Equation 2.8, we get

$$\frac{1}{k_{eq}} = \frac{1}{k_1} + \frac{1}{k_2} \tag{2.12}$$

Equation 2.12 can be generalized to the case of n springs in series

$$\frac{1}{k_{eq}} = \frac{1}{k_1} + \frac{1}{k_2} + \cdots + \frac{1}{k_n} \tag{2.13}$$

We shall illustrate these principles through a few solved problems here.

Problem 2.7

A hoisting drum carrying a steel wire rope is mounted at the end of a cantilever beam of length b (Fig. 2.12). Determine the spring constant of the system when the suspended length of wire rope is l, the diameter of the rope is d, and Young's modulus of the beam is E. The cross-section of the beam is $a \times t$ where a is the width.

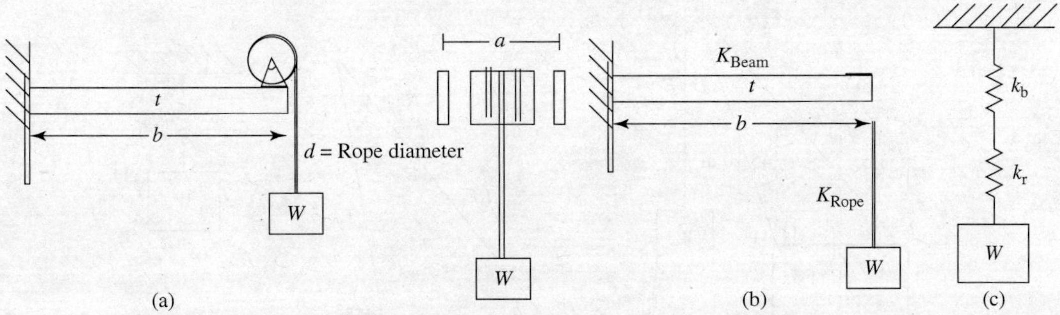

Figure 2.12 Hoisting Rope

Solution:
We assume that the beam is massless. The spring constant of the beam is

$$k_b = \frac{3EI}{b^3} = \frac{3E}{b^3}\left(\frac{1}{12}at^3\right) = \frac{Eat^3}{4b^3} \tag{1}$$

The stiffness of the wire rope which is subjected to axial loading is given by $k_r = \dfrac{AE}{l}$

This is because the stress in wire rope $\sigma = W/A =$ strain $xE = E \times \delta l/l$ and $W/\delta l = k_r$

From this we get

$$k_r = \frac{\prod d^2 E}{4l} \tag{2}$$

It may be understood that both the cantilever and the rope experience the same load as shown in the free-body diagram [Fig. 2.12(b)]. Hence, they can be considered and modelled as springs in series. The equivalent spring stiffness k_{eq} is given by

$$\frac{1}{k_{eq}} = \frac{1}{k_1} + \frac{1}{k_2} \tag{3}$$

where $k_1 = k_b$ and $k_2 = k_r$. Substituting these in Equation 3, we get,

$$k_{eq} = \frac{\prod Eat^3 d^2}{4(\prod d^2 b^3 + lat^3)}$$

Problem 2.8

Figure 2.13 shows a crane having boom AB made of uniform steel bar of length 10 m and cross-sectional area 2500 mm^2. In the stationary condition a weight W is suspended as shown. The cable $CDEBF$ is made of steel (area 100 mm^2). Find the spring constant of the system in the vertical direction.

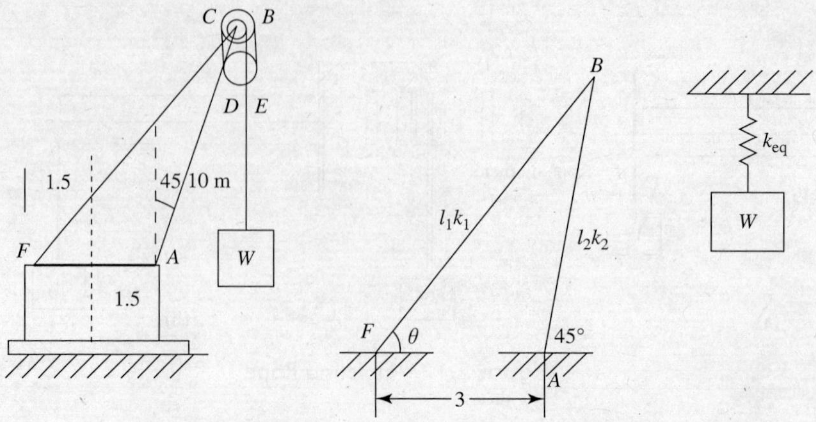

Figure 2.13 Schematic Arrangement of a Crane

Solution:

We shall see that it is not necessary to give the numerical value of W as stiffness depends only upon elastic and geometrical properties. Let us assume that the base is rigid. Note that PE of the two systems FB and AB are equal. The effect of cable $CDEB$ is negligible. Hence W can be considered to be acting through B.

The vertical displacement y of point B will cause the spring k_2 (boom) to deform by an amount $y_2 = y \cos 45$ and spring k_1 (cable) deform by an amount $y_1 = y \cos(90 - \theta)$. The length l_1 of cable FB, is given by

$$l_1^2 = 3^2 + 10^2 - 2.3 \times 10 \cos 135 = 151, \text{ giving } l_1 = 12.3 \text{ m}.$$

Also since $$l_1^2 = 3^2 - 2l_1 \cdot 3 \cos\theta = 10^2, \ \theta = 35°$$

The total PE (U) stored in the springs k_1 and k_2 is

$$U = \frac{k_1(y\cos 45)^2}{2} + \frac{k_2(y\cos(90-\theta))^2}{2}$$

where $k_1 = A_1E/l_1$. Here area is 100×10^{-6} cm^2, $E = 207 \times 10^9$ N/m, and $l_1 = 12.3$ m. Substituting these values, we get,

$$k_1 = 1.69 \times 10^6 \text{ N/m}.$$

Similarly $k_2 = A_2E/l_2$. Here the area is 2500×10^{-6} and $l_2 = 10$. Using these values we obtain

$$k_2 = 5.2 \times 10^6.$$

We have assumed the value of E.

Since the equivalent spring in the vertical direction undergoes a deformation y, the PE of the equivalent spring U_{eq} is given by

$$U_{eq} = \tfrac{1}{2}(k_{eq}\, y^2) = U$$

Using this and after calculations,

we get, $k_{eq} = 26 \times 10^6$ N/m.

(2) Mass or inertia element: The mass or inertia element is assumed to be a rigid-body that can only lose or gain kinetic energy. The product of mass and acceleration of vibration ($\ddot{x}$) is equal to the force, and the force multiplied by the displacement in the direction of the force is the work done. The work done on a mass is stored as kinetic energy.

In vibration analysis, we are required to appropriately model the mass. Once the model is chosen, the mass or inertia element can be easily identified. For example, consider the cantilever system loaded at free end shown in Fig. 2.8(a). In reality, it is a continuous system. However, for quick and reasonable accuracy, the mass and the damping (discussed later in this chapter) of the beam can be disregarded and the system can be modelled as a spring/mass system as shown in Fig. 2.8(b), where stiffness of the spring is $3EI/L^3$ and the tip mass m represents the mass element.

Yet another example can be seen in Fig. 2.14(a), which shows a multi-storey building subjected to earthquake forces. In modelling this structure, we assume that the supporting structure has negligible mass compared to the masses of the floor but has elasticity (spring) to support the masses of the floors. Thus, the building shown in Fig. 2.14(a) is modelled as a multi-degrees-of-freedom system comprising of four masses m_1, m_2, m_3, and m_4 (which represent masses of the floors) and supported springs k_1, k_2, k_3, and k_4, which represent the elasticity (spring) of the vertical support members. In a strict sense, This model, by no means an exact, is good enough for making reasonably accurate analysis of the vibration of the building. Let us illustrate the concept through some practical cases.

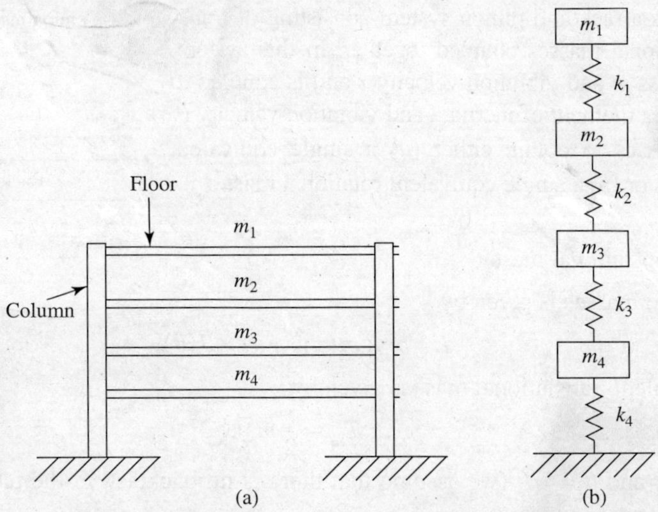

Figure 2.14 Spring–Mass Model of a Building

Problem 2.9

Determine the equivalent mass of the system shown in Fig. 2.15. The bar can be considered as rigid and massless.

Solution:

The equivalent mass can be assumed to be located anywhere along the bar as long as the bar has only elasticity and no mass. Let us assume that the location of the equivalent mass is at point A. Let x_1, x_2, and x_3 be the displacements of masses m_1, m_2, and m_3. The corresponding velocities are $\dot{x}_1$, $\dot{x}_2$, and $\dot{x}_3$. Assuming small angular

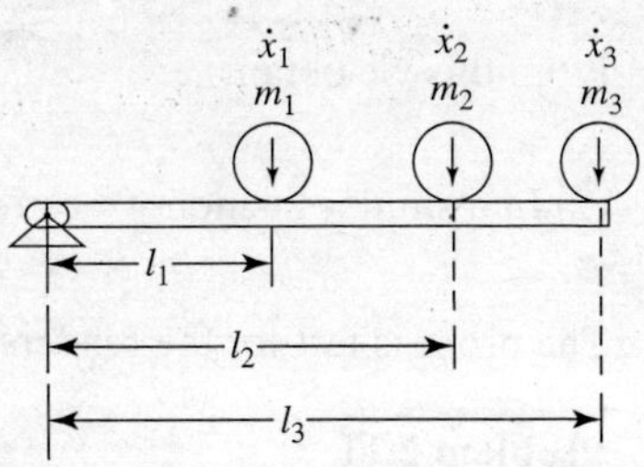

Figure 2.15 Massless Bar with External Loads

displacements of the bar, the velocities of masses m_2 and m_3 can be expressed in terms of velocity of mass m_1 as

$$\dot{x}_2 = \frac{l_2}{l_1}\dot{x}_1, \quad \dot{x}_3 = \frac{l_3}{l_1}\dot{x}_1 \tag{2.14}$$

and

$$\dot{x}_{eq} = \dot{x}_1 \tag{2.15}$$

For the system shown in Fig. 2.16 to be equivalent to the system shown in Fig. 2.15, we must equate the *KE* of both systems, that is,

$$m_1(\dot{x}_1)^2 + m_2(\dot{x}_2)^2 + m_3(\dot{x}_3)^2 = m_{eq}(\dot{x}_{eq})^2 \tag{2.16}$$

Using Equation 2.14 in the Equation 2.16, we get,

$$m_{eq} = m_1 + (l_2/l_1)^2 m_2 + (l_3/l_1)^2 \tag{2.17}$$

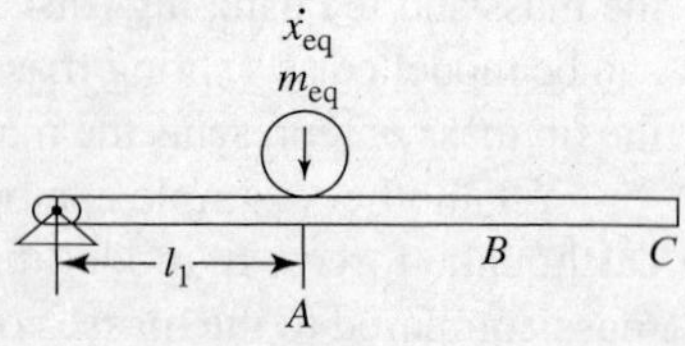

Figure 2.16

The importance of the concept lies in the fact that when the vibration excitation force is applied, say at mass m_1, the other masses m_2 and m_3 also participate in the vibration response and thus m_1, $(l_2/l_1)^2 m_2$ and $(l_3/l_1)^2 m_3$ can be considered as participating masses and the equivalent mass m_{eq} comprises of contributions from these masses.

Problem 2.10

Figure 2.17 shows a rack and pinion system consisting of translational and rotational masses coupled together. In this system, the rack has a mass m and vibration velocity x and is coupled to pinion having mass moment of inertia J and vibration velocity $\dot{\theta}$. Combine the masses to obtain either (*a*) a single equivalent translational mass or (*b*) a single equivalent rotational mass J_{eq}.

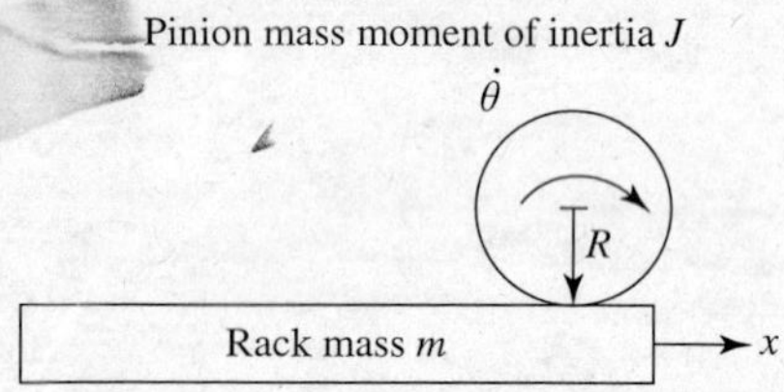

Figure 2.17 Rack–Pinion System

Solution:

(1) Equivalent translational mass

The *KE* of the two masses is given by

$$T = \tfrac{1}{2} \times m(\dot{x})^2 + \tfrac{1}{2} \times J(\dot{\theta})^2 \tag{2.18}$$

The *KE* of equivalent translational mass is given by

$$T_{eq} = \frac{1}{2} m_{eq}(\dot{x}_{eq})^2 \tag{2.19}$$

Since $x_{eq} = x$ and $\theta = x/R$ (we assume that there is no backlash in the pinion/rack system), the equivalence of T and T_{eq} gives

$$\tfrac{1}{2} \cdot m_{eq}(\dot{x})^2 = \tfrac{1}{2} \cdot m(\dot{x})^2 + \tfrac{1}{2} \cdot J(\dot{\theta})^2$$

From this we obtain

$$m_{eq} = m + J/R \tag{2.20}$$

(2) In a similar fashion we can show that a single-equivalent rotational mass J_{eq} is given by

$$J_{eq} = J + mR^2 \tag{2.21}$$

The proof is left for the readers to assess.

Problem 2.11

Figure 2.18 shows the valve-operating mechanism used in the internal combustion engines. The rotary motion of the shaft is converted into the reciprocating motion of the valve by the cam mechanism.

The cam follower system shown in the Fig. 2.18 consists of a push rod of mass m_p, rocker arm of mass m_r and mass moment of inertia J_r about CG (G), a valve of mass m_v, and a valve spring of negligible mass. Find the equivalent mass m_{eq} at (1) point A and (2) point C.

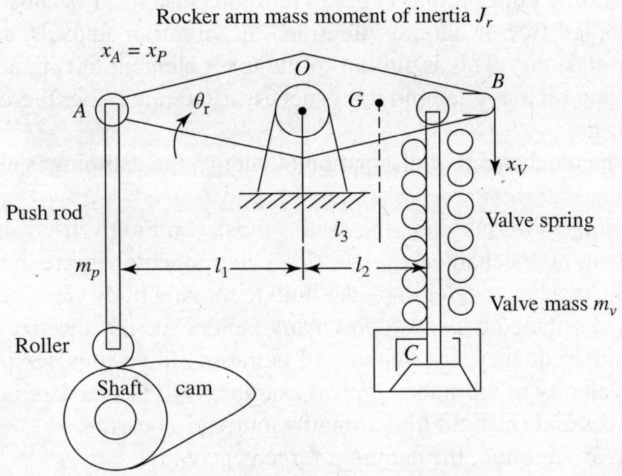

Figure 2.18 Cam Follower System

Solution:

The equivalent mass of the cam follower system can be determined using equivalence of the kinetic energy of the two systems. Due to vertical movement x of the push rod, the rocker arm rotates by an angle $\theta_r = x/l_1$, the valve moves downward by $x_v = x(l_2/l_1)$, and the CG of the rocker arm moves downward by $x_r = \theta l_3 = x(l_3/l_1)$.

The *KE* of the system is then

$$2T = m_p(\dot{x}_p)^2 + m_v(\dot{x}_v)^2 + J_r(\dot{\theta})^2 + m_r(\dot{x}_r)^2 \tag{1}$$

where $\dot{\theta}$ is the angular velocity of the rocker arm.

If the *KE* and *TE* of the equivalent mass put at A with velocity $\dot{x}_{eq} = \dot{x}$ are equated, we get,

$$2T = m_p(\dot{x}_p)^2 + m_v(\dot{x}_v)^2 + J_r(\dot{\theta})^2 + m_r(\dot{x}_r)^2 = m_{eq}(\dot{x}_{eq})^2.$$

After substituting the value of various velocities in this and simplifying, we get,

$$m_{eq} = m_p + J_r/(l_1)^2 + m_v(l_2/l_1)^2 + m_r(l_3/l_1)^2.$$

In a similar manner we can find the equivalent mass placed at point C. This is left as an exercise for the students.

Thus, we have seen the procedure for modelling structures or assemblies into an equivalent single spring and mass system. Although such a modelling can never be considered as very rigorous, it does effectively serve the purpose of estimating (approximately) the vibration behaviour to a known excitation force. This aspect of estimating the vibration behaviour is discussed later in this chapter. We shall now turn our attention to the third component of vibration system, that is the damper.

(3) Damping element: In many practical systems, the vibration energy is dissipated into heat or sound energy. Due to reduction in the energy, the response such as vibration displacement of the system, when disturbed from its position of equilibrium, gradually decreases. The mechanism by which energy dissipation takes place is known as damping. In absence of damping (which really never happens), the system set into vibration, by disturbing it from its position of equilibrium, will keep on vibrating forever. Conversely, to sustain the vibrations, the system must have continuous action of disturbing force. In case the system has continuous action of disturbing force, it is considered as having forced vibration. In absence of continuously acting disturbing force, the system still vibrates for a certain amount of time but eventually these vibrations disappear because of energy dissipation. These vibrations are called free or natural vibrations. In vibration analysis, a damper is assumed to have neither mass nor elasticity. This is similar to the mass element having no elasticity/damping or the spring element having no mass/damping. It is necessary to remember these concepts and assumptions in vibration analysis.

Depending upon the mechanism of dissipation of energy, the damping is modelled as one or more of the following types.

(i) Viscous damping: Viscous damping is the most commonly used damping mechanism in vibration analysis. When the mechanical system or its components vibrate in a fluid medium such as air, gas, water or oil, the resistance offered by the fluid to moving body causes energy to be dissipated. The amount of energy dissipated depends upon many factors such as the size and shape of the body, the viscosity of the fluid, frequency, and velocity of vibration. In viscous damping, the damping force is proportional to the velocity of vibration. Typical examples of viscous damping include (i) fluid film between moving surfaces and (ii) fluid film around a journal in bearing.

If $\dot{x}$ is the velocity of vibration, the damping force is given by

$$F_{\mathrm{d}} = f\dot{x} \tag{2.22}$$

where f is termed the damping coefficient. Recalling that $\dot{x}$= velocity of vibration $= x_{\mathrm{p}}\omega$, where x_{p} is the amplitude of vibration displacement and ω is the circular frequency.
Thus,

$$F_{\mathrm{d}} = x_{\mathrm{p}}\omega f \tag{2.23}$$

(ii) Coulomb or dry-friction damping: In this case, the damping force is constant in magnitude but opposite to the direction of motion of the vibrating body. It is caused by friction between the rubbing surfaces that are either dry or having insufficient lubrication. Friction dampers in the form of damping ring or pins between the blades of steam/gas turbine are extensively used in high speed turbo-machinery.

(iii) Material or solid or hysteretic damping: When materials are deformed the friction between internal planes which slip or slide induces friction. In many practical cases of vibration, the vibrations are associated with stretching or unstretching due to the action of alternating forces. For example, the alternating bending and twisting of turbine blades cause friction between the internal planes of the blades. This friction results into dissipation of energy. These forms of damping as well as Coulomb damping are discussed later in this chapter.

We now illustrate the method of finding the stiffness of the equivalent spring and equivalent damping coefficient of the complex structure. The aim in doing so is to idealize the complex structure as a simple system comprising of a mass, spring and a damper. We shall now show in the forthcoming paragraphs that such an idealization gives a fairly accurate estimation of the response of the system to various excitation forces.

Problem 2.12

The steam turbine–generator unit of a power plant is on four shock mounts as shown in Fig. 2.19. The elasticity and damping of each shock mount can be modelled as a spring and a viscous damper as shown in Fig. 2.19(b). Find the equivalent spring constant k_{eq} and equivalent damping constant f_{eq} of the unit in terms of spring constants and damping constants (k_i and f_i) of the mounts.

Solution:

Figure 2.19 shows the free-body diagrams of the springs and the dampers. It is assumed that CG of the machine is located symmetrically with respect to four springs and dampers. Thus, all springs will be subjected to same displacement x and all dampers will be subjected to same relative velocity x of the centre of mass of the generating set. Denoting the spring force and the damping force as $F_s(i)$ and $F_d(i)$, we can write

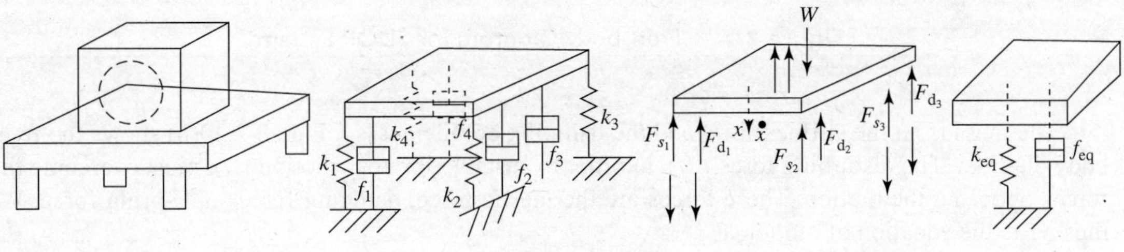

Figure 2.19 Turbine–Generator on Shock Mounts

$$F_s(i) = k_i x_i, \quad i = 1, 2, 3, 4 \quad \text{and} \quad F_d(i) = f_i \dot{x}_i, \quad i = 1, 2, 3, 4$$

Total spring force

$$F_s = F_s(1) + F_s(2) + F_s(3) + F_s(4) = k_{eq} x$$

Total damping force

$$F_d = F_d(1) + F_d(2) + F_d(3) + F_d(4) = f_{eq} \dot{x}$$

and $F_s + F_d = W =$ total vertical force including inertia forces acting on the machine. The inertia forces are caused by the centrifugal forces due to rotation of moving parts of the turbine/generator.

Thus, we can see

$$k_{eq} = \sum_1^4 k_i \quad \text{and} \quad f_{eq} = \sum_1^4 f_i.$$

Usually it is customary to use same design mounts. In such a case

$$k_{eq} = 4k \quad \text{and} \quad f_{eq} = 4f.$$

2.3 EQUATION OF MOTION FOR SINGLE DEGREE-OF-FREEDOM SYSTEM (SDOF)

We have discussed the elements of the single degree-of-freedom system comprising of a spring, mass and a damper. We have also discussed the procedure of developing a single degree-of-freedom vibration system model for complex bodies and structure as the first-level approximation.

Let us now write the governing equation of motion of a spring, mass, and damper system subjected to harmonic excitation force (Fig. 2.20).

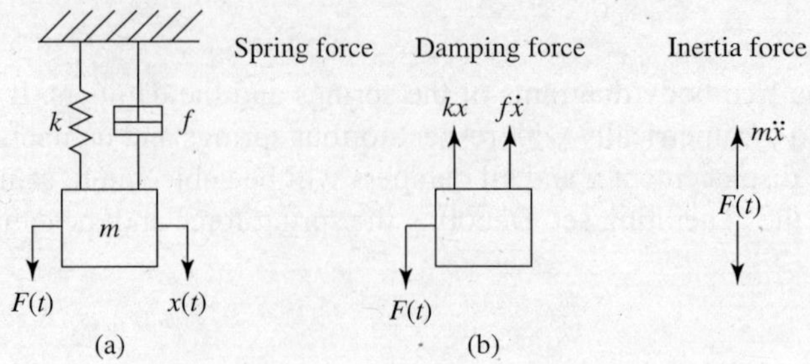

Figure 2.20 Free-body Diagram for SDOF System

The mass is m, the stiffness is k, and the damping coefficient is f. Figure 2.20(b) shows the free-body diagram. The disturbing force $F(t)$, in order to sustain vibratory motion $x(t)$, must overcome the forces opposing the motion. These forces are the inertia force, damping force, and spring force. We thus write the equation of motion as

$$m\ddot{x} + f\dot{x} + kx = F(t). \tag{2.24}$$

The equation assumes that the dynamic deflections are small enough to consider the spring element as linear ($F/x = k$) and also damping is viscous so that the damping force is linearly proportional to the velocity $\dot{x}$.

In case there is no continuous force $F(t)$ on the system shown in Fig. 2.20 and the system is momentarily disturbed from its position of static equilibrium by a force which acts only for a very short duration Δt, the system will be set into vibrations. However, these will cease, sooner or later depending upon the magnitude of damping in the system, after sometime. Such vibrations are called free vibrations. The study of free vibrations is extremely important since it only reveals certain important characteristics of the spring, mass, and damper system. These are very important in analyzing the vibrations occurring when the system is acted upon by force $F(t)$. In fact, quite a significant part of vibration analysis pertains to analysis of free vibrations.

2.3.1 Free Vibrations of Single Degree-of-Freedom System—Viscous Damping

Consider again the SDOF system shown in Fig. 2.20(a) but without any force $F(t)$ acting on it. As mentioned earlier, the system has been momentarily disturbed from its position of static equilibrium and is set into vibrations.

The governing equations for these vibrations can be obtained by making the right-hand side of Equation 2.24 as 0.

Thus, for free vibrations, the governing equation becomes

$$m\ddot{x} + f\dot{x} + kx = 0 \tag{2.25}$$

Since m, f, and k are constants, Equation 2.25 is a second-order ordinary differential equation with constant coefficients. We assume the solution of Equation 2.25 as

$$x(t) = Ce^{st} \tag{2.26}$$

where C and s are constants. Inserting Equation 2.26 in Equation 2.25, we get the characteristic equation

$$ms^2 + fs + k = 0. \tag{2.27}$$

The roots of Equation 2.27 are

$$s_{1,2} = \frac{-f \pm \sqrt{f^2 - 4mk}}{2m} = -\frac{f}{2m} \pm \sqrt{\left\{\left(\frac{f}{2m}\right)^2 - \frac{k}{m}\right\}} \tag{2.28}$$

These roots give two solutions

$$x_1(t) = C_1 e^{s_1 t}, \quad x_2(t) = C_2 e^{s_2 t} \tag{2.29}$$

The general solution of Equation 2.25 is given by the combination of the two solutions $x_1(t)$ and $x_2(t)$

$$x(t) = C_1\, e^{s_1 t} + C_2\, e^{s_2 t}$$

$$= C_1 e^{\left\{-f/2m + \sqrt{\left\{(f/2m)^2 - k/m\right\}}\right\}t} + C_2 e^{\left\{-f/2m - \sqrt{\left\{(f/2m)^2 - k/m\right\}}\right\}t} \tag{2.30}$$

C_1 and C_2 are arbitrary constants which are determined from the initial conditions of the system.

Critical Damping Constant and Damping Ratio

The critical damping factor f_c is defined as the value of the damping constant f for which the radical in Equation 2.28 becomes zero.

$$(f_c/2m)^2 - k/m = 0, \quad \text{or}$$

$$f_c = 2m\sqrt{k/m} = 2\sqrt{km} \tag{2.31}$$

We shall revert to Equation 2.31 after some analysis of the system in absence of damping, that is, we consider the free vibrations of the undamped system ($f = 0$).
For the system with no damping, the governing equation becomes

$$m\ddot{x} + kx = 0. \tag{2.32}$$

The solution of Equation 2.32 is given by the Equation 2.30 by putting $f = 0$, that is

$$x(t) = \mathrm{C}_1\, \mathrm{e}^{\sqrt{-k/m \cdot t}} + \mathrm{C}_2 \mathrm{e}^{-\sqrt{-k/m \cdot t}} \tag{2.33}$$

Let $$\sqrt{k/m} = \omega_n \tag{2.34}$$

Then Equation 2.33 becomes

$$x(t) = \mathrm{C}_1 \mathrm{e}^{i\omega_n t} + \mathrm{C}_2 \mathrm{e}^{-i\omega_n t}, \quad (I = \sqrt{-1})$$

We know that $e^{i\theta} = \cos\theta + i\sin\theta$ and $\mathrm{e}^{-i\theta} = \cos\theta - i\sin\theta$.

Using this in above equation, we get,

$$x(t) = \mathrm{A}_1 \cos\omega_n t + \mathrm{A}_2 \sin\omega_n t \tag{2.35}$$

where, A_1 and A_2 are new constants which can be determined from the initial conditions. The initial conditions to be used are

$$x(t = 0) = \mathrm{A}_1 = x_0$$

$$\dot{x}(t = 0) = \omega_n \mathrm{A}_2 = \dot{x}_0.$$

Thus,

$$A_1 = x_0 \quad \text{and} \quad A_2 = \dot{x}_0/\omega_n \tag{2.36}$$

Equation 2.35, then becomes

$$x(t) = x_0 \cos \omega_n t + (\dot{x}_0/\omega_n) \sin \omega_n t \tag{2.37}$$

Equation 2.37 can be written as

$$x(t) = A \cos(\omega_n t - \Phi) \quad [2.37(a)]$$

such that

$$A \cos \Phi = x_0 \quad [2.37(b)]$$

and

$$A \sin \Phi = \dot{x}_0/\omega_n \quad [2.37(c)]$$

These relationships imply

$$A = \sqrt{\{x_0^2 + (\dot{x}_0/\omega_n)^2\}} \tag{2.38}$$

and

$$\Phi = \tan^{-1}(\dot{x}_0/x_0\, \omega_n) \tag{2.39}$$

Thus, Equation 2.35 which describes free vibration behaviour of undamped system can be rewritten as

$$x(t) = [\sqrt{\{x_0^2 + (\dot{x}_0/\omega_n)^2\}}] \cos(\omega_n t - \Phi)$$

with

$$\Phi = \tan^{-1}(\dot{x}_0/x_0\, \omega_n) \tag{2.40}$$

Equation 2.40 shows that the undamped system, once disturbed, will go on vibrating forever and the frequency of vibration is ω_n. This frequency is termed as the natural frequency of the undamped system. Equation 2.40 graphically appears as shown in Fig. 2.21.

The natural frequency $\omega_n (= \sqrt{k/m})$ is the characteristic of the undamped system. This frequency plays a major role in the analysis of vibration behaviour of systems and we shall discuss this many a times in later chapter.

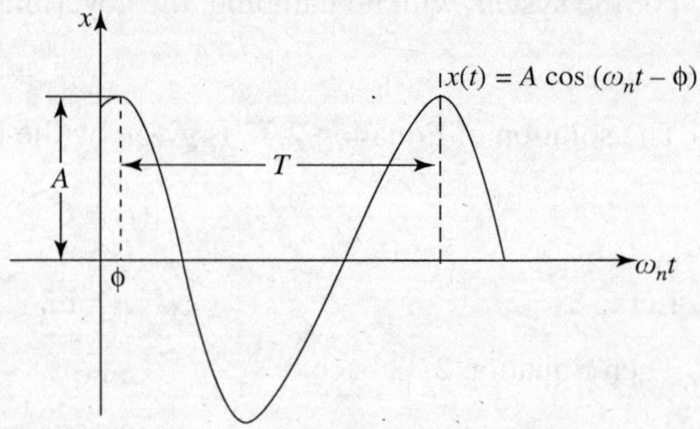

Figure 2.21 Vibration Waveform at Natural Frequency

We now revert back to Equation 2.31 which defines the critical damping in the SDOF spring, mass, and damper system. Equation 2.31 shows critical damping f_c as given by

$$f_c = 2m\sqrt{k/m} = 2\sqrt{km} = 2m\omega_n \tag{2.41}$$

where $\sqrt{k/m} = \omega_n$ = natural frequency of undamped system, that is, when damper is absent. A strong reason for the appearance of the parameter ω_n in vibration analysis is that both stiffness and mass are the inherent properties of the component/structure/system. The damper (viscous or any other) is usually outside the system or external to the system. For any damped system, the damping ratio ξ is defined as the ratio of the damping constant to the critical damping constant, that is,

$$\xi = f/f_c \tag{2.42}$$

Using Equations 2.41 and 2.42, we can write,

$$f/2m = (f/f_c)\cdot(f_c/2m) = \xi\ \omega_n \tag{2.43}$$

Hence,

$$s_{1,2} = [-\xi \pm \sqrt{\ \xi^2 - 1})]\ \omega_n \tag{2.44}$$

Thus, the solution given by Equation 2.30 can now be written as

$$x(t) = C_1 e^{\left[-\xi+\sqrt{(\xi^2-1)}\right]\omega_n t} + C_2 e^{\left[-\xi-\sqrt{(\xi^2-1)}\right]\omega_n t} \tag{2.45}$$

The nature of roots s_1 and s_2 and hence the behaviour of the solution, of Equation 2.45, solely depend upon ξ (i.e. damping). It can be seen that the case $\xi = 0$ leads to undamped vibrations as discussed previously. Hence, for further analysis we assume that damping (ξ) is not zero and consider the following three cases.

Case 1: ***Underdamped system [(ξ < 1 or f < f_c or f/2m < √K/m)]***

For this case $(\xi^2 - 1)$ is –ve, which makes $\sqrt{(\xi^2 - 1)}$ a complex number.
The roots s_1 and s_2 for this case are

$$s_1 = [-\xi + i\sqrt{\ (1 - \xi^2)}]\ \omega_n \quad \text{and} \quad s_2 = [-\xi - i\sqrt{\ (1 - \xi^2)}]\ \omega_n \tag{2.45(a)}$$

Using these in Equation 2.45, we get,

$$\begin{aligned}
x(t) &= C_1 e^{[-\xi + i\sqrt{(1-\xi^2)}]\omega_n t} + C_2 e^{[-\xi - i\sqrt{(1-\xi^2)}]\omega_n t} \\
&= e^{-\xi\omega_n t}\{C_1 e^{i\sqrt{(1-\xi^2)}\omega_n t} + C_2 e^{-i\sqrt{(1-\xi^2)}\omega_n t} \\
&= e^{-\xi\omega_n t}\{(C_1 + C_2)\cos\sqrt{(1-\xi^2)}\omega_n t + i\,(C_1 - C_2)\sin\sqrt{(1-\xi^2)}\omega_n t\} \\
&= e^{-\xi\omega_n t}\ \{c_1 \cos\sqrt{\ (1-\xi^2)}\omega_n t + c_2 \sin\sqrt{\ (1-\xi^2)}\omega_n t\ \} \\
&= X\, e^{-\xi\omega_n t} \sin\,[\sqrt{\ (1-\xi^2)}\omega_n t + \Phi]
\end{aligned} \tag{2.46}$$

or

$$= X_0\ e^{-\xi\omega_n t} \cos\,[\sqrt{\ (1-\xi^2)}\omega_n t - \Phi_0] \tag{2.47}$$

where (c_1, c_2), (X, Φ), and (X_0, Φ_0) are arbitrary constants which can be determined from the initial conditions. The initial conditions are

$$x(t = 0) = x_0, \qquad \dot{x}(t = 0) = \dot{x}_0.$$

Using these, we find,

$$c_1 = x_0 \quad \text{and} \quad c_2 = \{\dot{x}_0 + \xi\omega_n x_0\}/\ (1 - \xi^2)\omega_n \tag{2.48}$$

Hence the solution becomes,

$$x(t) = e^{-\xi\omega_n t}\left[x_0 \cos\sqrt{1-\xi^2}\,\omega_n t + \frac{(\dot{x}_0 + \xi\omega_n x_0)}{\sqrt{1-\xi^2}\,\omega_n}\sin\sqrt{1-\xi^2}\,\omega_n t\right] \tag{2.49}$$

The constants (X, Φ) and (X_0, Φ_0) can be expressed as

$$X = X_0 = \sqrt{\ (c_1^{\,2} + c_2^{\,2})} = \{x_0^{\,2} + [\{\dot{x}_0 + \xi\omega_n x_0\}^2/\{\omega_n^{\,2} - \omega_n^{\,2}\,\xi^2\}]\}^{1/2} \tag{2.50}$$

$$\Phi = \tan^{-1}\left\{\frac{x_0\sqrt{1-\xi^2}\,\omega_n}{\dot{x}_0 + \xi\omega_n x_0}\right\} \tag{2.51}$$

$$\Phi_0 = \tan^{-1} [-(\dot{x}_0 + \xi\omega_n x_0)/\{x_0 \sqrt{((1-\xi^2)\,\omega_n}\}] \tag{2.52}$$

It may be remembered that the response $x(t)$ can be expressed either by Equation 2.46 or 2.47.

The motion described by Equation 2.49 is a harmonic motion of angular frequency $\sqrt{(1-\xi^2)}\omega_n$, but because of the factor $e^{-\xi\omega nt}$, the amplitude decreases exponentially with time as shown in Fig. 2.22.

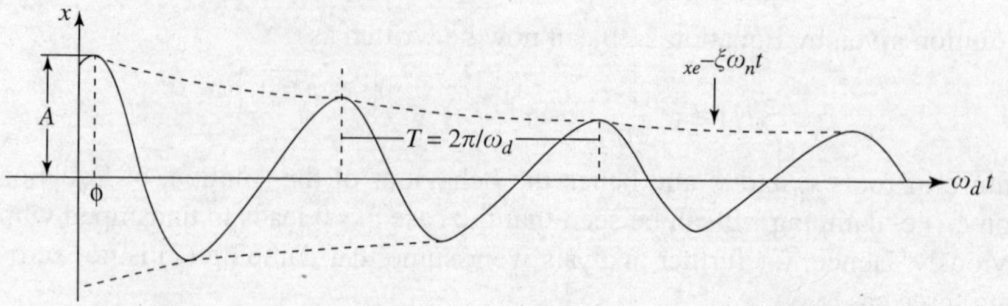

Figure 2.22 Free Vibrations of Damped System

$$\omega_d = \sqrt{(1-\xi^2)}\,\omega_n \tag{2.53}$$

The frequency quantity is called natural frequency of damped vibrations. It can be seen that after certain number of cycles the vibrations totally stops.

Before proceeding further, let us consider the cases of critically damped and overdamped systems.

Case 2: *Critically damped systems*

In this case $\xi = 1$ or $f = f_c$ which in other words means that $f_c/2m = \sqrt{K/m} = \omega_n$. In this case the two roots s_1 and s_2 in Equation 2.44 are equal.

$$s_1 \text{ and } s_2 = -\omega_n \tag{2.54}$$

Examination of Equation 2.49 reveals that as $\xi \to 1$, $(1-\xi^2)$ ½ $\omega_n \to 0$. This makes $\cos\omega_d \to 1$ and $\sin\omega_d \to \omega_d$.

Hence, we get, $x(t) = e^{-\omega nt}(c_1 + c_2\,\omega_d\,t) = (c_1 + c_2 t)\,e^{-\omega nt}$

with new constants c_1 and c_2.

The application of initial conditions, $x(t=0) = x_0$ and $\dot{x}$ $(t=0) = \dot{x}_0$ gives

$$c_1 = x_0, \quad c_2 = \dot{x}_0 + \omega_n x_0 \tag{2.55}$$

Thus,

$$x(t) = [x_0 + (\dot{x}_0 + \omega_n x_0)t]\,e^{-\omega nt} \tag{2.56}$$

It can be seen that Equation 2.56 does not show any vibratory motion. The mass comes to rest with an increase in t (Fig. 2.23). This motion is aperiodic. Figure 2.24 shows variation of (ω_d/ω_n) that is, ratio of damped natural frequency and underdamped frequency. Point A corresponds to undamped case, while point B corresponds to critically damped case.

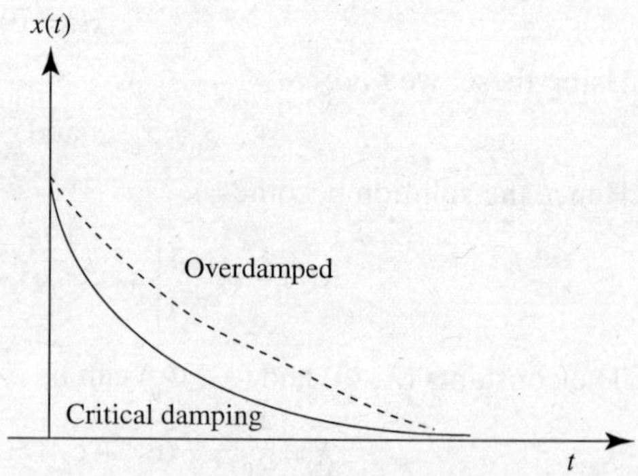

Figure 2.23 Response of Critically Damped and Overdamped Systems

The fact that in critical damping the system once disturbed comes back to rest quickly is used in measuring instruments. In these instruments, the measuring needles are critically damped so that the instrument within a short time shows the correct result and there are no oscillations of the pointer.

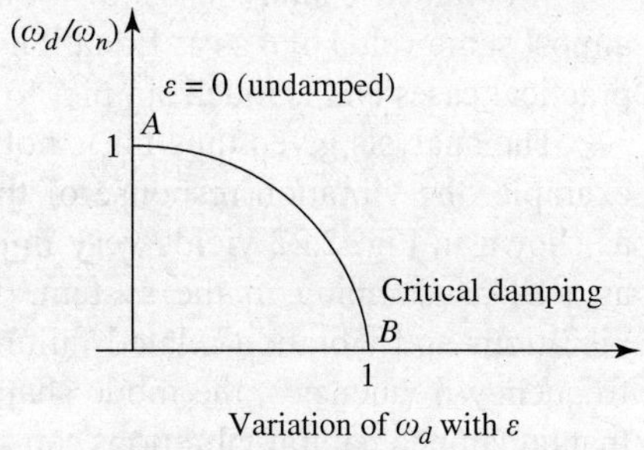

Figure 2.24

Case 3: ***Overdamped systems***

In these systems $\xi > 1$ or $f > f_c$ or $f/2m > K/m\ (= \omega_n^2)$

As $\sqrt{(\xi^2 - 1)} > 0$, equation 2.44 shows the roots to be real numbers contrary to the case when $\sqrt{(\xi^2 - 1)} \leq 0$. The roots are

$$s_1 = [-\xi + \sqrt{(\xi^2 - 1)}]\omega_n < 0, \quad s_2 = [-\xi - \sqrt{(\xi^2 - 1)}]\,\omega_n < 0 \quad \text{with} \quad s_2 << s_1$$

The solution is

$$x(t) = c_1\, e^{(-\xi + \sqrt{(\xi 2 - 1)\omega nt}} + c_2\, e^{(-\xi - \sqrt{(\xi 2 - 1)\omega nt}} \tag{2.57}$$

The initial conditions are

$$x(t = 0) = x_0\ \dot{x}\ (t = 0) = \dot{x}_0$$

Using these, we find,

$$c_1 = \{x_0\, \omega_n(\xi + \sqrt{\xi^2 - 1}) + \dot{x}_0/2\omega_n\sqrt{(\xi^2 - 1)}\},\ c_2 = \{-x_0\, \omega_n(\xi - \sqrt{(\xi^2 - 1)} - \dot{x}_0/2\omega_n\sqrt{(\xi^2 - 1)}\}, \tag{2.58}$$

Study of Equation 2.57 shows that with increase in t, $x(t)$ diminishes and ultimately becomes zero. One can also see that Equation 2.57 does not have any sine or cosine terms. Therefore, clear that with $\xi > 1$, no vibrations are possible. The response of the overdamped system is shown in Fig. 2.23 by a dotted curve.

It is now clear that free vibrations can occur only in the underdamped systems and undamped conditions. The free vibrations cease only after a certain period of time (as shown in Fig. 2.22) for underdamped systems; while they will go on and on when damping is zero (undamped condition).

Examination of Fig. 2.22 reveals that in the case of free vibrations of underdamped systems, the amplitudes of vibrations progressively decrease. This decrement is indicated by a parameter called **logarithmic decrement,** which is defined as a natural logarithm of the ratio of any two successive amplitudes. Let t_1 and t_2 denote the times corresponding to two consecutive amplitudes measured one cycle apart, this is shown in Fig. 2.22. Use of Equation 2.47 or 2.46 gives

$$X_1/X_2 = X_0\, e^{-\xi\omega nt1} \cos(\omega_d t_1 - \Phi_0)/X_0\, e^{-\xi\omega nt2} \cos(\omega_d t_2 - \Phi_0) \tag{2.59}$$

But $t_2 = t_1 + T_d$ where $T_d = 2\pi/\omega_d$ is the period of vibration.

As a result $\cos(\omega_d t_1 - \Phi_0) = \cos(\omega_d t_2 - \Phi_0)$ and hence after some simplification, we get,

$$X_1/X_2 = e^{\xi\omega Td} \tag{2.60}$$

Logarithmic decrement δ can now be obtained by taking logarithms on both sides

$$\delta = \log_e(X_1/X_2) = \xi\,\omega_n\,T_d = \xi\,\omega_n\,2\pi/\sqrt{(1 - \xi^2)} = 2\pi\,\xi/\sqrt{(1 - \xi^2)} = (2\pi/\omega_d)\cdot(f/2m) \tag{2.61}$$

For **small damping**, Equation 2.61 can be approximated as

$$\delta = \sim 2\pi\xi \qquad (\text{if } \xi \leq 1) \tag{2.62}$$

A parametric study has shown that when ξ is equal to 0.3 or less than 0.3, Equation 2.62 gives almost same value of δ as as Equation 2.61 would give. Thus Equation 2.62 is applicable to almost all practical cases and is much simpler to use.

The analysis given thus far is not just for academic purpose but has important applications. For example, the vibration response of the system when disturbed from its state of static equilibrium, as shown in Fig. 2.22 yields very important information about the natural frequency of the system as well as damping in the system. These parameters are of utmost importance while analyzing vibrations and vibration-related failures. Special methods have been devised to evaluate the natural frequency/frequencies, the mode shape of vibrations, and the system damping. It may also be seen that undamped natural vibrations can also be estimated once damped natural frequency and the damping ratio ξ are evaluated by tests. The undamped natural frequency solely depends upon the mass and elasticity distribution, which are very important in the analysis and solution of vibration problems. We shall now illustrate these principles through a few examples.

Problem 2.13

The column of a water tank, (Fig. 2.25), is 300 ft tall and is made of R.C.C with a tubular cross-section (**ID** 8 and **OD** 10 ft, respectively). The tank with full water content weighs 600000 lbs. Assuming E = 4000000 psi, find (*a*) natural frequency and the period of vibration waveform, (*b*) find the response of the tank in terms of velocity and acceleration during an earthquake activity when the water tank displaces by 10 in. Consider tank along its contents as a point mass added to the cantilever of 300 ft length.

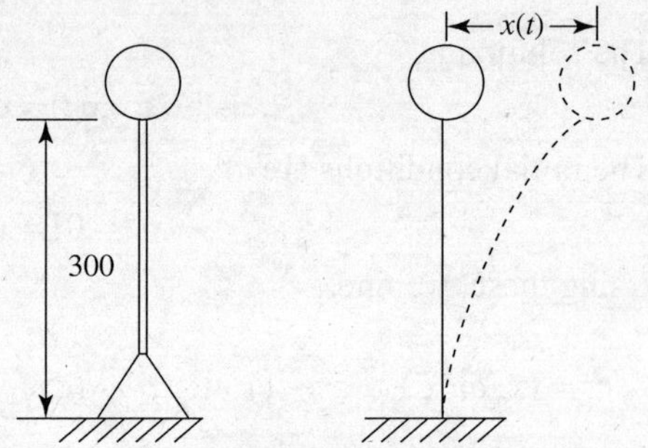

Figure 2.25 Column-mounted Water Tank

Solution:

The transverse-static deflection δ due to the transverse load on a cantilever is given by

$$\delta = PL^3/3EI \text{ or}$$

$$k = P/\delta = 3EI/L^3,\ L = 3600 \text{ in.}$$

$$E = 4000000 \text{ psi},\ I = \pi/64\ (120^4 - 96^4) = 6000000$$

Substituting these values and remembering that m = 600000/386.4, we get,

$$K = 3 \times 4000000 \times 6000000/3600^3 = 1545 \text{ lb/in.}$$

And $$K/M = \omega_n^2 = 0.98 \text{ or } \omega_n = 0.99 \text{ rad/s}$$

Also, T = period = $2\pi/\omega_n = 2\pi/0.99 = 6.3$ s.

Initial displacement $x_0 = 10$, we assume initial velocity = $\dot{x}_0 = 0$.

The response is $A = [x_0^2 + (\dot{x}_0/\omega_n)^2]^{1/2} = 10$. Since velocity is zero, phase angle $\Phi = \pi/2$

$$x(t) = 10 \sin(0.99t + \pi/2)$$

Velocity = $\dot{x}(t) = 0.99 \times 10 \times \cos(0.99t + \pi/2)$

Maximum velocity = 9.9 in./s

Acceleration = $\ddot{x} = -0.99^2 \times 10 \times \sin(0.99t + \pi/2)$

Maximum acceleration = $10 \times 0.99^2 = 9.96$ in./s^2

Dynamic load = Mass (600000/386.4) × Acceleration (9.96) = 1460000 lbs

Problem 2.14

A massless cantilever beam carries a mass M at the free end (Fig. 2.26). *A* mass m falls from a height h on the mass M and adheres to it without rebounding. Determine the resultant transverse vibration.

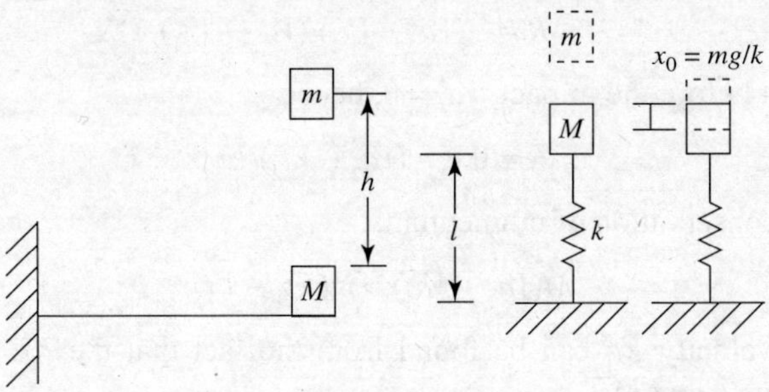

Figure 2.26 Transverse Vibration of Beam Due to Striking Mass

Solution:

Let k be the spring constant of the beam. The dropping mass m will cause this spring to compress by an amount equal to $x = mg/k$ if it does not rebound. Once the mass adheres to the mass M there is a transfer of momentum. If v is the velocity of mass m, then $mv = (m + M)V$, where V = initial velocity of spring, mass $(m + M)$. Thus we have

$$m\sqrt{2gh} = (m + M)V$$

or

$$V = [m/(m + M)]\sqrt{2gh}$$

The stiffness of the beam spring $k = 3EI/L^3$.

The free vibration system consists of mass $(m + M)$ spring k with initial displacement $x_0 = mg/K$ and initial velocity $\dot{x}_0 = V = [M/(M + m)]\sqrt{2gh}$.

The resulting transverse vibrations are $x(t) = A\cos(\omega_n t - \Phi)$ where

$$A = \{x_0{}^2 + (\dot{x}_0/\omega_n)^2\}^{1/2} \text{ and } \Phi = \tan^{-1}(\dot{x}_0/x_0\omega_n) \text{ and } \omega_n = \sqrt{3EI/(m + M)L^3}.$$

It can be noticed the fact if $m <<< M$, ω_n will correspond to natural frequency of the beam. Repeated application of mass m striking the beam will cause the beam to oscillate at its natural frequency at the same rate as that of striking of mass m. Thus, the principle behind the process is called shot pinning, which is used for introducing compressive residual stresses in beam (or any other structure) and improving its strength.

Problem 2.15

The anvil of a forging hammer weighs 5000 N and is mounted on a rigid foundation having a stiffness 5000000 N/m and viscous-damping constant 10000 N s/m. During the forging operation, the tup (falling weight or hammer) weighing 1000 N is made to fall from a height of 2 m on the anvil (Fig. 2.27). Assume that the anvil is at rest before

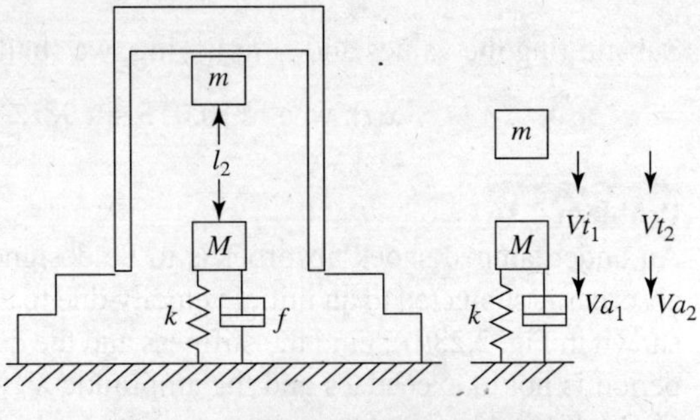

Figure 2.27 Forging Hammer

the impact takes place. Determine the response of the anvil after the impact. The coefficient of restitution between the anvil and the tup is 0.04.

Solution:

Let the velocities of the tup before and after the impact be Vt_1 and Vt_2. Let the velocities of the anvil before and after the impact be Va_1 and Va_2. As per the definition, the coefficient of restitution R is given by

$$R = -(Va_2 - Vt_2)/(Va_1 - Vt_1) \qquad (1)$$

Since the anvil is at rest before the impact, $Va_1 = 0$, hence

$$R = 0.4 = (Va_2 - Vt_2)/Vt_1 \qquad (2)$$

Using the principle of conservation of momentum,

$$M(Va_2 - Va_1) = m(Vt_1 - Vt_2) \qquad (3)$$

But $Va_1 = 0$. The velocity Vt_1 can be found from the fact that the KE of the tup is developed through the PE of the tup (= mgh).

Thus, $\quad (\frac{1}{2})mVt_1^2 = mgh$ or $Vt_1 = \sqrt{2gh} = \sqrt{2x9.81x_2} = 6.26$ m/s

Substituting this in Equation 3

$$\frac{5000}{9.81}Va_2 = \frac{1000}{9.81}(6.26 - Vt_2) \text{ or}$$

$$510.2Va_2 = 638.8 - 102\ Vt_2 \qquad (4)$$

Simplification of Equation 2 gives

$$0.4\ Vt_1 = Va_2 - Vt_2 = 25.04 \qquad (5)$$

Solving Equations 4 and 5, we get $Va_2 = 1.46$ and $Vt_2 = -1.04$ m/s

Thus the initial conditions for anvil are at $t = 0$, $x_0 = 0$ and $\dot{x}_0 = 1.46$

The damping coefficient $\xi = \dfrac{f}{2\sqrt{KM}} = 1000/2(\sqrt{5000000} \times 50000/9.81) = 0.099$

The undamped and the damped natural frequencies of the anvil are given by

$$\omega_n = \sqrt{K/M} = \sqrt{5000000/(5000/9.81)} \approx 99 \text{ rad/s}$$

$$\omega_d = \omega_n\sqrt{(1 - \xi^2)} = 98 \text{ rad/s}$$

Thus, the displacement response of the anvil is

$$x(t) = e^{-\xi\omega nt}\{x_0 \cos \omega_d + ([\dot{x}_0 + \xi\omega_n x_0]/\omega_d)\sin \omega_d t\}$$

Substituting the values and simplifying, we shall obtain,

$$x(t) \simeq e^{-9.8t}\{0.015 \sin 98t\}. \text{ (The details have been omitted.)}$$

Problem 2.16

An underdamped shock absorber is to be designed for a motor bike of mass 200 kg. When the shock absorber is subjected to an initial velocity due to a road bump, the resulting displacement time is to be as shown in Fig. 2.28(b). Find the stiffness and the damping coefficient of the shock absorber if the damped period is not to exceed 2 s and the amplitude X_1 is reduced to at least one-fourth in one half-cycle, i.e., ($X_{1.5} = X_1/4$). Also, find the minimum initial velocity that leads to maximum displacement 250 mm.

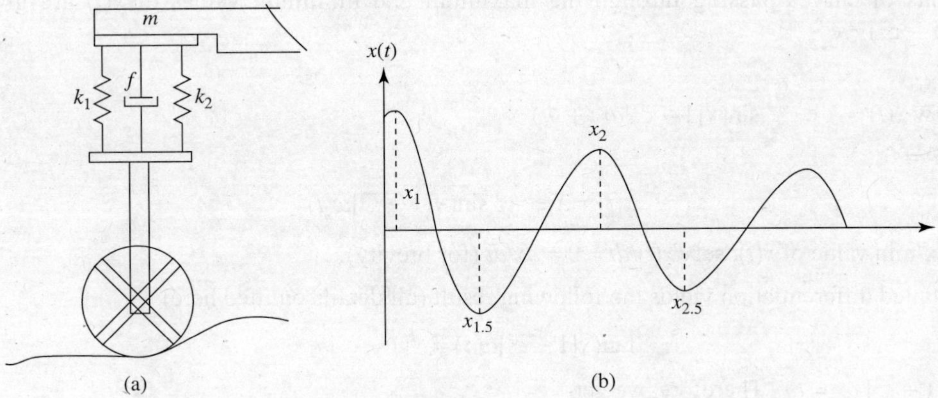

Figure 2.28 Shock Absorber for Motor Cycle

Solution:

This example reveals the process of arriving at the conceptual design of shock absorbers. The readers should note the requirement of an underdamped shock absorber. Overdamping may prove to be injurious to the rider and thus should never attempted. This is similar to closing the door smoothly instead of with a bang.

Since $X_{1.5} = X_1/4$. $X_2 = X_{1.5}/4$.

The logarithmic decrement is

$$\delta = \log (X_1/X_2) = \log 16 = 2.77$$

$$\delta = 2.77 = 2\pi/\sqrt{(1-\xi^2)}$$

This gives $\xi = 0.43$. The period of vibration is 2 s. Hence, $2 = T = 2\pi/\omega_d$.

From this we get $\omega_d = \pi$

Also, undamped natural frequency $\omega_n = \omega_d/\sqrt{(1-\xi^2)}$. From this we obtain, $\omega_n = 3.43$ rad/s

The critical damping constant $f_c = 2m\,\omega_n = 2 \times 200 \times 3.43 = 1373$ N s/m. Therefore, the damping constant is $f = 0.43 \times 1373 = 554.5$ N s/m

Since $\sqrt{(K/m)} = \omega_n$, $k = m\,(\omega n)^2 = 200 \times 3.43^2 = 2358$ N/m

Thus, we have designed the shock absorber.

Also, the displacement of the mass will attain its maximum value at time t_1 given by

$$\sin \omega_d t_1 = \sqrt{(1-\xi^2)}, \quad \sin \pi t_1 = \sqrt{(1-0.403^2)} = 0.915$$

This gives $t_1 = 0.3678$ s.

For the remaining part, see the next problem.

Problem 2.17

Assuming that the phase angle is zero, show that the response $x(t)$ of an underdamped single degree-of-freedom system reaches a maximum value when $\sin \omega_d = -(1-\xi^2)$. Also, show that equations of curves passing through the maximum and minimum values of $x(t)$ are given by $x = \pm(1 - \xi^2)^{1/2}\, e^{-\xi\omega nt}$.

Solution:

We know $x(t) = X\, e^{-\xi\omega nt} \sin(\sqrt{[1 - \xi^2]}\omega_n t + \Phi)$

Since $\Phi = 0$,

$$x(t) = X\, e^{-\xi\omega nt} \sin(\sqrt{[1 - \xi^2]}\omega_n t) \tag{1}$$

For max/min value of $x(t)$, set $dx(t)/dt = 0 = dx/dt$ (for brevity).

The detailed differentiation yields the following result (all details omitted here).

$$\tan(\sqrt{[1 - \xi^2]}\omega_n t) = \sqrt{[1 - \xi^2]}/\xi$$

But $\sqrt{[1 - \xi^2]}\, \omega_n = \omega_d$. Therefore, we get,

$$\tan \omega_d t = \sqrt{[1 - \xi^2]}/\xi$$

$$1 + \tan^2 \omega_d t = 1 + (1 - \xi^2)/\xi^2 = 1/\xi^2 = \sec^2 \omega_d t.$$

With this, we obtain

$$\sec \omega_d t = 1/\xi \text{ or } \cos \omega_d t = \xi.$$

From this we obtain

$$\sin^2 \omega_d t = 1 - \xi^2 \text{ or}$$

$$\sin \omega_d t = \pm\sqrt{(1 - \xi^2)} \tag{2}$$

The (+) sign in Equation 2 corresponds to maximum value of $x(t)$ while (–) sign corresponds to a minimum value. We now substitute these values in Equation 1

$$x_{max} = X\, e^{-\xi\omega nt} \sin(\sqrt{[1 - \xi^2]}\omega_n t) = X\, e^{-\xi\omega nt} \sin \omega_d t = X\, e^{-\xi\omega nt} \sqrt{(1 - \xi^2)}$$

$$x_{max} = \sqrt{(1 - \xi^2)}\, X\, e^{-\xi\omega nt} \tag{3}$$

$$x_{min} = -\sqrt{(1 - \xi^2)}\, X\, e^{-\xi\omega nt} \tag{4}$$

The value of the above equations mean that the envelope passing through the maximum points is given by

$$x = \sqrt{(1 - \xi^2)} X\, e^{-\xi\omega nt}$$

For Problem 2.16, we substitute $x = 0.25$, $\xi = 0.403$, $\omega_n = 3.43$ to obtain $X = 0.45$

For velocity, Equation 1 can be differentiated to obtain initial velocity $= X\omega_d$. Using the maximum value of X (= 0.45), we can obtain the initial velocity as 1.43 m/s. The readers should solve it further.

We shall now deal with yet another design-related problem where the concepts that we have discussed earlier are applied.

Problem 2.18

It is required to design a support system for a large cannon schematic of which has been shown in Fig. 2.29. It is necessary that after the shell is delivered the recoil is without oscillations. The gun barrel and recoil mechanism have a mass of 500 kg. The recoil spring has a stiffness of 10000 N/m. The gun recoil is set to be 0.4 m upon firing. Find (*a*) the damping coefficient, (*b*) initial recoil velocity, and (*c*) time taken by the gun to return to a position 0.1 m from its initial position.

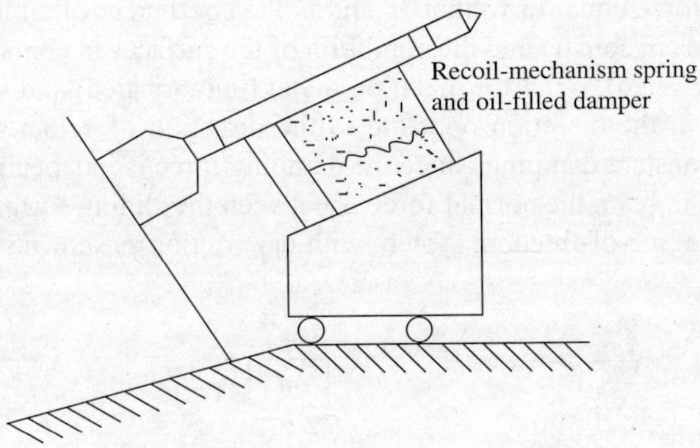

Figure 2.29 Recoil of Cannon

Solution:

(i) Here, we need to put a critical damper such that the gun does not oscillate. The undamped natural frequency of the system $\omega_n = \sqrt{(K/m)} = \sqrt{(10000/500)} = 4.47$ rad/s

Critical damping coefficient $f_c = 2m\omega_n = 2 \times 500 \times 4.47 = 4470$ N s/m

(ii) We know that the response of the critically damped system is given as

$$x(t) = (c_1 + c_2 t)\, e^{-\omega nt} \quad \text{where} \quad c_1 = x_0 \text{ and } c_2 = \dot{x}_0 + \omega_n x_0$$

Differentiating, we get $\dot{x}(t) = c_2 e^{-\omega nt} - \omega_n(c_1 + c_2 t)\, e^{-\omega nt}$. For maximum displacement, $\dot{x}(t) = 0$

Hence, $\dot{x}(t) = 0$ gives $t_1 = (1/\omega_n) - c_1/c_2$

In this case $c_1 = x_0 = 0$ and hence $t_1 = (1/\omega_n)$.

Since the maximum value of $x(t = t_1) = 0.4$,

$$x(t = t_1) = 0.4 = c_2 t_1 e^{-\omega_n t_1} = \frac{\dot{x}_0 e^{-1}}{\omega_n} = \frac{\dot{x}_0}{e\omega_n}$$

From this, we get initial velocity $\dot{x}_0(t = 0) = 0.4 \times 4.47 \times 2.718 = 4.86$ m/s.

(iii) If t_2 denotes the time taken by the gun to return to a position 0.1 m from its initial position, we have $0.1 = c_2 t_2\, e^{-\omega_n t_2} = 4.86\, t_2 e^{-4.47 t_2}$. Solution to this equation gives $t_2 = 0.82$.

We shall now deal with Coulomb and hysteretic damping.

2.3.2 Free Vibrations of Single Degree-of-Freedom System—Coulomb and Hysteretic Damping

Coulomb Damping

In many mechanical systems, components slide or move relative to each other. The relative motion causes frictional forces, which provide resistance to motion. The frictional resistance thus acts as a damping force. The steam- or gas-turbine blades employ damping ring or damping pins, which provide frictional resistance to the vibrating blades and thus control the vibrations of the rotating blades.

The dry friction or Coulomb damping appears internally. Coulomb's law of dry friction states that, when two bodies are in contact, the force required to provide sliding is proportional to the normal force acting in the plane of contact. Thus, the friction force F is given by

$$F = \mu N = \mu W = \mu mg \tag{2.63}$$

where N is the normal force equal to weight W and is the coefficient of sliding or kinetic friction μ depending upon materials in contact and the condition of the surfaces in contact. For example, $\mu = 0.1$ for metal on metal (lubricated), $= 0.3$ for metal on metal (unlubricated) and ≈ 1 for rubber on metal. The friction force acts in the direction opposite to the direction of velocity. Coulomb damping is sometimes also called constant damping, since the damping force is independent of the displacement and velocity; it depends only on the normal force N between the sliding surfaces.

Consider a single degree-of-freedom system with dry friction as shown in [Fig. 2.30(a)].

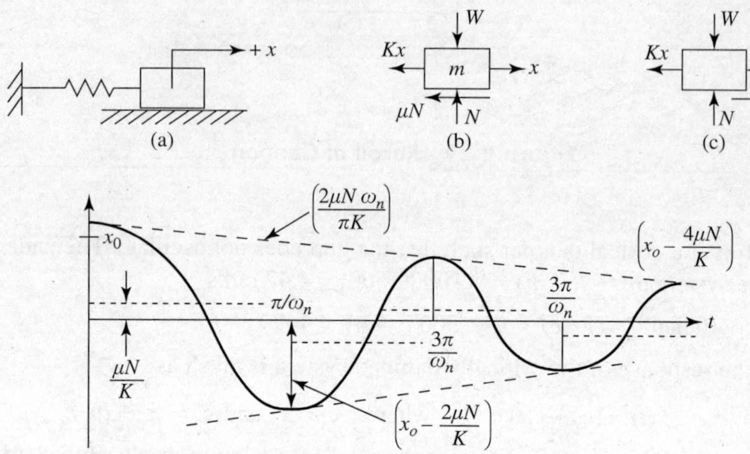

Motion of mass with Coulombs

(d)

Figure 2.30 SDOF with Dry Friction

Since friction varies with the direction of velocity, we need to consider two cases. These include case 1 when x is +ve and dx/dt is +ve or when x is −ve and dx/dt is −ve that is, for the half cycle during which the mass moves from left to right [Fig. 2.30(b)], and Case 2 when x is +ve and dx/dt is −ve or when x is −ve and dx/dt is −ve that is, for the half cycle during which the mass moves from right to left [Fig. 2.30(c)].

For Case 1, [Fig. 2.30(b)], equation of equilibrium is

$$m\ddot{x} + kx = -\mu N \tag{2.64}$$

This is a second-order non-homogeneous differential equation. The solution of this equation (proof omitted) is

$$x(t) = A_1 \cos \omega_n t + A_2 \sin \omega_n t - \mu N/k \tag{2.65}$$

where $\omega_n = \sqrt{k/m}$ A_1 and A_2 are constants the values of which depend upon the initial conditions of the half-cycle.

For the Case 2 (Fig. 2.30), the equation of equilibrium is

$$m\ddot{x} + kx = \mu N$$

The solution of this equation is

$$x(t) = A_3 \cos \omega_n t + A_4 \sin \omega nt + \mu N/k \tag{2.66}$$

where A_3 and A_4 are constants to be obtained from the initial conditions of this half-cycle.

The term $\mu N/k$ appearing in Equations 2.65 and 2.66 is a constant representing the virtual displacement of the spring under force μN, if it were applied as a static force. Equations 2.65 and 2.66 indicate that in each half-cycle, the motion is harmonic and the equilibrium position changes from $\mu N/k$ to $-\mu N/k$ every half-cycle as shown in [Fig. 2.30(d)]. Equations 2.65 and 2.66 can be expressed as a single equation (using $N = mg$)

$$m\ddot{x} + \mu mg\, sgn(\dot{x}) + kx = 0 \tag{2.67}$$

where $sgn(\dot{x})$ is called signum function, the value of which is defined as 1 for $x > 0$, -1 for $x < 0$ and 0 for $x = 0$. The Equation 2.67 is a non-linear differential equation for which a simple close form solution does not exist. The solution can be found only from using numerical methods. However, be solved if we break the time axis into segments separated by $\dot{x} = 0$ (i.e. time intervals with different directions of motion). The readers may refer to the book Mechanical Vibrations by Singiresu S. Rao (Pearson Education 2004).

The important characteristics of a system with Coulomb damping are

- Equation of motion is nonlinear with Coulomb damping, whereas it is linear when the damping is linear.
- Natural frequency of the system is unaltered with addition of Coulomb damping while it reduces with addition of viscous damping.
- The motion is periodic with Coulomb damping while depending upon the amount of damping, the motion can be aperiodic in viscous damping (overdamped system).
- The system comes to rest after sometime with Coulomb damping whereas the motion takes place for a longer duration with viscous damping.
- Amplitude reduces linearly with Coulomb damping whereas it decreases exponentially with viscous damping.

Hysteretic Damping

When the material is deformed due to stresses, energy is absorbed and dissipated by the material. This effect is due to friction between the internal planes, which slip or slide as the deformation takes place. When a body having such a damping (material damping) is subjected to vibration it experiences alternating stresses and strains. The alternate stressing and straining gives rise to what is called hysteresis loop as shown in Fig. 2.31.

The area under the loop denotes the energy lost per unit volume of the body per cycle due to damping. If we represent the viscous damping as shown in Fig. 2.31, then

$$F = kx + f\dot{x} \tag{2.68}$$

Let $\qquad x = X\sin\omega t$

$$F(t) = kX\sin\omega t + \omega f X\cos\omega t = kx \pm \sqrt{(X^2 - [X\sin\omega t]^2)}$$

Then

$$= kx \pm f\omega\sqrt{(X^2 - x^2)} \tag{2.69}$$

When F versus x is plotted, Equation 2.69 shows a closed loop as shown in Fig. 2.31. Area of the loop is equal to the energy dissipated in damper $= \oint F dX = \pi\omega f X^2$ (2.70)

Proof: Energy dissipated $= \Delta W = \int^{2\pi/\omega} (kX\sin\omega t + fX\omega cs\omega t)(\omega X\cos\omega t)dt = \pi\omega f X^2$ (2.71)

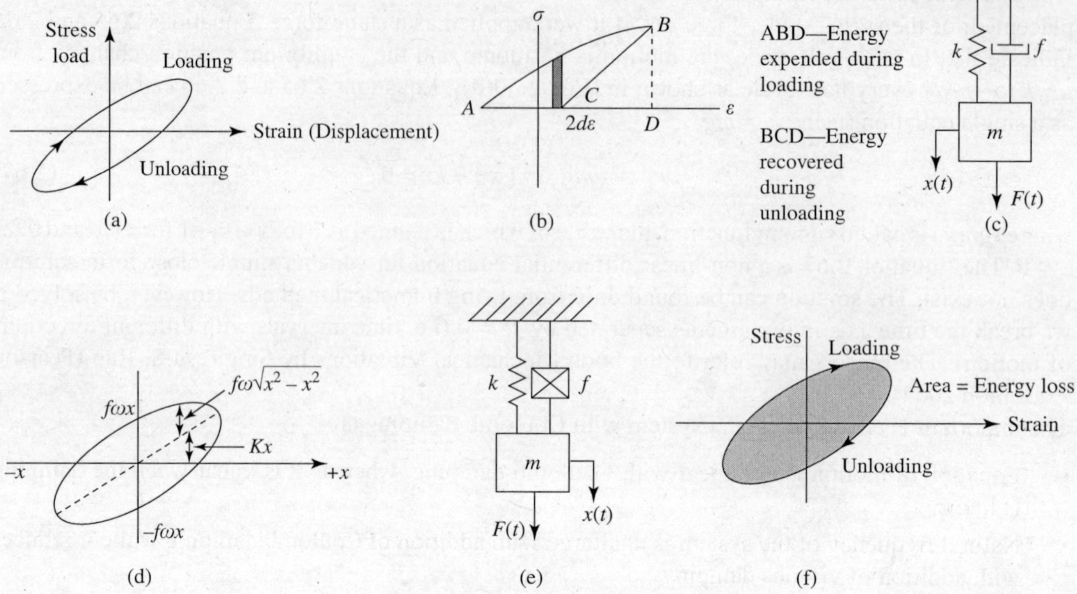

Figure 2.31 Hysteresis Loops

As stated earlier, the damping caused by the friction between the internal planes that slip or slide as the material deforms is called the hysteretic/solid or structural damping. This causes a hysteresis loop to be formed as shown in Fig. 2.31. The energy loss in one cycle of loading and unloading is equal to the area enclosed by the hysteresis loop. The similarity between loop figures can be used to define the hysteretic damping constant. The experimental data shows that the energy loss per cycle in hysteretic damping is independent of the frequency but approximately proportional to the square of the amplitude. In order to achieve this observed behaviour from Equation 2.70, the damping coefficient f is assumed to be inversely proportional to the frequency as

$$f = h/\omega \tag{2.72}$$

In Equation 2.72, h is called the hysteretic damping coefficient.
Combining Equations 2.70 and 2.72, we get,

$$\Delta W = \pi h X^2 \tag{2.73}$$

Complex stiffness: In Fig. 2.31, the spring and the damper are connected in parallel and for a general harmonic motion $x = X\,e^{i\omega t}$, the force is given by

$$F = kX\,e^{i\omega t} + f\omega iX\,e^{i\omega t} = (k + ifw)X\,e^{i\omega t} \tag{2.74}$$

Similarly, if a spring and hysteretic damper are connected in parallel, as shown in Fig. 2.31, the force–displacement relation is expressed as

$$F = (k + ih) \tag{2.75}$$

where $$(k + ih) = k(1 + ih/k) = k(1 + i\beta) \tag{2.76}$$

is called the complex stiffness of the system and $\beta = h/k$ is a constant indicating dimensionless measure of damping.

Response: In terms of β, the energy loss per cycle can be expressed as

$$\Delta W = \pi k \beta X^2 \tag{2.77}$$

As ΔW is very small under hysteretic damping, the motion is nearly harmonic, and the decrease in amplitude per cycle can be determined using energy balance.
For example, the energies at point P and Q (separated by half-cycle) in Fig. 2.32 are related $\frac{1}{2}(kX_J^2) - \pi k\beta X_J^2/4 - \pi k \beta X_{J+0.5}^2 = \frac{1}{2}(X_{J+0.5}^2)$

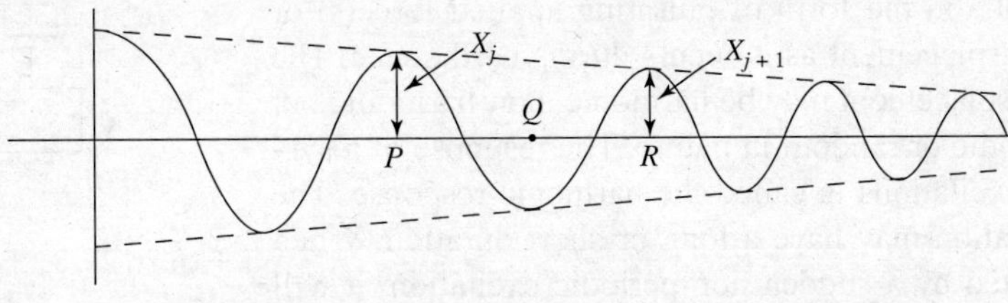

Figure 2.32 Decrement in Hysteretic Damping

or
$$X_J/X_{J+0.5} = [(2+\pi\beta)/(2-\pi\beta)]^{1/2} \tag{2.78}$$

Similarly energies at Q and R are related as

$$X_{J+0.5}/X_{J+1} = [(2+\pi\beta)/(2-\pi\beta)]^{1/2} \tag{2.79}$$

Equations 2.78 and 2.79 give

$$X_J/X_{J+1} = [(2+\pi\beta)/(2-\pi\beta)] = 1 + \pi\beta = \text{constant} \tag{2.80}$$

Hysteretic logarithmic decrement can now be defined as

$$\delta = \log(X_J/X_{J+1}) = \log(1+\pi\beta) \approx \pi\beta \tag{2.81}$$

Since the motion is assumed to be nearly harmonic the corresponding natural frequency is

$$\Omega = \sqrt{k/m} \tag{2.82}$$

The equivalent damping ratio ξ_{eq} can be found by equating the relation for the logarithmic decrement δ

$$\delta \approx 2\xi_{eq} = \beta/2 = h/2k \tag{2.83}$$

The equivalent damping constant f_{eq} is given by

$$f_{eq} = f_c\,\xi_{eq} = 2\sqrt{mk} \times \beta/2 = \beta\sqrt{mk} = \beta k/\omega = h/\omega \tag{2.84}$$

Hysteretic damping is of considerable interest when vibrations are associated with cyclic deformations.

We have learnt about free undamped vibration here viscous damping was considered. In practicality such cases are not many but a large number of engineering structures/components such as aircraft wing, turbine/compressor/pump impellers etc. exhibit Coulomb and hysteretic damping but their analyses are quite complex. Therefore, it is customary to consider the damping as viscous damping to facilitate analysis. The error involved in this approach is not significant and for all practical purposes treating vibrations system as consisting of viscous damping is acceptable.

Now, we turn our attention towards forced vibrations of the single degree-of-freedom vibration system consisting of a spring, mass, and damper subjected to (1) harmonically varying force $F = F_0 \cos \omega t$, (2) a general periodically varying force.

2.4 FORCED VIBRATIONS OF SINGLE DEGREE-OF-FREEDOM SYSTEM TO HARMONIC EXCITATION FORCE

A mechanical or structural system experiences forced vibration whenever external energy is supplied to the system (Fig. 2.33). The input energy may be in the form of pulsating applied force(s) or imposed ground displacement as it occurs during earthquake. The applied force or displacement may be harmonic, non-harmonic but periodic, non-periodic or random in nature. The response of a system to harmonic excitations is called the harmonic response. The non-periodic excitation may have a long or short duration. When the system is excited by a sudden non-periodic excitation, it will exhibit a vibration response which is called transient response.

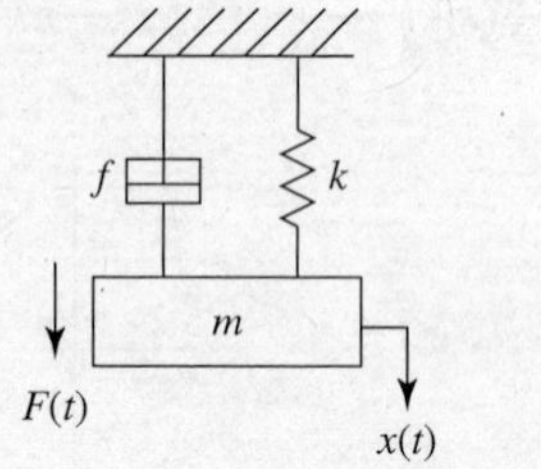

Figure 2.33 Forced Vibration of SDOF System

We shall now consider the dynamic response of a single degree-of-freedom system subjected to a harmonic excitation force described by $F(t) = F_0 e^{i(\omega t + \Phi)}$ or $F(t) = F_0 \sin(\omega t + \Phi)$, or $F(t) = F_0 \cos(\omega t + \Phi)$. F_0 is the amplitude of the force, ω is the frequency of the force, and Φ is the phase angle of the force, which depends upon the value of $F(t)$ at $t = 0$. Usually it is considered to be zero.

It is natural to expect that if the frequency of the excitation force is ω, the response will also be harmonic. The response is in the form of vibration displacement, which results into stressing and unstressing of the elastic part (spring) of the body. This causes fatigue of the vibrating body. Depending upon the magnitude of the pulsating stresses and time duration of their action, fatigue-induced failure takes place. It is for this reason that we are concerned about the response of the mechanical or structural system.

Let us first consider a spring, mass, and viscous damper system subjected to excitation force $F(t)$. The equation of motion (Equation 2.24) is rewritten as

$$m\ddot{x} + f\dot{x} + kx = F(t) \tag{2.85}$$

The solution of this equation is given by

$x(t)$ = Complimentary function (CF) + Particular integral (PI)/(Solution).

The complimentary function (CF), x_{CF} is the solution of the homogeneous equation

$$m\ddot{x} + f\dot{x} + kx = 0 \tag{2.86}$$

This equation represents the free vibration of the system. As seen in previous section, the free vibration dies down rapidly with time, instantaneously or slowly depending upon the nature of damping in the system, that is, overdamped/critically damped or underdamped. Thus, in the case of damped system, the solution of Equation 2.85 consists of particular integral (solution) only as the steady-state solution. The solution of the equation of motion of an undamped system subjected to harmonic excitation force (Equation 2.85 with $f = 0$) will comprise for both CF and PI. We shall therefore, first consider the case of forced vibrations of the undamped system.

2.4.1 Response of an Undamped System to Harmonic Excitation Force

Let a force $F(t) = F_0 \cos \omega t$ act on the system comprising of spring and mass only (Fig. 2.34).

The governing equation is

$$m\ddot{x} + kx = F_0 \cos \omega t \qquad (2.87)$$

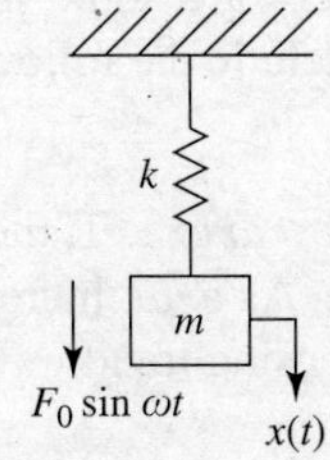

Figure 2.34 Undamped Forced Vibration of SDOF System

The solution of this equation is

$$x(t) = \text{CF} + \text{PI} = x_n(t) + x_p(t)$$

where $x_n(t)$ is the solution of the homogeneous equation

$$m\ddot{x} + kx = 0 \qquad (2.88)$$

As mentioned earlier, $x_n(t)$ can be written as

$$x_n(t) = C_1 \cos \omega_n t + C_2 \sin \omega_n t \qquad (2.89)$$

where C_1 and C_2 are constant and $\omega_n = \sqrt{(k/m)}$ = natural frequency.

Because the exciting force $F(t)$ is harmonic the particular solution is also harmonic and has the same frequency ω. Thus, we assume a solution in the form

$$x_p(t) = X \cos \omega t \qquad (2.90)$$

By substituting Equation 2.90 in Equation 2.87, and solving for X, we get,

$$X = F_0/(k - m\omega^2) = (F_0/k)/[1 - (m/k)\omega^2] = \delta_{st}/[1 - (\omega/\omega_n)^2] \qquad (2.91)$$

where $\delta_{st} = F_0/k$ denotes the static deflection of the mass under force F_0 (constant). Thus, the total solution of Equation 2.87 becomes

$$x(t) = C_1 \cos \omega_n t + C_2 \sin \omega_n t + \frac{F_0}{k - m\omega^2} \cos \omega t \qquad (2.92)$$

The initial conditions $x(t = 0) = x_0$ and $\dot{x}(t = 0) = \dot{x}_0$.

Putting these in Equation 2.92, we get,

$$C_1 = x_0 - \frac{F_0}{k - m\omega^2}, \quad C_2 = \frac{\dot{x}_0}{\omega_n}$$ and hence Equation 2.92 becomes

$$x(t) = \left(x_0 - \frac{F_0}{k - m\omega^2} \right) \cos \omega_n t + \left(\frac{\dot{x}_0}{\omega_n} \right) \sin \omega_n t + \frac{F_0}{k - m\omega^2} \cos \omega t \qquad (2.93)$$

The maximum amplitude of X in Equation 2.91 can be expressed as

$$\frac{X}{\delta_{st}} = \frac{1}{1 - \left(\frac{\omega}{\omega_n} \right)^2} \qquad (2.94)$$

The quantity $\frac{X}{\delta_{st}}$ represents the ratio of dynamic to the static amplitude of motion and is called the magnification factor or amplification factor or amplitude ratio. The variation of the amplitude ratio $\frac{X}{\delta_{st}}$ with the frequency ratio $r = \omega/\omega_n$ (Equation 2.94) is shown in Fig. 2.35.

There are some important cases to be considered. They include the following.

Case 1:

When $0 < \omega/\omega_n < 1$, the denominator in Equation 2.94 is positive. As ω/ω_n increases, the magnification factor has an increasing trend.

Case 2:

When $\omega/\omega_n > 1$, the denominator in Equation 2.04 is negative. The magnification factor has a decreasing trend. Thus,

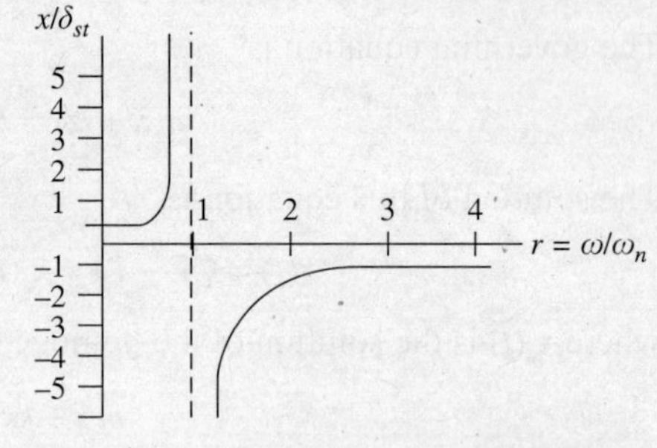

Figure 2.35 Variation of Amplitude Ratio with Frequency Ratio

$$X = \frac{\delta_{st}}{(\omega/\omega_n)^2 - 1} \tag{2.95}$$

Figure 2.36 show the force/time and response/time distributions for Cases 1 and 2.

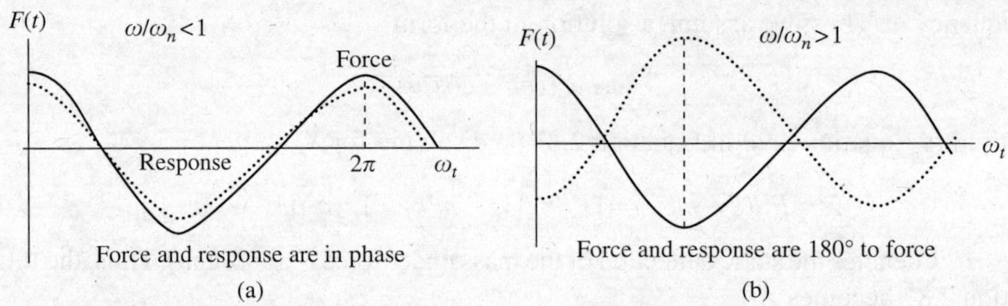

Figure 2.36 Force/Time vs Response Time

It can be seen that when $(\omega/\omega_n) \to$ infinity, $X \to 0$ (Equation 2.95).

Case 3:

When $\omega/\omega_n = 1$, the amplitude X (Equation 2.95) becomes infinite. This condition, that is, the excitation frequency ω is equal to the natural frequency ω_n which is called resonance. The infinite value of X means very large stresses and deformations in the component/structure causing their failures. It is for this reason it is very important to know the natural frequency of the system, which depends upon the stiffness (elasticity) and the mass of the system.

Let us now find out the total response for this condition, that is, resonance. For this purpose we rewrite Equation 2.93 as

$$x(t) = x_0 \cos\omega_n t + \frac{\dot{x}}{\omega_n}\sin\omega_n t + \delta_{st}\left[\frac{[\cos\omega t - \cos\omega_n t]}{1-(\omega/\omega_n)^2}\right] \tag{2.96}$$

At $\omega = \omega_n$, we know that the response becomes infinite. Let us see as to how the resonance condition is reached as $\lim(\omega/\omega_n) \to 1$. For this purpose, we apply L' Hospital's rule to evaluate this limit

$$\lim_{\omega\to\omega_n}\left\{[\cos\omega t - \cos\omega_n]/(\omega/\omega_n)^2\right\} = \lim_{\omega\to\omega_n}\left[\frac{d/d\omega(\cos\omega t - \cos\omega_n t)}{d/d\omega(1-\omega^2/\omega_n^2)}\right]$$

$$\lim_{\omega\to\omega_n}\frac{t\sin\omega t}{2\omega/\omega_n} = \frac{\omega_n t\sin\omega_n t}{2} \tag{2.97}$$

Thus at $\omega \to \omega_n$, the response of the system becomes

$$x(t) = x_0 \cos\omega_n t + \frac{\dot{x}_0 \sin\omega_n t}{\omega_n} + \frac{\delta_{st}\omega_n t}{2}\sin\omega_n t \tag{2.98}$$

Equation 2.98 shows that at resonance, $x(t)$ increases indefinitely. The last term, $\delta_{st}\omega_n t/2$ is shown in Fig. 2.37 from which the amplitude of the response can be seen to increase linearly with time. The total response of the system, Equation 2.92 or 2.93 can also be expressed as

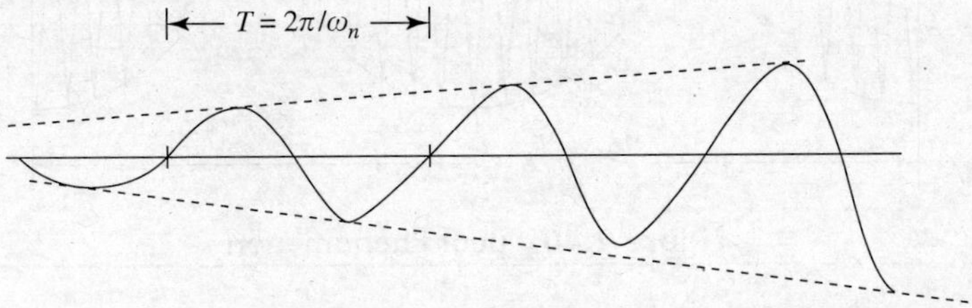

Figure 2.37 Forced Vibration Response of Undamped System

$$x(t) = A\cos(\omega_n t + \phi I) + \frac{\delta_{st}}{1-(\omega/\omega_n)^2}\cos\omega t, \quad \text{for } \frac{\omega}{\omega_n} < 1 \quad \text{and} \tag{2.99(a)}$$

$$x(t) = A\cos(\omega_n t - \phi) - \frac{\delta_{st}}{1-(\omega/\omega_n)^2}\cos\omega t, \quad \text{for } \frac{\omega}{\omega_n} > 1 \tag{2.99(b)}$$

Thus, the complete motion can be expressed as a sum of two cosine curves of different frequencies, ω and ω_n. The responses are shown in Fig. 2.38 and 2.39.

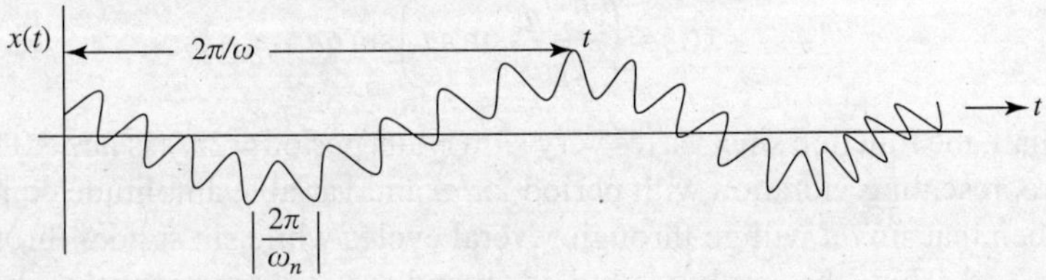

Figure 2.38 Total Response (ω/ω_n <1)

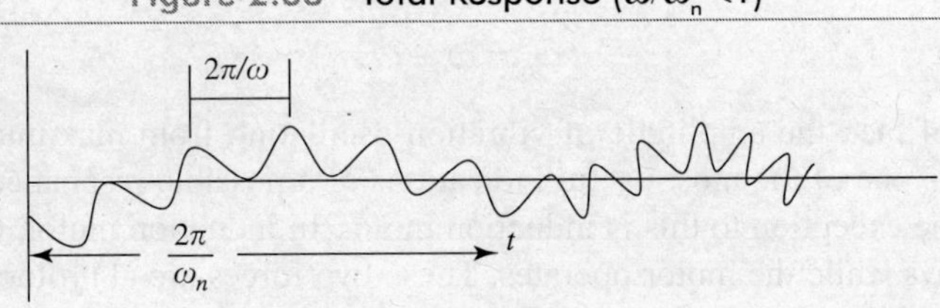

Figure 2.39 Total Response (ω/ω_n >1)

Response of the System whose Natural Frequency is Close to Forcing Frequency

There are situations where the natural frequency does not exactly match with the forcing frequency but is very close to it. In other words,

$$\omega_n - \omega = 2\varepsilon \text{ and } \omega_n + \omega \approx 2\omega \tag{2.100 and 2.101}$$

In such situations, the amplitude builds up and then diminishes in a regular pattern as shown in Fig. 2.40.

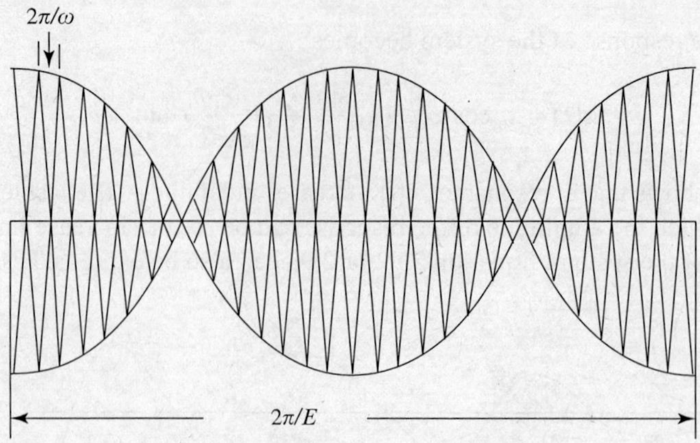

Figure 2.40 Beat Phenomenon

This pattern called beating can be explained by considering the solution given by Equation 2.93. Consider $x_0 = \dot{x}_0 = 0$.

Then Equation 2.93 becomes

$$x(t) = \frac{F_0/m}{\omega_n^2 - \omega^2}(\cos \omega t - \cos \omega_n t).$$

This then reduces to

$$x(t) = \frac{F_0/m}{\omega_n^2 - \omega^2}\left[\frac{2\sin(\omega_n + \omega)}{2} * \sin(\omega - \omega_n)\right]$$

This further reduces to

$$x(t) = \left(\frac{F_0/m}{2\varepsilon\omega}\sin \varepsilon.t\right)\sin \omega t \tag{2.102}$$

Since ε is small, the function sinεt varies very slowly; its period = $2\pi/\varepsilon$ is large. The Equation 2.102 may be seen as representing vibration with period $2\pi/\omega$ and variable amplitude equal to $[(F_0/m)/2\varepsilon\omega]$ sinεt. It can be seen that sin ωt will go through several cycles while sin εt goes through a single cycle as shown in Fig. 2.45. Thus, the amplitude builds up and reduces continuously. The period of beat is $T = 2\pi/2\varepsilon = 2\pi/(\omega_n - \omega)$ and the beat frequency is

$$\omega_b = \omega_n - \omega = 2\varepsilon \tag{2.103}$$

As a matter of fact, the amplitude of vibration oscillating from maximum to minimum (sometimes zero also), is one of the most useful indications of a possible resonance phenomenon in rotating machinery. One exception to this is induction motor. In induction motor, there are two excitation forces acting always while the motor operates. These two forces are (1) rotor unbalance force which occurs at a speed equal to the speed corresponding to line frequency minus the slip speed. For a 50 Hz two-pole induction motor having, say, 3% slip, the rotating speed is 3000 – 3000.3/100 = 2910 rpm. Thus, the frequency of unbalance force is 2910/60 = 48.33 Hz. (2) electrical excitation force at 50 Hz. These two excitation forces, closely spaced in their frequencies of action produce beat-like vibration behaviour. For a well-balanced rotor, these beats will be insignificant, though always present. In case the rotor has higher unbalance or say, a rotor bar has cracked, these beats will be very significant and will always occur at slip frequency.

We shall now solve a few numerical problems for clarifying the concepts discussed so far.

Problem 2.19

A 150 lb reciprocating engine is mounted on a 100 in. long steel beam (both ends fixed) having a cross-section of 20 in. width and 0.5 in. thickness (Fig. 2.41). Find (*a*) resonant frequency of the structure, (*b*) amplitude of vibration if the engine provides a pulsating force $F(t) = 50 \cos 62.832t$, (*c*) find the response if the engine speed reduces causing the pulsating force to change to $F(t) = 40 \cos 54t$, (*d*) if the engine is required to run continuously at reduced speed and generates a disturbing force $F(t) = 40 \cos 54t$, find the response. In case resonance condition is seen, find out changes required to be implemented.

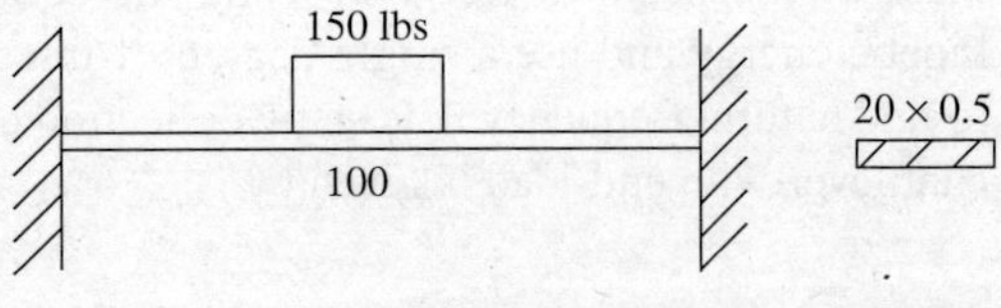

Figure 2.41 Engine Mounted on Fixed Beam

Solution:

(*a*) The area moment of inertia of the beam (massless) $I = 1/12 \times 20 \times 0.5^3 = 0.2083$ in^4.

The bending stiffness $k = 192EI/l^3 = 192 \times 30 \times 10^6 \times 0.2083/100^3 = 1200$ lb/in.

$$\omega_n = \sqrt{k/m} = \sqrt{1200/(150/386.4)} = 56 \text{ rad/s}$$

$$(b)\quad X = \frac{F_0}{k - m\omega^2} = \frac{50}{1200 - (150/386.4)62.83^2} = -0.1504 \text{ in}$$

This is a expected result since $\omega/\omega_n > 1$

$$(c)\quad X = \frac{F_0}{k - m\omega^2} = \frac{40}{1200 - (150/386.4)54^2} = 0.44 \text{ in}$$

(*d*) Since the margin between the natural frequency (56 rad/s) and excitation frequency (54 rad/s) is very small, it is advisable to increase this margin as the dynamic displacement (0.44 in.) of the order of thickness of the beam will induce very high dynamic stresses and the fatigue failure of the beam. Since nothing can be done to alter the frequency of the excitation frequency, we can increase the margin between the natural frequency of the beam and the frequency of excitation force by stiffening the beam and increasing the natural frequency of the beam. Readers may verify that increasing the thickness of the beam from 0.5 to 0.75 in. will make the natural frequency of the beam ~70 rad/s. In many instances, it is impossible to alter the frequency of the excitation force. Thus, the only way left is to alter the natural frequency of the structure/component. This requires alteration of the stiffness either by stiffening the member or making it more flexible. There are instances when the mass of the beam cannot be neglected while finding the natural frequency. We shall describe the procedure to take the mass of the beam into account through a solved example.

Problem 2.20

A cantilever beam of length l carries a weight $P(= Mg)$ at its free end (Fig. 2.42). Find the natural frequency of the beam by including the mass of the beam. A dynamic force $F = F_0 \sin 0.9\omega t$ acts on the beam. Under this condition, evaluate the effect of the self-weight of the beam on the response. $\omega = \sqrt{kg/P}$, where k is the stiffness of the beam.

Solution:

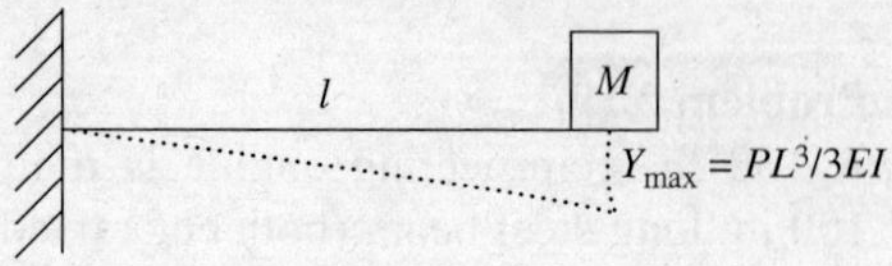

Figure 2.42 Cantilever Beam with End Mass

To include the mass of the beam, we find the equivalent mass of the beam at free end using the equivalence of kinetic energy and use a single degree-of-freedom model to find natural frequency. The static deflection curve of the cantilever with end-load is given by

$$y(x) = \frac{Px^2(3l-x)}{6EI} = \frac{y_{max}x^2(3l-x)}{2l^3} = \frac{y_{max}(3x^2l-x^3)}{2l^3} \tag{1}$$

The maximum *KE* of the beam itself (T_{max}) is given as

$$T_{max} = \frac{1}{2}\int \frac{m}{l}\{\dot{y}(x)\}^2\,dx \tag{2}$$

where m = total mass of the beam, m/l is the mass per unit length.

Equation 1, in terms of velocity can be written as

$$\dot{y}(x) = \frac{\dot{y}_{max}x^2(3l-x)}{2l^3} \tag{3}$$

$$\therefore T_{max} = \frac{m}{2l}\left(\frac{\dot{y}_{max}}{2l^3}\right)^2 \int_o^l (3x^2l-x^3)^2\,dx = \frac{1}{2}\frac{m}{l}\frac{\dot{y}_{max}^2}{4l^6}\frac{33l^7}{35} = \frac{1}{2}\left(\frac{33}{140}m\right)\dot{y}_{max}^2 \tag{4}$$

If m_{eq} denotes the equivalent mass of the beam at the free end, its *KE* is given by

$$T_{max} = \tfrac{1}{2}m_{eq}\dot{y}_{max}^2 \tag{5}$$

Equating Equations 4 and 5, we get

$$m_{eq} = \frac{33}{140}m \tag{6}$$

Total effective mass at the end of the beam is $M_{eff} = M + m_{eq}$. (7)

Natural frequency considering mass of the beam $= \sqrt{k/M_{eff}} = \sqrt{k\Big/\left(M + \frac{33}{140}m\right)} = \omega_n$

$(\omega_n)^2 = \dfrac{k}{M + \frac{33}{140}m}$ and $(\omega')^2 = \dfrac{k}{M}$ = natural frequency neglecting the mass of the beam.

Thus $\omega' > \omega_n$.

Let r be the mass ratio m/M. In that case

$$\omega'^2 - \omega_n^2 = \frac{k}{m} - \frac{k}{M + \frac{33}{140}m} = \frac{33mk/140}{M(M+33m/140)} = \frac{(33mk/140M)}{M+33m/140} = \frac{33\omega'^2 m/140}{M+33m/140}$$

Neglecting $33m/140$ in the above equation, we get,

$$\omega'^2 - \omega_n^2 = (33/140)r\,\omega'^2 \quad \text{or}$$

$$(\omega'^2 - \omega_n'^2)/\omega^2 = (33/140)r$$

Since ω' and ω_n are close to each other

$$2\frac{\omega'\Delta\omega}{\omega'^2} = \frac{33}{140}r$$

Therefore,

$$\frac{\Delta\omega}{\omega'} = \frac{33}{280}r$$

If the disturbing frequency is $0.9\omega'$, $\Delta\omega/\omega' = 0.9$. Since at resonance, $\Delta\omega/\omega' = (33/280)r$, the present force is unlikely to cause any resonance.

We shall now consider the response of a damped system under harmonic excitation force.

2.4.2 Response of a Damped System Under Harmonic Force

If the forcing function is $F(t) = F_0 \cos \omega t$, the equation of motion is

$$m\ddot{x} + f\dot{x} + kx = F_0 \cos \omega t \qquad (2.104)$$

As discussed earlier, the solution comprises of CF $[x_{CF}(t)]$ and the particular solution (PI). Since the CF $[x_{CF}(t)]$ is of transient nature and vanishes after sometime, the steady-state solution comprises of only particular integral.

The particular solution is also expected to be harmonic. We assume it in the form

$$x_p(t) = X\cos(\omega t - \Phi) \qquad (2.105)$$

where the amplitude X and the phase angle Φ are the unknowns and required to be determined. Substituting Equation 2.105 in Equation 2.104, we get,

$$X[(k - m\omega^2)\cos(\omega t - \Phi) - f\omega \sin(\omega t - \Phi)] = F_0 \cos \omega t \qquad (2.106)$$

We know

$$\cos(\omega t - \Phi) = \cos \omega t \cos \Phi + \sin \omega t \sin \Phi$$

$$\sin(\omega t - \Phi) = \sin \omega t \cos \Phi - \cos \omega t \sin \Phi$$

Substituting the above relations in Equation 2.106, we get

$$X[(k - m\omega^2)\cos \Phi + f\omega \sin \Phi] = F_0$$

$$X[(k - m\omega^2)\sin \Phi - f\omega \cos \Phi] = 0 \qquad (2.107)$$

Solution of Equation 2.107 gives

$$X = F_0 / \sqrt{(k - m\omega^2)^2 + f^2\omega^2} \qquad (2.108)$$

$$\Phi = \tan^{-1}\left(\frac{f\omega}{k - m\omega^2}\right) \qquad (2.109)$$

It can be seen that the phase difference Φ between the excitation force and the displacement x_p depends upon the damping constant f. If $f = 0$, the force $F(t)$ and x_p are in phase. Figure 2.43 gives pictorial representation of the response.

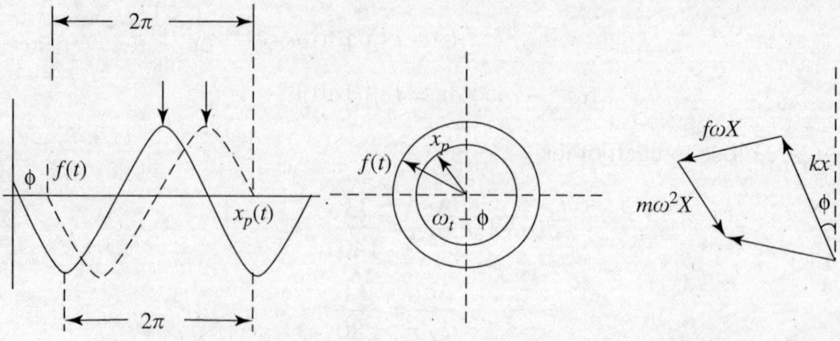

Figure 2.43 Phase Between Excitation Force and Response

We know $\omega_n = \sqrt{k/m}$ = natural frequency of the undamped system. Also $\xi = f/f_c = f/2m\omega_n = f/2\sqrt{mk}$, $f/m = 2\xi\omega_n$ where f_c is a critical damping coefficient. $\delta_{st} = F_0/k$ = deflection under F_0 applied statically, $r = \omega/\omega_n$ = frequency ratio.

Using there relations in Equations 2.108 and 2.109, we get,

$$\frac{x}{\delta_{st}} = 1/\sqrt{(1-\{\omega/\omega_n\}^2 + (2\xi\omega/\omega_n)^2} = 1/\sqrt{(1-r^2)^2 + (2\xi r)^2} \tag{2.110}$$

$$\Phi = \tan^{-1}\{(2\xi\omega/\omega_n)/(1-[\omega/\omega_n]^2\} = \tan^{-1}(2\xi r/1-r^2) \tag{2.111}$$

As mentioned earlier, the quantity $M = X/\delta_{st}$ is called the magnification factor or amplification factor or amplitude ratio. Figure 2.44(a) and 2.44(b) show the behaviour of M and Φ for various frequency ratio r and damping factor r.

The following important characteristics should be noted.

- For an undamped system ($\xi = 0$), $M \to \infty$ as $r \to 1$.
- Any amount of damping ($\xi > 0$) reduces the magnification factor for all values of forcing frequencies.
- For a given r, higher is the damping, lesser is the magnification factor M. Even at resonance ($\omega/\omega_n = 1$), higher is the damping lesser is M.
- As $r \to \infty$, $M \to 0$ which means for $r \gg 1$, the amplification factor gets reduced drastically. Thus at forcing frequencies higher than the natural frequency, vibratory response or magnification factor reduces.
- For $0 < \xi < 1/\sqrt{2}$, the maximum value of M occurs when $r = \sqrt{(1 - 2\xi^2)}$ or $\omega = \omega_n\sqrt{1 - 2\xi^2})$, which can be seen to be lower than the undamped natural frequency ω_n and damped natural frequency $\omega_d = \omega_n\sqrt{1 - \xi^2})$.
- The maximum value of X, when $r = \sqrt{1 - 2\xi^2}$ is given by

$$(X/\delta_{st})_{max} = 1/[2\,\xi\sqrt{(1 - 2\xi^2)}. \tag{2.112}$$

The value of X at $\omega = \omega_n$ is given by

$$(X/\delta_{st})_{max} = \frac{1}{2\xi} \tag{2.113}$$

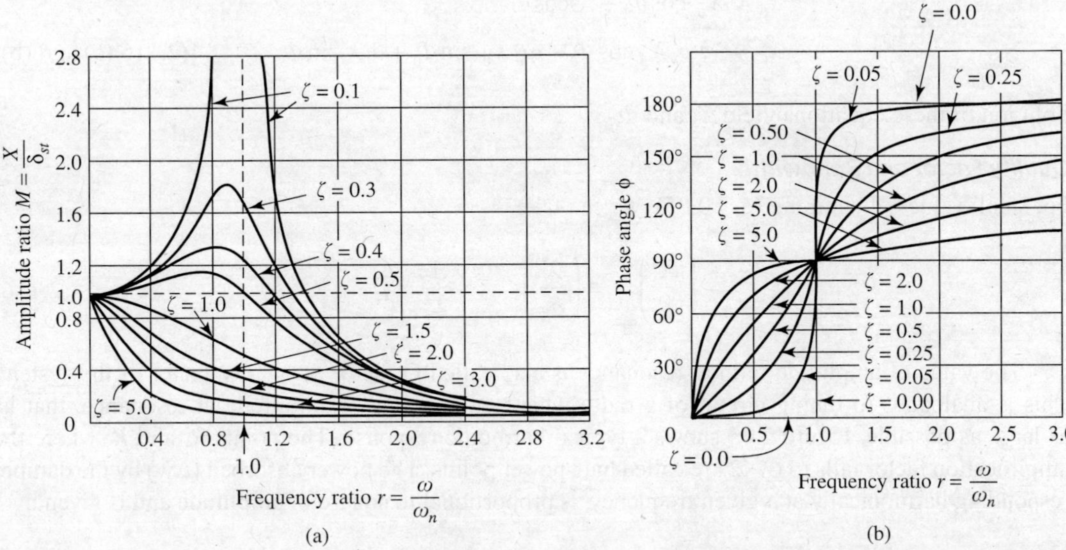

Figure 2.44 Forced Vibrations—Amplitude Ratio/Phase Angle versus Frequency-Ratio Curve

Equation 2.113 has a special significance. It is used for estimating the damping present in the system. In the vibration test, if the maximum amplitude of response X_{max} is measured, the damping in the system can be found out using Equation 2.112. This requires that the stiffness k of the spring is known *a priori* to enable determination of δ_{st}. It may be recalled that a simpler method to determine ξ is to measure the logarithm decrement in the free vibration test.

- For undamped systems ($\xi = 0$), the phase angle is zero for $0 < r < 1$ and 180° for $r > 1$. This means that the excitation and the response are in phase when for $0 < r < 1$ and out of phase for $r > 1$, $\xi = 0$.
- For $\xi > 0$ and $0 < r < 1$, the phase angle is given by $0 < \Phi < 90°$ implying that the response lags the excitation.
- For $\xi > 0$ and $r > 1$, the phase angle is given by $90 < \Phi < 180°$ implying that the response leads the excitation. For very large r, the phase angle approaches 180°.
- For $\xi > 0$ and $r = 1$, the phase angle $F = 90°$ implying that the phase difference between the excitation and response is 90°. This is a very important observation and the litmus test for identifying the resonant frequency. At around the resonance, there is a phase shift of 90° when damping is present and a phase shift of ~180° when there is no or very little damping. This is because while making the resonance test, sometimes the location of the vibration transducer may be a point of 0 or minimum vibration displacement. Thus, if resonance is to be determined only through observing frequency at which the response is maximum, we may fail to identify the resonance condition. Noting the phase reversal is, therefore, the correct method of identifying resonance condition.

The total response $x(t) = x_{cf} + x_p(t)$, where x_{cf} is the response during few cycles of vibration which ultimately die out and $x_p(t)$ is the steady-state response. Thus

$$x(t) = X_0 e^{-\xi\omega_n t} \cos(\omega_d - \Phi) + X \cos(\omega t - \Phi) \tag{2.114}$$

X and Φ are given by Equations 2.110 and 2.111, while X_0 and Φ_0 can be found out from initial conditions $x(t = 0) = x_0$ and $\dot{x}(t = 0) = \dot{x}_0$. These conditions when put in Equation 2.114, gives

$$x_0 = X_0 \cos \Phi_0 + X \cos \Phi$$
$$\dot{x}_0 = -\xi \omega_n' X_0 \cos \Phi_0 + \omega_d X_0 \sin \Phi_0 + \omega X \sin \Phi \quad [(2.115\text{ (a) and (b)}]$$

Solution of these equations yield X_0 and Φ_0.

Quality Factor and Bandwidth

For small values of damping ($\xi < 0.05$), we can take

$$\left(\frac{X}{\delta_{st}}\right)_{max} \equiv \left(\frac{X}{\delta_{st}}\right)_{\omega=\omega_n} = \frac{1}{2\xi} = Q \quad (2.116)$$

The value of amplitude ratio at resonance is also called Q factor or quality factor of the system. This is analogous to tuning circuit of a radio where interest lies in amplitude at resonance, that is, as large as possible. Figure 2.45 shows a typical harmonic response. The points R_1 and R_2 where the amplification factor falls to $Q/\sqrt{2}$ are called half-power points. The power absorbed (Δw) by the damper responding harmonically at a given frequency, is proportional to square of amplitude and is given by

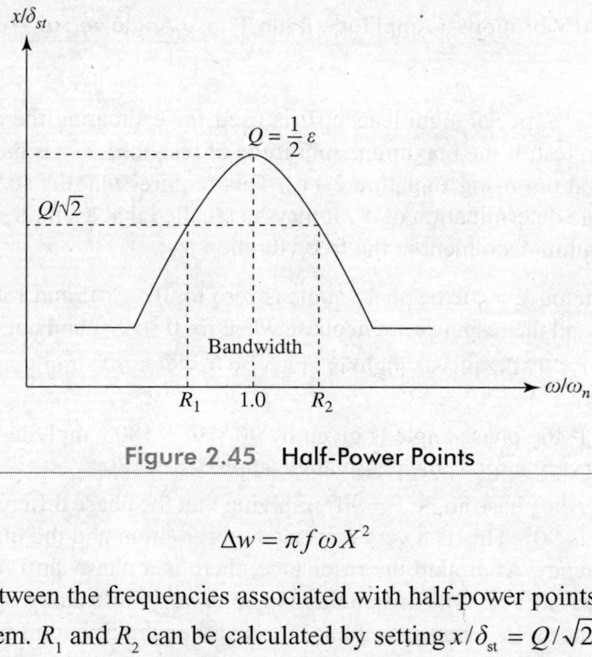

Figure 2.45 Half-Power Points

$$\Delta w = \pi f \omega X^2 \quad (2.117)$$

The difference between the frequencies associated with half-power points R_1 and R_2 is called the bandwidth of the system. R_1 and R_2 can be calculated by setting $x/\delta_{st} = Q/\sqrt{2} = 1/2\sqrt{2}\xi$ in Equation 2.110, so that

$$1/\sqrt{(1-r^2)^2 + (2\xi rr)^2} = \frac{Q}{\sqrt{2}} = \frac{1}{2\sqrt{2}\xi} \quad \text{or}$$

$$r^4 - r^2(2-4\xi^2) + (1-8\xi^2) = 0 \quad (2.118)$$

Solution of this equation is

$$r_1^2 = 1 - 2\xi^2 - 2\xi\sqrt{1+\xi^2}$$
$$r_2^2 = 1 - 2\xi^2 + 2\xi\sqrt{1+\xi^2} \quad (2.119)$$

For small ξ, Equation 2.119 can be approximated as

$$(r_1)^2 = (R_1)^2 = (\omega_1/\omega_n)^2 \simeq 1 - 2\xi,\ (r_2)^2 = (R_2)^2 = (\omega_2/\omega_n)^2 \simeq 1 + 2\xi \quad (2.120)$$

where $\omega_1 = \omega_{R1}$ and $\omega_2 = \omega_{R2}$.

From Equations 2.120

$$(\omega_2)^2 - (\omega_1)^2 = [(R_2)^2 - (R_1)^2]\,\omega_n 2 = 4\xi\omega_n^{\ 2} \quad (2.121)$$

But

$\omega_1 + \omega_2 = 2\omega_n$ and using Equation 2.121, bandwidth is

$$\Delta\omega = \omega_2 - \omega_1 \approx 2\xi\omega_n \quad (2.122)$$

Combining Equation 2.116 and 2.122, we get,

$$Q \approx \frac{1}{2}\xi \approx \omega_n/(\omega_2 - \omega_1) \quad (2.123)$$

The above analysis is useful from the point of view of vibration testing. The quality factor can be used for estimating the viscous damping in the mechanical system. It is easy to understand that sharper is the resonant frequency lesser is the damping in the system. On the other hand, if the resonant peak is associated with a fat curve, the system has a good damping. These observations are of great value in diagnosis of vibration problems.

Problem 2.21

A single degree-of-freedom system comprises of m = 10 kg, f = 20 N s/m, k = 4000 N. The initial conditions are X_0 = 0.01 m, x_0 0. Find total response of the system when an external force $F(t)$ = 100 cos 10t is applied. Also find the time duration after which the system response will be fully a steady-state response.

Solution:

$$\omega_n = \sqrt{k/m} = \sqrt{4000/10} = 20\ \text{rad/s},$$

$$\delta_{\text{st}} = F_0/k = 100/4000 = 0.025\,\text{m}$$

$$\xi = f/f_c = f/2\sqrt{km} = 20/2\sqrt{4000.10} = 0.05$$

$$\omega_d = \sqrt{(1-\xi^2)}\omega_n = \sqrt{(1-0.05^2)}.20 = 19.98\,\text{rad/s}$$

$$r = \omega/\omega_n = 10/20 = 0.5$$

$$X = \frac{\delta_{\text{st}}}{\sqrt{(1-r^2)+(2\xi r)^2}} = \frac{0.025}{\sqrt{(1-0.05^2)^2+(2.0.5.0.05)^2}} = 0.033$$

$$\text{Tan}\,\Phi = \left(\frac{2\xi r}{1-r^2}\right) = \left(\frac{2.0.05.0.5}{1-0.5^2}\right)$$

Therefore,

$$\Phi = 3.81^\circ,\ \cos 3.81^\circ = 0.997$$

We know that the total response of the system is given by

$$x(t) = X_0 e^{-\xi\omega_n t}\cos(\omega_d - \Phi_0) + X\cos(\omega t - \Phi)$$

We also have initial conditions, $x_0 = 0.01$ m, $\dot{x}_0 = 0$.

These conditions are used in the equations $x_0 = X_0 \cos \Phi_0 + X \cos \Phi$

$$\dot{x}_0 = -\xi\omega_n X_0 \cos \Phi_0 + \omega_d X_0 \sin \Phi_0 + \omega X \sin \Phi$$

Using these, we obtain

$$0.01 = X_0 \cos \Phi_0 + (0.033)(0.997) \text{ or } X_0 \cos \Phi_0 = -0.0231 \quad (1)$$

$$0 = -(0.05)(20)\, X_0 \cos \Phi_0 + X_0(19.97) \sin \Phi_0 + (0.033)(10) \sin (3.81) \quad (2)$$

Substituting value of $X_0 \cos \Phi_0$ from Equation 1, we get

$$X_0 \sin \Phi_0 = -0.022 \quad (3)$$

Solution of Equations 1 and 3 gives

$$X_0 = \sqrt{\{(X_0 \cos \Phi_0)^2 + (X_0 \sin \Phi_0)^2\}} = \sqrt{0.0231^2 + 0.0022^2)} = 0.232$$

$$\text{Tan } \Phi_0 = X_0 \sin \Phi_0 / X_0 \cos \Phi_0 = 0.0978$$

This gives $\Phi_0 = 5.586°$. Thus the total response is given by

$$x(t) = 0.0233\mathrm{e}^{-(0.05)(20)t} \cos(19.97t - 5.586) + 0/033 \cos(10t - 3.81)$$

The transient part of this is

$$x_{cf}(t) = 0.0233\mathrm{e}^{-(0.05)(20)t} \cos(19.97t - 5.586) = 0.0233\mathrm{e}^{-4} \cos(19.97t - 5.586)$$

Let t_1 be the time when

$$x_{cf}(t) = 0$$

or

$$0 = 0.0233\mathrm{e}^{-tj} \cos(19.97t_1 - 5.586).$$

Since $\cos(19.97t_1 - 5.586)$ cannot be zero $0 = 0.0233\ \mathrm{e}^{-tj}$.

We find the solution by taking logarithm on both sides. This gives $t_1 = 0.61$s.

Problem 2.22

Figure 2.46 shows a spring–mass–damper system mounted on a base which undergoes a displacement $x_1(t)$. Determine the response of the system. The base can be considered as massless and undergoes a rigid-body harmonic motion $x_1(t) = X_1 \sin \omega t$.

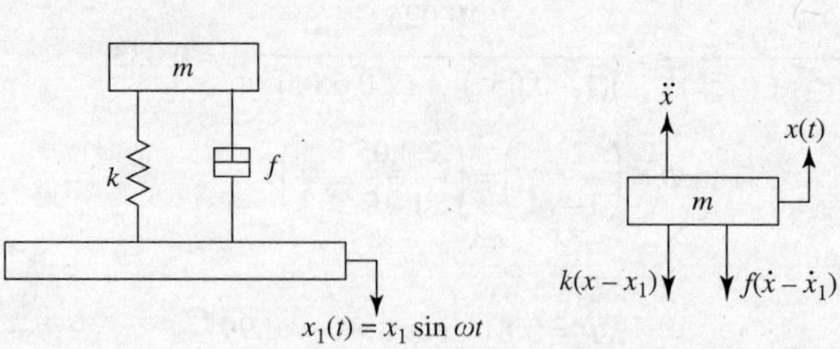

Figure 2.46 SDOF System with Base Excitation

Solution:

Let the displacement of the mass m be $x(t)$. At the outset, the system appears to be a two degrees-of-freedom system. However, with proper analysis, we can show that it is in reality a SDOF system.

The base motion is $x_1(t)$. Consider the free-body diagram [Fig. 2.49(b)]. By considering the equilibrium of the body, we get

$$m\ddot{x} = -k(x - x_1) - f(\dot{x} - \dot{x}_1) \quad \text{or}$$

$$m\ddot{x} + k(x - x_1) + f(\dot{x} - \dot{x}_1) = 0 \quad \text{or}$$

$$m\ddot{x} + kx + f\dot{x} = f\dot{x}_1 + kx_1 = (f\omega)X_1 \cos\omega.t + kX_1 \sin\omega.t \tag{1}$$

or

This can also be written as

$$m\dot{x} + f\dot{x} + kx = A\sin(\omega.t - a)$$

$$A \cos a = kX_1, A \sin a = fX_1 \tag{2, 3}$$

Thus,

$$A = X_1\sqrt{k^2 + (f\omega)^2}, \ \tan a = \left\{-\frac{f\omega}{k}\right\}$$

These equations show that the system shown in Fig. 2.46 is indeed a single degree-of-freedom system. The excitation at the base is equivalent to application of force of magnitude A to the mass. By using the solution indicated earlier

$$x_p(t) = \frac{A}{\sqrt{((k - m\omega^2 + (f\omega)^2)}} \sin(\omega.t - \Phi_1 - a) \quad \text{or}$$

$$x_p(t) = \frac{X_1\sqrt{(k^2 + f^2\omega^2)}}{\sqrt{((k - m\omega^2 + (f\omega)^2)}} \sin(\omega.t - \Phi_1 - a) \tag{4}$$

where $a = \tan^{-1}(-f\omega/k)$ and $\Phi_1 = \tan^{-1}(f\omega/(k - m\omega^2)$ (5, 6)

Using trigonometric identities, the above equations can be written as

$$x_p(t) = X\sin(\omega t - \Phi) \tag{7}$$

where X and Φ are given by

$$\frac{X}{X_1} = \left[\frac{k^2 + f^2\omega^2}{(k - m\omega^2 + (f\omega)^2}\right]^{1/2} = \left[\frac{1 + (2r\xi)^2}{(1 - r^2)^2 + (2r\xi)^2}\right]^{1/2}, \ r = \omega/\omega_n \tag{8}$$

$$\Phi = \tan^{-1}\left[\frac{mf\omega^3}{(k - m\omega^2 + (f\omega)^2}\right] = \tan^{-1}\left[\frac{2\xi.r^3}{1 + (4\xi^2 - 1)r^2}\right] \tag{9}$$

Transmissibility

Displacement transmissibility: The amplitude ratio X/X_1 (i.e. amplitude of response $x_p(t)$/amplitude of base motion $x_1(t)$) is called the displacement transmissibility T_d

$$T_d = \frac{X}{X_1} = \left[\frac{k^2 + f^2\omega^2}{(k - m\omega^2) + (f\omega)^2}\right]^{1/2} = \left[\frac{1 + (2r\xi)^2}{(1 - r^2)^2 + (2r\xi)^2}\right]^{1/2} \tag{2.124}$$

Force transmissibility: In Fig. 2.46, a force F is transmitted to the base or support due to the reaction from the spring and the dashpot.

$$F = -m\ddot{x} = k(x - x_1) + f(\dot{x} - \dot{x}_1) \tag{2.125}$$

From Equation 7 (Problem 2.22), Equation 2.125 can be written as

$$F = m\omega^2 X \sin(\omega t - \Phi) = F_T \sin(\omega t - \Phi) \tag{2.126}$$

where F_T is amplitude or maximum value of force transmitted to the base given by

$$\frac{F_T}{kX_1} = r^2\left[\frac{1+(2r\xi)^2}{(1-r^2)+(2r\xi)^2}\right]^{1/2} \tag{2.127}$$

The ratio $\frac{F_T}{kX_1}$ is known as force transmissibility T_F. It may be noted that the transmitted force is in phase with the motion of the mass $x(t)$. The use of the concept of transmissibility is extensively done in the design of vibration isolation systems.

Study of Equations 2.124 and 2.127 reveals the following important deductions. These are very useful in the designing of vibration-isolation systems.

- The value of $T_d = 1$ at $r = 1$ and close to unity for small values of r $(= \omega/\omega_n)$.
- For undamped systems $(\xi = 0)$, $T_d \to \infty$ at $r = 1$.
- $T_d < 1$ for values $r > \sqrt{2}$ for any amount of damping.
- $T_d = 1$ for all values of ξ at $r = \sqrt{2}$.
- For $r < \sqrt{2}$, smaller damping ratio leads to larger value of T_d.
- T_d attains a maximum for $0 < \xi < 1$ at frequency ratio $r = r_m < 1$ given by $r_m = \frac{1}{2\xi}[\sqrt{1+8\xi^2} - 1]^{1/2}$ (differentiate Equation 8 of previous problem with respect to r and set it equal to zero)

This analysis is highly relevant for evaluating what is called mounted resonance of accelerometer. Accelerometer is an acceleration-sensitive transducer used for vibration measurements. The accelerometer consists of mass(es), which are small and supported by springs. The accelerometer is firmly fixed to the vibrating component/structure. The acceleration of the mass produces force, which is converted into electrical signal by a piezoelectric device. This electrical signal is then converted to acceleration of vibration/velocity/displacement using appropriate electronic circuits. The generalized model of accelerometer is shown in Fig. 2.47.

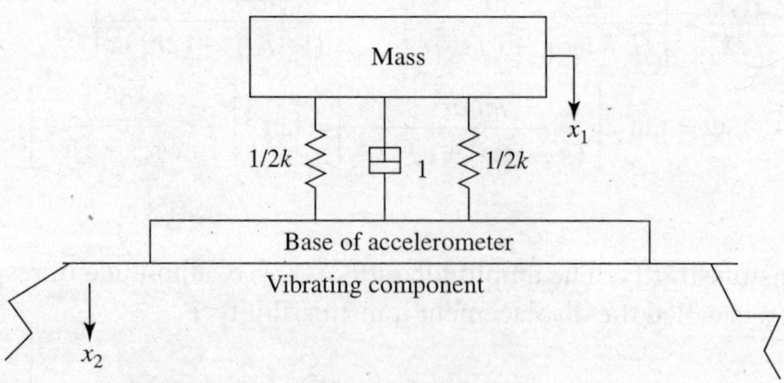

Figure 2.47 Model of Accelerometer

The forces acting on the mass are spring force = $k(x_1 - x_2)$ and damping force = $f(\dot{x}_1 - \dot{x}_2)$. Assuming $x_1 > x_2$, $\Sigma F = ma$ gives $m\ddot{x} = -k(x_1 - x_2) - f(\dot{x}_1 - \dot{x}_2)$.

Let $x_1 - x_2 = x$ and $(\dot{x}_1 - \dot{x}_2) = \dot{x}$ so that $x_1 = x + x_2$ and $\dot{x}_1 = \dot{x} + \dot{x}_2$.

The equation of motion takes the form

$$m(\ddot{x} + \ddot{x}_2) + f\dot{x} + kx = 0, \quad \text{or}$$

$$m\ddot{x} + f\dot{x} + kx = -m\ddot{x}_2 \tag{2.128}$$

But $x_2 = A \sin \omega t$ is the vibration (to be measured) of the body.

Equation 2.128 then becomes

$$m\ddot{x} + f\dot{x} + kx = -m\ddot{x}_2 = mA\omega^2 \sin \omega t \tag{2.129}$$

Solution of this equation is

$$x_p = \frac{F_0}{\sqrt{[(k - m\omega^2)^2 + (f\omega)^2]}} \sin(\omega.t - \Phi)$$

$$\Phi = \tan^{-1}\left(\frac{f\omega}{k - m\omega^2}\right)$$

Substituting $F_0 = mA\omega^2$, $(\omega_n)^2 = k/m$, $\xi = f/2m\omega_n$, $r = \omega/\omega_n$

$$x_p = \left[\frac{Ar^2}{\sqrt{(1-r^2)^2 + (2r\xi)^2}}\right] \sin(\omega.t - \Phi) \tag{2.130}$$

In this case, ω_n = natural frequency of the pick-up (accelerometer) and ω is the frequency of vibrations. If ω_n is large (by having very stiff spring or by having very small mass), the ratio $r = \omega/\omega_n$ is quite small. The expression for x_p becomes

$$x_p = \left[\frac{Ar^2}{\sqrt{(1-r^2)^2 + (2r\xi)^2}}\right] \cong \frac{Ar^2}{\sqrt{1-0+0}} = Ar^2 = A\left(\frac{\omega}{\omega_n}\right)^2 = A'\omega^2 \tag{2.131}$$

The design-mounted resonance frequency ω_n is very large compared to the frequency of vibration to be measured. Hence, for a large frequency range, the transducer response for a given base acceleration is as shown in Fig. 2.48.

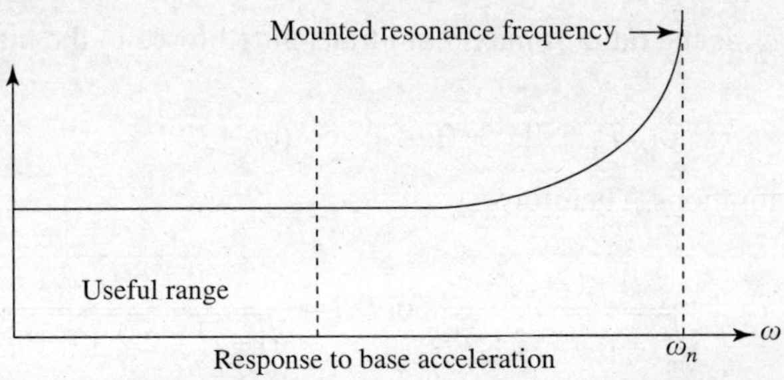

Figure 2.48 Response of Accelerometer to Base Accelerations

Figure 2.48 shows that for a large frequency range (away from the mounted resonant frequency), the response is flat and the measurements are accurate. The usual frequency range selected is 0–$\frac{1}{3}$ of ω_n for reliable results. Thus for an accelerometer whose mounted resonant frequency is 15 KHz, the measurements done in frequency range 0–5 KHz will be highly reliable. In general, accelerometers are the most-favoured transducers for vibration measurements. We shall discuss this aspect in details in the chapter of diagnosis.

Problem 2.23

A centrifugal fan weighs 100 lbs and has a rotating unbalance of 20 lb in. when dampers having damping factor $\xi = 0.2$ are used. Specify the springs for mounting such that only 10% of unbalance force is transmitted to the floor. Also, determine the magnitude of transmitted force. The fan runs at a constant speed of 1000 rpm.

Solution:

If a disc is mounted on the shaft in such a way that its CG does not coincide with the centre of rotation and is at a distance e from the centre of rotation (Fig. 2.49) O, the rotation of the disc at angular speed ω, produces a centrifugal force $me\omega^2$. This force is called unbalance force and produces harmonic forces $me\omega^2 \cos \omega t$ and $me\omega^2 \sin \omega t$ in horizontal and vertical directions, respectively. The quantity (me) is defined as rotating unbalance and has units lb in., kg m, etc.

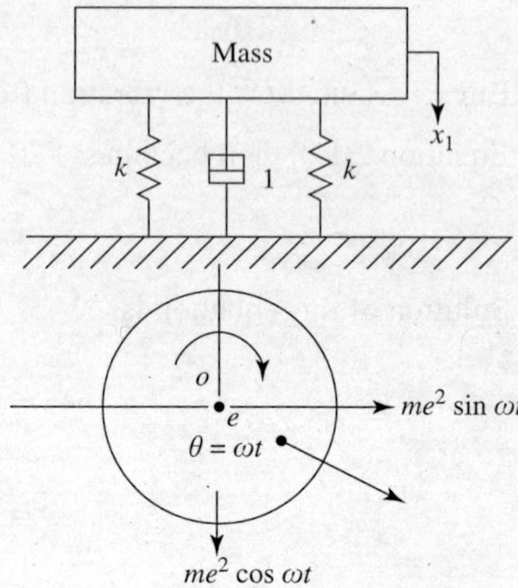

Figure 2.49 Centrifugal Fan on Vibration Mounts

The total force transmitted to the rigid foundation is the sum of reactions at the fixed end of the spring and dashpot;

$$F_T = kx + f\dot{x}.$$

Under steady-state conditions, amplitude of vibrations is

$$x_P = \frac{F_0/k}{\sqrt{(1-r^2)^2 + (2r\xi)^2}} \sin(\omega t - \Phi) = A\sin(\omega t - \varphi)$$

$$\dot{x}_P = A\omega\cos(\omega t - \varphi)$$

$$\therefore F_T = kA \sin(\omega.t - \Phi) + A\omega.f \cos(\omega.t - \varphi)$$

It may be remembered that the spring force is maximum when velocity is zero (or displacement maximum), while damping force is maximum when displacement is zero (or velocity maximum). Since the phase angle between the spring force and damping force is 90°, the resultant maximum force transmitted is $A[k^2 + (f\omega)^2]^{1/2}$. Then $F_T = A\,[k^2 + (f\omega)^2]^{1/2} \cos(\omega t + \Phi)$, $\Phi = \tan^{-1}(f\omega/k)$.

Transmissibility, T_R, is the ratio of maximum transmitted force to the maximum impressed force and is thus given by

$$T_R = A[k^2 + (f\omega)^2]^{1/2} \cos(\omega t + \Phi)/F_0$$

But $x = x_p$ for steady vibrations. Therefore,

$$T_R = \frac{\sqrt{1+(2r\xi)^2}}{\sqrt{(1-r^2)+(2r\xi)^2}} \text{...or.} 0.1 = \frac{\sqrt{1+0.16r^2}}{\sqrt{1+r^4-2r^2+0.16r^2}}$$

Solving the is equation, we get, $r = 3.7$.

$$\omega = 2\,\pi \times 1000/60 = 105 \text{ rad/s}, \quad r = \omega/\omega_n = 3.7.$$

From this we get, $\omega_n = 105/3.7 = 28.4$ and $k = m\omega_n^2$.

The mass is 100/32.

From this we get, $k = 100.28.4^2/32.2.12 = 210$ lb/in.

Amplitude of force transmitted $= 0.1\, me\omega^2 = 0.1.20.105^2/32.2.12 = 57$ lbs.

It may be understood that Problem 2.23 dealt with rigid foundation where the force (F_T) transmitted to the foundation was through spring and dash pot. This force is given by

$$F_T = kx(t) + f\dot{x}(t) = kX\cos(\omega t - \Phi) - f\omega X\sin(\omega t - \Phi) \tag{2.132}$$

where $$X = \frac{F_0}{\sqrt{(k - m\omega^2)^2 + f^2\omega^2}}, \quad \Phi = \tan^{-1}\left(\frac{f\omega}{k - m\omega^2}\right)$$

Since the spring and damping force have a phase difference of 90° (Fig. 2.50), F_T can also be written as $F_T = \{(kx)^2 + (f\dot{x})^2\}^{1/2} = X\sqrt{k^2 + f^2\omega^2}$

$$\therefore F_T = \frac{F_0\sqrt{k^2 + f^2\omega^2}}{\sqrt{(k - m\omega^2)^2 + f^2\omega^2}}$$

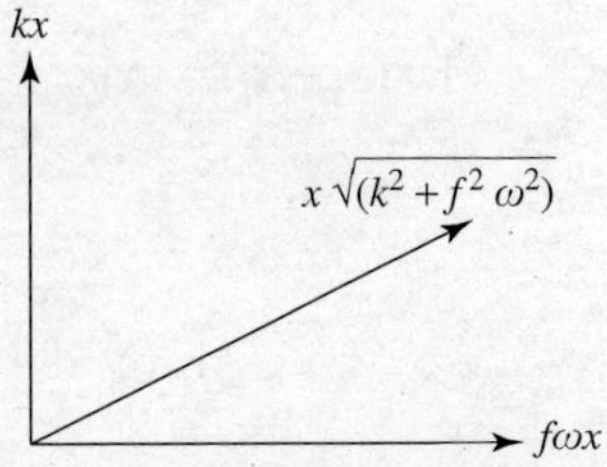

Figure 2.50 Spring and Damping Force

From this we obtain

$$\left(\frac{F_T}{F_0}\right)^2 = \left\{\frac{1 + f^2\omega^2/k^2}{(1 - m\omega^2/k)^2 + f^2\omega^2/k^2}\right\} \tag{2.133}$$

We know $\xi = f/f_c$ and $f_c = 2m\omega_n$. Using this and $r = \omega/\omega_n$, we obtain $2\xi r = f\omega/k$.

Thus, we obtain,

$$T_R = \left(\frac{F_T}{F_0}\right) = \left\{\frac{1 + (2r\xi)^2}{(1 - r^2)^2 + (2r\xi)^2}\right\}^{½} = T_R \tag{2.134}$$

Thus, Equation 2.134 gives the transmissibility ratio for spring–mass–damper system (excited by a force $F_0 \sin \omega t$) mounted on rigid foundation, while Equation 2.127 gives the transmissibility ratio when the base is excited by $y = y_0 \sin \omega t$. It is important to note this difference. Later, we shall deal with whether force transmissibility of harmonically excited spring–mass–damper system and the foundation is flexible.

Problem 2.24

The instruments used in the cockpit of the aircraft are required to be vibration-isolated to a significant extent. The instrument boards are mounted on isolators. If the damping used in isolators is negligible and isolator deflects 0.125 in. under a weight of 50 lbs, find the percentage of motion transmitted to the instrument board if the frequency of vibration of the aircraft is 2000 CPM.

Solution:

We know that transmissibility is given by

$$\frac{F_T}{F_0} = \left\{\frac{1 + (2r\xi)^2}{(1 - r^2)^2 + (2r\xi)^2}\right\}^{½} = T_R . r = \omega/\omega_n \quad \text{and} \quad \xi = f/2\text{m}. \text{But } \xi = 0.\omega_n = \sqrt{k/m}$$

Since isolator deflects 0.125 in. under a load of 50 lbs $k = 50/0.125$ $(\omega_n)^2 = k/m = 50 \times 32.2 \times 12/0.125.50 = 3056$.

This gives $\omega_n = 55.4$ rad/s. The frequency corresponding to this $= 55.4/2\,\pi x = 8.85$ CPS or 530 CPM.

The ratio $r = 2000/530 = 3.773$. Substituting this and remembering that $\xi = 0$, we get from $T_R = I/(1 - r^2)$, $T_R = 0.076$.

Hence, the isolator transmits only 7.6% of the vibratory motion of the aircraft.

Problem 2.25

A simplified spring–mass vibration pick-up shown in Fig. 2.51 is used to measure the vertical acceleration of the train caused by wheel-hammer-blow process. The frequency of hammer blow is 10 rad/s. The vibration pick-up comprises of a mass weighing 3.86 lbs and spring of stiffness 100 lb/in. The amplitude of the relative vibration of the mass is recorded as 0.05 in. Find the maximum vertical acceleration and amplitude of vibration of the train.

Solution:

The natural frequency $\omega_n = \sqrt{100 \times 12 \times 32.2/3.86} = 100$ rad/s, $\omega = 10$ rad/s

$$\text{Max}(x_2 - x_1) = \text{max. } x = x_\text{p} = \left[\frac{Ar^2}{\sqrt{(1-r^2)^2 + (2r\xi)^2}}\right] \text{ since } \xi = 0$$

$$x_\text{p} = \frac{Ar^2}{1-r^2} = \frac{A\omega^2/\omega_n^{\,2}}{1-\omega^2/\omega_n^{\,2}} \text{ or}$$

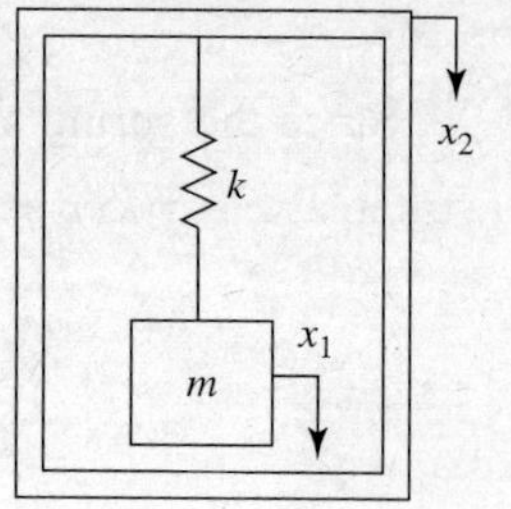

Figure 2.51 Vibration Pick-Up

$A\omega^2 = x_\text{p}(\omega_n^{\,2} - \omega^2) = 0.05[(100 \times 386/3.86) - 100]$ = maximum vertical acceleration of the train = 495 in./s^2.

Amplitude of vibration of train $A = 495/100 = 4.95$ in.

Problem 2.26

Figure 2.52 shows a simplified model of a motor vehicle which travels on a rough road. The mass of the vehicle is 1200 kg. The suspension system consists of a 400 kN/m spring and dampers with a damping ratio of 0.5. During the travel, the vehicle meets a rough road with road surface varying sinusoidally with amplitude of $Y = 0.05$ m and wavelength of 6 m. If the speed of the vehicle is 20 km/h, determine the displacement amplitude of the vehicle.

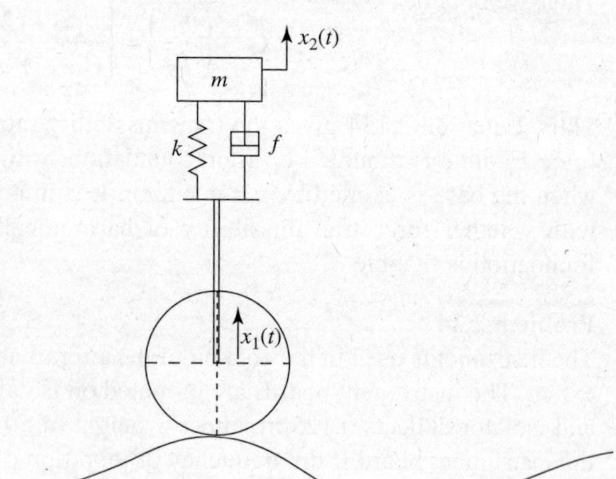

Figure 2.52 Motor Vehicle on Rough Road

Solution:

The equivalent spring–mass–damper system for the motor vehicle is shown in Fig. 2.53. The speed of vehicle is 20 km/h.

Since the wavelength is 6 m, time taken to cross the rough T is = 6/(20000/3600) seconds. This T is the period of vibration. Therefore, the frequency of vibration is $\omega = 2\pi/T = (2\pi \times 20 \times 1000)/3600 \times 6 = 5.818$ rad/s.

Natural frequency of the vehicle $\omega_n = \sqrt{k/m} = (400 \times 10^3/1200)^{½} = 18.25$ rad/s. Therefore, in the frequency ratio $r = 5.818/18.25 = 0.3186$.

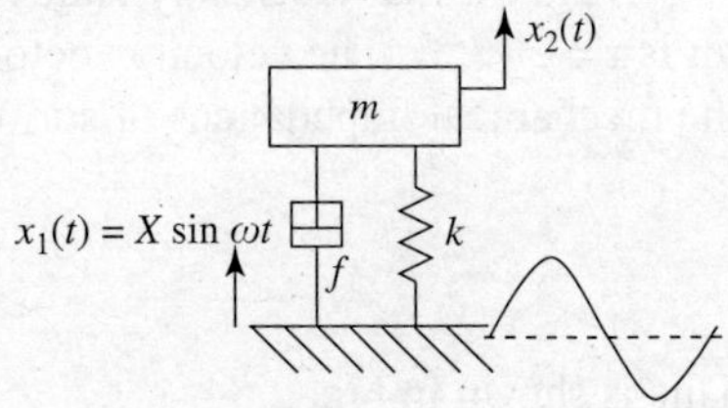

Figure 2.53 SDOF Model of Vehicle

$$\text{Amplitude ratio}\frac{X_2}{X_1} = \left[\frac{k^2 + f^2\omega^2}{(k - m\omega^2) + (f\omega)^2}\right]^{1/2}$$

$$= \left[\frac{1 + (2r\xi)^2}{(1 - r^2)^2 + (2r\xi)^2}\right]^{1/2}$$

$$= \frac{1 + (2 \times 0.5 \times 0.3186)^2}{(1 - 0.318^2)^2 + (2 \times 0.5 \times 0.3186^2)^2} = 1.47$$

Thus, the displacement amplitude $X_2 = 1.47 \times 0.05 = 0.0735$ m.

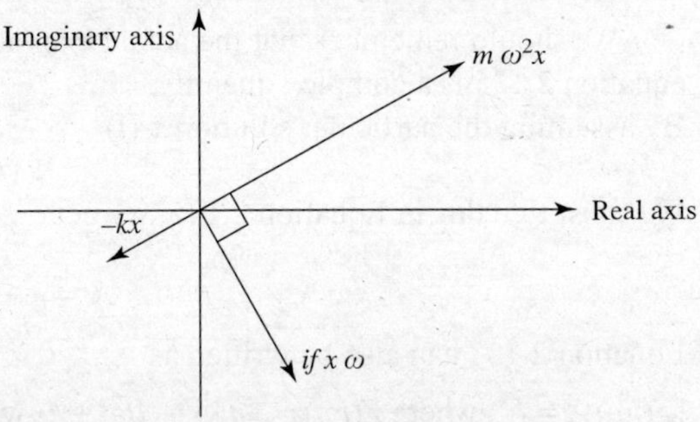

Figure 2.54 Impedance Method

Problem 2.27

A heavy machine weighing 3000 N is mounted on a flexible foundation. The static deflection of the foundation due to the machine is 7.5 cm. The measured vibration levels on the machine are 1 cm when the foundation is subjected to harmonic oscillations at the undamped natural frequency with an amplitude of 0.25 cm. Find the damping constant, dynamic force at the base, and amplification of displacement at the machine relative to the base.

Solution:

Stiffness of the foundation = 3000/(7.5/100) = 40000 N/m

At resonance $r = 1$, $X = 0.01$ m, $X_1 = 0.0025$ m, $X/X_1 = 4 = \left[\frac{1 + 2\xi^2}{(2\xi)^2}\right]$

since $r = 1$

From the above equation $\xi = 0.1291$

$$f = \xi f_c = 2\xi \sqrt{km} = 0.1291 \times 2 \times \sqrt{40000 \times 3000 / 9.81} = 903 \text{ N s/m}$$

$$\text{Dynamic force} = F_T = kX_1\left[\frac{1 + 4\xi^2}{4\xi^2}\right]^{½} = kX = 40000 \times 0.01 = 400 \text{ N.}$$

2.4.3 Mechanical-Impedance Method

In the determination of the steady-state vibration of the system, the mechanical-impedance method is simple and straight-forward compared to other methods. This method is based upon the vectorial representation of harmonic functions. Let the force vector be $F = e^{i\omega t}$.

We know that the steady-state response lags behind the excitation force i.e. the displacement vector is $x = e^{ic(\omega t - \Phi)}$. The velocity vector is $\dot{x} = i\omega . x$ and the acceleration vector is $\ddot{x} = -(\omega^2 x)$. Therefore, the mechanical impedances of spring, mass, and damper are

$$\text{Spring} = k$$
$$\text{Mass} = -m\omega^2$$
$$\text{Damper} = if\omega$$

This is shown in Fig. 2.54.

Let the harmonic-forcing function be expressed as $F(t) = F_0 e^{i\omega t}$ so that the equation of motion becomes

$$m\ddot{x} + f\ddot{x} + kx = F_0 e^{i\omega t} \tag{2.135}$$

We should remember that the actual excitation is given only by the real part of $x(t)$ where $x(t)$ in equation 2.135 is a complex quantity.

By assuming the particular solution $x_p(t)$

$$x_p(t) = Xe^{i\omega t} \tag{2.136}$$

We substitute this in Equation 2.135, we get

$$X = \frac{F_0}{(k - m\omega^2) + if\omega} \tag{2.137}$$

Equation 2.137 can also be written as

$[Z(i\omega)\, X = F_0$, where $Z(i\omega) = -m\omega^2 + if\omega + k]$ which is the impedance of the system as defined above.

Multiplying the numerator and denominator on RHS of Equation 2.137 by $[(k - m\omega^2) + if\omega]$ and separating real and imaginary parts, we get,

$$X = F_0\left[\frac{k - m\omega^2}{(k - m\omega^2)^2 + f^2\omega^2} - i\frac{f\omega}{(k - m\omega^2)^2 + f^2\omega^2}\right] \tag{2.138}$$

We know that $x + iy = Ae^{i\Phi}$ where $A = (x^2 + y^2)$, $\tan \Phi = y/x$, Equation 2.138 can be expressed as

$$X = \frac{F_0}{[(k - m\omega^2)^2 + f^2\omega^2]^{1/2}} e^{-i\Phi} \tag{2.139}$$

where

$$\Phi = \tan^{-1}\left(\frac{f\omega}{k - m\omega^2}\right) \tag{2.140}$$

The steady-state solution, Equation 2.136, becomes

$$x_P(t) = \frac{F_0}{\sqrt{[(k - m\omega^2)^2 + (f\omega)^2}} e^{i(\omega t - \Phi)} \tag{2.141}$$

These manipulations are done to bring out a very important concept in vibration measurements and analysis, known as frequency-response function. Equation 2.137 can be rewritten as

$$\frac{kX}{F_0} = \frac{1}{1 - r^2 + i2r\xi} \equiv H(i\omega) \tag{2.142}$$

where $H(i\omega)$ is known as complex frequency response of the system. The absolute value of $H(i\omega)$ is given by

$$|H(i\omega)| = \left|\frac{kX}{F_0}\right| = \frac{1}{\sqrt{(1 - r^2)^2 + (2r\xi)^2}} \tag{2.143}$$

This denotes the magnification factor described earlier.
Recalling that

$$e^{i\Phi} = \cos \Phi + i \sin \Phi,$$

We can show that

$$H(i\omega) = |H(i\omega)|\, e^{-i\Phi} \tag{2.144}$$

where $\Phi = \tan^{-1}(2r/(1 - r^2))$ (2.145)

Thus, Equation 2.141 can be written as

$$x_P(t) = \frac{F_0}{K}|H(i\omega)|\, e^{i(\omega t-\Phi)} \tag{2.146}$$

It is important to note that the complex response function, $|H(i\omega)|$ contains both magnitude and the phase of the steady-state response. The use of this function is made use of in the experimental determination of system parameters ($m, f,$ and k).

If $F(t) = F_0 \cos \omega t$, the corresponding steady-state solution is given by the imaginary part of Equation 2.141

$$x_p(t) = \text{image}\left[\frac{F_0}{K}|H(i\omega)|\, e^{i(\omega t-\Phi)}\right] \tag{2.147}$$

Figure 2.54 shows the vector plot of the harmonic motion.

2.4.4 Forced Vibrations with Coulomb Damping

Equation of motion Figure 2.55 (a and b) show free body diagram of SDOF with Coulomb damping for the system (Fig. 2.55) is given by

$$m\ddot{x} + kx \pm \mu NF(t) = F_0 \sin \omega t \tag{2.148}$$

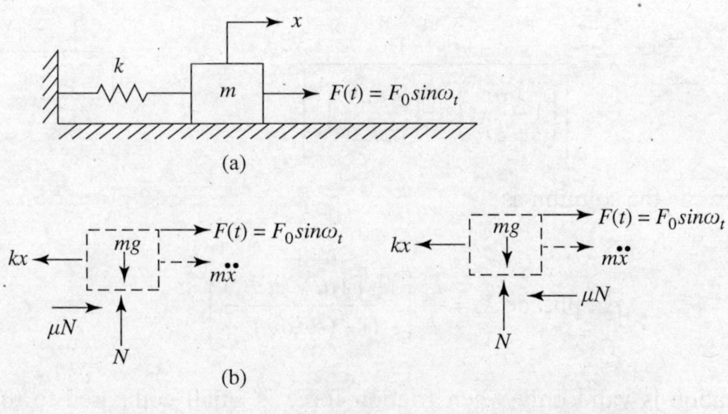

Figure 2.55 Forced Vibrations with Coulomb Damping

where the sign of friction force ($\mu N = \mu mg$) is +ve when mass moves from left to right and is negative when it is from right to left. The solution of Equation 2.148 is very complex. The motion of the mass will be discontinuous if the friction is very large (dry-friction damping force). We, in this book, are not interested in this case. On the other hand, if the dry-friction force is small compared to the amplitude of the excitation force, the system will vibrate almost in a harmonic manner. This typically happens in steam- and gas-turbine rotating blades having damper ring or damping pins. In this case, we find an approximate solution of Equation 2.148 by introducing the concept of equivalent viscous damping.

The equivalent viscous-damping ratio is found by equating the energy dissipation due to dry friction to the energy dissipated by an equivalent viscous damper during full-cycle of motion. If the amplitude of motion is denoted as *X*, the energy dissipated by friction force μN in a quarter of cycle is μNX. Hence, in a full-cycle, the energy dissipation by dry friction damping is

$$\Delta W = 4\mu NX \tag{2.149}$$

If the equivalent viscous-damping coefficient is denoted by f_{eq}, the energy dissipation in a full cycle will be

$$\Delta W = \pi f_{\text{eq}}.\omega X^2 \tag{2.150}$$

Proof of Equation 2.150:

$$\Delta W = \int_0^{2\pi/\omega d} f_{\text{eq}}.\dot{x}^2\,dt = \pi.f\omega_d X^2$$

where ω_d is damped natural frequency
Equating Equations 2.149 and 2.150, we obtain,

$$f_{\text{eq}} = \frac{4\mu.N}{\pi.\omega.X} \tag{2.151}$$

The steady-state response is given by $x_{\text{p}}(t) = X\sin(\omega t - \Phi)$
As in previous cases (spring, mass, and damper in SDOF systems as shown in Fig. 2.56)

$$X = \frac{F_0}{\sqrt{(k - m\omega^2)^2 + f_{\text{eq}}^2\omega^2}} = \frac{F_0/k}{\sqrt{(1-(\omega/\omega_n)^2)^2 + (2\xi_{\text{eq}}.\omega/\omega_n)^2}} \tag{2.152}$$

where $\xi_{\text{eq}} = f_{\text{eq}}/f_{\text{c}} = f_{\text{eq}}/2m\omega_n = 4\mu N/2m\omega_n\pi\omega X = 2\mu N/m\omega_n\pi\omega X$.
Thus, we obtain,

$$X = \frac{(F_0/k)}{\left[\left(1-\frac{\omega^2}{\omega_n}\right)^2 + \left(\frac{4\mu.N}{\pi.kX}\right)^2\right]}$$

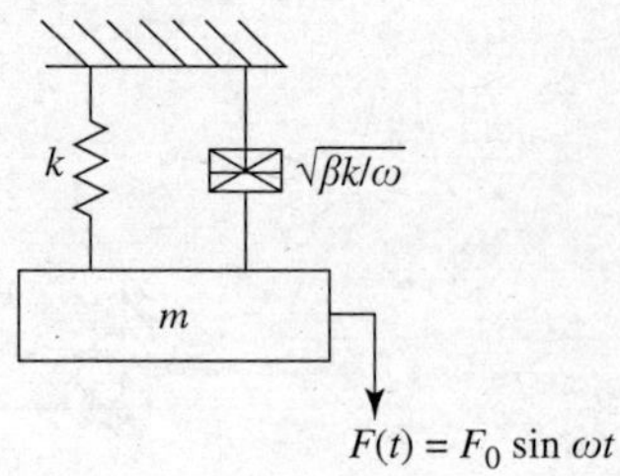

Figure 2.56 SDOF with Coulomb Damping

Thus, we can write the solution as

$$\text{Amplitude } X = \frac{F_0}{k}\left[\frac{1-(4\mu.N/\pi.F_0)}{(1-(\omega/\omega_n)^2}\right]^{1/2} \tag{2.153}$$

This equation is valid only when friction force is small compared to the excitation force F_0. Analysis of Equation 2.153 shows that the solution *X* exists only when the numerator is positive, that means

$$F_0/\mu N > 4/\pi \tag{2.154}$$

The phase angle is given by

$$\Phi = \tan^{-1}\left(\frac{f_{\text{eq}}\omega}{k - m\omega^2}\right) = \tan^{-1}\left[\frac{2\xi.\omega/\omega_n}{1-\omega^2/\omega 2_n}\right] = \tan^{-1}\left[\frac{4\mu.N/\pi.kX}{1-\omega^2/\omega_n^2}\right] \tag{2.155}$$

Substituting the value of X in the Equation 2.155, we get,

$$\Phi = \tan^{-1}\left[\frac{4\mu.N/\pi.F_0}{\{1-(4.\mu.N/\pi.F_0)\}^{1/2}}\right] \tag{2.156}$$

Study of Equation 2.153 shows that friction serves to limit amplitude of vibration for $\omega/\omega_n \neq 1$. However at $\omega/\omega_n = 1$ (resonance), the amplitude of vibration increases enormously (this does not happen in the case of viscous damping where with damping the amplitude at resonance decreases). This happens because resonance more energy is directed into the system than the energy dissipated. The proof briefly given here. Energy directed into the system is given by

$$\Delta W' = \int_0^{\tau=2\pi/\omega} F dx = \int_{\text{cycle}} F_0 \sin\omega t.(\omega X \cos(\omega t - \Phi)dt)$$

At resonance, the phase angle between the force and response is 90°.

Put $\Phi = 90°$ at resonance. This equation reduces to $\Delta W' = \pi F_0 X$. Energy dissipated in friction $\Delta W = 4\mu NX$ (Equation 2.149). Thus, we get $\Delta W', > \Delta W$ or more energy is directed into the system than is dissipated in a cycle. This extra energy is used to build up amplitude of vibration. For non-resonance condition,

$$\Delta W'' = \omega.F_0 X \int_{\text{cycle}} \sin\omega t \cos(\omega t - \Phi)dt$$

$$= \pi F_0 X \sin\Phi$$

Due to the presence of the term sin Φ, the energy input coincides with energy dissipated and thus no vibration build-up takes place under this condition.

2.4.5 Forced Vibration with Hysteretic Damping

In this case, as shown previously, the damping ratio = $h/\omega = \beta k/\omega$.

The equation of motion is

$$m\ddot{x} + \beta(k/\omega)\,\dot{x} + kxF_0 \sin\omega t \tag{2.157}$$

The response of the system can be worked out according to the steps explained earlier. The final solution is

$$X = \frac{F_0}{k(1-\left(\frac{\omega^2}{\omega_n^2}\right) + \beta^2)^{1/2}} \tag{2.158}$$

$$\Phi = \tan^{-1}\left[\frac{\beta}{1-\frac{\omega^2}{\omega_n^2}}\right] \tag{2.159}$$

Thus, at resonance, the amplitude ratio is

$$\frac{X}{F_0/k} = \frac{1}{\beta} \tag{2.160}$$

Structural damping being small, at resonance the amplification factor is very large even though it does not become infinite.

2.4.6 Response of SDOF Systems Subjected to a General Periodic Force

We have already considered SDOF systems which are harmonically excited (i.e. vibratory force $F(t)$ is sinusoidal). Such harmonic excitations take place when the structure or component under consideration experiences vibratory forces caused by the rotating masses. One such system is shown in Figure 2.57. It shows a rotor supported on bearings which are mounted on a foundation block. The bearings can be hydrodynamic bearings or antifriction bearings as shown in Fig. 2.59.

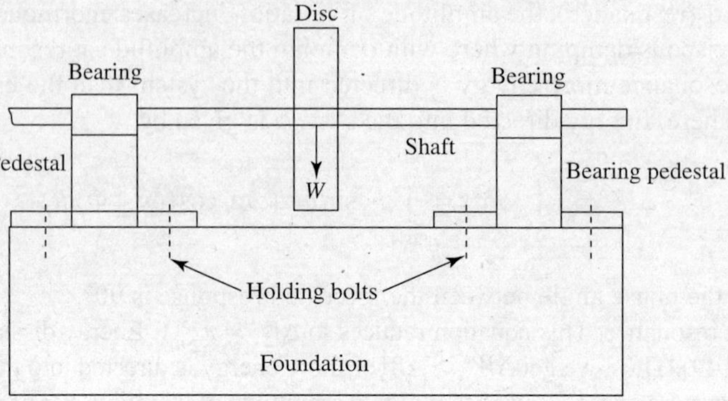

Figure 2.57 Rotor Mounted in Bearings

The shaft carries a disc of weight W. The CG of the disc does not coincide with the centre of rotation. In practice this happens on account of various reasons—one of them being manufacturing errors in machining of the shaft having integral disc or improper fit of the disc on the shaft or both. Due to this non-coincidence, also called the eccentricity, the disc while rotating at frequency ω (= $2\pi n/60$, where N is rpm of the rotor) experiences unbalance forces. The assembly of disc(s) on the shaft is termed as rotor.

The rotor (Fig. 2.58) having eccentricity e produces a centrifugal force $F = \frac{W}{g}e\omega^2$. The bearings, which support the rotor, experience a force proportional to $F = \frac{W}{g}e\omega^2$ and have components $F_Y = \frac{W}{g}e\omega^2 \cos\theta$ in vertical direction and $F_X = \frac{W}{g}e\omega^2 \sin\theta$ in the horizontal direction. Since $\theta = \omega t$, where t is time, the forces acting on the bearing are $F_Y = \frac{W}{g}e\omega^2 \cos\omega.t$ and $F_X = \frac{W}{g}e\omega^2 \sin\omega t$ and these cause the bearing and the bearing pedestal to vibrate in X and Y directions at a frequency ω (= $2\pi N/60$, where N is rpm of the rotor). The shaft also oscillates/vibrates at the same frequency in space with respect to or relative to bearings. Fig. 2.59 shows how these vibrations are measured.

In actual practice, it is rare that system will get excited by only a single excitation or perturbation force such as unbalance of the rotor. There are other excitation/perturbation forces caused by

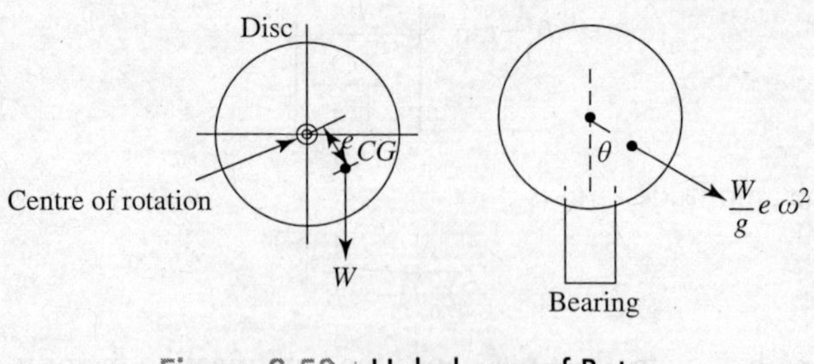

Figure 2.58 Unbalance of Rotor

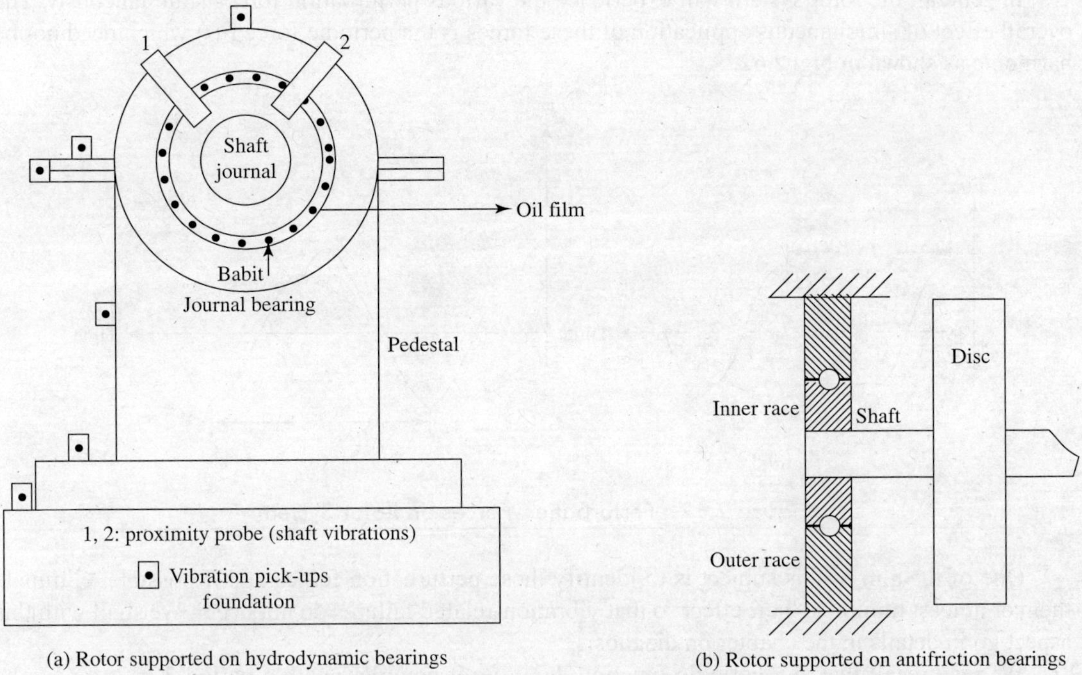

Figure 2.59 Vibration Measurement on Rotor bearing System

misalignment of coupled rotors (Fig. 2.60) or other forces such as unsteady pressure oscillation in the space between moving blades and the guide blades (Fig. 2.61).

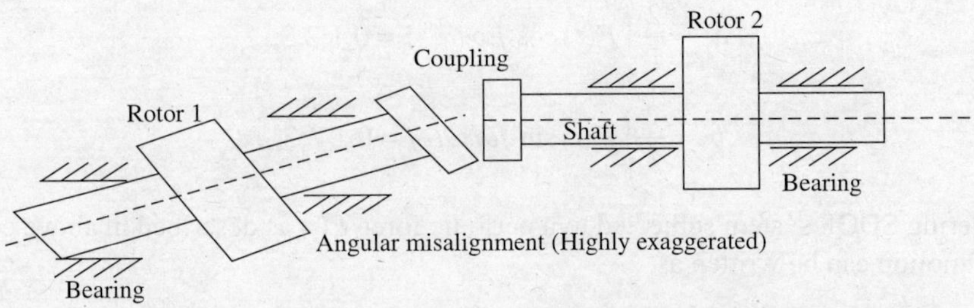

Figure 2.60 Misaligned Rotor System

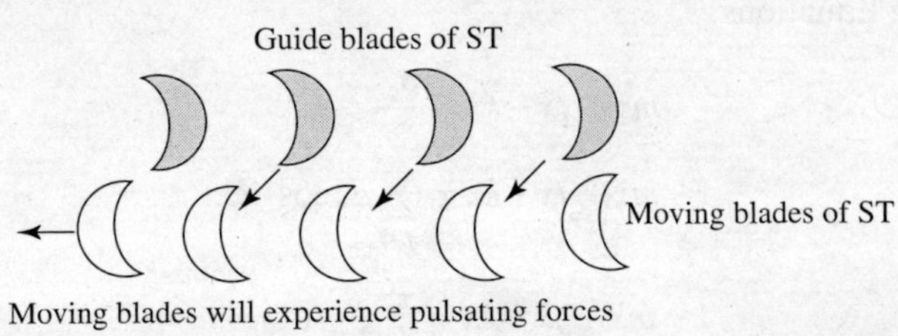

Figure 2.61 Flow-Induced Pressure-Pulsation Forces on Steam-Turbine Blades

In general, the rotor system will experience the various perturbation forces simultaneously. The overall effect of simultaneous application of these forces is the periodic force $F(t)$ which need not be harmonic as shown in Fig. 2.62.

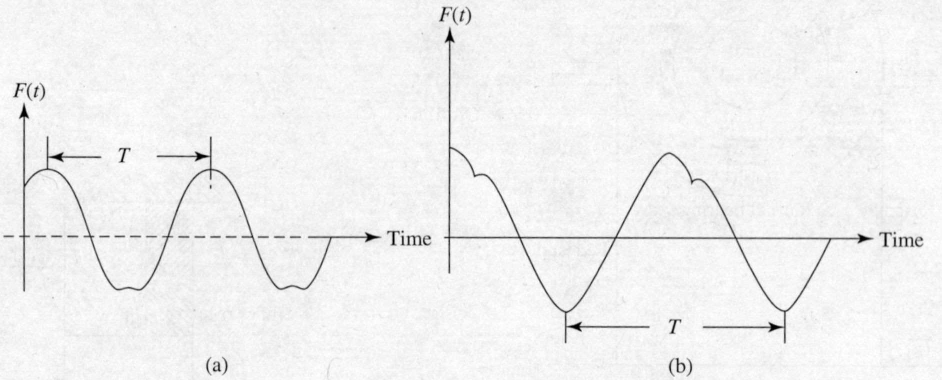

Figure 2.62 Perturbation Forces on Rotor System

One of the aim of this subject is to identify these perturbation forces and if possible, eliminate them or at least minimize their effect so that vibration-related failures do not occur. We shall with this aspect go in details in the chapter on diagnosis.

We may recall that any periodic function, howsoever complex, with a period $T = 2\pi/\omega$ can be expressed as a Fourier series such as

$$F(t) = \frac{a_0}{2} + \sum_{j=1}^{\infty} a_j \cos(j\omega t) + \sum_{j=1}^{\infty} b_j \sin(j\omega t) \tag{2.161}$$

where

$$a_j = \frac{2}{T}\int_0^T F(t)\cos j\omega.tdt \quad j = 0,1,2,3,..... \tag{2.162}$$

$$b_j = \frac{2}{T}\int_0^T F(t)\sin j\omega.tdt \quad j = 0,1,2,3,..... \tag{2.163}$$

Considering SDOF system subjected to a periodic force $F(t)$ as described in above equations, the equation of motion can be written as

$$m\ddot{x} + f\dot{x} + kx = F(t) = \frac{a_0}{2} + \sum_{j=1}^{\infty} a_j \cos j\omega.t + \sum_{j=1}^{\infty} b_j \sin j\omega.t \tag{2.164}$$

The right-hand-side of this equation is a constant plus a sum of harmonic functions. Using the principle of superposition, the steady-state solution of Equation 2.164 is the sum of steady-state solutions of the following Equations.

$$m\ddot{x} + f\dot{x} + kx = \frac{a_0}{2} \tag{2.165}$$

$$m\ddot{x} + f\dot{x} + kx = \sum_{j=1} a_j \cos j\omega.t \tag{2.166}$$

$$m\ddot{x} + f\dot{x} + kx = \sum_{j=1}^{\infty} b_j \sin j\omega.t \tag{2.167}$$

The solution of Equation 2.165 is

$$x_p = \frac{a_0}{2k} \tag{2.168}$$

Solutions of Equations 2.166 and 2.165 are

$$x_p(t) = \frac{(a_j/k)}{\sqrt{(1-j^2r^2)^2+(2\xi.r)}}\cos(j\omega.t-\Phi_j) \tag{2.169}$$

$$x_p(t) = \frac{(b_j/k)}{\sqrt{(1-j^2r^2)^2+(2\xi.r)}}\sin(j\omega.t-\Phi_j) \tag{2.170}$$

where

$$\Phi_j = \tan^{-1}\left(\frac{2\xi.jr}{1-j^2r^2}\right), \qquad r = \frac{\omega}{\omega_n} \tag{2.171}$$

(One may note that there are several equations in above three equations.) Therefore, the complete solution is,

$$X_P(t) = \frac{a_0}{2k} + \sum_{j=1}^{\infty}\frac{(a_j/k)}{\sqrt{(1-j^2r^2)+(2\xi.jr)^2}}\cos(j\omega t-\Phi_j)$$
$$\times\sum\frac{(b_j/k)}{\sqrt{(1-j^2r^2)+(2\xi.jr)^2}}\sin(j\xi-\Phi_j) \tag{2.172}$$

It can be seen from Equation 2.172 that the amplitude and phase shift corresponding to jth term depend on j. If $j\omega = \omega_n$ (jth harmonic equals natural frequency or very near natural frequency), the amplitude of the corresponding harmonic will be very large in case ξ is small and also if j is small. On the other hand, as j becomes larger and larger, the amplitude becomes smaller and smaller. Thus, the first few frequency terms may be sufficient to obtain the response with reasonable accuracy.

This analysis is given for steady-state response only. The transient part is usually not a major concern.

We shall illustrate this procedure for a component such as turbine blade subjected to pressure pulsations such as those appearing in steam turbines or gas turbines. In order to understand the principle, let us assume that the component can be idealized as spring–mass–damper system i.e. SDOF system. (In reality, the blades are subjected to centrifugal force, gas-bending force, twisting moment, pressure pulsation, etc. and hence they have to be considered as complex three-dimensional body with very complex end fixity conditions.) Let the pressure be represented by curve as shown in Fig. 2.63.

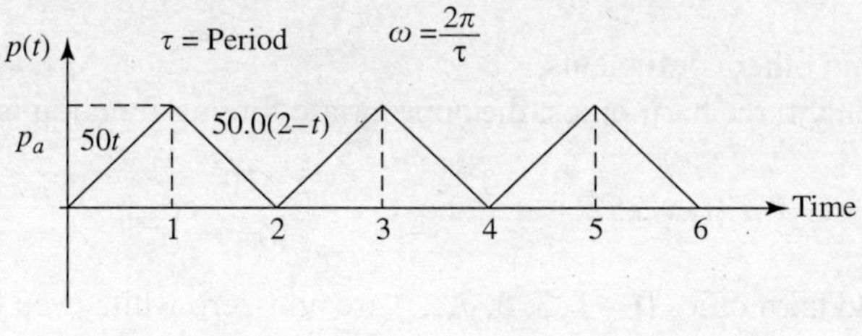

Figure 2.63 Pressure Pulsations in ST/GT

The forcing function is

$$F(t) = Ap(t) \tag{2.173}$$

where A is the cross-sectional area of blade normal to pressure.

Let us assume $A = 1$. The system and the disturbing force are shown in Fig. 2.64.

$$F(t) = 50t \le t < \tau$$
$$= 50(2-t) \text{ for } t/2 \le \tau \quad (2.174)$$

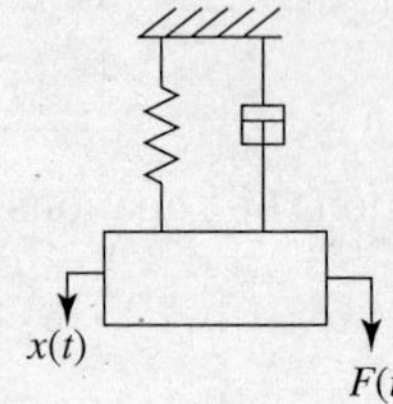

Figure 2.64 SDOF Model for ST/GT Blade

Let us denote

$$F(t) = \frac{a_0}{2} + a_1 \cos\omega.t + a_2 \cos 2\omega t + \cdots + b_1 \sin\omega t + b_2 \sin 2\omega t \quad (2.175)$$

where

$$\omega = \pi \quad (2.176)$$

The next step is to evaluate the constants a_j and b_j, we know

$$a_j = \frac{2}{\tau}\int_0^{\tau} F(t)\cos j\omega.tdt,\ j = 1,2,3,\ldots\ldots \quad (2.177a)$$

$$b_j = \frac{2}{\tau}\int_0^{\tau} F(t)\sin j\omega.tdt,\ j = 1,2,3,\ldots\ldots \quad (2.177b)$$

It may be remembered that the system equation is

$$m\ddot{x} + f\dot{x} = kx + F(t) = \frac{a_0}{2} + \sum_{j=1}^{\infty} a_j \cos j\omega.t + \sum_{j=1}^{\infty} b_j \sin j\omega.t$$

Using above equations, we get

$$a_0 = \frac{2}{2}\left[\int_0^1 50tdt + \int_1^2 50(92-t)dt\right] = 50 \quad (2.178)$$

$$a_1 = \frac{2}{2}\left[\int_0^1 50\cos\pi.tdt + \int_1^2 50(2-t)\cos\pi.tdt\right] = -\frac{2\times10^5}{\pi^2}\Bigg] \quad (2.179)$$

$$b_1 = \frac{2}{2}\left[\int_0^1 50t\sin\pi.tdt + \int_1^2 50(2-t)\sin\pi.tdt\right] = 0\Bigg] \quad (2.180)$$

Similarly we can show that $b_2 = 0$

We can also show that (2.181)

$$a_3 = -\frac{2\times10^5}{9\pi^2} \quad (2.182)$$

Likewise, we can obtain other coefficients.
For considering, say only three harmonics, the appropriate forcing function is given by

$$F(t) = 25 - \frac{2\times10^5}{\pi^2}\cos\omega.t - \frac{2\times10^5}{9\pi^2}\cos 3\omega.t \quad (2.183)$$

Note that only odd harmonics ($j = 1, 3, 5, 7, \ldots$) are non-zero while even harmonics are zero. The steady-state response of the system is then

$$x_p(t) = \frac{25}{k} - \frac{2\times10^5/k\pi^2}{\sqrt{(1-r^2)^2 + (2\xi.r)^2}}\cos(\omega t - \Phi_1) - \frac{2\times10^5/9\pi^2}{\sqrt{(1-r^2)^2 + (6\xi r)^2}}\cos(3\omega t - \Phi_3) \quad (2.184)$$

Additional terms comprising of 5ω, 7ω, 9ω, etc. can be evaluated but as can be noted from Equation 2.184, their contributions will be very small and thus can be neglected. Equation 2.184 will also enable us to formulate a diagnostic rule that if the spectrum of $x_p(t)$ shows odd harmonics (1, 3, 5,…, x times rotating speed), the perturbation force is most likely to be of the type shown in Fig. 2.63 and can be appropriately related to a physical process in the flow path.

Natural frequency (undamped) assuming m = 2500 N/m and mass m = 0.25 kg. f = 10 N s/m

$$\omega_n = \sqrt{\frac{k}{m}} = \sqrt{\frac{2500}{0.25}} = 100 \text{ rad/s.}$$

The forcing frequencies $\omega = \pi\ 5\omega$, etc

The frequency ratio for first frequency = $\pi/100 = 0.0314$.

Damping ratio $\xi = f/f_c = f/2m\omega_n = 10/2 \times 0.25 \times 100 = 0.2$.

The phase angles are given by

$$\Phi_1 = \tan^{-1}\frac{2r\xi}{1-r^2} = 0.012 \text{ rad}$$

$$\Phi_3 = \tan^{-1}\left\{\frac{6r\xi}{1-r^2}\right\} = 0.03 \text{ rad}$$

Finally, we can show the response as

$$x_p(t) = 0.019 - 0.0159\cos(\omega t\ 0.012) - 0.00018\cos(3\omega t).$$

We shall illustrate this procedure by yet another example.

Consider periodic excitation of the base of the system as shown in Fig. 2.65.

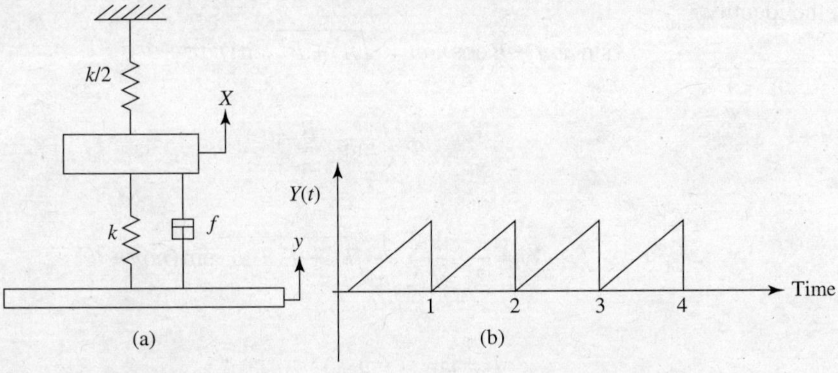

Figure 2.65 Periodic Excitation of Base SDOF System

The base motion $y(t)$ causes the mass m to have motion $x(t)$. Applying force equilibrium at mass m, we get

$$m\ddot{x} = -f(\dot{x} - \dot{y}) - k(x - y)$$

or

$$m\ddot{x} + f\dot{x} + kx = f\dot{y} + ky \tag{1}$$

As explained earlier, the periodic motion $y(t)$ can be expressed as a Fourier series

$$y(t) = \sum_{n=0}^{\infty}(a_n \cos n\omega t + b_n \sin n\omega t) \text{ i.e.}$$

$$y(t) = a_0 + a_1 \cos \omega t + a_2 \cos 2\omega t + \cdots\cdots + b_1 \sin \omega t + b_2 \sin 2\omega t + \cdots\cdots$$

where

$$a_0 = \frac{1}{T}\int_0^T y(t)dt,\ a_n = \frac{2}{T}\int_0^T y(t)\cos n\omega t dt,\ b_n = \frac{2}{T}\int_0^T y(t)\sin n\omega t dt$$

For the given waveform, $y(t) = t$ and $T = 1$ and thus

$$a_0 = \frac{1}{T}\int_0^{T=1} t dt = \frac{1}{2},\ a_n = \frac{2}{T}\int_0^{T=1} t\cos n\omega t dt = 0,\ b_n = \frac{2}{T}\int_0^{T=1} t\sin n\omega t dt = -\frac{1}{n\pi}.$$

Thus,

$$y(t) = \frac{1}{2} - \left(\frac{1}{\pi}\right)\sum_{n=1}^{\infty}\frac{1}{n}\sin n\omega.t$$

From this, we get,

$$\dot{y}(t) = -\sum_{n=1}^{\infty}\left(\frac{\omega}{\pi}\right)\cos n\omega.t$$

And the equation of motion (1) becomes

$$m\ddot{x} + f\dot{x} + kx = f\left[-\sum_{1}^{\infty}\left(\frac{\omega}{\pi}\right)\cos n\omega.t\right] + k\left[\frac{1}{2} - \frac{1}{\pi}\sum_{1}^{\infty}\frac{1}{n}\sin n\omega.t\right]$$

or

$$m\ddot{x} + f\dot{x} + kx = \frac{k}{2} - \sum_{1}^{\infty}\left\{\frac{k}{n\pi}\sin n\omega.t + \frac{f\omega}{\pi}\cos n\omega t\right\}$$

Since from the identity

$$A\sin n\omega t + B\cos n\omega t = \sqrt{A^2 + B^2}\,\sin(n\omega t + \Phi)$$

and

$$\Phi = \tan^{-1}\frac{B}{A}$$

then

$$fy + ky = \frac{k}{2} - \frac{1}{\pi}\sum_{n=1}^{\infty}\frac{1}{n}\sqrt{k^2 + f^2 n^2\omega^2}\,\sin(n\omega t + \Phi)$$

where

$$\Phi = \tan^{-1}(f\omega/k)$$

Also,

$$\omega_n^{\ 2} = k/m \text{ and } \xi = f/2m\omega_n$$

$$fy + ky = \frac{k}{2} - \frac{1}{\pi}\sum_{n=1}^{\infty}\left(\frac{k}{n}\right)\sqrt{1 + (2\xi.rn)^2}\,\sin(n\omega.t + \Phi)$$

This is the impressed force and the steady-state response.

According to earlier discussions

$$x_p(t) = \frac{1}{2} - \left(\frac{1}{\pi}\right)\sum_{n=1}^{\infty}\left(\frac{1}{n^2}\right)\sqrt{\frac{1 + (2\xi.rn)^2}{(1 - n^2r^2)^2 + (2\xi.rn)^2}}\,\sin(n\omega.t - \varphi)$$

where

$$\varphi = \tan^{-1}\left(\frac{2nr\xi}{1-n^2r^2} - \Phi\right)$$

In this particular case, $\omega_n = \sqrt{40/10} = 2$ rad/s, $r = \frac{1}{2} = 0.5$, $\xi = f/2m\,\omega_n = 20/2 \times 10 \times 2 = 0.5$, $x_p = 0.5 - 0.396 \sin(6.24t - 7.1) - 0.225 \sin(12.48t - 26.50) - \cdots\cdots$

Thus, the response consists of harmonics of $\omega = 6.24$ (1, 2, 3…)

This example illustrates as to how diagnostic rules can be framed for a variety of disturbing-force pattern. It should be noted that each of the disturbing force can be identified to some known malfunction in the system such as unbalance, misalignment, pressure pulsations in the flow path, etc. The example cited is typical of mechanical looseness in the system. It is clear that the analysis leads to framing of the diagnostic rules.

CONCLUSION

In this chapter, we have carried out a detailed analysis of single degree-of-freedom systems. In the beginning, we have discussed how to determine the number of degrees-of-freedom of the vibratory system. We also dealt with rigid-body oscillations and individual elements of vibratory system, that is, the mass, spring, and damper. We also analyzed the vibrations of spring, mass system (no damper) and presented the concept of natural frequency of system having elasticity and inertia (spring and mass). This is followed by analyzing systems with damping. We discussed the cases of free-vibration response of overdamping, critical damping and underdamping. We showed that vibrations can occur only in the case of damping lesser than critical damping. We discussed the free vibration response of an underdamped system and method of utilizing the information of logarithmic decrement to obtain information about the damping ratio in the system. We also discussed other damping mechanisms such as Coulomb or dry-friction damping and hysteretic damping.

We analyzed the response of the one degree-of-freedom system to harmonic and non-harmonic periodic-excitation forces. In the end, we discussed as to how diagnostic rules can be framed from the detailed analysis of vibration response to a variety of excitation forces.

In the forthcoming chapter, we shall discuss the two degrees-of-freedom systems subjected to vibratory forces.

EXERCISES

2.1 For the system shown in the figure, find the equivalent mass. Assume that the cylinder with mass m_c does not slip and the pulley has mass moment of inertia J_p and the links 1 and 2 having lengths l_1 and l_2, respectively are rigid. (*Hint:* Consider the kinetic energy of each element such as the pulley, cylinder, and the mass m. The pulley with R as radius will have only angular motion θ_p, while the cylinder will have linear motion x_2 as well. $\Theta_p = x/R$. $x_2 = \theta_p \times l_1$)

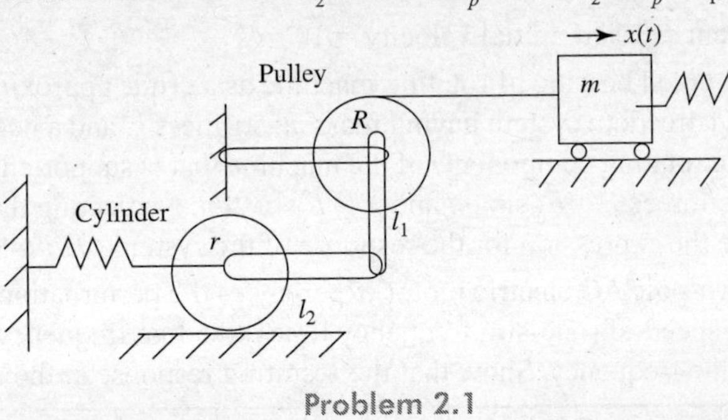

Problem 2.1

2.2 A machine of mass 500 kg is mounted on a simply supported beam of length 2 m. The beam cross-section is rectangular with a depth of 10 cm and a width of 12 cm. The modulus of elasticity of the material of beam is 2.06×10^{11} N/m². It is required to limit the central deflection of the beam by 25% by providing a spring of stiffness k at midpoint of the beam. Find k.

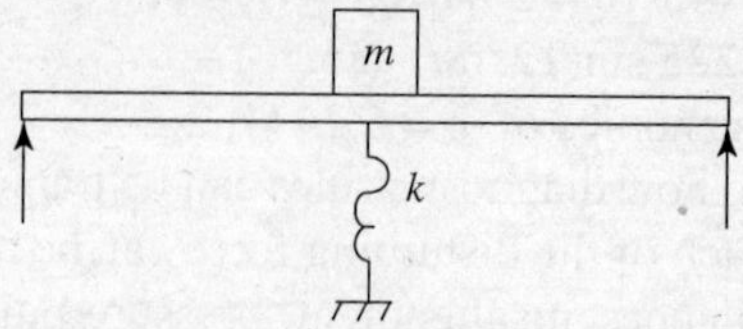

Problem 2.2

If the same beam is put such that the depth of the beam is 12 cm and width is 10 cm, find k. Which of these arrangements is stiffer?

(Answer: $k = 37 \times 10^7$ N/m)

2.3 Find the equivalent mass of the system shown.

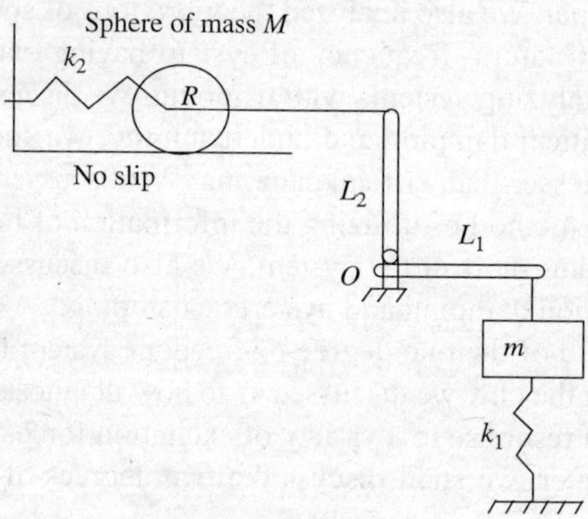

Problem 2.3

2.4 Consider a spring–mass system with $k = 4000$ N/m and $m = 10$ kg subjected to harmonic excitation force $F(t) = 400 \cos 10t$ N. Find and plot the total response under the following initial conditions:

(1) $x_0 = 0.1$ m, initial velocity = 0

(2) $x_0 = 0$, initial velocity 10 m/s

(3) Initial displacement = 0.1 m, initial velocity = 10 m/s

2.5 The pedestal-mounted bearing of rotating machine as a crude approximation can be considered as single degree-of-freedom system having mass m, stiffness k, and a negligible damping. The force transmitted by the rotating components of the machine that is supported on two such bearings comprises of harmonic forces $F = F_0 \sin \omega t$ and $F = F_0 \sin 2\omega t$. Neglecting the phases of these perturbation forces, derive the expression for the response of the system. (*Hint:* See the section 2.4.5.)

2.6 The rotor of a two-pole AC electric motor experiences the perturbation forces (1) unbalance force due to rotational speed of (one-slip frequency times) the line frequency and electrical forces corresponding to the line frequency. Show that the resulting response at the bearings of the motor will show beat characteristics. You may consider the amplitudes of both forces experienced at the bearings to be equal, and bearings can be considered as single degree-of-freedom systems. What will happen if the rotor of the motor has higher unbalance? (*Hint:* Consider action of two harmonic forces, one at line frequency and other at a frequency corresponding to running speed of the motor. The concepts explained in section 2.4.5 may then be used.)

2.7 An aircraft engine has a rotating unbalanced mass m at a radius r, and two of such engines are mounted on the wings at a distance L from the fuselage body on both wings. The wing has a uniform cross-section $a \times b$. Assuming that both the engines are running at N rpm, find out the maximum deflection of the wing of total length L_1. Neglect the effects of damping and the mass of wings.

2.8 A centrifugal pump weighing 900 N, operating at 1500 rpm, is mounted on six springs of stiffness 6000 N/m each. Find the maximum permissible unbalance in order to limit the steady-state deflection to 500 microns. (1 micron = 10^{-6} m.)

2.9 A fixed-fixed steel beam of length 5 m, thickness 0.1 m, and width 0.5 m carries an electrical motor of mass 75 kg at speed 1500 rpm at its mid-span. A rotating unbalance of magnitude $F_0 = 6000$ N is developed due to unbalance in the rotor of the motor. Find the amplitude of steady-state vibration. Neglect the weight of the beam.

2.10 Assume that one of the fixed supports in Example 2.9 fails. What will be the steady vibration response of the structure? What corrective measure are required so that the motor continues to run?

2.11 An electric motor weighing 750 lbs and running at a speed of 1800 rpm is supported on four helical springs each of which has eight active coils with a wire diameter of 0.25 in. and coil diameter of 3 in. The rotor has a weight of 100 lbs with its centre of mass located at a distance of 0.01 in. from the centre of rotation. Find the amplitude of vibration of the motor and the force transmitted through the springs to the base.

2.12 A single-storey building frame can be modelled by a rigid floor of mass m and columns of stiffness k as shown in the figure. It is proposed that a damper shown in this figure is attached to absorb vibrations due to horizontal ground motion $y(t) = Y \cos \omega t$. Derive an expression for the damping constant of the damper that absorbs maximum power.

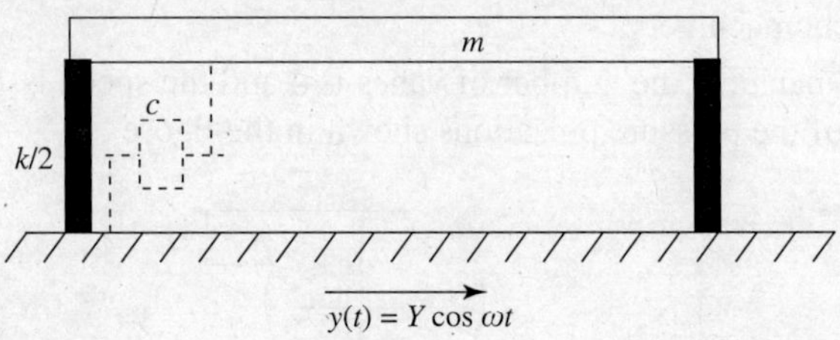

Problem 2.12

(*Answer:* $c = (k - m\omega^2)/\omega$)

2.13 For the building shown in Problem 2.12, assume that ground acceleration is 100 sin ωt. Find the horizontal displacement of the floor. Assume $m = 2000$ kg, $k = 0.1$ MN/m, $\omega = 25$ rad/s, and all the initial velocities and displacements for ground as well as the building are zero.

(*Answer:* 0.334 sin 25 t)

2.14 An automobile is modelled as a single degree-of-freedom system vibrating in vertical direction. It is driven along a road whose elevation varies sinusoidally. The distance from peak to trough is 0.2 m and the distance along the road between the peaks is 35 m. If the natural frequency of the automobile is 2 Hz and the damping ratio of the shock absorbers is 0.15, determine the amplitude of vibration of the automobile at a speed of 60 km/h. Find the most unfavourable speed of the automobile from the point of view of passenger's comfort.

(*Answer:* $X = 0.106$ m, $s = 247$ km/h)

2.15 Show that for small values of damping, the damping ratio ζ can be expressed as $\zeta = (\omega_2 - \omega_1)/(\omega_2 + \omega_1)$.

2.16 An air compressor of mass 100 kg is mounted on an elastic foundation. It has been observed that when a harmonic force of amplitude 100 N is applied to the compressor, the maximum steady-state displacement of 5 mm occurred at a frequency of 300 rpm. Determine the equivalent stiffness and damping coefficient of the foundation.

(*Answer:* $k = 1.006 \times 10^5$ N/m, $c = 633.3$ N-s/m)

2.17 A precision grinding machine is supported on an isolator having a stiffness of 1 MN/m and viscous-damping constant of 1 kN-s/m. The floor on which it is mounted is subjected to a harmonic disturbance due to the operation of unbalanced engine in the near vicinity of the grinding machine. Find the maximum-acceptable displacement amplitude of the floor, if the resulting amplitude of the grinding wheel is to be restricted to 1 micron. For this evaluation, we may consider the grinding machine and the wheel as rigid bodies of weight 5000 N.

(*Answer:* 170 microns)

2.18 An automobile having empty weight of 1000 lb and fully loaded weight of 3000 lb vibrates in a vertical direction while travelling at 55 mph on a rough road having a sinusoidal waveform with an amplitude Y ft and a period of 12 ft. Assuming that the automobile can be modelled as a single degree-of-freedom system with stiffness of 30,000 lb/ft and damping ratio $\zeta = 0.2$, determine the amplitude of vibration of the automobile when it is (*a*) empty and (*b*) fully loaded.

2.19 In a centrifugal rotating machine such as the runner of the hydraulic turbine, the fluid at any point is subjected to periodic impulses each time a blade passes the point. The frequency of these impulses (which manifest in the form of pressure pulsations in the flow path of machine) is given by the product of the frequency of rotation and the number of blades. This frequency is called the blade-passing frequency or vane-passing frequency. These impulses are experienced by all the components of the turbomachinery.

In a typical fluid machine, the number of vanes is 4 and the speed is 100 rpm. Determine the first three harmonics of the pressure pulsations shown in the figure.

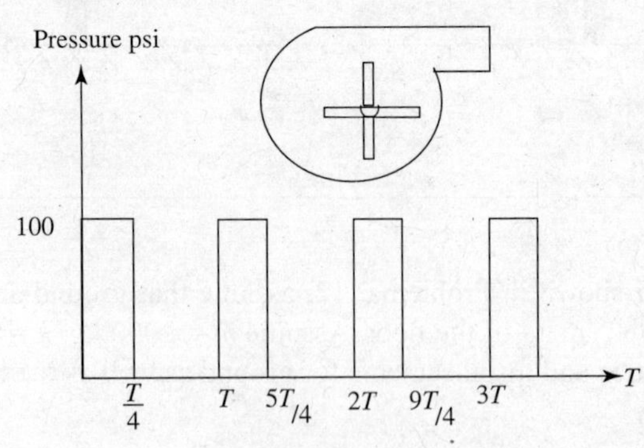

Problem 2.19

2.20 A turbine-generator rotor system can be considered as a single degree-of-freedom system having mass m, stiffness k, and negligible damping. Derive an equation for vibration response of the system to an unbalance force $F = \sin \omega t$, where ω represents the angular speed at any given instant of time. The machine starts from zero rpm and approaches its critical speed at a certain rate. Show how vibration build-up takes place and explain how you will prevent the likely damage around the critical speed (natural frequency) by quickly moving away from the critical speed.

Hint: See the derivation of Equation 2.98.

2.21 For Problem 2.20, consider some damping element in the system and derive the equation for the total response of the system which starts from rest and approaches its normal operating speed in certain time duration.

3 Two Degrees-of-Freedom Systems

In the previous chapter, we considered single degree-of-freedom vibration systems. We showed that many complex systems that one encounters in real life can be approximated as a single degree-of-freedom system to obtain a first-level assessment of vibration problem. We now consider more complex idealization of the real-world systems. The real-world systems usually are multi-degrees-of-freedom systems. To begin with, we shall now deal with a two-degrees-of-freedom system.

The systems that require two independent coordinates to describe their motion are called two degrees-of-freedom systems. Figs. 3.1–3.3 show some typical systems, which belong to the category of two degrees-of-freedom systems.

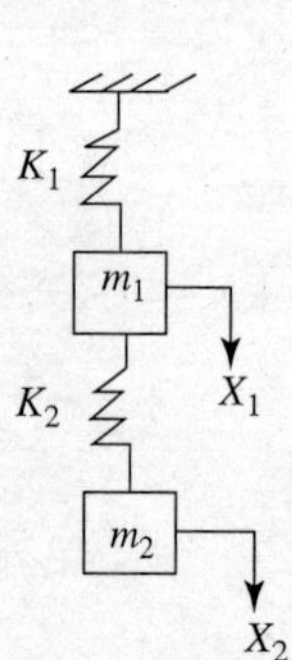

Figure 3.1 Two Degrees-of-Freedom Spring–Mass–Damper System

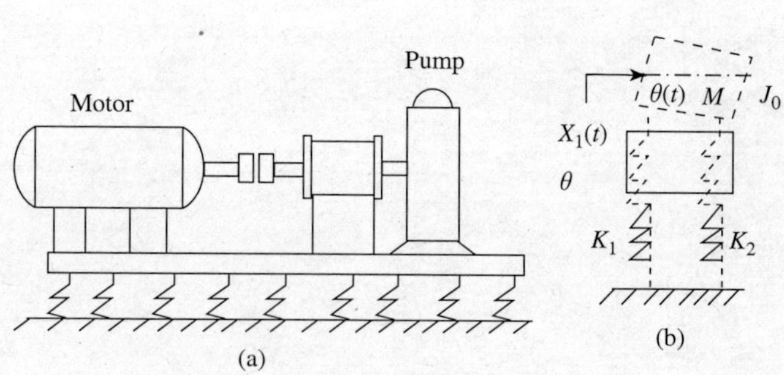

Figure 3.2 Motor Pump Assembly

The system shown in Fig 3.1 is constrained to move in the vertical direction. Each mass has its own motion and in order to specify the position of the system at any given instant of time, we need to specify x_1 and x_2.

Figure 3.2 shows a pump motor assembly supported on a spring system. This can be idealized as a bar of mass m and mass moment of inertia J_0 supported on two springs k_1 and k_2. The displacement of the system at any time can be specified by a linear coordinate $x(t)$, indicating the vertical displacement of the CG of the mass and an angular coordinate $\theta(t)$, denoting the rotation of the mass m about CG. Instead of $x(t)$ and $\theta(t)$, we can also use $x_1(t)$ and $x_2(t)$ as independent coordinates to specify the motion of the system. Thus, the system has two degrees-of-freedom. It is very important to note that in this case, the mass m is not treated as a point mass as shown in Fig. 3.1, but as a rigid body having two possible types of motion.

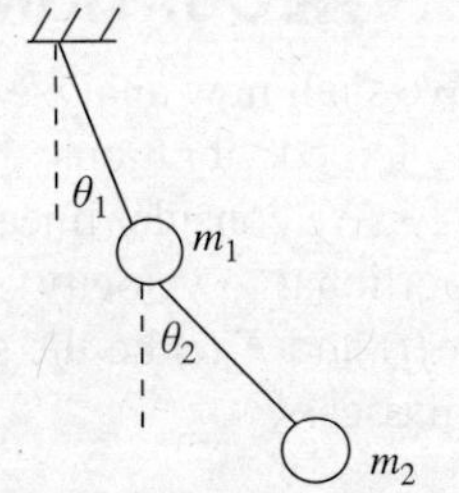

Figure 3.3 Double Pendulum

Figure 3.3 shows a double pendulum. It is necessary to specify θ_1 (for mass m_1) and θ_2 (for mass m_2) to specify the motion. Hence, this is also a two degrees-of-freedom system.

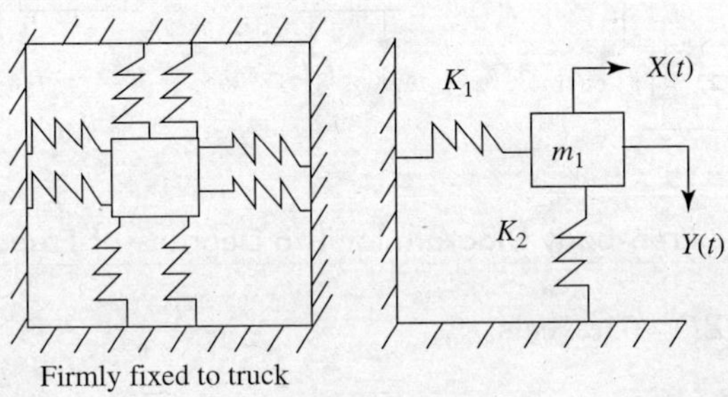

Figure 3.4 Two Degrees-of-Freedom Idealization of Packaging

Figure 3.4 shows packaging of mass m, usually done while transporting or shipping. The mass (m) is confined to motion in x and y directions. The system then can be modelled as a mass supported by springs in the x and y directions. This is a case where we have one mass (point mass) and two degrees-of-freedom because of motions in the x and y directions. In fact, the strict definition of number of degrees-of-freedom is

$$N = \text{Number of masses} \times \text{Number of possible types of motion for each mass.}$$

There are two equations of motion for a two degrees-of-freedom system, one for each mass (more precisely, for each degree-of-freedom). They are generally in the form of coupled differential equations, which means that each of the equation involves all coordinates.

The system has two natural frequencies. If we give suitable initial excitation (disturbance), the system vibrates at one of the natural frequencies. During free vibrations, at one of the natural frequencies, the amplitudes of the two degrees-of-freedom (coordinates) are related in a specific manner and that configuration (of motion) is called normal mode/principal mode or natural mode of vibration. Thus, the two degrees-of-freedom system has two normal/principal/natural modes of vibration corresponding to the two natural frequencies. If given an arbitrary excitation to the system, the resulting free vibrations will be a superposition of the two normal modes of vibration. However, if a forced excitation is given to the system, the system vibrates at the frequency of excitation force.

Under harmonic excitation, resonance occurs when the excitation frequency matches one of the natural frequencies.

The systems shown in Figs. 3.1–3.4 are required to be specified by a set of independent coordinates such as length, angle, that is, x_1, x_2, x, θ etc. Any such set of coordinates are called **generalized coordinates**. As we shall see in the forthcoming paragraphs, the equations of motion are coupled so that each equation involves all the coordinates; it is always possible to find a particular set of coordinates such that each equation of motion contains only one coordinate. The equations of motion are then uncoupled and can be solved independent of each other. Such set of coordinates, which lead to an uncoupled set of equations is called **principal coordinates**.

3.1 EQUATIONS OF MOTION

We shall now analyze the vibratory motion of a two mass, three springs and three dampers subjected to forced vibrations.

We consider linear springs (constant stiffness) and viscous dampers (damping force linearly proportional to velocity). Let $x_1(t)$ and $F_1(t)$ be the displacement and the excitation force at mass m_1 and $x_2(t)$ and $F_2(t)$ be the same at location of mass m_2. Figure 3.5 shows the free-body diagram for both masses.

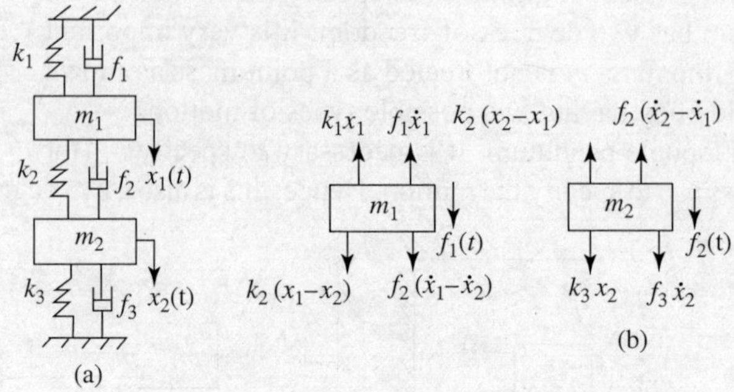

Figure 3.5 Free-body Diagram for Two Degrees-of-Freedom System

The equilibrium equation $\Sigma F = 0$, reveals

$$m_1\ddot{x}_1 + k_1x_1 + k_2(x_1 - x_2) + f_1\dot{x}_1 + f_2(\dot{x}_1 - \dot{x}_2) = F_1(t)$$

$$m_1\ddot{x}_1 + (k_1 + k_2)x_1 + (f_1 + f_2)\dot{x}_1 - k_2x_2 - f_2\dot{x}_2 = F_1(t) \tag{3.1}$$

Similarly, we get,

$$m_2\ddot{x}_2 + (k_2 + k_3)x_2 + (f_1 + f_3)\ddot{x}_2 - k_2x_1 - f_2\dot{x}_1 = F_2(t) \tag{3.2}$$

Examination of Equation 3.1 reveals that there are terms containing $x_2, \dot{x}_2$, whereas Equation 3.2 contains terms, $x_1, \dot{x}_1$; hence, Equations 3.1 and 3.2 represent a system of two coupled, second-order differential equations. It is easy to recognize that motion of mass m_1 will influence motion of mass m_2 and vice versa.

We can also write the Equations 3.1 and 3.2 in the matrix form

$$\begin{bmatrix} m_1 & 0 \\ 0 & m_2 \end{bmatrix}\begin{Bmatrix} \ddot{x}_1 \\ \ddot{x}_2 \end{Bmatrix} + \begin{bmatrix} f_1 + f_2 & -f_2 \\ -f_2 & f_2 + f_3 \end{bmatrix}\begin{Bmatrix} \dot{x}_1 \\ \dot{x}_2 \end{Bmatrix} + \begin{bmatrix} k_1 + k_2 & -k_2 \\ -k_2 & k_2 + k_3 \end{bmatrix}\begin{Bmatrix} x_1 \\ x_2 \end{Bmatrix} = \begin{Bmatrix} F_1 \\ F_2 \end{Bmatrix} \tag{3.3}$$

where

$$[\mathrm{m}] = \begin{bmatrix} m_1 & 0 \\ 0 & m_2 \end{bmatrix} \tag{3.4}$$

$$[\mathrm{f}] = \begin{bmatrix} (f_1 + f_2) & -f_2 \\ -f_2 & (f_2 + f_3) \end{bmatrix} \tag{3.5}$$

$$[\mathrm{k}] = \begin{bmatrix} (k_1 + k_2) & -k_2 \\ -k_2 & (k_2 + k_3) \end{bmatrix} \tag{3.6}$$

It can be seen that the matrices [m], [k], and [f] are symmetric matrices. $\begin{Bmatrix} \ddot{x}_1 \\ \ddot{x} \end{Bmatrix}$ is the acceleration vector, $\begin{Bmatrix} \dot{x}_1 \\ \dot{x}_2 \end{Bmatrix}$ is the velocity vector, and $\begin{Bmatrix} x_1 \\ x_2 \end{Bmatrix}$ is the displacement vector.

$$\begin{Bmatrix} F_1 \\ F_2 \end{Bmatrix} \text{ is the force vector} \tag{3.7}$$

Since the mass matrix [m], the damping matrix [f], and the spring or stiffness matrix are symmetrical, the transpose of these matrices are equal to the original matrices.
Thus,

$$[m]^T = [m],\ [f]^T = [f], \text{ and } [k]^T = [k] \tag{3.8}$$

It is easy to notice that Equations 3.1 and 3.2 of motion are coupled and can become uncoupled, that is, independent of each other, only when f_2 and k_2 are zero. This means that the masses m_1 and m_2 are not physically connected. In such a case the mass matrix [m], damping matrix [f], and the stiffness matrix become diagonal. This of course is an absurd case.

The solution of Equations 3.1 and 3.2 for any arbitrary forces $F_1(t)$ and $F_2(t)$ is difficult to obtain, mainly because of coupling of variables $x_1(t)$ and $x_2(t)$. The solution involves four constants of integration (two for each equation). Usually, initial displacements and velocities are specified as $\mathrm{x}_1(\mathrm{t}=0)=\mathrm{x}_1(0)$ and $\dot{x}_1(t=0)=\dot{x}_1(0), X_2(t=0)=X_2(0)$ and $\dot{x}_2(t=0)=\dot{x}_2(0)$. With these initial conditions, let us first solve Equations 3.1 and 3.2 with no applied excitation force, that is free vibrations. We divide this exercise into two parts: undamped-free vibrations and then damped-free vibrations. First we will consider two degrees-of-freedom system with no damping and no external force applied.

3.1.1 Analysis of Free Vibrations of an Undamped System

Equations 3.1 and 3.2 for this case become

$$m_1\ddot{x}_1 + (k_1 + k_2)x_1 - k_2x_2 = 0 \tag{3.9}$$

$$m_2\ddot{x}_2 + (k_2 + k_3)x_2 - k_2x_1 = 0 \tag{3.10}$$

Let us assume that the harmonic motion of both masses takes place at the same frequency and the phase angles. Their amplitudes, however, may be different. Let us assume

$$x_1 = X_1\cos(\omega.t + \Phi), x_2 = X_2\cos(\omega.t + \Phi) \tag{3.11}$$

where X_1 and X_2 denote amplitudes of x_1 and x_2, respectively, and Φ is the phase angle.
Using Equation 3.11 in Equations 3.9 and 3.10, we get,

$$[\{-m_1\omega^2 + (k_1 + k_2)\}X_1 - k_2X_2]\cos(\omega.t + \Phi) = 0 \tag{3.12(a)}$$

$$[-k_2X_1 + \{-m_2\,\omega^2 + (k_2 + k_3)\}X_2]\cos(\omega.t + \Phi) = 0 \tag{3.12(b)}$$

Equations 3.12 represent two simultaneous homogeneous algebraic equations in X_1 and X_2;.one solution of Equations 3.12(a) and (b) is X_1 and $X_2 = 0$, which implies that there is no vibration and thus is a trivial solution.

For non-trivial solution of X_1 and X_2, the determinant of the coefficients of X_1 and X_2 must be zero. Thus,

$$\text{Determinant}\begin{vmatrix}\{-m_1\omega^2+(k_1+k_2)\} & -k_2 \\ -k_2 & \{-m_2\omega^2+(k_2+k_3)\}\end{vmatrix}=0 \tag{3.13}$$

The expansion of the determinant gives

$$(m_1m_2)\omega^2-\{(k_1+k_2)m_2+(k_2+k_3)\mathrm{m}_1\}\omega^2+\{(k_1+k_2)(k_2+k_3)-k_2^2\}=0 \tag{3.13}$$

Equation 3.13 is called frequency or characteristic equation because the solution of this equation yields the frequencies or characteristic values of the system. The roots of Equation 3.13 are

$$\omega_1^2,\omega_2^2=\frac{1}{2}\left\{\frac{(k_1+k_2)m_2+(k_2+k_3)m_1}{m_1m_2}\right\}\mp\frac{1}{2}\left[\left\{\frac{(k_1+k_2)m_2+(k_2+k_3)m_1}{m_1m_2}\right\}^2 -4\left\{\frac{(k_1+k_2)(k_2+k_3)-k_2^2}{m_1m_2}\right\}\right]^{1/2} \tag{3.14}$$

This shows that with these values of ω_1 and ω_2, we get a non-trivial solution of Equation 3.12. ω_1 and ω_2 are the natural frequencies of the system.

The values of X_1 and X_2 are still undetermined. These values depend upon the frequencies ω_1 and ω_2. Let X_1^1, X_2^1 be the amplitudes of X_1 and X_2 at ω_1 and $X_1^{(2)}, X_2^{(2)}$ be the amplitudes of X_1 and X_2 the same at ω_2. Further, since Equation 3.12 is homogeneous, only ratios $r_1=\frac{X_2^1}{X_1^1}$ and $r_2=\frac{X_2^{(2)}}{X_1^{(2)}}$, can be found.

$$r_1=\frac{X_2^1}{X_1^1}=\frac{-m_1\omega_1^2+(k_1+k_2)}{k_2}=\frac{k_2}{-m_2\omega_1^2+(k_2+k_3)} \tag{3.15}$$

$$r_2=\frac{X_2^{(2)}}{X_1^{(2)}}=\frac{-m_1\omega_2^2+(k_1+k_2)}{k_2}=\frac{k_2}{-m_2\omega_2^2+(k_2+k_3)} \tag{3.16}$$

Examination of Equations 3.15 and 3.16 reveal that the two ratios given for each r_i ($i = 1, 2$) are identical. The normal modes of vibration corresponding to ω_1^2 and ω_2^2 can be expressed as

$$\{X\}^1=\begin{Bmatrix}X_1^1\\X_2^1\end{Bmatrix}=\begin{Bmatrix}X_1^1\\r_1X_1^1\end{Bmatrix} \tag{3.17(a)}$$

$$\{X\}^{(2)}=\begin{Bmatrix}X_1^{(2)}\\X_2^{(2)}\end{Bmatrix}=\begin{Bmatrix}X_1^{(2)}\\r_2X_1^{(2)}\end{Bmatrix} \tag{3.17(b)}$$

The vectors $\{X\}^1$ *and* $\{X\}^{(2)}$ which denote the normal modes of vibration are known as modal vectors of the system since they describe the mode shapes at natural frequencies.

The free-vibration solution or the motion in time can be expressed using Equation 3.11 as

$$\{X\}^1=\begin{Bmatrix}x_1^1(t)\\x_2^1(t)\end{Bmatrix}=\begin{Bmatrix}X_1^1\cos(\omega_1t+\Phi_1)\\r_1X_1^1\cos(\omega_1t+\Phi_1)\end{Bmatrix}=\text{First mode} \tag{3.18(a)}$$

$$\{X\}^{(2)}=\begin{Bmatrix}x_1^{(2)}(t)\\x_2^{(2)}(t)\end{Bmatrix}=\begin{Bmatrix}X_1^{(2)}\cos(\omega_2t+\Phi_2)\\r_2X_1^{(2)}\cos(\omega_2t+\Phi_2)\end{Bmatrix}=\text{Second mode} \tag{3.18(b)}$$

where constants X_1^{1}, $X_1^{(2)}$, Φ_1 and Φ_2 are determined by the initial conditions as follows

$$x_1(\mathrm{t}=0) = X_1^{(i)} = \text{some constant}, \ \dot{x}_1(t=0)=0$$

$$x_2(\mathrm{t}=0) = r_i X_1^{(i)} \text{ and, } \dot{x}_2(t=0)=0, \ \dot{x}_2(t=0)=\dot{x}_2(0) \tag{3.21}$$

However, for any general initial condition, both modes will get excited. The resulting motion, which is given, by the general solution of Equations 3.9 and 3.10, can be obtained by a linear superposition of the two normal modes (Equation 3.18)

$$\{x\} = \mathrm{c}_1\{x_1\} + \mathrm{c}_2\{x_2\} \tag{3.19}$$

c_1 and c_2 are constants. Since $\{\dot{x}_1\}$ and $\{\dot{x}_2\}$ already involve unknown constants $\{X_1^1\}$ and $\{X_1^2\}$ (Equation 3.18), we can choose $\mathrm{c}_1 = \mathrm{c}_2 = 1$ with no loss of generality. Thus using Equation 3.19 with $\mathrm{c}_1 = \mathrm{c}_2 = 1$ and Equation 3.18, the components of vector $\{\mathrm{x}\}$ can be expressed as,

$$x_1(t) = x_1^1(t) + x_1^{(2)}(t) = X_1^1\cos(\omega_1 t + \Phi_1) + X_1^{(2)}\cos(\omega_2 t + \Phi_2) \tag{3.20(a)}$$

$$x_2(t) = x_2^1(t) + x_2^{(2)}(t) = r_1 X_1^1\cos(\omega_1 t + \Phi_1) + r_2 X_1^{(2)}\cos(\omega_2 t + \Phi_2) \tag{3.20(b)}$$

The unknown constants are $X_1^1, X_1^{(2)}$, Φ_1 and Φ_2 can be determined from the initial conditions

$$x_1(t=0) = x_1(0) \text{ and, } \dot{x}_1(t=0) = \dot{x}_1(0), x_2(t=0) = x_2(0) \tag{3.21}$$

Substitution of Equation 3.21 into Equation 3.20 gives

$$\begin{aligned}
x_1(0) &= X_1^1\cos\Phi_1 + X_1^{(2)}\cos\Phi_2 \\
\dot{x}_1(0) &= -\omega_1 X_1^1\sin\Phi_1 - \omega_2 X_1^{(2)}\sin\Phi_2 \\
x_2(0) &= r_1 X_1^1\cos\Phi_1 + r_2 X_1^{(2)}\cos\Phi_2 \\
\dot{x}_2(0) &= -\omega_1 r_1 X_1^1\sin\Phi_1 - \omega_2 r_2 X_1^{(2)}\sin\Phi_2
\end{aligned} \tag{3.22}$$

Equation 3.22 comprises four algebraic equations with the unknowns $X_1^1\cos\Phi_1, X_1^{(2)}\cos\Phi_2$, $X_1^1\sin\Phi_1$, and $X_1^{(2)}\sin\Phi_2$ The solution of Equation 3.22 is

$$X_1^1\cos\Phi_1 = \left\{\frac{r_2 x_1(0) - x_2(0)}{r_2 - r_1}\right\}, X_1^{(2)}\cos\Phi_2 = \left\{\frac{-r_1 x_1(0) - x_2(0)}{r_2 - r_1}\right\}, \tag{3.23}$$

$$X_1^1\sin\Phi_1 = \left\{\frac{-r_2\dot{x}_1(0) - \dot{x}_2(0)}{\omega_1(r_2 - r_1)}\right\}, X_1^{(2)}\sin\Phi_2 = \left\{\frac{r_1\dot{x}_1(0) - \dot{x}_2(0)}{\omega_2(r_2 - r_1)}\right\}$$

From these equations, we obtain,

$$X_1^1 = \left[\left(X_1^1\cos\Phi_1\right)^2 + \left(X_1^1\sin\Phi_1\right)^2\right]^{1/2} = \frac{1}{r_2 - r_1}\left[\{r_2 x_1(0) - x_2(0)\}^2 + \frac{\{-r_2\dot{x}_1(0) + \dot{x}_2(0)\}^2}{\omega_1^2}\right]^{1/2} \cdot \tag{3.23(a)}$$

Similarly,

$$X_1^{(2)} = \frac{1}{r_2 - r_1}\left[\{-r_1 x_1(0) + x_2(0)\}^2 + \frac{\{r_1\dot{x}_1(0) + \dot{x}_2(0)\}^2}{\omega_2^2}\right]^{1/2} \tag{3.23(b)}$$

$$\Phi_1 = \tan^{-1}\left\{\frac{X_1^1 \sin \Phi_1}{X_1^1 \cos \Phi_2}\right\} = \tan^{-1}\left\{\frac{-r_2\dot{x}_1(0) + \dot{x}_2}{\omega_1\left[r_2 x_1(0) - x_2(0)\right]}\right\} \quad [3.23(c)]$$

and

$$\Phi_2 = \tan^{-1}\left\{\frac{r_1\dot{x}_1(0) - \dot{x}_2}{\omega_2\left[-r_1 x_1(0) + x_2(0)\right]}\right\} \quad [3.23(d)]$$

Let us illustrate these procedure through a few solved problems.

Problem 3.1

Figure 3.6 shows two degrees-of-freedom system with two masses and three springs. If the mass m_1 is displaced 1 cm from its static equilibrium position and released, determine the displacements $x_1(t)$ and $x_2(t)$ of the masses. Also, show how individual modes can be excited. Assume both masses and springs as equal.

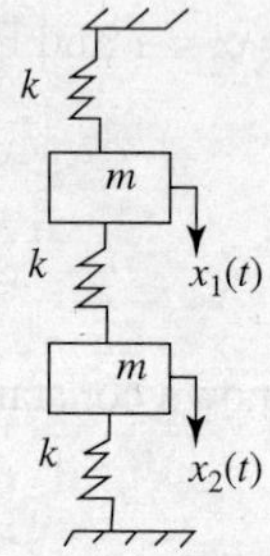

Figure 3.6 TDOF Undamped System

Solution:

The system equations are

$$\begin{aligned} m_1\ddot{x}_1 &= -kx_1 - k(x_1 - x_2) \\ m_2\ddot{x}_2 &= -kx_2 - k(x_2 - x_1) \end{aligned} \quad (1)$$

Assume that the motion is periodic and is composed of harmonic components of various amplitudes and frequencies. Let one of the components be $x_1 = A \cos (\omega t + \Phi)$ and $x_2 = B \cos (\omega t + \Phi)$. Putting these in Equation 1, we get,

$$\begin{aligned} (2k - m\omega^2)\,\mathrm{A} - k\mathrm{B} &= 0 \\ -kA + (2k - m\omega^2) &= 0 \end{aligned} \quad (2)$$

For a non-trivial solution of Equation 2, the determinant of the above Equation 2 must be zero.

$$\begin{vmatrix} (2k - m\omega^2) & -k \\ -k & (2k - m\omega^2) \end{vmatrix} = 0 \quad (3)$$

The expansion of the determinant gives the frequency equation

$$\omega^4 - \frac{4k}{m}\omega^2 + 3\left(\frac{k}{m}\right)^2 = 0 \quad (4)$$

From these we obtain,

$$\omega_1 = \sqrt{\frac{k}{m}} \text{ and } \omega_2 = \sqrt{\frac{3k}{m}} \quad (5)$$

Solving Equation 2, we get,

$$r_1 = \frac{X_2^1}{X_1^1} = \frac{A_1}{B_1} = \frac{k}{(2k - m\omega_1^2)} = 1 \quad (6)$$

$$r_2 = \frac{X_2^{(2)}}{X_1^{(2)}} = \frac{A_2}{B_2} = \frac{k}{(2k - m\omega_2^2)} = -1$$

Hence, the motion of the masses is given by

$$x_1(t) = A_1 \cos(\sqrt{(k/m)}\, t + \Phi_1) + A_2 \cos(\sqrt{(3k/m)}\, t + \Phi_2)$$

$$x_2(t) = A_1 \cos(\sqrt{(k/m)}\, t + \Phi_1) - A_2 \cos(\sqrt{(3k/m)}\, t + \Phi_2) \quad (7)$$

Four constants to be determined are A_1, A_2, Φ_1, and Φ_2. The initial conditions are

$$x_1(0)=1,\ x_2 = 0, \dot{x}_1(0) = \dot{x}_2(0) = 0$$

$$\therefore 1 = A_1 \cos\Phi_1 + A_2 \cos\Phi_2 \quad (8)$$

$$0 = A_1 \cos\Phi_1 - A_2 \cos\Phi_2 \quad (9)$$

$$0 = -\omega_1 A_1 \sin\Phi_1 - \omega_2 A_2 \sin\Phi_2 \quad (10)$$

$$0 = -\omega_1 A_1 \sin\Phi_1 - \omega_2 A_2 \sin\Phi_2 \quad (11)$$

Solving Equations 8 and 9, we get $A_1 = 1/(2 \cos \Phi)$. Solving Equations 10 and 11, we get $\sin \Phi_1 = \sin F_2 = 0$ or $\Phi_1 = \Phi_2 = 0$. Hence $A_1 = A_2 = 1/2$.
Thus, the motion is given by

$$x_1(t) = \frac{1}{2} \cos\sqrt{(k/m)t} + \frac{1}{2} \cos\sqrt{(3k/m)t} \quad (12)$$

$$x_2(t) = \frac{1}{2} \cos\sqrt{(k/m)t} - \frac{1}{2} \cos\sqrt{(3k/m)t} \quad (13)$$

The natural modes are given by the above equations (12 & 13).

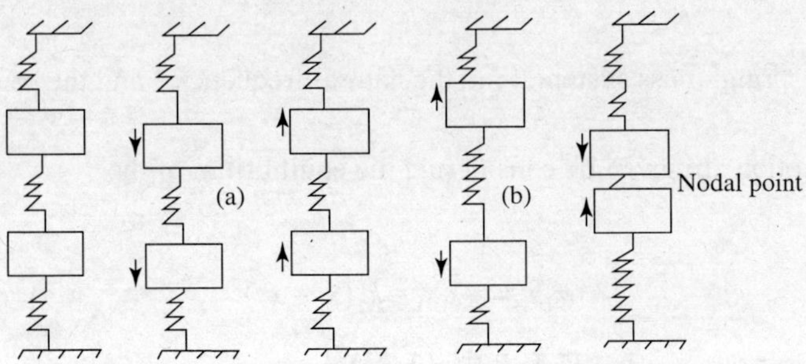

(a) Both masses moving together Equal amplitude (b) Masses moving in opposite direction

Figure 3.7 Mode Shapes

In the first mode (ω_1), both masses move together [Fig. 3.7(a)], whereas in the second mode (ω_2), the masses move in opposite directions [Fig. 3.7(b)] with equal amplitudes. Thus, the midpoint of the coupling spring is stationary due to symmetry of the system. The stationary point is called the node. This mode shape is an important point in the analysis of vibrations. Later, we shall see the importance of this concept in solving practical vibration problems.

Let us now find the initial conditions to be imposed on the system shown in Fig. 3.6 so that it vibrates only in the first mode or second mode.

Equations 12 and 13 show the displacements $x_1(t)$ and $x_2(t)$ when both modes are simultaneously excited, that is,

$$x_1(t) = \frac{1}{2}\cos\sqrt{(k/m)t} + \frac{1}{2}\cos\sqrt{(3k/m)t}$$

$$x_2(t) = \frac{1}{2}\cos\sqrt{(k/m)t} - \frac{1}{2}\cos\sqrt{(3k/m)t}$$

and the initial conditions were

$$x_1(0)=1,\ x_2 = 0, \dot{x}_1(0) = \dot{x}_2(0) = 0$$

If only first mode is to be excited, $A_2 = 0$ (in Equations 6 and 7), which is possible only when $x_1(0) = x_2(0) = 1$ and $\dot{x}_1 = \dot{x}_2 = 0$. With these initial conditions, the displacements are

$$x_1(t) = \frac{1}{2}\cos\sqrt{(k/m)t}$$

$$x_2(t) = \frac{1}{2}\cos\sqrt{(k/m)t}$$

Thus, both masses must be given initial displacement of 1.

If only second mode is to be excited, then $A_1 = 0$. This is possible only when $x_1(0)=1$, $x_2(0) = -1, \dot{x}_1(0) = -\dot{x}_2(0)$. Thus, both masses must be given equal displacement but in opposite directions.

We shall now illustrate these concepts through solution of another problem.

Problem 3.2

Figure 3.8 shows a spring–mass system. Find the natural frequencies and the principal coordinates.

Solution:

The equations of motion are given by considering the equilibrium of the masses

$$m_1\ddot{x}_1 = -k_1x_1 - k_2(x_1 - x_2)$$

$$m_2\ddot{x}_2 = -k_2(x_2 - x_1)$$

or

$$m_1\ddot{x}_1 + (k_1 + k_2)x_1 - k_2x_2 = 0$$

$$m_2\ddot{x}_2 + k_2(x_2 - x_1) = 0$$

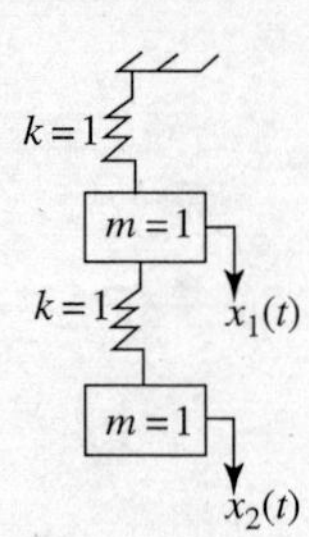

Figure 3.8 TDOF System

Assuming

$$x_1 = A \sin(\omega.t + \Psi)$$

$$x_2 = B \sin(\omega.t + \Psi)$$

and substituting in the equations of equilibrium and simplifying, we get,

$$(k_1 + k_2 - m_1 \omega^2)A - k_2 B = 0 \tag{1}$$

$$k_2 A + (k_2 - \omega^2)B = 0 \tag{2}$$

Equations 1 and 2 have non-trivial solution (a = b ≠ 0) only when the determinant of Equations 1 and 2 is zero, that is,

$$\begin{vmatrix} k_1 + k_2 - m_1\omega^2 & -k_2 \\ -k_2 & k_1 + k_2 - m_1\omega^2 \end{vmatrix} = 0.$$

This gives

$$\omega^4 - \left(\frac{k_1 + k_2}{m_1} + \frac{k_2}{m_2}\right)\omega^2 + \frac{k_1 k_2}{m_1 m_2} = 0.$$

The two roots are

$$\omega^2 = \frac{k_1 + k_2}{2m_1} + \frac{k_2}{2m_2} \pm \sqrt{\frac{1}{4}\left\{\frac{k_1 + k_2}{m_1} + \frac{k_2}{m_2}\right\}^2 - \frac{k_1 k_2}{m_1 m_2}}$$

But

$$k_1 = k_2 = k = 1, m_1 = m_2 = m = 1$$

$$\therefore \omega^2 = 1 + \frac{1}{2} \pm \sqrt{\frac{1}{4}(2+1)^2 - 1} = 1.5 \pm \sqrt{1.25}$$

This gives

$$\omega_1 = 1.62 \text{ and } \omega_2 = 0.63$$

Thus, the general solution is

$$x_1 = \mathrm{A}_1 \sin(\omega_1 t + \Psi_1) + \mathrm{A}_2 \sin(\omega_2 t + \Psi_2)$$

$$x_2 = \mathrm{B}_1 \sin(\omega_1 t + \Psi_1) + \mathrm{B}_2 \sin(\omega_2 t + \Psi_2)$$

where A_i and B_i are arbitrary constants (i = 1, 2).

The amplitude ratio are given by

$$\frac{A_1}{B_1} = \frac{k_2}{k_1 + k_2 - m_1\omega_1^{\,2}} = \frac{k_2 - m_1\omega_1^{\,2}}{k_2} = r_1$$

$$\frac{A_2}{B_2} = \frac{k_2}{k_1 + k_2 - m_1\omega_2^{\,2}} = \frac{k_2 - m_2\omega_2^{\,2}}{k_2} = r_2$$

But

$$k_1 = k_2 = k = 1, m_1 = m_2 = m = 1.$$

This gives $r_1 = 0.6$ and $r_2 = -1.6$.

Hence, the general solution is

$$x_1 = A_1 \sin(\omega_1 t + \Psi_1) + A_2 \sin(\omega_2 t + \Psi_2)$$

$$x_2 = \frac{1}{r_1} A_1 \sin(\omega_1 t + \Psi_1) + \frac{1}{r_2} A_2 \sin(\omega_2 t + \Psi_2)$$

The constants A_1, A_2, Ψ_1, and Ψ_2 are determined from the initial conditions

$$x_1(t = 0) = x_1(0) \text{ and } \dot{x}_1(t = 0) = \dot{x}_1(0), X_2(t = 0) = X_2(0)$$

Thus, the motion is given by

$$x_1(t) = A_1 \sin(0.63t + \Psi_1) + A_2 \sin(1.62t + \Psi_2) \tag{3}$$

$$x_2(t) = 1.6A_1 \sin(0.63t + \Psi_1) - 0.63A_2 \sin(1.62t + \Psi_2) \tag{4}$$

Let us now define a new set of coordinates y_1 and y_2 such that

$$y_1 = A_1 \sin(0.63t + \Psi_1) \tag{5}$$

$$y_2 = A_2 \sin(1.62t + \Psi_2) \tag{6}$$

Note that y_1 and y_2 are harmonic motions; their corresponding equations of motion are found as follows.

Differentiating Equation 5 twice, we get $\ddot{y}_1 = -0.4A \sin(0.63t + \Psi_1)$ and when we add $0.4A_1 y_1$, the sum is zero, that is,

$$\ddot{y}_1 + 0.4y_1 = 0 \tag{7}$$

Likewise, we get,

$$\ddot{y}_2 + 2.6y_1 = 0 \tag{8}$$

A close examination of Equations 7 and 8 reveal that they represent a two degrees-of-freedom system with natural frequencies $\omega_1 = 0.63$ and $\omega_2 = 1.62$ and both equations are uncoupled that is, Equation 7 contains only y_1 and no term in y_2 and vice versa. Therefore, the quantities y_1 and y_2 are the principal coordinates.

$$x_1 = y_1 + y_2 \tag{9}$$

$$x_2 = 1.62\, y_1 - 0.63y_2 \tag{10}$$

From these equations, we obtain the principal coordinates,

$$y_1 = 0.28x_1 + 0.45x_2, \tag{11}$$

and

$$y_2 = 0.72x_1 - 0.45x_2 \tag{12}$$

We follow the same procedure in the next problem.

Problem 3.3

For the system shown in Fig. 3.6, determine the principal coordinates if both masses are equal.

Solution:

Refer to Equations 6 and 7 in Problem 3.1

$$x_1(t) = A_1 \cos(\sqrt{(k/m)t} + \Phi_1) + A_2 \cos(\sqrt{(3k/m)t} + \Phi_2) \tag{1}$$

$$x_2(t) = A_1 \cos(\sqrt{(k/m)t} + \Phi_1) - A_2 \cos(\sqrt{(3k/m)t} + \Phi_2) \tag{2}$$

Define a new set of coordinates y_1 and y_2 such that

$$y_1 = A_1 \sin\left(\sqrt{\frac{k}{m}}t + \Psi_1\right) \tag{3}$$

$$y_2 = A_2 \sin\left(\sqrt{\frac{3k}{m}}t + \Psi_2\right) \tag{4}$$

Since y_1 and y_2 are harmonic motions, the following equations hold good (readers may verify).

$$\ddot{y}_1 + \left(\frac{k}{m}\right) y_1 = 0 \tag{5}$$

$$\ddot{y}_2 + \left(\frac{3k}{m}\right) y_2 = 0 \tag{6}$$

These equations represent two degrees-of-freedom system with natural frequencies $\omega_1 = \sqrt{\frac{k}{m}}$ and $\omega_2 = \sqrt{\frac{3k}{m}}$. Equations 5 and 6 are uncoupled equations as described in the previous problem. The y_1 and y_2 are therefore the principal coordinates given as follows. We can verify that

$$x_1 = y_1 + y_2 \quad \text{and} \quad x_2 = y_1 - y_2.$$

Hence, we get principal coordinates as

$$y_1 = \frac{1}{2}(x_1 + x_2)$$

$$y_2 = \frac{1}{2}(x_1 - x_2)$$

Thus it is possible to uncouple the equations of motion by finding the principal coordinates.

Problem 3.4

Figure 3.9 shows a schematic of a motor car weighing 4000 lbs and having 4.5 ft of radius of gyration about the CG. The front suspensions have a stiffness of 270 lb/in. Neglect damping in the system, find the principal modes of vibrations of the car.

Solution:

We assume that the motion of the car is in the plane of the paper. Consequently, the motion can be described by two parameters, namely, $x(t)$ as translational displacement and $\theta(t)$ as angular displacement. The system therefore is a two degrees-of-freedom vibration system. The equations of motion are given by

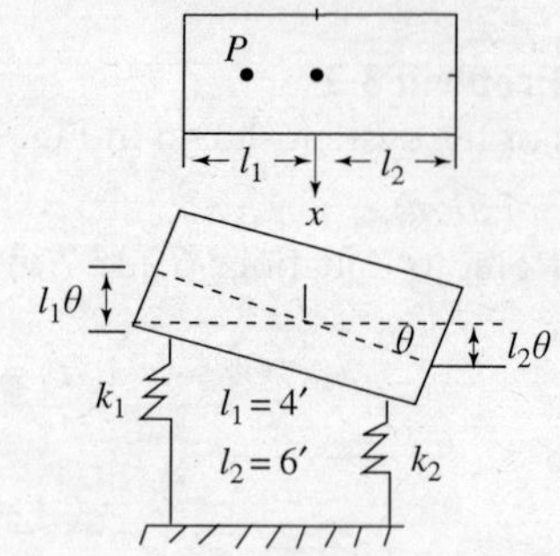

Figure 3.9 Idealization of Car

$$m\ddot{x} = -k_1(x - l_1\theta) - k_2(x + l_2\theta)$$

$$J\ddot{\theta} = k_1(x - l_1\theta)l_1 - k_2(x - l_2\theta)l_2$$

This assumes that θ is small so that $\sin\theta \approx \theta$. Also, $J = mK^2$, where m = mass and K is the radius of gyration. Rearranging the equations, we get,

$$m\ddot{x} + (k_1 + k_2)x - (k_1l_1 - k_2l_2)\theta \tag{1}$$

$$J\ddot{\theta} + (k_1l_1^2 + k_2l_2^2)x - (k_1l_1 - k_2l_2)x \tag{2}$$

These equations are coupled equations with $(k_1l_1 - k_2l_2)$ as static coupling terms. If this term is zero, that is, $k_1l_1 = k_2l_2$, the equations become two uncoupled equations

$$m\ddot{x} + (k_1 + k_2)x = 0$$

$$J\ddot{\theta} + (k_1l_1^2 + k_2l_2^2)\theta = 0$$

Substituting in Equations 1 and 2, the values of k_i and l_i (i = 1, 2), $J = mK^2$ (m = 4000/32.2 and K = 4), we get,

$$\ddot{x} + 50.12x + 532\theta = 0 \tag{3}$$

$$\ddot{\theta} + 64.2\theta + 0.54x = 0 \tag{4}$$

Let $x = \text{A}\sin(\omega t + \Psi)$, $\theta = \text{B}\sin(\omega t + \Psi)$. Substituting these in Equations 3 and 4, we get

$$(50.12 - \omega^2)\,\text{A} + 532\text{B} = 0 \tag{5}$$

$$0.54\text{A} + (64.22 - \omega^2) = 0 \tag{6}$$

By equating the determinant of the above homogeneous equations to zero, we get the frequency from the equation

$$\omega^4 - 114.2\omega^2 + 2923 = 0 \tag{7}$$

Solution of Equation 7 is ω_1 = 5.5 and ω_2 = 8.9 rad/s. The principal modes are found from the amplitude ratios

$$\frac{A_1}{B_1} = \frac{532}{(50.12 - 5.5^2)} = 26.1$$

$$\frac{A_2}{B_2} = \frac{532}{(50.12 - 8.9^2)} = -18.6$$

These modes are shown in Fig. 3.10.

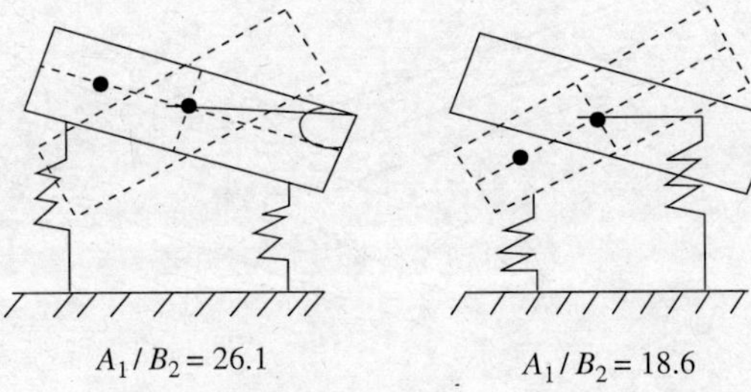

Figure 3.10 Principal Modes

In the preceding example, we chose $x(t)$ and $\theta(t)$ as the coordinates (of CG motion) to describe the vibratory motion of the body. The resulting equations were coupled equations with coupling term $(k_1 l_1 - k_2 l_2)$. If this term happens to be zero, the equations of motion would be uncoupled. The equations of motion can also be written in terms of

- Deflection x_1 and x_2 at ends.
- Deflection x of CG and rotation as we did in previous example.
- Deflection x of one end and rotation θ.
- Deflection y of point P located at e from CG and rotation θ (Fig. 3.9)

Depending upon the choice of generalized coordinates, the resulting equation will be coupled, for example, with x and θ, we had coupled equations with coupling term $(k_1 l_1 - k_2 l_2)$ called static or elastic coupling. Suppose, now we choose y, θ that is, displacement of point P at a distance e from the CG and the rotation θ of the body (Fig. 3.9), the equations of motion can be written (Fig. 3.11)

$$m\ddot{y} = -k_1(y_1 - l_1'\theta) - k_2(y + l_2'\theta) - me\ddot{\theta}$$

$$J_P\ddot{\theta} = -k_1(y - l_1'\theta)l_1' - k_2(y - l_2'\theta)l_2' - me\ddot{y}$$

These equations, after rearranging, can be written

$$\begin{bmatrix} m & me \\ me & J_P \end{bmatrix}\begin{Bmatrix} \ddot{y} \\ \ddot{\theta} \end{Bmatrix} + \begin{bmatrix} (k_1 + k_2) & (k_2 l_2' - k_1 l_1') \\ -(k_1 l_1' + k_2 l_2') & k_1 l_1'^2 + k_2 l_2'^2 \end{bmatrix}\begin{Bmatrix} y \\ \theta \end{Bmatrix} = \begin{Bmatrix} 0 \\ 0 \end{Bmatrix}$$

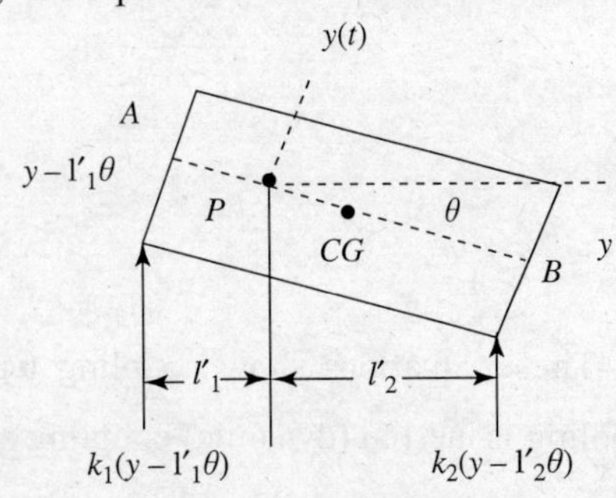

Figure 3.11 Changed Reference Axes (Problem 3.4)

Both of these equations of motion contain y, θ. They contain the static (elastic) as well as dynamic (mass) coupling term $m\ddot{e}$.

The following example illustrates this approach.

Problem 3.5

Figure 3.12 shows a weightless (assumed) stiff rod with two equal masses m attached to each end of a cantilever with torsional stiffness K and bending stiffness k. Derive the equation of motion and identify the coupling terms.

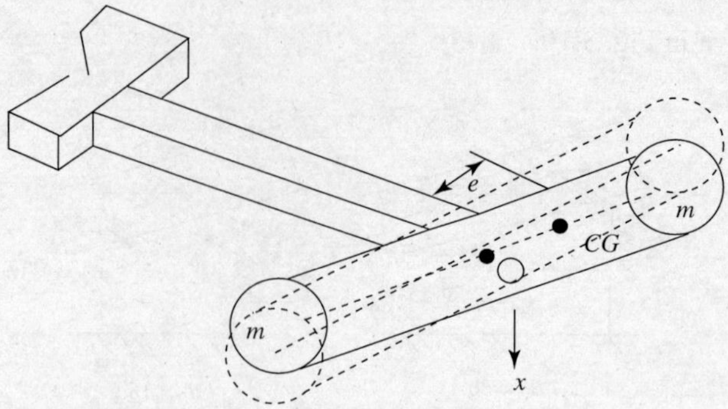

Figure 3.12 Structure for Problem 3.5

Solution:
The CG is located at a distance e from the end of the cantilever. We assume that the downward movement x of CG, when the stiff rod rotates through a small angle θ, is $e\theta$.

Thus, the acceleration is $e\ddot{\theta}$ in the x direction. The total accelerating force in the x direction is $2m\ddot{x} + 2m\ddot{\theta}$. The spring force is kx.

Thus, the equation of motion in the x direction is

$$2m\ddot{x} + 2m\ddot{\theta} + \mathrm{kx} = 0 \tag{1}$$

The second equation of motion comes from the moment balance $\Sigma M = 0$.

$2m\ddot{x}$ force in x direction produces a moment $2m\ddot{x}\mathrm{e}$ because of eccentricity e. The torsional stiffness of the bar is K, giving restoring twisting moment $K\theta$. Let J be the moment of inertia of the rod with respect to point O. Then, we get $J\ddot{\theta} + 2me\ddot{x} + K\theta = 0$, which is another equation of motion.

Thus, the two equations of motion are

$$\ddot{x} + \frac{k}{2m}x + e\ddot{\theta} = 0$$

$$\ddot{\theta} + \frac{K}{J}\theta + \frac{2me}{J}\ddot{x} = 0$$

These equations show coupling terms involving mass and moment of inertia. Therefore the coupling is inertial (dynamic) coupling and the coupling terms are $e\ddot{\theta}$ and $\frac{2me}{J}\ddot{x}$.

3.1.2 Lagrange's Equations

The Lagrange's equation in its fundamental form for generalized coordinates q_i is given by

$$\frac{d}{dt}\frac{\partial(KE)}{\partial \dot{q}_1} - \frac{\partial(KE)}{\partial q_1} + \frac{\partial(PE)}{\partial q_1} + \frac{\partial(DE)}{\partial \dot{q}_1} = Q_i. \tag{3.24}$$

where

q_i = generalized coordinates

$$\dot{q}_1 = \frac{dq_1}{dt}$$

KE = kinetic energy of the system say $\frac{1}{2}m\dot{x}^2$ for $q = x$

PE = potential energy of the system say $\frac{1}{2}kx^2$

DE = dissipation energy (damping) say $\frac{1}{2}f\dot{x}^2$

Q_i = generalized external force on the system

For a conservative system (which means that there is no dissipation of energy or no damping in the system), the Lagrange's equation can be written as

$$\frac{d}{dt}\frac{\partial L}{\partial \dot{q}_1} - \frac{\partial L}{\partial q_1} = 0 \tag{3.25}$$

where $L = KE - PE$ is called the Lagrangian.

The use of Lagrange's equation will directly yield as many equations of motion as the number of degrees-of-freedom of the system when the basic energy expressions of the system are known. The derivation of the equation can be found in advanced textbooks on dynamics. Here we are more interested in application of this concept in the analysis of two or more degrees-of-freedom systems. We shall study the application of Lagrange's equation through a few solved problems.

Problem 3.6

Figure 3.13 shows a coupled pendulum. Derive the equations of motion for the system.

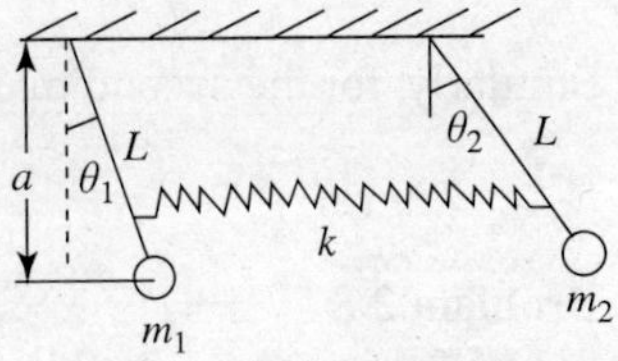

Figure 3.13 Coupled Pendulums

Solution:

It is possible to write the equations of motion for this system using Newton's laws of motion. We, however, apply Lagrange's equation to illustrate the ease with which these can be derived. Since there is no damping in the system, the system is conservative.

For conservative systems without excitation ($Q_i = 0$), the Lagrange's equation is written as

$$\frac{d}{dt}\frac{\partial (KE)}{\partial \dot{q}_i} - \frac{\partial (KE)}{\partial q_i} + \frac{\partial (PE)}{\partial q_i} = 0$$

For the system shown in Fig. 3.13, the generalized coordinates are θ_1 and θ_2. Thus, the energy expression is

$$KE = \frac{1}{2}m_1L^2\dot{\theta}_1^2 + \frac{1}{2}m_2L^2\dot{\theta}_2^2$$

$$PE = m_1gL(1-\cos\theta_1) + m_2gL(1-\cos\theta_2) + \frac{1}{2}k(a\theta_2 - a\theta_1),$$

where $(a\theta_2 - a\theta_1)$ = stretch of spring

$$\frac{d}{dt}\frac{\partial (KE)}{\partial \dot{\theta}_1} = m_1L^2\ddot{\theta}_1, \frac{\partial (KE)}{\partial \theta_1} = 0, \quad \frac{\partial (PE)}{\partial \theta_1} = m_1gL\sin\theta_1 - ka(a\theta_2 - a\theta_1).$$

Hence the equation of motion is

$$m_1L^2\ddot{\theta}_1 + m_1gL\sin\theta_1 - a^2k(\theta_2 - \theta_1) = 0 \quad (1)$$

Similarly,

$$m_2L^2\ddot{\theta}_2 + m_2gL\sin\theta_2 + ka^2(\theta_2 - \theta_1) = 0 \quad (2)$$

Problem 3.7

For the system shown in Fig. 3.14, formulate the Equations of motion. The friction can be neglected.

Solution:

$$\text{KE} = \frac{1}{2}(m_1\dot{x}_1^2 + m_2\dot{x}_2^2),\ \text{PE} = \frac{1}{2}k(x_2 - x_1)^2$$

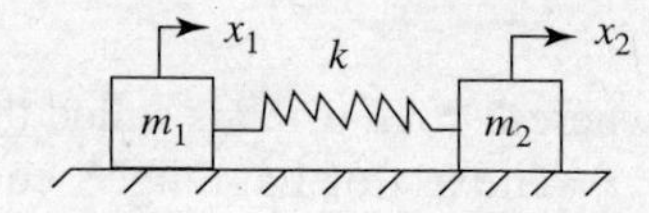

Figure 3.14 Semi-definite System

Since the system is a conservative system,

$$\frac{d}{dt}\frac{\partial(KE)}{\partial\dot{q}_1} - \frac{\partial(KE)}{\partial q_1} + \frac{\partial(PE)}{\partial q_1} = 0$$

$$\frac{d}{dt}\frac{\partial(KE)}{\partial\dot{x}_1} = m_1\ddot{x}_1,\ \frac{\partial(KE)}{\partial x_1} = 0,\ \frac{\partial(PE)}{\partial x_1} = -k(x_2 - x_1).$$

Therefore, the equation of motion is $m_1\ddot{x}_1 + k(x_1 - x_2) = 0$ (mass m_1)

Similarly, for the second mass, we can show $m_2\ddot{x}_2 + k(x_2 - x_1) = 0$.

Problem 3.8

Figure 3.15 shows a double pendulum. Derive the equation of motion.

Solution:

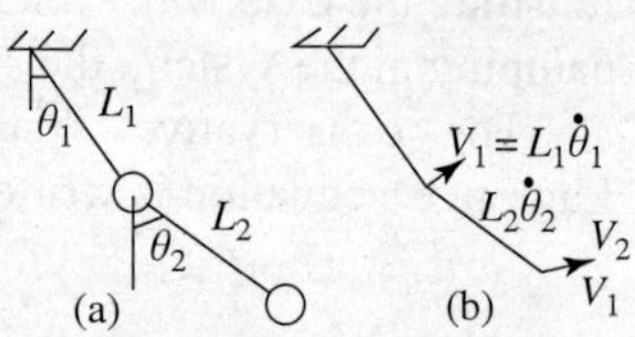

Figure 3.15 Double Pendulum

KE of pendulum $= \frac{1}{2}(m_1v_1^2 + m_2v_2^2)$.

For small oscillations, $v_1^2 = (L_1\theta_1^2)$

Similarly v_2 is given by $v_2^2 = (L_1\dot{\theta}_1)^2 + (L_2\dot{\theta}_2)^2 + 2L_1L_2\dot{\theta}_1\dot{\theta}_2\cos(\theta_2 - \theta_1)$

Note that v_2 is a vector sum. Thus, we have the following expressions for KE

$$KE = \frac{1}{2}\left\{m_1L_1^2\dot{\theta}_1^2 + [m_2L_1^2\dot{\theta}_1^2 + m_2L_2^2\dot{\theta}_2^2 + 2m_2L_1L_2\dot{\theta}_1\dot{\theta}_2\cos(\theta_2 - \theta_1)]\right\}$$

$$\text{PE} = m_1gL_1(1 - \cos\theta_1) + m_2g[L_1(1 - \cos\theta_1) + L_2(1 - \cos\theta_2)]$$

Note the contribution of PE of m_2.

Lagrange's equation is written as

$$\frac{d}{dt}\frac{\partial(KE)}{\partial q_i} - \frac{\partial(KE)}{\partial q_i} + \frac{\partial(PE)}{\partial q_i} = 0$$

Now, we evaluate individual elements of this equation

$$\frac{d}{dt}\frac{\partial(KE)}{\partial\dot{\theta}_1}=\frac{d}{dt}\left[\frac{1}{2}m_1L_1^2\dot{\theta}_1^2+\frac{1}{2}m_2\{L_1^2\theta_1^2+2L_1L_2\dot{\theta}_2\dot{\theta}_1\cos(\theta_2-\theta_1)\}\right]$$

$$=m_1L_1^2\ddot{\theta}_1+m_2L_1^2\ddot{\theta}_1+m_2L_1L_2\ddot{\theta}_2$$

where $\sin\theta\approx\theta, \cos(\theta_2-\theta_1)\approx 1, \text{d}(\cos(\theta_2-\theta_1))/\text{dt}=0$

$$\frac{\partial(PE)}{\partial\dot{\theta}_1}=m_1gL_1\sin\theta_1+m_2gL_1\sin\theta_1, \frac{\partial(KE)}{\partial\theta_1}=0$$

The first equation of motion can then be derived, after some simplifications and rearrangements as

$$(m_1+m_2)L_1\ddot{\theta}_1+m_2L_2\ddot{\theta}_2+m_2L_2\ddot{\theta}_2+(m_1+m_2)g\theta_1=0 \quad (1)$$

Similarly, we can show,

$$\frac{d}{dt}\frac{\partial(KE)}{\partial\dot{\theta}_2}=m_2L_2^2\ddot{\theta}_2+m_2L_1L_2\ddot{\theta}_1, \frac{\partial(KE)}{\partial\theta_2}=0 \quad \text{and} \quad \frac{\partial(PE)}{\partial\theta_2}=m_2gL_3\sin\theta_2=m_2gL_2\theta_2$$

Therefore, second equation of motion is

$$L_2\ddot{\theta}_2+g\theta_2+L_1\ddot{\theta}_1=0. \quad (2)$$

Note that motion of the second mass is independent of the mass m_2.

Problem 3.9

Figure 3.16 shows two identical rotors connected by a spring of stiffness $k = 5$, the torsional stiffness $K = 90$, and the polar mass moment of inertia $J = 1$. The spring is located at a distance $a = 2$. Find the frequency equation.

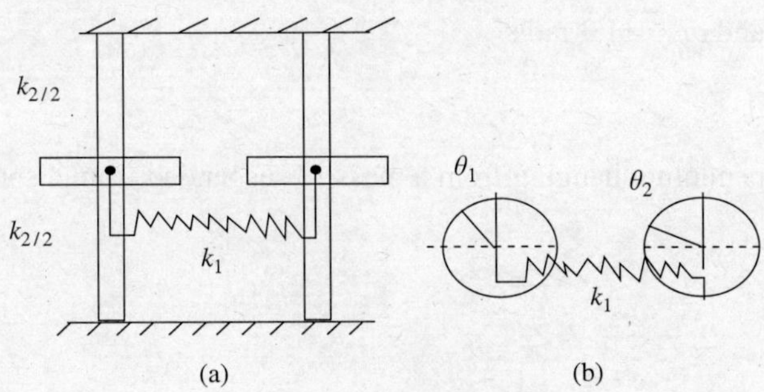

Figure 3.16 TDOF System

Solution:

Although we shall deal with torsional vibration in a separate chapter, this example is given with an intention to explain the method used in the analysis of two degrees-of-freedom systems. Let θ_1 and θ_2 represent the angular displacements of the rotors. The energy expression can be written as

$$KE=\frac{1}{2}J\dot{\theta}_1^2+\frac{1}{2}J\dot{\theta}_2^2, PE=\frac{1}{2}\{K(\theta_1^2+\theta_2^2)+Ka^2(\theta_1-\theta_2)^2\}.$$

The second term is spring stretch-induced *PE*. With this, we obtain,

$$\frac{d}{dt}\frac{\partial(KE)}{\partial\dot{\theta}_1} = J\ddot{\theta}_1, \frac{\partial(KE)}{\partial\theta_1} = 0, \frac{\partial(PE)}{\partial\theta_1} = K\theta_1 + ka^2(\theta_1 - \theta_2).$$

With this we obtain the first equation of motion as

$$J\ddot{\theta}_1 + K\theta_1 + ka^2(\theta_1 - \theta_2) = 0 \quad (1)$$

In a similar fashion, we obtain the second Equation of motion as

$$J\ddot{\theta}_2 + K\theta_2 + ka^2(\theta_2 - \theta_1) = 0 \quad (2)$$

Substituting the values, we get,

$$\ddot{\theta}_1 + 110\theta_1 - 20\theta_2 = 0$$
$$\ddot{\theta}_2 + 110\theta_2 - 20\theta_1 = 0$$

Assuming that the motion is periodic and is composed of harmonic motion of various amplitudes and frequencies, Let sin $\theta_1 = A$ sin $(\omega.t + \Psi)$ and $\theta_2 = B$ sin $(\omega.t + \Psi)$. Substituting these in equations (1) and (2) simplifying, we get,

$$(110 - \omega^2)\,\mathrm{A} - 20\mathrm{B} = 0$$
$$-20\mathrm{A} + (110 - \omega^2)\,\mathrm{B} = 0$$

Therefore, the frequency equation is

$$(110 - \omega^2)^2 - 20^2 = 0.$$

This gives $\omega_1 = 9.43$ and $\omega_2 = 11.3$ rad/s

Problem 3.10

Figure 3.17 shows a pendulum hanging from a mass *M* suspended from a spring. Find the equations of motion.

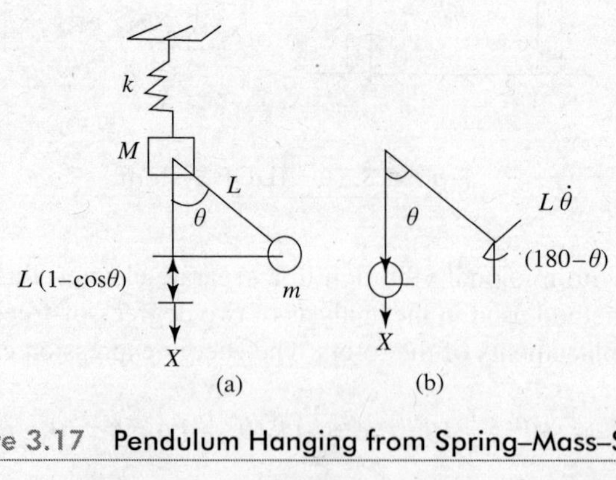

Figure 3.17 Pendulum Hanging from Spring–Mass–System

Solution:
The KE of the system is

$$KE = \frac{1}{2}M\dot{x}^2 + \frac{1}{2}m(\dot{x}^2 + L^2\dot{\theta}^2 - 2L\dot{x}\dot{\theta}\cos(180-\theta))$$

$$\therefore KE = \frac{1}{2}M\dot{x}^2 + \frac{1}{2}m(\dot{x}^2 + L^2\dot{\theta}^2 - 2L\dot{x}\dot{\theta})$$

$$PE = \frac{1}{2}kx^2 + mgL(1-\cos\theta).$$

Hence

$$\frac{d}{dt}\frac{\partial(KE)}{\partial\dot{x}} = (M+m)\ddot{x} + mL\ddot{\theta}\cos(180-\theta) - mL\sin\theta$$

$$\frac{\partial(KE)}{\partial x} = 0$$

$$\frac{\partial(PE)}{\partial x} = kx$$

For small angles of oscillations, $\sin\theta \approx \theta$ and $\cos(180-\theta) = -1$.

Therefore, we get, $(M+m)\ddot{x} - mL\ddot{\theta} + kx = 0$, neglecting term involving $\dot{\theta}^2$.

Similarly, we can show $L\ddot{\theta} + g\theta + \dot{x} = 0$.

Problem 3.11

Figure 3.18 shows idealization of inner race and the rollers in the anti-friction bearings. The cylindrical roller of radius r and mass m rolls without slipping in a semi-circular groove of radius R. The inner race M is supported by the outer race modelled as a spring of stiffness k. The inner race can move only in vertical direction. Derive the equation of motion.

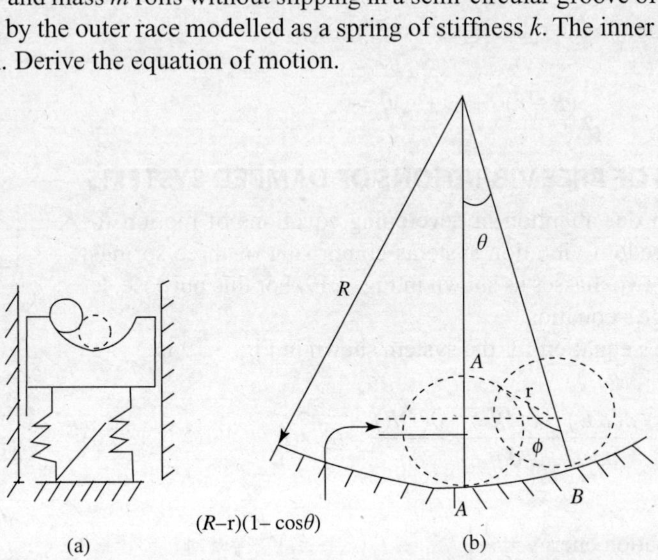

Figure 3.18 Modelling of Anti-friction Bearing

Solution:

Since the angular displacement of roller = $\Phi - \theta$, the kinetic energy is $KE = \frac{1}{2}M\dot{x}^2 + \frac{1}{2}mv^2 + \frac{1}{2}J(\dot{\Phi} - \dot{\theta})^2$, where J = mass moment of inertia = (½)mr^2.

Also, arc AB = $r\Phi = R\theta$ because the roller rolls without slipping. Thus, we get, $\Phi = R\theta/r$.

The velocity of the centre of cylindrical roller is

$$v = \sqrt{\left\{(R-r)\dot{\theta}\right\}^2 + \dot{x}^2 - 2[(R-r)\dot{\theta}\dot{x}\cos(90+\theta)]}$$

$$v^2 = [(R-r)\dot{\theta}]^2 + \dot{x}^2 + 2(R-r)^2\dot{\theta}\dot{x}\sin\theta$$

$$KE = \frac{1}{2}M\dot{x}^2 + \frac{1}{4}m\dot{\theta}^2(R-r)^2 + \frac{1}{2}m[\dot{x}^2 + (R-r)^2\dot{\theta}^2 + 2\dot{x}\dot{\theta}(R-r)\sin\theta]$$

$$PE = \frac{1}{2}kx^2 + mg(R-r)(1-\cos\theta).$$

Assuming small angles of oscillations, $\sin\theta \sim \theta$, $\cos\theta \sim 1$ and neglecting higher order terms, we have,

$$\frac{d}{dt}\frac{\partial(KE)}{\partial\dot{x}} = M\ddot{x} + m\ddot{x} + m\ddot{\theta}(R-r)\sin\theta + m\dot{\theta}^2(R-r)\cos\theta = (m+M)\ddot{x} + m(R-r)\ddot{\theta}$$

$$\frac{\partial(KE)}{\partial\dot{x}} = 0, \quad \frac{\partial(PE)}{\partial\dot{x}} = kx$$

Therefore, The first equation of motion is,

$$(m+M)\ddot{x} + kx + m\theta(R-r)\ddot{\theta}$$

Similarly, we can obtain the other equation of motion as,

$$\frac{3}{2}(R-r)\ddot{\theta} + (\ddot{x} + g)\theta = 0$$

3.2 ANALYSIS OF FREE VIBRATIONS OF DAMPED SYSTEMS

We shall now turn our attention in developing equations of motion for two degrees-of-freedom vibration systems comprising of three springs, three dampers and two masses as shown in Fig. 3.19. For this purpose, let us use the Lagrange's equation.

The Lagrange's equation for the system shown in Fig. 3.19 is

$$\frac{d}{dt}\frac{\partial(KE)}{\partial\dot{q}_i} - \frac{\partial(KE)}{\partial q_i} + \frac{\partial(PE)}{\partial q_i} + \frac{\partial(DE)}{\partial\dot{q}_i} = Q_i = 0 \tag{3.26}$$

where DE = dissipation energy = $\frac{1}{2}\left\{f_1\dot{x}_1^2 + f_2\left(\dot{x}_1 - \dot{x}_2\right)^2 + f_3\dot{x}_3^2\right\}$

= energy dissipated in dampers.

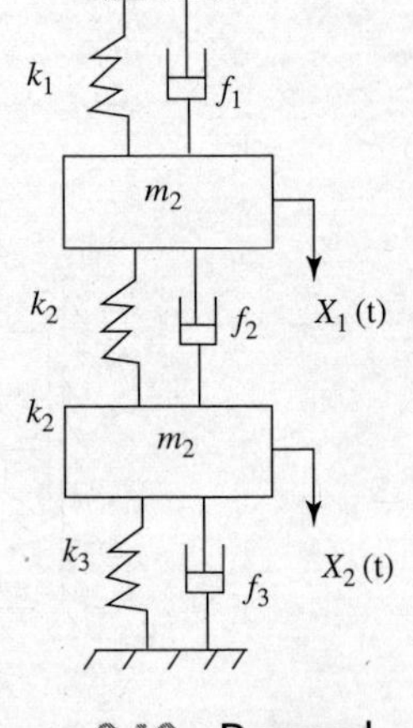

Figure 3.19 Damped TDOF System

$$KE = \frac{1}{2}[m_1\dot{x}_1^2 + m_2\dot{x}_2^2 \quad \text{and} \quad \text{PE} = \frac{1}{2}[k_1x_1^2 + k_2(x_1 - x_2)^2 + k_3x_3^2]$$

$$\frac{d}{dt}\frac{\partial(KE)}{\partial\dot{x}_1} = m_1\dot{x}_1,$$

$$\frac{\partial(KE)}{\partial x_1} = k_1x_1 + k_2(x_1 - x_2),$$

$$\frac{\partial(DE)}{\partial\dot{x}_1} = f_1\dot{x}_1 + f_2(\dot{x}_1 - \dot{x}_2)$$

Thus, the first equation of motion is

$$m_1\ddot{x}_1 + (f_1 + f_2)\dot{x}_1 - f_2\dot{x}_2 + (k_1 + k_2)x_1 - k_2x_2 = 0 \tag{3.27}$$

We can similarly obtain the other equations of motion as

$$m_2\ddot{x}_2 + (f_2 + f_3)\dot{x}_2 - f_2\dot{x}_1 + (k_2 + k_3)x_2 - k_2x_1 = 0 \tag{3.28}$$

These equations can also be derived using Newton's laws of motion. Since the components of vibration for a damped system are non-periodic (depending upon the damping factor) or oscillatory with diminishing amplitudes, let $x_1 = Ae^{st}$, $x_2 = Be^{s}$ be the solutions.

Substituting these expressions into the differential equations and dividing throughout by e^{st}, we get,

$$\{m_1s^2+(f_1+f_2)s + (k_1 + k_2)\}A - (f_2s + k_2)B = 0 \tag{3.29}$$

$$\{m_2s^2+(f_2+f_3)s + (k_2 + k_3)\}B - (f_2s + k_1)A = 0 \tag{3.30}$$

Equations 3.29 and 3.30 will have a non-trivial solution only when the determinant of these equations is zero; i.e.

$$\begin{vmatrix} m_1s^2 + (f_1 + f_2)s + (k_1 + k_2) & -(f_2s + k_2) \\ -(f_2s + k_1) & m_2s^2 + (f_2 + f_3)s + (k_2 + k_3) \end{vmatrix} = 0 \tag{3.31}$$

Expansion of this equation is termed as characteristic equation of the system. The solution of the characteristic equation will yield four values of s. Therefore, the complete general solution (motion) of the system can be expressed as

$$x_1(t) = A_1\, e^{s1t} + A_2e^{s2t} + A_3e^{s3t} + A_4e^{s4t} \tag{3.32}$$

$$x_2(t) = B_1\, e^{s1t} + B_2e^{s2t} + B_3e^{s3t} + B_4e^{s4t} \qquad [3.33(a)]$$

There are only four unknown constants, say, A1, A2, A3, and A4 to be determined. These and the other constants, Bi = λ_iAi (i = 1, 2, 3, and 4) are to be found out from the four initial conditions, namely, $x_1(0)$, $x_2(0)$, $\dot{x}_1(0)$, and $\dot{x}_2(0)$. The amplitude ratios λ_i are found from the algebraic equations with coefficients Ai and Bi (i = 1, 2, 3, and 4).

$$\frac{A_i}{B_i} = \frac{f_2 s_i + k_2}{m_1 s_i^2 + (f_1 + f_2)s_i + (k_1 + k_2)} = \frac{m_2 s_i^2 + (f_2 + f_3)s_i + (k_2 + k_3)}{f_2 s_i + k_1} = \frac{1}{\lambda_i} \quad (3.33(b))$$

We know that Equations 3.32 and 3.33 will represent vibration only when the roots of characteristic equation are complex. Suppose the roots of the characteristic Equation are complex, then there must be complex conjugate roots, i.e.,

$$s_1 = -(r + id) \quad \text{and} \quad s_2 = -(r - id).$$

Thus,

$$\begin{aligned} A_1 e^{-(r+id)t} + A_2 e^{-(r-id)t} &= e^{-rt}(A_1 e^{-idt} + A_2 e^{idt}) \\ &= e^{-rt}[A_1(\cos dt - i\sin dt) + A_2(\cos dt + i\sin dt)] \\ &= Ce^{-rt}\sin(dt + \Psi). \end{aligned}$$

Similarly,

$$B_i e^{s_1 t} + B_2 e^{s_2 t} = De^{-rt}\sin(dt + \Psi).$$

Therefore, the general motion is

$$\begin{aligned} x_1(t) &= Ce^{-rt}\sin(dt + \Psi) + A_3 e^{s_3 t} + A_4 e^{s_4 t} \\ x_2(t) &= De^{-rt}\sin(dt + \Psi) + B_3 e^{s_3 t} + B_4 e^{s_4 t} \end{aligned} \quad (3.34)$$

These are oscillatory with diminishing amplitude and aperiodic. We shall demonstrate the procedure through a few worked examples.

Problem 3.12

Figure 3.20 shows a system of masses m_1 and m_2 attached to a rigid and massless bar supported by springs k_1 and k_2 and dashpot f. Assuming the motion of the system only in the plane of the paper, determine the equations of motion.

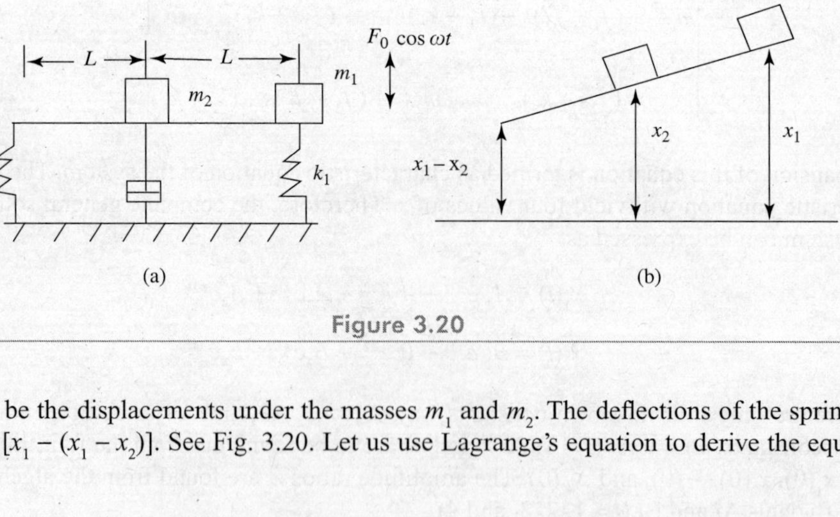

Figure 3.20

Solution:

Let x_1 and x_2 be the displacements under the masses m_1 and m_2. The deflections of the springs k_1 and k_2 are x_1 and $[x_1 - (x_1 - x_2)]$. See Fig. 3.20. Let us use Lagrange's equation to derive the equations of motion.

$$KE = \frac{1}{2}[m_1\dot{x}_1^2 + m_2\dot{x}_2^2]$$

$$PE = \frac{1}{2}[k_1x_1^2 + k_2(2x_2 - x_1)^2]$$

$$DE = \frac{1}{2}f\dot{x}_2^2$$

The Lagrange's equation for this system is given by

$$\frac{d}{dt}\frac{\partial(KE)}{\partial\dot{x}_1} - \frac{\partial(KE)}{\partial x_1} + \frac{\partial(PE)}{\partial x_1} + \frac{\partial(DE)}{\partial\dot{x}_1} = F_1 = \text{Force at } m_1$$

$$\frac{d}{dt}\frac{\partial(KE)}{\partial\dot{x}_2} - \frac{\partial(KE)}{\partial x_2} + \frac{\partial(PE)}{\partial x_2} + \frac{\partial(DE)}{\partial\dot{x}_2} = F_2 = \text{Force at mass } m_2 = 0 \text{ as given.}$$

Let us now evaluate the derivatives

$$\frac{d}{dt}\frac{\partial(KE)}{\partial\dot{x}_1} = m_1\ddot{x}_1,$$

$$\frac{\partial(KE)}{\partial x_1} = 0, \frac{\partial(PE)}{\partial x_1} = k_1x_1 - k_2(2x_2 - x_1)$$

$$\frac{d}{dt}\frac{\partial(KE)}{\partial\dot{x}_2} = m_2\ddot{x}_2,$$

$$\frac{\partial(KE)}{\partial x_2} = 0, \frac{\partial(PE)}{\partial x_2} = 2k_2(2x_2 - x_1).$$

$$\frac{\partial(DE)}{\partial x_2} = fx_2$$

Thus, we get the equations of motion as

$$m_1\ddot{x}_1 + k_1x_1 - k_2(2x_2 - x_1) = F_0\cos\omega.t = m_1\ddot{x}_1 + (k_1 + k_2)x_1 - 2k_2x_2 \tag{1}$$

$$m_2\ddot{x}_2 + 4k_2x_2 + f\dot{x}_2 - 2k_2x_1 = 0 \tag{2}$$

Note that there is no *DE* term for the Equation 1 as there is no damper element at the place of first mass.

Before we move on to forced vibrations of two degrees-of-freedom system, we shall discuss about an important concept called orthogonality principle.

3.2.1 Orthogonality Principle

This states that the principal modes of vibration of multi-degrees-of-freedom systems are orthogonal. The important property that the principal modes are vibrations along mutually perpendicular straight lines is extremely useful for the computation of natural frequencies.

The orthogonality principle for two degrees-of-freedom systems can be written as

$$m_1A_1A_2 + m_2B_1B_2 = 0 \tag{3.35}$$

where, A_1, A_2, B_1, and B_2 are the amplitudes of the two coordinates for the first and second mode of vibrations.

Proof: In general, the motion of two degrees-of-freedom system is given by

$$x_1(t) = A_1 \sin(\omega_1 t + \Psi_1) + A_2 \sin(\omega_2 t + \Psi_2)$$
$$x_2(t) = B_1 \sin(\omega_1 t + \Psi_1) + B_2 \sin(\omega_2 t + \Psi_2) \tag{3.36}$$

where, A_1, A_2, B_1, and B_2 are the amplitudes of vibration of the two masses, Ψ_1 and Ψ_2 are the phase angles, and ω_1 and ω_2 are the natural frequencies.

The kinetic energy of the system is

$$KE = \frac{1}{2}[m_1 v_1^2 + m_2 v_2^2]$$

Therefore,
$$(KE)_{MAX} = \frac{1}{2}[m_1(\dot{x}_1)^2_{\max} + m_2(\dot{x}_2)^2_{\max}]$$
$$= \frac{1}{2}[m_1(A_1\omega_1 + A_2\omega_2)^2 + m_2(B_1\omega_1^2 + B_2\omega_2)^2] \tag{3.37}$$

$$= \frac{1}{2}\{m_1[(A_1\omega_1)^2 + (A_2\omega_2)^2 + 2A_1A_2\omega_1\omega_1] + m_2[(B_1\omega_1)^2 + (B_2\omega_2)^2 + 2B_1B_2\omega_1\omega_2]\}$$

When the system is having principal modes of vibration, then,

$$KE_{MAX} = \frac{1}{2}\left\{m_1(A_1\omega_1)^2 + m_2(B_1\omega_1)^2 + m_1(A_2\omega_2)^2 + m_2(B_2\omega_2)^2\right\} \tag{3.38}$$

Since the Kinetic energy given by Equations 3.37 and 3.38 must be equal,

$$m_1A_1\omega_1A_2\omega_2 + m_2B_1\omega_1B_2\omega_2 = 0.$$

Since ω_1 and ω_2 cannot be zero, we have,

$$m_1\,(A_1A_2) + m_2\,(B_1B_2) = 0 \tag{3.39}$$

Thus, we have proved the orthogonality principle. Let us now see the application of this principle through a few examples.

Problem 3.13

Figure 3.21 shows a two degrees-of-freedom system consisting of equal masses and springs of equal stiffness. Examine the principal modes of vibrations.

Solution:

The equations of motion can be shown to be

$$m\ddot{x}_1 + 2kx_1 - kx_2 = 0 \tag{1}$$
$$m\ddot{x}_2 + kx_2 - kx_1 = 0 \tag{2}$$

We can assume solution as $x_1(t) = \text{A} \sin(\omega.t + \Psi)$, $x_2(t) = \text{B} \sin(\omega.t + \Psi)$.

Substituting in Equations 1 and 2 and simplifying, we get,

$$(2k - m\omega^2)\text{A} - Kb = 0$$

and
$$-k\text{A} + (k - m\omega^2) = 0.$$

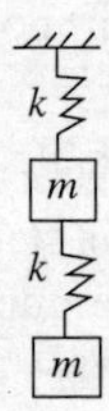

Figure 3.21

These equations have a non-trivial solution only when the determinant of the Equations is zero.

$$\begin{vmatrix} (2k - m\omega^2) & -k \\ -k & (k - m\omega^2) \end{vmatrix} = 0.$$

The expansion of this determinant gives the frequency equation

$$(2k - m\omega^2)(k - m\omega^2) - k^2 = 0 \quad \text{or} \quad \omega^4 - \frac{3k}{m}\omega^2 + \frac{k^2}{m^2} = 0.$$

The solution of this equation is

$$\omega_1 = 0.62(k/m)^{½} \quad \text{and} \quad \omega_2 = 1.62(k/m)^{½}$$

The amplitude ratios are given by

$$\frac{A_1}{B_1} = \frac{k}{2k - m\omega_1^2} = \frac{k - m\omega_1^2}{k} = 1 - \frac{\omega_1^2}{(k/m)} = 1 - 0.62^2 = 0.63 = \frac{1}{\lambda_1}$$

$$\frac{A_2}{B_2} = \frac{k}{2k - m\omega_2^2} = \frac{k - m\omega_2^2}{k} = 1 - \frac{\omega_2^2}{(k/m)} 1 - 1.62^2 = -1.62 = \frac{1}{\lambda_2}$$

Thus, the solution can be written as

$$x_1(t) = A_1 \sin[0.62(k/m)^{1/2}t + \Psi_1] + A_2 \sin[1.62(k/m)^{1/2}t + \Psi_2]$$

$$x_2(t) = 1.62 \sin[0.62(k/m)^{1/2}t + \Psi_1] - 0.63\, A_2 \sin[1.62(k/m)^{1/2}t + \Psi_2]$$

$$\therefore A_1A_2 + B_1B_2 = A_1A_2 + (1.62A_1)(-0.63A_2) = 0$$

Therefore, the principal modes of vibration are orthogonal.

Problem 3.14

A mass m is connected by springs $k_1, k_2, k_3, \ldots, k_n$. Assume that the motion of the mass is in the plane of paper and there is no damping. If the mass is displaced from its position of equilibrium O (Fig. 3.22), find the principal modes.

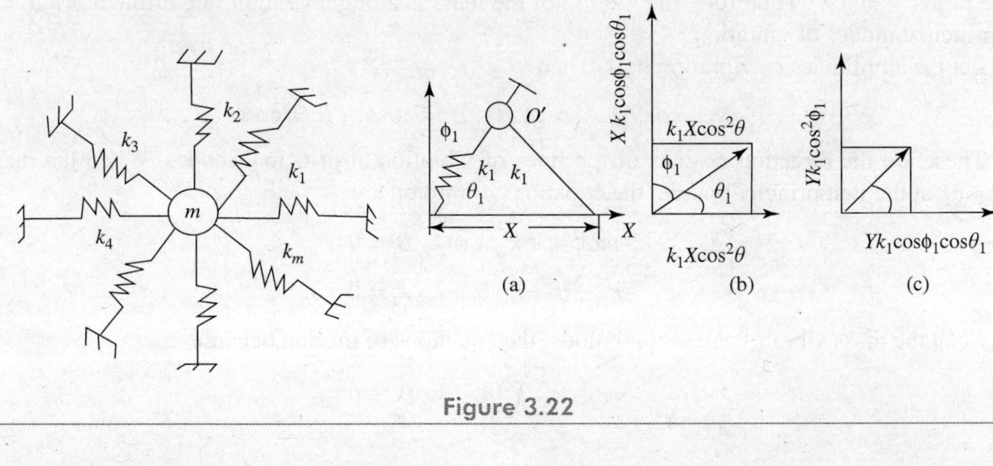

Figure 3.22

Solution:

The system shown in Fig. 3.22 is a two degrees-of-freedom system as the motion of the mass can be described by its x, y coordinates. When appropriate initial conditions are applied to the mass, it will vibrate at the first principal mode with a natural frequency ω_1 or at the second mode with a natural frequency ω_2.

Let the position of the mass be O′. The spring k_1 is deformed to $X \cos \theta_1$ as shown in Fig. 3.22(a). The x-component of the spring force is $k_1 X \cos \theta_1 \times \cos \theta_1 = Xk_1 \cos^2 \theta_1$. The y-component of this spring force is $X k_1 \cos \theta_1 \cos \Phi_1$. Similarly, the x-component of springs $k_2, k_3 \dots k_n$ are $Xk_2 \cos^2 \theta_2$, $Xk_3 \cos^2 \theta_3, \dots Xk_n \cos {}^2\theta_n$ and the y-components of the spring force are $Xk_2 \cos \theta_2 \cos \Phi_2$, $Xk_3 \cos \theta_3 \cos \Phi_3, \dots\dots Xk_n \cos \theta_{n\,n}$.

Thus, we can get

$$F_X = X(k_1 \cos^2 \theta_1 + k_2 \cos^2 \theta_2 + \dots + k_n \cos^2 \theta_n) = k_{xx} X$$

and

$$F_Y = X(k_1 \cos \theta_1 \cos\Phi_1 + k_2 \cos \theta_2 \cos\Phi_2 + \dots + k_n \cos\theta_n \cos\Phi_n) = k_{yx} X$$

These have been depicted as shown in Fig 3.22(b). In a similar manner, k_1 is deformed by $Y \cos \Phi_1$ when the mass is displaced due to the displacement Y from its position of equilibrium. The y-component of this force is $Yk_1 \cos^2{}_1$ and the x-component is $Yk_1 \cos\Phi_1 \cos \theta_1$ (Fig. 3.22(c)). Therefore, the x-component of all spring forces due to the displacement Y of the mass m in the y-direction is $k_{xy}Y, = Y(k_1 \cos\Phi_1 \cos \theta_1 + k_2 \cos\Phi_2 \cos \theta_1 + \dots + k_n \cos\Phi_n \cos\theta_n)$ and the y-component is $k_{yy} Y = Y(k_1 \cos^2\Phi_1 + k_2 \cos^2 \Phi_2 + \dots + k_n \cos^2 \Phi_n)$.

Therefore, the equations of motion are

$$m\ddot{X} + k_{xx} X + k_{xy} Y = 0$$

$$m\ddot{Y} + k_{yy} Y + k_{yx} X = 0$$

As mentioned, with appropriate initial conditions applied to the mass, it will vibrate either in the first mode with natural frequency ω_1 or second mode with natural frequency ω_2. In other words, this means that

$$x(t) = \mathrm{A}_1 \sin(\omega_1 t + \Psi_1),\ y(t) = \lambda_1 \mathrm{A}_1 \sin(\omega_1 t + \Psi_1)$$

$$x(t) = \mathrm{A}_2 \sin(\omega_2 t + \Psi_2),\ y(t) = \lambda_2 \mathrm{A}_2 \sin(\omega_2 t + \Psi_2)$$

Thus therefore, the amplitudes x and y of the mass vibrating in the principal modes are in the same ratios λ_1 and λ_2. Therefore, the motion of the mass is along a straight line through origin O for the principal modes of vibration.

Let the amplitudes of vibration be 1. Then

$$\mathrm{A}_1 = \cos\theta'_1,\ \mathrm{A}_2 = \cos\theta'_2,\ \mathrm{B}_1 = \cos\phi'_1,\ B_2 = \cos\phi'_2.$$

These are the direction cosines of the lines of vibration at principal modes. When the mass is vibrating at the first principal mode, the equations of motion are

$$-m\omega_1^2 A_1 + k_{xx} A_1 + k_{xy} B_1 = 0 \tag{1}$$

$$-m\omega_1^2 B_1 + k_{yy} B_1 + k_{yx} A_1 = 0 \tag{2}$$

and when the mass vibrates at a second mode, the equations of motion become

$$-m\omega_2^2 A_2 + k_{xx} A_2 + k_{xy} B_2 = 0 \tag{3}$$

$$-m\omega_2^2 B_2 + k_{yy} B_2 + k_{yx} A_2 = 0 \tag{4}$$

Multiplying Equation 3 by ($-A_2$), second by ($-B_2$), third by (A_1), and Equation (4) by (B_1). Adding them together, we get,

$$m\omega_1^2 A_1 A_2 - k_{xx} A_1 A_2 - k_{xy} A_2 B_1 = 0 = m\omega_1^2 B_1 B_2 - k_{yy} B_1 B_2 - k_{yx} A_1 B_2$$

$$m\omega_2^2 A_1 A_2 - k_{xx} A_1 A_2 - k_{xy} A_1 B_2 = 0 = m\omega_2^2 B_1 B_2 + k_{yy} B_1 B_2 + k_{yx} A_2 B_1$$

From these equations, we get,

$$m(A_1 A_2 + B_1 B_2)(\omega_2^2 - \omega_1^2) = 0\,.$$

Since $\quad k_{xy} = k_{yx}, m \quad$ and $\quad (\omega_2{}^2 - \omega_1{}^2) \neq 0$,

we get, $$A_1 A_2 + B_1 B_2 = 0 = \cos\theta_1' \cos\theta_3' + \cos\phi_1^1 \cos\phi_2' \tag{3}$$

These equation means, that θ_1', ϕ_1' line and θ_2', ϕ_2' line are perpendicular to each other. Thus, the principal modes are orthogonal.

3.3 SEMI-DEFINITE SYSTEM

Sometimes, when one of the roots of frequency equation of a vibration system is equal to zero, one of the natural frequencies is equal to zero. Such systems are called semi-definite systems. One of the natural frequencies being zero physically means that the system will move as rigid body without any deformation of the springs and zero relative velocity for the damper. For example, consider the system shown in Fig. 3.23.

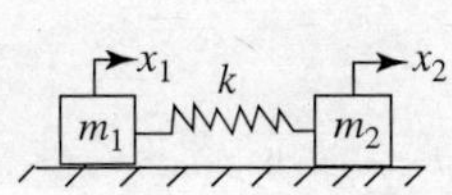

Figure 3.23 Semi-definite System

The equations of motion for the system are

$$m_1 \ddot{x}_1 + k(x_1 - x_2) = 0 \tag{3.40}$$

$$m_2 \ddot{x}_2 + k(x_2 - x_1) = 0 \tag{3.41}$$

Let the solution be $x_1 = A \sin(\omega.t + \Psi)$, $x_2 = B \sin(\omega.t + \Psi)$

Substituting these in Equations 3.40 and 3.41 and simplifying

$$(k - m_1\omega^2)A - kB = 0 \tag{3.42}$$

$$-kA + (k - m\omega^2)B = 0 \tag{3.43}$$

The frequency equation is obtained by equating the determinant of these equations to zero.

$$m_1 m_2 \omega^4 - k(m_1 + m_2)\omega^2 = 0$$

or $$\omega_1 = 0, \quad \omega_2 = \sqrt{\frac{k(m_1 + m_2)}{m_1 m_2}}\ \text{rad/s} \tag{3.44}$$

The first natural frequency of the system being zero means that the system is not vibrating or oscillating. In other words, this means that the two masses move as a whole unit and have no relative motion and no spring deformation.

By substituting the value of ω_2 into Equations 3.42 and 3.43, we can verify that $X_1^{(2)}$ and $X_2^{(2)}$ are opposite in phase. This means that there is a node at the middle of the spring.

Problem 3.15

Figure 3.24 shows a simple pendulum hinged at the centre of mass *M* sliding on a frictionless surface. Show that this is a semi-definite system. Also, find the natural frequencies.

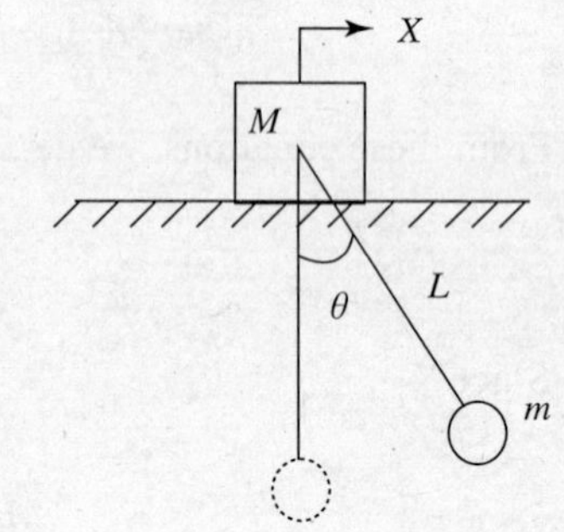

Figure 3.24

Solution:

For the system, the generalized coordinates are x and θ. The energies in the system are

$$KE = \frac{1}{2}\{M\dot{x}^2 + m[\dot{x}^2 + (L\dot{\theta})^2 + 2\dot{x}L\dot{\theta}\cos\theta]\}$$

$$PE = mgL(1 - \cos\theta)$$

Assuming small oscillations such that $\sin\theta \approx \theta$, $\cos\theta \approx 1 - \theta^2/2$ (because $\cos\theta = \cos^2\theta/2 - \sin^2\theta/2$), we get,

$$KE = \frac{1}{2}[M\dot{x}^2 + m(\dot{x} + L\dot{\theta})^2$$

$$PE = \frac{1}{2}mgL\theta^2$$

Using Lagrange's equation for the above, we get

$$(m+M)\ddot{x} + mL\ddot{\theta} = 0$$

$$\ddot{x} + L\ddot{\theta} + g\theta = 0$$

Let x = A cos (ωt+Ψ) and θ = B cos (ωt + Ψ). Substituting these in the equations above and simplifying, we get,

$$-\omega^2(m+M)A - mL\omega^2 B = 0$$

$$-\omega^2 A - (\omega^2 L - g)B = 0$$

As explained in previous examples, we set the determinant of the above equations to zero to obtain the frequency equation

$$\begin{vmatrix} -(m+M)\omega^2 & -mL\omega^2 \\ -\omega^2 & -(\omega^2 L - g) \end{vmatrix} = 0$$

$$\therefore \omega^2[(m+M)(\omega^2 L - g) - mL\omega^2] = 0$$

Therefore, $\omega_1 = 0, \omega_2 = \sqrt{\dfrac{g(M+m)}{ML}}$ rad/s

Since one of the natural frequencies is zero, the system shown in Fig. 3.24 is a semi-definite system.

3.4 FORCED VIBRATION OF TWO DEGREES-OF-FREEDOM SYSTEM

We shall first deal with a two degrees-of-freedom system consisting of only mass and spring elements. The analysis can, later, be extended to systems comprising of damper element also.

Figure 3.25 shows a typical undamped two degrees-of-freedom system subjected to a forced excitation. The equations of motion are

$$m_1\ddot{x}_1 + (k_1+k_2)x_1 - k_2x_2 = F_0 \cos\omega.t \tag{3.45}$$

$$m_2\ddot{x}_2 + (k_2+k_3)x_2 - k_2x_1 = 0 \tag{3.46}$$

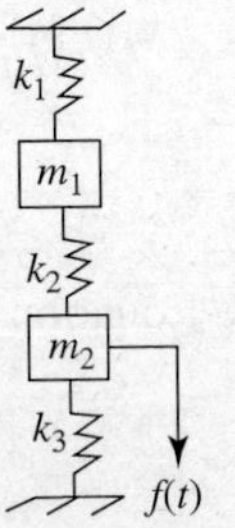

Figure 3.25 Forced Vibration TDOF System

For systems comprising of dampers, equations can be developed using Newton's laws of motion or using Lagrange's equation. See for example, Equation 3.3. Similarly equations can be developed for semi-definite system. The method of solution will be similar to the method we describe for the system shown in Fig. 3.25.

Assume $x_1 = A\cos(\omega.t+\Phi)$, $x_2 = B\cos(\omega.t+\Phi)$, as the steady-state forced responses. Substituting these in Equations 3.45 and 3.46, we get

$$(k_1+k_2-m_1\omega^2)A - k_2B = F_0$$

$$-k_2A + (k_2+k_3-m_2\omega^2)B = 0$$

Using Cramer's rule, we find

$$A = \frac{\begin{vmatrix} F_0 & -k_2 \\ 0 & k_2+k_3-m_2\omega^2 \end{vmatrix}}{(k_1+k_2-m_1\omega^2)(k_2+k_3-m_2\omega^2)-k_2^2} = \frac{F_0(k_2+k_3-m_2\omega^2)}{(k_1+k_2-m_1\omega^2)(k_2+k_3-m_2\omega^2)-k_2^2} \tag{3.47}$$

$$B = -\frac{\begin{vmatrix} k_1+k_2-m_1\omega^2 & F_0 \\ -k_2 & 0 \end{vmatrix}}{(k_1+k_2-m_1\omega^2)(k_2+k_3-m_2\omega^2)-k_2^2} = \frac{F_0k_2}{(k_1+k_2-m_1\omega^2)(k_2+k_3-m_2\omega^2)-k_2^2} \tag{3.48}$$

Therefore, the response is

$$x_1(t) = \frac{F_0(k_2+k_3-m_2\omega^2)}{(k_1+k_2-m_1\omega^2)(k_2+k_3-m_2\omega^2)-k_2^2}\cos\omega.t \tag{3.49}$$

$$x_2(t) = \frac{F_0k_2}{(k_1+k_2-m_1\omega^2)(k_2+k_3-m_2\omega^2)-k_2^2}\cos\omega.t \tag{3.50}$$

and the phase angle Φ may be zero or 180°

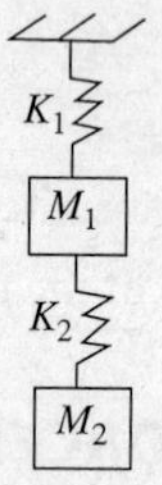

Figure 3.26 TDOF System

We can also use impedance method. For example, consider the system shown in Fig. 3.26. We have two masses M_1 and M_2, two springs K_1 and K_2. Let x_1 and x_2 be the displacements. For this case, equations of motion are

$$M_1\ddot{x}_1 + (K_1 + K_2)x_1 - K_2x_2 = F_0 \sin \omega.t, (say) \qquad (3.51)$$

$$M_2\ddot{x}_2 + K_2x_2 - K_2x_1 = 0 \qquad (3.52)$$

In the impedance method, we substitute $F_0e^{i\omega.t}$ for F_0 sin ωt, $X_1e^{i\omega.t}$ for x_1, and $X_2e^{i\omega.t}$ for x_2. Using these, Equations (3.51) and (3.52) become

$$M_1i^2\omega^2X_1e^{i\omega.t} + (K_1 + K_2)X_1e^{i\omega.t} - K_2X_2e^{i\omega.t} = F_0e^{i\omega.t} \qquad (3.53)$$

$$M_2i^2\omega^2X_2e^{i\omega.t} + K_2X_2e^{i\omega.t} - K_2X_1e^{i\omega.t} = 0.---(i = \sqrt{-1}) \qquad (3.54)$$

From the Equations (3.53) and (3.54) we obtain,

$$(K_1 + K_2 - M_1\omega^2)X_1 - K_2X_2 = F_0 \qquad (3.55)$$

$$-K_2X_1 + (K_2 - M_2\omega^2)X_2 = 0 \qquad (3.56)$$

Solving Equations (3.55) and (3.56) using Cramer's rule, we get,

$$X_1 = \frac{\begin{vmatrix} F_0 & -K_2 \\ 0 & K_2 - M_2\omega^2 \end{vmatrix}}{(K_1 + K_2 - M_1\omega^2)(K_2 - M_2\omega^2) - k_2^2} = \frac{F_0(K_2 - M_2\omega^2)}{M_1M_2\omega^4 - (M_1K_2 + M_2K_2 + M_2K_1)\omega^2 + K_1K_2} \qquad (3.57)$$

$$B = -\frac{\begin{vmatrix} (K_1 + K_2 - M_1\omega^2) & F_0 \\ -K_2 & 0 \end{vmatrix}}{(K_1 + K_2 - M_1\omega^2)(K_2 - M_2\omega^2) - k_2^2} = \frac{F_0k_2}{M_1M_2\omega^4 - (M_1K_2 + M_2K_2 + M_2K_1)\omega^2 + K_1K_2} \qquad (3.58)$$

Since the forcing function is F_0 sin $\omega.t$ = Imag(F_0 $e^{i\omega.t}$), x_1 = Imag(X_1 $e^{i\omega.t}$)

or $$x_1 = \text{Im}(X_1e^{i\phi})(e^{i\omega.t}) = \text{Im}(X_1e^{i(\omega.t+\phi)}) = X_1 \sin(\omega.t + \phi),$$

and $$x_2 = X_2 \sin(\omega.t + \phi)\cdot$$

But (cos sin)$x_1 = X_1$ (cos ϕ + i sin ϕ) and expression X_1 contains real quantities only. Thus, $\Phi = 0$ or 180°. This means that the motion of the masses is either in phase or out of phase with excitation. Therefore, $X_i = x_i$. Thus, the steady-state response is given by

$$x_1(\text{t}) = \frac{F_0(K_2 - M_2\omega^2)}{M_1M_2\omega^4 - (M_1K_2 + M_2K_2 + M_2K_1)\omega^2 + K_1K_2}\sin\omega\text{t} \qquad (3.59)$$

$$x_2(\text{t}) = \frac{F_0k_2}{M_1M_2\omega^4 - (M_1K_2 + M_2K_2 + M_2K_1)\omega^2 + K_1K_2}\sin\omega\text{t} \qquad (3.60)$$

The example cited is for explaining the concept of 'Tuned Absorber', which is extensively used to control the vibration in the equipment.

3.4.1 Tuned Absorber

Suppose we have very high vibrations in mass M_1 (Fig. 3.27) in SDOF system because of the natural frequency ω_n (= K/M)^{½} = ω (frequency of the excitation force) or in the near neighbourhood. Suppose, we attach a spring of stiffness K_2 and attach a mass M_2 (shown by a dotted line) such that K_1/M_1 = K_2/M_2, then the system becomes a two degrees-of-freedom system and Equations 3.59 and 3.60 apply. Under the condition, $K_1/M_1 = K_2/M_2$, Equation 3.59 becomes zero since $K_2/M_2 = \omega^2$, or $K_2 - M_2\,\omega^2 = 0$. Under this condition, vibrations on M_1 are arrested

or $$x_1(t) = 0 \qquad (3.61)$$

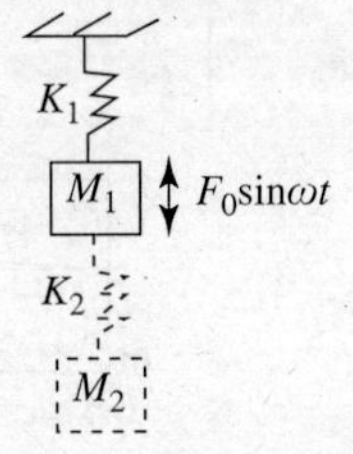

Figure 3.27 Tuned Absorber

In order to understand the physical phenomena, let us rewrite Equations 3.59 and 3.60 in a slightly different form

$$x_j = X_j \sin \omega.t, \qquad j = 1, 2 \qquad (3.62)$$

$$X_1(\text{t}) = \frac{(K_2 - M_2\omega^2)F_0}{(K_1 + K_2 - M_1\omega^2)(K_2 - M_2\omega^2) - k_2^2} \qquad (3.63)$$

$$X_2(\text{t}) = \frac{F_0 k_2}{(K_1 + K_2 - M_1\omega^2)(K_2 - M_2\omega^2) - k_2^2} \qquad (3.64)$$

As mentioned earlier, absorber is designed such that $K_1/M_1 = K_2/M_2 = \omega^2$ and the operation of the original machine will be free of vibration. Let us define $\delta_{st} = F_0/K_1$, $\omega_1 = (K_1/M_1)^{1/2}$ as the natural frequency of the main system and $\omega^2 = (K_2/M_2)^{1/2}$ as the natural frequency of the absorber or auxiliary system. Then we can write

$$\frac{X_1}{\delta_{st}} = \frac{1-(\omega/\omega_2)^2}{\{[1+(K_2/K_1)-(\omega/\omega_1)^2][1-(\omega/\omega_2)^2]-K_2/K_1\}} \qquad (3.65)$$

$$\frac{X_2}{\delta_{st}} = \frac{1}{\{[1+(K_2/K_1)-(\omega/\omega_1)^2][1-(\omega/\omega_2)^2]-K_2/K_1\}} \qquad (3.66)$$

$$\text{At } \omega = \omega_1, X_1 = 0$$

At this frequency $$X_2 = -\frac{K_1}{K_2}\delta_{st} = -F_0/K_2 \qquad (3.66)$$

This shows that the force exerted by the absorber or auxilliary spring is opposite to the impressed force ($K_2X_2 = -F$) and neutralizes it, thus $X_1 = 0$. The size of the dynamic (tuned) absorber can be found from Equations 3.67 and 3.68. Fig. 3.28 shows the vibration characteristics of the total system. We have now two natural frequencies Ω_1 and Ω_2. The values of Ω_1 and Ω_2.can be found by equating the denominator of Equation 3.65 to zero.

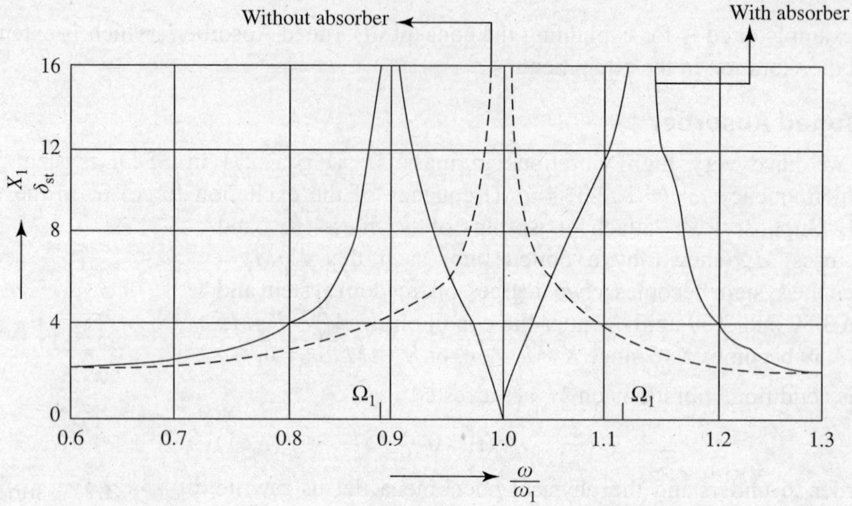

Figure 3.28 Response of Machine with Tuned Absorber (Mass Ratio 0.05)

Noting that,

$\frac{K_2}{K_1} = \frac{K_2}{M_2}\cdot\frac{M_2}{M_1}\cdot\frac{M_1}{K_1} = \frac{M_2}{M_1}\left(\frac{\omega_2^2}{\omega_1^2}\right), \frac{K_2}{M_2} = \omega_2^2, \frac{K_1}{M_1} = \omega_1^2$ (say) and setting the denominator of Equation 3.65 to zero leads to

$$\left(\frac{\omega}{\omega_2}\right)^4\left(\frac{\omega_2}{\omega_1}\right)^2 - \left(\frac{\omega}{\omega_2}\right)^2\left[1+\left(1+\frac{M_2}{M_1}\right)\left(\frac{\omega_2}{\omega_1}\right)^2\right] + 1 = 0 \tag{3.67}$$

The two roots are given by $(\Omega_1/\omega_2)^2$ and $(\Omega_2/\omega_2)^2$

$$\left.\begin{matrix}(\Omega_1/\omega)^2\\ (\Omega_2/\omega)^2\end{matrix}\right\} = \frac{\left[1+\left(1+\frac{M_2}{M_1}\right)\left(\frac{\omega_2}{\omega_1}\right)^2\right] \mp \left\{\left[1+\left(1+\frac{M_2}{M_1}\right)\left(\frac{\omega_2}{\omega_1}\right)^2\right]^2 - 4\left(\frac{\omega_2}{\omega_1}\right)^2\right\}^{1/2}}{2\left(\frac{\omega_2}{\omega_1}\right)^2} \tag{3.68}$$

We can conclude following from this analysis that

- Ω_1 is less than and Ω_2 is greater than the operating speed, which is equal to the natural frequency ω_1 of the machine. Thus the machine must pass through Ω_1 during start up and stopping. At these speeds, the amplitude is very large.
- Since the tuned absorber is tuned to one-excitation frequency ω, the steady-state amplitude of the machine is almost zero at that frequency.

CONCLUSION

In this chapter, we discussed the method of arriving at system equations for systems having two degrees-of-freedom. We also discussed the method of evaluating the natural frequencies and the mode shapes by putting the appropriate boundary conditions. Concept of coordinate coupling has been discussed. We also discussed the application of Lagrange's equation for finding the natural frequencies. We discussed the orthogonality principle for the principal modes.

We discussed the method to evaluate forced-vibration response of two degrees-of-freedom system and explained the principle behind the tuned absorbers.

Exercises

3.1 For the system shown here, find the vibration response for the initial conditions: at $t = 0$, displacement of first mass = 1, velocity of first mass = 0, velocity of second mass = displacement of second mass.

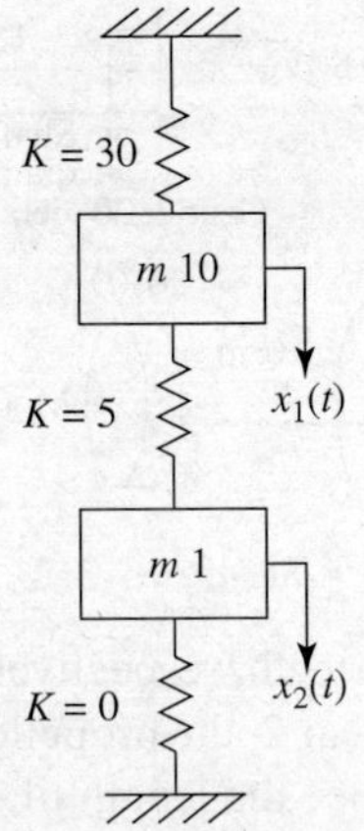

Problem 3.1

(*Answer:* $\omega_1 = 1.58, \omega_2 = 2.45$)

$$x_1(t) = \frac{5}{7}\cos 1.581t + \frac{2}{7}\cos 2.45t$$

$$x_2(t) = \frac{10}{7}\cos 1.581t - \frac{10}{7}\cos 2.45t$$

3.2 The figure here shows a two degrees-of-freedom torsional system. Find the natural frequencies and the mode shapes. $J_1 = J_0$, $J_2 = 2J_0$, $K1 = K2 = K3 = K$

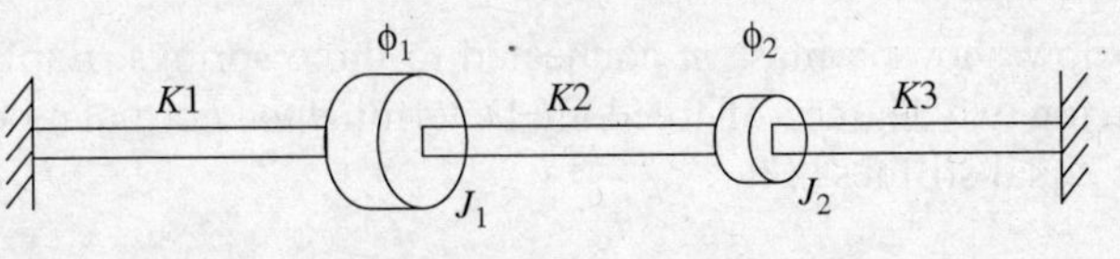

Problem 3.2

(*Hint:* Torsional system can be analysed exactly the same manner as spring–mass system. The mass moment of inertia J of the disc is analogous to mass and the torsional stiffness of the shaft is analogous to spring stiffness. The method of arriving at the governing equation is similar to that used in spring, mass, and damper system.

(*Answer:* $\omega_1 = (0.22K/J_0)^{1/2}$, $\omega_2 = (2.3K/J_0)^{1/2}$)

3.3 The figure (on the next page) shows schematic of a marine engine coupled to a propeller through gears. The mass moments of the flywheel, engine, gear 1, gear 2, and the propeller are 9000, 1000, 250, 150, and 2000 kg-m^2, respectively. Assume that the flywheel on account of its very high moment of inertia is fixed and does not vibrate. Also assume that the engine and the gears can be considered as a single disc (rotor) because of very short length of the shaft on which they are mounted.

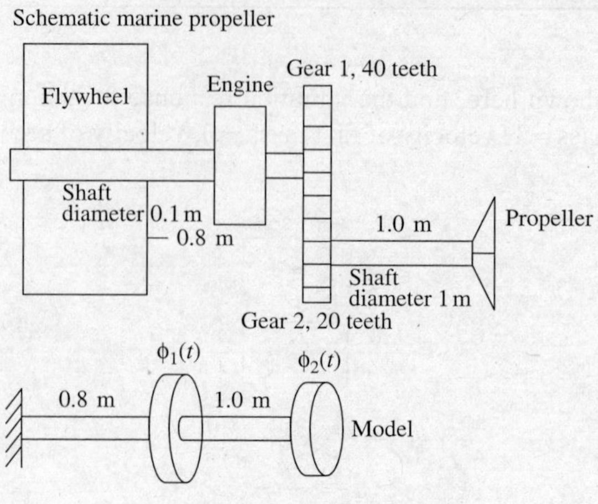

Problem 3.3

Hint: The gears 1 and 2 have 40 and 20 teeth, respectively. As a result, shaft 2 rotates at twice the speed of shaft 1. Because of the MI of gear 2, the propeller will be $J_{GEQ} = 2^2 \times 150 = 600$ kg m^2 and $J_{PEQ} = 2^2 \times 2000 = 8000$ kg m^2. Also, since the length of shaft between the engine and the gears is very small, all of them can be replaced by a single disc (rotor) of $MI = 1000 + 250 + 600 = 1850$ kg m^2. Thus as shown in the model, we have one disc having mass moment of inertia equal to 1850 kg m^2 and the other disc having the mass moment of inertia equal to 8000 kg m^2. Assuming the shear modulus for the shaft as 80×10^9 N/m^2, calculate K_1 and K_2 using the formula $k = (G/l)\,(\pi d^4/32)$. Now for the model in the figure, calculate the natural frequencies and the mode shapes.

(Answer: $\omega_1 = 9.23$ rad/s, $\omega_2 = 55.6$ rad/s)

Mode 1 displacement ratio 1/1.207

Mode 2 displacement ratio 1/-0.1916

3.4 Explain the basic concept behind the principle coordinates. Find the principle coordinates of the system as shown in figure of Problem 3.1 but having mass as m and $2m$ with springs having stiffness k and $2k$, respectively.

3.5 The figure below shows a mass m connected to three springs. Explain why this system is considered to be having two degrees-of-freedom. Determine the normal modes of the system when all the springs have equal stiffness (k).

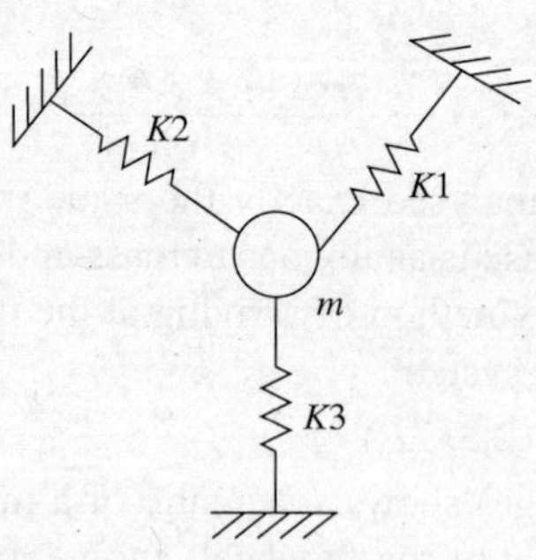

Problem 3.5

Answer: Natural frequencies are $\omega_1 = (\dot{x}_1)^{1/2}$ and $\omega_2 = (2k/m)^{1/2}$

3.6 A heavy machine having a mass m = 1000 kg and a mass moment of inertia of J_0 = 300 kg-m^2 is supported on elastic supports having stiffnesses k_1 = 3000 N/m and k_2 = 2000 N/m. The figure here shows their locations. Find the natural frequencies and the mode shapes.

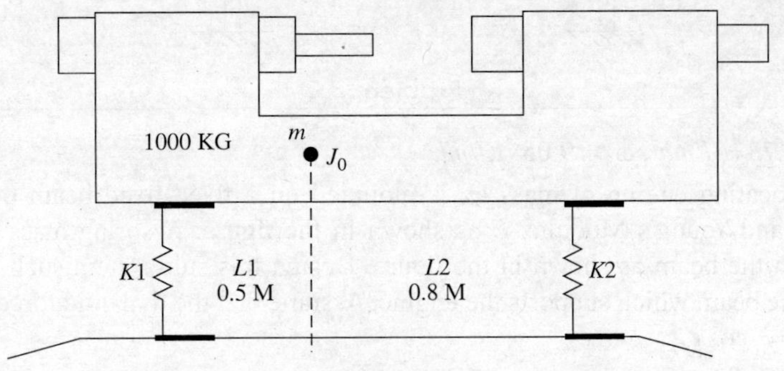

Problem 3.6

Hint: Follow the procedure outlined in Solved Problem 3.4 solved in the main text.

3.7 Figure below shows schematic arrangement of an electric-overhead travelling crane, consisting of a girder, trolley, and a wire rope. The girder has a flexural rigidity (EI) 6×10^{12} lb-in.^{2} and a span L of 30 ft. The weight of the trolley is 8000 lb and the load lifted is 2000 lb. Design the rope such that the fundamental natural frequency is greater than 20 Hz (1200 cycles per min).

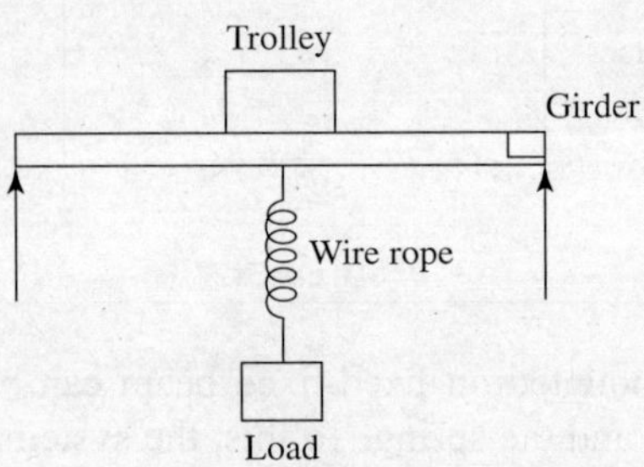

Problem 3.7

Hint: Consider the girder as massless and as a simply supported beam. Using the formula for the deflection of beam, find the spring constant of the spring that represents the girder. We now have a two degrees-of-freedom system with girder spring and the mass of the trolley as one system to which the wire rope as another spring which carries the load of 2000 lbs is attached. We thus model the crane as a two degrees-of-freedom system. The system has two natural frequencies, the lowest being >20 Hz. Use this information to calculate the stiffness of wire rope and hence the area of the wire rope.

(*Answer:* 1.1 in^2.)

3.8 A two-storey building frame is modelled as shown in the figure. Assume that the girders are rigid and the columns have flexural rigidities EI, The stiffness of the columns is given by $k = 24\ EI/h^3$. Assuming that both the girders have same weight, determine the natural frequencies and the mode shapes.

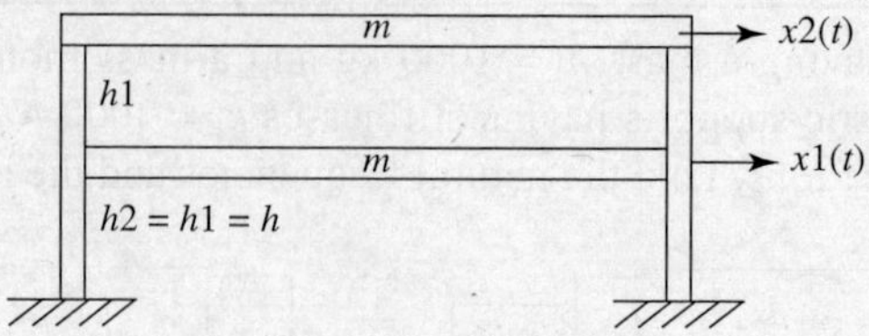

Problem 3.8

Answer: $\omega_1 = 3.75\sqrt{EI/mh^3}$, $\omega_2 = 9.05\sqrt{EI/mh^3}$

3.9 A reciprocating engine of mass *m*, is mounted on a fixed-fixed beam of length *l*, width *a*, thickness *t*, and Young's Modulus *E* as shown in the figure. A spring-mass system (*k*, *m*) is suspended from the beam as shown in the figure. Design this sub-system such that there are no vibrations on the beam which supports the engine. Assume that the dynamic force produced by the engine is $F(t) = F \cos \omega t$.

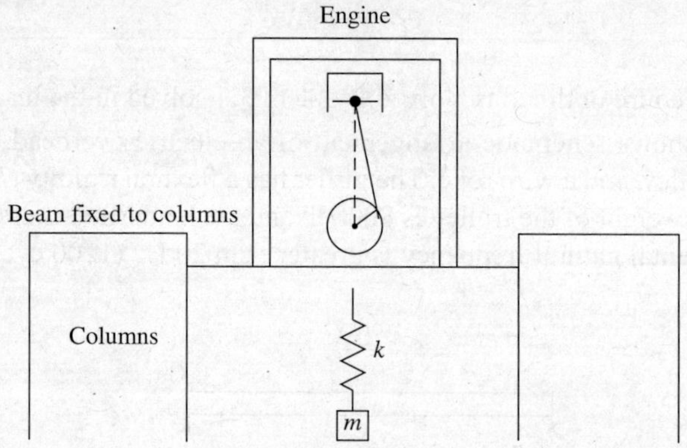

Problem 3.9

Hint: The system of engine mounted on fixed-fixed beam can be modelled as a single degree-of-system comprising of mass and the spring. To this, the systems *m* and *k* are attached. We thus have two degrees-of-freedom system. The above problem shows the principle behind the tuned absorber. Refer to the text.

3.10 The figure shows a steam turbine coupled to the generator through speed-reduction gears having 20 and 30 teeth. The mass moment of inertia of the turbine, generator, gear 1 and gear 2 are given, respectively, by 3000, 2000, 500, and 1000 kg m². Shafts 1 and 2 are made of steel and have diameters 30 cm and 10 cm and lengths 2 cm and 1.0 m, respectively. Find the natural frequencies of the system.

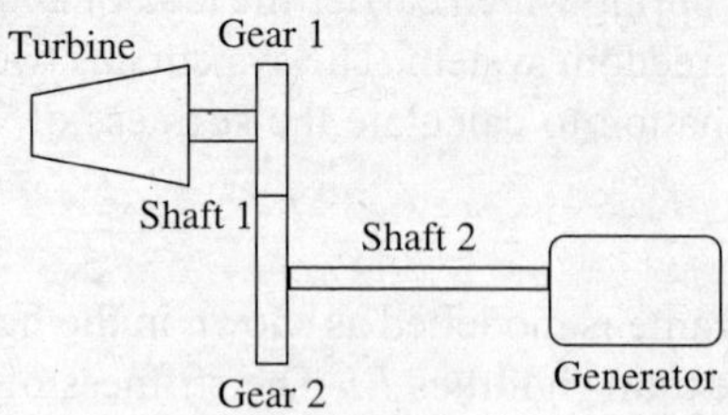

Problem 3.10

Hint: The gear assembly must be reduced to a single disc. The moment of inertia of the gear 2 as well as the generator as referred to the engine will be $(20/30)^2 \times MI$ of gear 2 and $(20/30)^2 \times MI$ of generator. The system then becomes a routine two degrees-of-freedom torsional system.

3.11 In the system shown here the mass m_1 is excited by a harmonic force having amplitude of 50 N and a frequency of 2 Hz. Find the forced amplitude of each mass where $m_1 = 10$ kg, $m_2 = 5$ kg, $k_1 = 8000$ N/m and $k = 2000$ N/m.

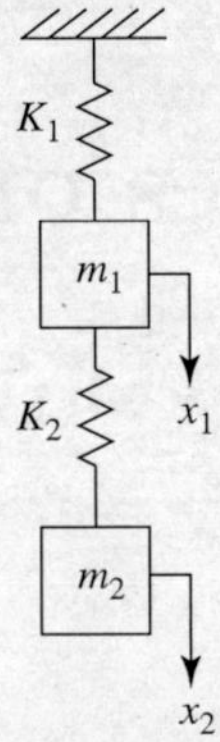

Problem 3.11

Answer: $X_1 = 0.009\ 77$, $X_2 = 0.0161$

4

Multi Degrees-of-Freedom Systems

4.1 INTRODUCTION

Most engineering systems are continuous and, have in theory, infinite degrees-of-freedom. However, as shown in previous chapters, many of them can be idealized as discrete spring–mass–damper system with reasonable accuracy. Whereas continuous systems require solution of complex partial differential Equations, modelling them by a single or two degrees-of-freedom system greatly simplifies the study of their behaviour/response to the given perturbation force. One may also note and appreciate the fact that no closed form solutions exist for many partial differential Equations.

We now extend the concepts of previous chapter to a higher degree of modelling so that we come quite close to the continuous system. In this approach also, the continuous system is modelled as an assemblage of many masses, springs, and dampers, and for such idealization, we write Equations of motion using Newton's laws of motion or by influence coefficients (which we shall discuss in this chapter) or preferably by using Lagrange's Equation.

The 'n' degrees-of-freedom system is associated with n natural frequencies, each associated with its mode shape. The method of determining the natural frequencies from the characteristic Equation obtained by equating the determinant of the system of Equations of motion, as shown in previous chapter, also applies to multi degrees-of-freedom systems. However, as the number of degrees-of-freedom increases, the solution of characteristic Equations becomes extremely complex. As explained in the previous chapter, the mode shapes exhibit a property known as Orthogonality; this often enables to simplify considerably the analysis of multi degrees-of-freedom systems.

4.2 MODELLING OF CONTINUOUS SYSTEMS

We have demonstrated in the previous chapter, the concept of replacing the mass or inertia and elasticity of the system by a finite number of lumped masses and springs. For example, as demonstrated in Problem 3.4, we idealized the car and the suspension system as a lumped mass (weight of the car) on two springs representing the suspension and treated it as two degrees-of-freedom system.

The roller-bearing system as described in Problem 3.11 was also idealized as two degrees-of-freedom system. In these analyses, the lumped masses are assumed to be connected by massless springs and damping members (having no mass or elasticity). Linear or angular coordinates are used to describe the motion of lumped masses. Such models are called lumped-parameter or lumped-mass or discrete-mass systems. The minimum number of coordinates necessary to describe the motion of lumped masses (or rigid bodies) defines the number of degrees-of-freedom of the system. It is needless to say that the larger the number of lumped masses used in the model, the higher is the accuracy of resulting analysis.

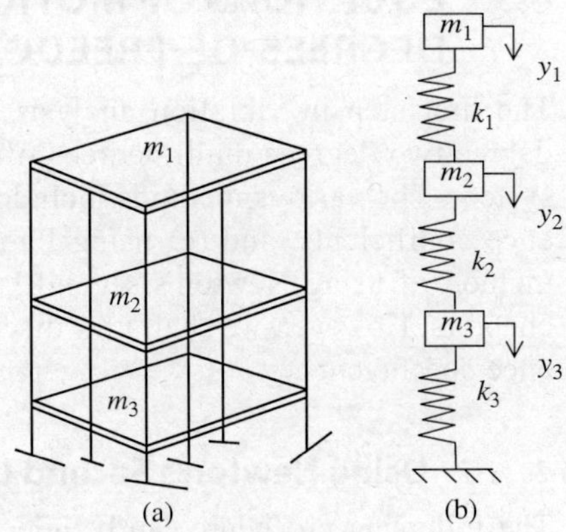

Figure 4.1 (a) Three-storey Building (b) Three-lumped-mass Model of the Building

Some problems automatically indicate the type of lumped parameter model to be used. For example, the three-storey building shown in Fig. 4.1(a), subjected to seismic disturbance, automatically suggests using a three-lumped mass model as shown in Fig. 4.1(b).

In this model, the inertia of the system is assumed to be concentrated at three-point masses located at the floor level and the elasticity of the columns are replaced by the springs. In this model, it is tacitly assumed that the floors represented by the masses m_1, m_2, and m_3 move as rigid bodies in y direction (displacements y_1, y_2, and y_3). On the other hand, if the vibration exciters exist on the floors, the present model of three masses and three springs is irrelevant and we have to go in for separate modelling, which accounts for flexing of the floors supported on columns. Such a model will be highly complex compared to the one shown in Fig. 4.1b. It must thus be remembered that the degree of modelling should be judged from what is required for the analysis. One can start with a very simple model to begin with and then go in for improvement when needed. To clarify this point, let us consider the modelling of a radial drilling machine [Fig. 4.2(a)].

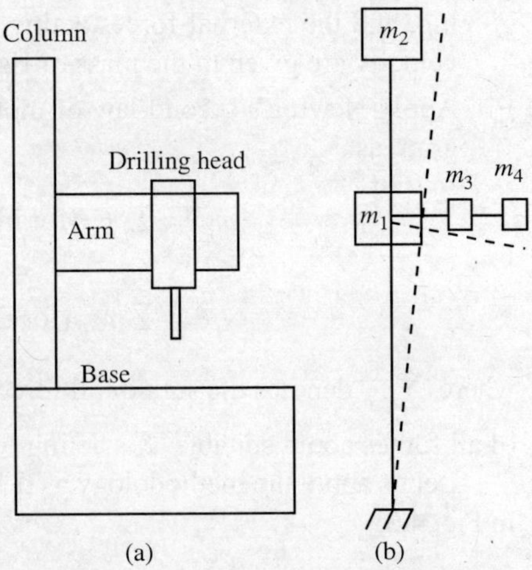

Figure 4.2 (a) Radial Drilling M/C (b) Lumped-mass-model of the Machine

In this, the column can be modelled as elastic (massless) cantilever beam with two lumped masses m_1 and m_2. The arm can be considered as an inertial element and thus represented as a lumped mass on a massless beam while the drilling head can be modelled as another lumped mass; thus the arm is modelled as a massless beam with two lumped masses m_4 and m_3. The dotted lines show the pattern of displacements. The axial deformation of the column is neglected since the column has enormous axial rigidity compared to transverse rigidity. By assuming simple solution within each element, the principles of compatibility and equilibrium are used to find an approximate (we may term it as first level) solution to the original system. This is the basic principle behind the finite-element method (FEM), which is one of the most popular method used for analysis of varieties of engineering problems including the stress and vibration analysis.

4.3 EQUATIONS OF MOTION FOR MULTI DEGREES-OF-FREEDOM SYSTEMS

The first step in vibration analysis (of course after correctly modelling the system as single degree/two degrees/multi degrees-of-freedom system) is to write the equations of motion of the system. The various methods include: (a) using Newton's second law of motion, (b) using influence coefficients, and (c) using Lagrange's equation. Of these, we have already described the method of using Newton's second law of motion and also Lagrange's equation in the previous chapters. However, we shall now discuss them in more detail and also discuss the method of influence coefficients.

4.3.1 Using Newton's Second Law of Motion

The following procedure can be adopted to derive the equations of motion of a multi degrees-of-freedom system,

- Set up suitable coordinates to describe the positions of various point masses and rigid bodies in the system. Assume suitable positive directions for the displacements, velocities, and acceleration of the masses and the rigid bodies.
- Determine the static-equilibrium configuration of the system and measure the displacement of the masses and rigid bodies from their respective static-equilibrium position.
- Draw the free-body diagram for each mass or rigid body in the system. Indicate the spring, damping, and the external forces acting on each mass or rigid body when positive displacement and velocity are given to the mass or rigid body.
- Apply Newton's second law of motion to each mass or rigid body shown by the free-body diagram as

$$m_i \ddot{y} = \Sigma_j F_{ij} \text{ (for mass } m_i) \tag{4.1}$$

or

$$J_i \ddot{\theta} = \Sigma_j M_{ij} \text{ (for rigid body of mass moment of inertia } J_i) \tag{4.2}$$

where $\sum_j F_{ij}$ denotes the sum of all forces acting on the mass m_i and $\sum M_{ij}$ indicates sum of moments of all forces about suitable axis acting on the rigid body of mass moment of inertia J_i.

Let us apply the methodology as described above to a multi degrees-of-freedom system as shown in Fig. 4.3.

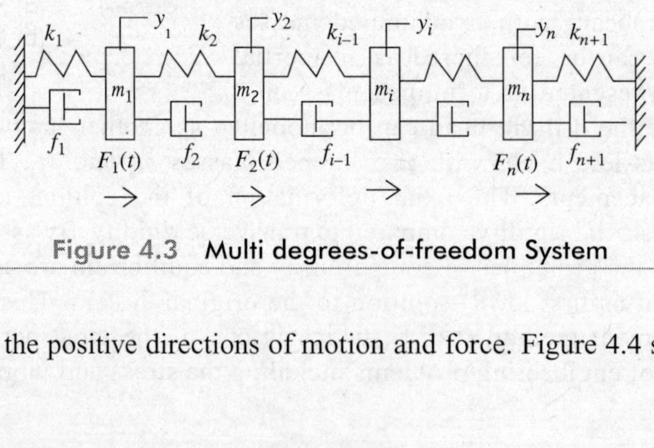

Figure 4.3 Multi degrees-of-freedom System

Figure 4.3 shows the positive directions of motion and force. Figure 4.4 shows a free-body diagram of the ith mass.

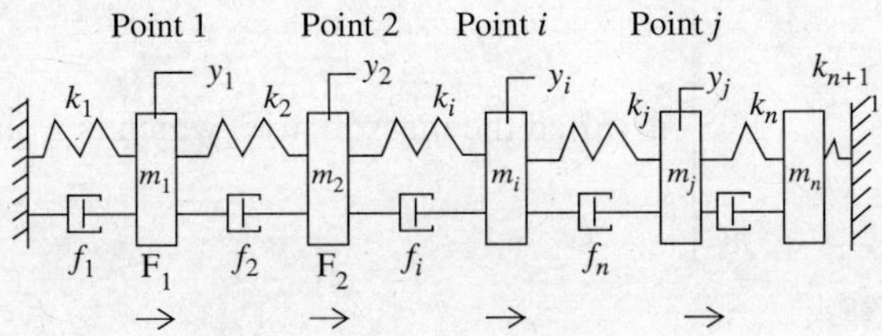

Figure 4.4 **Free-body Diagram**

The inertial force of the mass m_i is $m_i\ddot{y}_i$. Applying Newton's second law of motion, we get,

$$m_i\ddot{y}_i = -k_i(y_i - y_{i-1}) - f_i\,(\dot{y}_i - \dot{y}_{i-1}) + k_{i+1}(y_{i+1} - y_i) + f_{i+1}(\dot{y}_{i+1} - \dot{y}_i) + F_i(t), \quad \text{or}$$

$$m_i\ddot{y}_i + k_i(y_i - y_{i-1}) + f_i\,(\dot{y}_i - \dot{y}_{i-1}) - k_{i+1}(y_{i+1} - y_i) - f_{i+1}(\dot{y}_{i+1} - \dot{y}_i) = F_i(t).$$

This can be written as

$$m_i\ddot{y}_i - f_i\dot{y}_{i-1} + (k_i + k_{i+1})y_i - k_i y_{i-1} + (f_i + f_{i+1})\dot{y}_i - k_{i+1}y_{i+1} - f_{i+1}\dot{y}_{i+1} = F_i(t) \tag{4.3}$$

Equation 4.3 is a general equation valid for $i = 2, 3, 4, \ldots, n$. The equation of motion of mass m_1 and m_n can be derived from Equation 4.3 by setting $i = 1$ along with $y_0 = 0$ and $i = n$ along with $y_{n+1} = 0$, respectively. This means that

$$m_1\ddot{y}_1 + (f_1 + f_2)\dot{y}_1 - f_2\dot{y}_2 + (k_1 + k_2)y_1 - k_2 y_2 = F_1 \tag{4.4}$$

$$m_n\ddot{y}_n(f_n + f_{n+1})\dot{y}_n - f_n\dot{y}_{n-1} + (k_n + k_{n+1})y_n - k_n y_{n-1} = F_n \tag{4.5}$$

It is very convenient to write the equations of motion in matrix form. In order to be able to understand it, let us write the equations for few masses starting from the first mass

$$\begin{aligned}
m_1\ddot{y}_1 + (f_1 + f_2)\dot{y}_1 - f_2\dot{y}_2 + (k_1 + k_2)y_1 - k_2 y_2 &= F_1\\
m_2\ddot{y}_2 + (f_2 + f_3)\dot{y}_2 - f_2\dot{y}_1 - f_3\dot{y}_3 + (k_2 + k_3)y_2 - k_2 y_1 - k_3 y_3 &= F_2\\
m_3\ddot{y}_3 + (f_3 + f_4)\dot{y}_3 - f_3\dot{y}_2 - f_4\dot{y}_3 + (k_3 + k_4)y_3 - k_3 y_2 - k_4 y_3 &= F_3\\
\cdots\cdots\cdots\cdots\cdots\cdots\cdots\cdots\cdots\cdots\cdots\cdots&\\
m_n\ddot{y}_n + (f_n + f_{n+1})\dot{y}_n - f_n\dot{y}_{n-1} + (k_n + k_{n+1})y_n - k_n y_{n-1} &= F_n
\end{aligned} \tag{4.6}$$

Thus, it is very convenient to write the equations in matrix form.

$$\begin{bmatrix} m_1 & 0 & 0\cdot\cdot 0 \\ 0 & m_2 & 0\cdot\cdot 0 \\ 0 & 0 & m_3\cdot\cdot 0 \\ 0 & 0 & 0\cdot\cdot m_n \end{bmatrix}\begin{Bmatrix} \ddot{y}_1 \\ \ddot{y}_2 \\ \ddot{y}_3 \\ \ddot{y}_n \end{Bmatrix} + \begin{bmatrix} f_1 + f_2 & -f_2 & 0 & 0 \\ -f_2 & f_2 + f_3 & -f_3 & 0 \\ 0 & -f_3 & f_3 + f_4 & -f_4 \\ 0 & 0 & -f_n & f_n + f_{n+1} \end{bmatrix}\begin{Bmatrix} \dot{y}_1 \\ \dot{y}_2 \\ \cdot \\ \dot{y}_n \end{Bmatrix}$$

$$+\begin{bmatrix} k_1 + k_2 & -k_2 & 0 & 0 \\ -k_2 & k_2 + k_3 & -k_3 & 0 \\ 0 & -k_3 & k_3 + k_4 & -k_4 \\ 0 & 0 & -k_n & k_n + k_{n+1} \end{bmatrix}\begin{Bmatrix} y_1 \\ y_2 \\ y_3 \\ y_n \end{Bmatrix} = \begin{Bmatrix} F_1 \\ F_2 \\ \cdot \\ F_N \end{Bmatrix} \tag{4.7}$$

where $[m]=\begin{bmatrix} m_1 & 0 & 0 & & \\ & m_2 & & & \\ & & m & & 0 \\ & & & m_{n-1} & \\ & & & 0 & m_n \end{bmatrix}$ is called the mass matrix, which is a diagonal matrix with all the non-diagonal elements zero,

$$[f]=\begin{bmatrix} f_1+f_2 & -f_2 & 0 & 0 \\ -f_2 & f_2+f_3 & -f_3 & 0 \\ 0 & -f_3 & f_3+f_4 & \\ & & -f_n & f_n+f_{n+1} \end{bmatrix} \text{ a symmetric damping matrix, and}$$

$$[k]=\begin{bmatrix} k_1+k_2 & -k_2 & & \\ -k_2 & k_2+k_3 & -k_3 & \\ 0 & -k_3 & k_3+k_4 & -k_4 \\ & & -k_n & k_n+k_{n+1} \end{bmatrix} \text{ a symmetric stiffness matrix.} \quad (4.8)$$

These matrices are operated upon by the acceleration, velocity, and displacement vectors given by

$$\ddot{\mathbf{y}}=\begin{vmatrix} \ddot{y}_1 \\ \ddot{y}_2 \\ \\ \ddot{y}_{n-1} \\ \ddot{y}_n \end{vmatrix} = \text{acceleration vector, } \dot{\mathbf{y}}=\begin{vmatrix} \dot{y}_1 \\ \dot{y}_2 \\ \cdot \\ \dot{y}_{n-1} \\ \dot{y}_n \end{vmatrix} = \text{velocity vector, and } \mathbf{y}=\begin{vmatrix} y_1 \\ y_2 \\ \cdot \\ y_{n-1} \\ y_n \end{vmatrix}. \quad (4.9)$$

The force vector is $$\mathbf{F}=\begin{vmatrix} F_1 \\ F_2 \\ \\ F_{n-1} \\ F_N \end{vmatrix} \quad (4.10)$$

The system of equations given by Equations 4.6 can be written concisely as

$$[m]\ddot{\mathbf{y}}+[f]\dot{\mathbf{y}}+[k]\mathbf{y}=\boldsymbol{F} \quad (4.11)$$

For an undamped system, the damping matrix is zero. Thus the equations for undamped systems can be written as

$$[m]\ddot{\mathbf{y}}+[k]\mathbf{y}=\boldsymbol{F}. \quad (4.12)$$

For free-damped vibration the equations are

$$[m]\ddot{\mathbf{y}}+[f]\dot{\mathbf{y}}+[k]\mathbf{y}=\mathbf{0}. \quad (4.13)$$

For free vibrations of undamped system, we have

$$[m]\ddot{\mathbf{y}}+[k]\mathbf{y}=0 \quad (4.14)$$

The spring–mass–damper system described here is a particular case of general system, the matrices of which are as follows.

$$[m] = \begin{bmatrix} m_{11} & m_{12} & m_{13} & . & m_{1n} \\ m_{21} & m_{22} & m_{23} & . & m_{2n} \\ . & . & . & . & . \\ . & . & . & . & . \\ m_{n1} & m_{n2} & m_{n3} & . & m_{nn} \end{bmatrix}, \tag{4.15}$$

$$[f] = \begin{bmatrix} f_{11} & f_{12} & . & . & f_{1n} \\ . & . & . & . & f_{2n} \\ . & . & . & . & f_{3n} \\ . & . & . & . & . \\ f_{n1} & : & . & f_{n-1} & f_{nn} \end{bmatrix}, \tag{4.16}$$

$$[k] = \begin{bmatrix} k_{11} & . & . & . & k_{1n} \\ k_{21} & . & . & . & k_{2n} \\ . & . & . & . & . \\ . & . & . & . & . \\ k_{n1} & . & . & . & k_{nn} \end{bmatrix} \tag{4.17}$$

The differential Equations 4.6 (of spring, mass, and damper system) can be seen to be coupled, which means that each equation involves more than one coordinate. This means that these equations cannot be solved individually one at a time, but can only be solved simultaneously. In addition, the system can be seen to be statically coupled since stiffnesses are coupled. This means that the stiffness matrix has at least one non-zero off diagonal term. On the other hand, if the mass matrix has at least one off-diagonal term non-zero, the system is said to be dynamically coupled. Further, if both stiffness and mass matrices have non-zero off-diagonal terms, the system is said to be both statically and dynamically coupled.

4.3.2 Influence Coefficients

The equations of motion of a multi degrees-of-freedom system can also be written in terms of influence coefficients. The concept of influence coefficients is extensively used in structural engineering. Each of the matrices involved in the equations of motion are associated with one set of influence coefficients. The influence coefficients associated with the stiffness and mass are respectively known as stiffness and inertia influence coefficients. As will be explained later, at times, it is more convenient to rewrite the equations of motion using inverse of stiffness matrix (called flexibility matrix since stiffness = force/displacement. Therefore, flexibility = displacement/force or displacement per unit force) or inverse of mass matrix. The influence coefficients corresponding to the inverse-stiffness matrix are called flexibility-influence coefficients and those corresponding to inverse-mass matrix are called inverse-inertia coefficients.

Stiffness-Influence Coefficients

As mentioned here, for a linear spring, the force necessary to cause unit elongation is the stiffness of the spring. In a more complex system, we can express the relation between the displacement at a point and the forces acting at various other points of the system by means of stiffness-influence coefficients. Thus, the stiffness-influence coefficient denoted by k_{ij} is defined as the force at point i due to unit displacement at point j when all the points other than the point j are fixed (zero displacement). Thus for the system shown in Fig. 4.4, the total force at point i is the sum of all forces due to all displacements y_j ($j = 1, 2,\ldots, n$), that is,

$$F_i = \Sigma_{j=1}^{n} k_{ij} y_j,\ \mathrm{i} = 1, 2, 3,\ldots, n \tag{4.18}$$

Thus, Equation 4.18 can also be written as

$$\{F\} = \begin{bmatrix} F_1 \\ F_2 \\ \vdots \end{bmatrix} = \begin{pmatrix} k_{11} & k_{12}\cdots & k_{1n} \\ k_{21} & \ldots\ldots & k_{2n} \\ . & . & \cdots \end{pmatrix} \begin{bmatrix} y_1 \\ y_2 \\ . \\ y_n \end{bmatrix} \tag{4.19}$$

or simply,

$$\{\mathbf{F}\} = [\mathbf{k}]\mathbf{y} \tag{4.20}$$

where $[k]$ is given by

$$[k] = \begin{pmatrix} k_{11} & k_{12}\cdots & k_{1n} \\ k_{21} & \ldots\ldots & k_{2n} \end{pmatrix} \tag{4.21}$$

We shall now illustrate the concept of stiffness-influence coefficient. Consider three degrees-of-freedom system comprising of three springs and three masses as shown in Fig. 4.5(a).

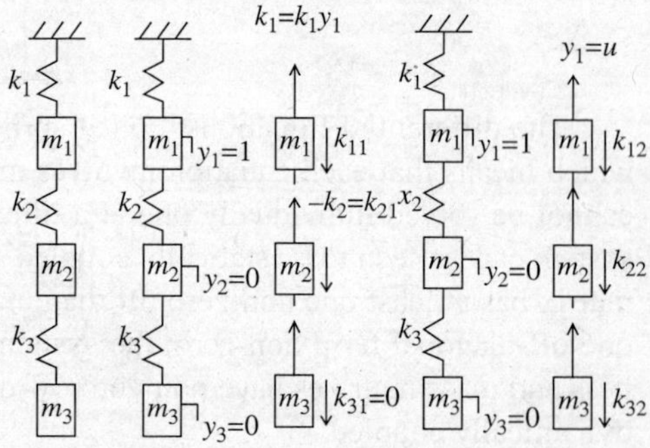

Figure 4.5 Stiffness-influence Coefficient

(i) Unit displacement y_1, y_2, and $y_3 = 0$.
Equilibrium of mass m_1 gives

$$k_1 = -k_2 + k_{11} \tag{A}$$

Equilibrium of m_2 gives

$$k_{21} = -k_2 \tag{B}$$

Equilibrium of m_3 gives $k_{31} = 0$ (C)

From the above three relations, we get

$$k_{11} = k_1 + k_2,$$
$$k_{21} = -k_2 \text{ and}$$
$$k_{31} = 0$$

(ii) Unit displacement $y_2 = 1, y_1 = y_3 = 0$
Equilibrium of mass m_1 gives

$$k_{12} + k_2 = 0 \tag{A1}$$

Equilibrium of m_2 gives

$$k_{22} - k_3 = k_2 \tag{B1}$$

Equilibrium of m_3 gives

$$k_{32} = -k_3 \tag{C1}$$

This gives $k_{12} = -k_2$, $k_{22} = k_2 + k_3$, $k_{32} = -k_3$

Likewise, we may now give $y_3 = 1$, $y_1 = y_2 = 0$. Following the same procedure, we get $k_{13} = 0$, $k_{23} = -k_3$ and $k_{33} = k_3$. Thus, the stiffness matrix is given by

$$[k] = \begin{pmatrix} (k_1 + k_2) & -k_2 & 0 \\ -k_2 & (k_2 + k_3) & -k_3 \\ 0 & -k_3 & k_3 \end{pmatrix} \tag{4.22}$$

Flexibility-Influence Coefficients

It is clear that computation of stiffness-influence coefficients requires application of principles of Statics. In fact, the generation of n stiffness coefficients $k_{1j}, k_{2j} \ldots k_{nj}$ for any specific j requires n simultaneous linear equations. Thus n sets of linear equations (n equations in each set) are to be solved to generate all the stiffness influence coefficients of an n degrees-of-freedom system. This, depending upon the size of the system, at times, involves a significant computational effort depending upon the value of n. The other option that is flexibility-influence coefficients is fairly simple and more convenient. For understanding the concept of flexibility coefficient, let us consider the spring–mass system shown in Fig. 4.6.

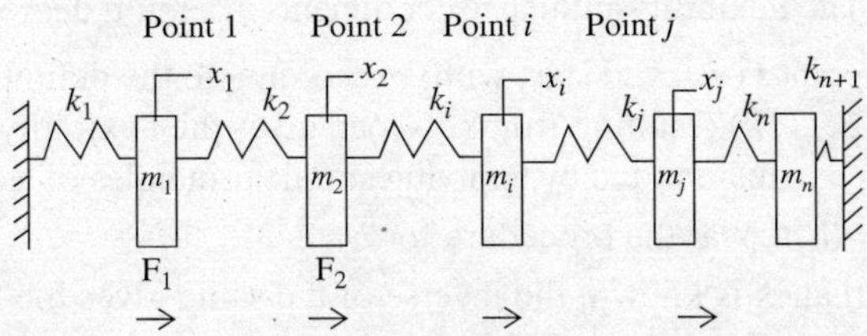

Figure 4.6 Flexibility-influence Coefficients

Let the system be acted upon by just one force F_j and let the displacement at point i (mass m_i) due to F_j be x_{ij}. The flexibility-influence coefficient denoted by a_{ij} is defined as the deflection at point i due to unit load at point j. Since the deflection varies linearly with load/force for linear springs, we have

$$x_{ij} = a_{ij} F_j \tag{4.23}$$

If, on the other hand, several forces F_j ($j = 1, 2, \ldots, n$) act at different points of the system, the total deflection at point i can be found by summing up the contribution of all forces F_j

$$x_i = \sum_{j=1}^{n} x_{ij} = \sum_{j=1}^{n} a_{ij} F_j \; (i = 1, 2, 3, \ldots, n) \tag{4.24}$$

Equation 4.24 can be written as

$$\{\mathbf{x}\} = [a]\,\{\mathbf{F}\} \tag{4.25}$$

where $\{\mathbf{x}\}$ and $\{\mathbf{F}\}$ are the displacement and force vectors defined in Equations 4.9 and 4.10 and $[a]$ is the flexibility matrix given by

$$a = \begin{pmatrix} a_{11} & a_{12} & a_{1n} \\ a_{21} & a_{22} & a_{2n} \\ \ldots & \ldots & \ldots \end{pmatrix} \tag{4.26}$$

The flexibility-influence coefficient matrix has the following important characteristics.

- Examination of Equations 4.25 and 4.20 reveals that the flexibility and stiffness matrices are related. If we substitute Equation 4.20 into Equation 4.25, we get,

$$\{\mathbf{x}\} = [a][k]\{\mathbf{x}\} \tag{4.27}$$

From this we obtain

$$[a][k] = [I] \tag{4.28}$$

In this equation, $[I]$ is a unit matrix. In other words, Equation 4.28 can also be written as

$$[k] = [a]^{-1}[I] = [a]^{-1} \quad \text{or} \quad [a] = [k]^{-1} \tag{4.29}$$

This means that stiffness and flexibility matrices are inverses of each other.

- Since the deflection at point i due to unit load at point j is the same as the deflection at point j due to the unit load at i, for a linear system (Maxwell's reciprocal theorem), $a_{ij} = a_{ji}$.
- The flexibility-influence coefficients of a torsional system can be defined in terms of unit torque and angular deflection it causes. In a multi-rotor torsional system, a_{ij} can be defined as an angular deflection of point i due to unit torque at point j (rotor j). We shall consider the torsional vibrations in a separate chapter later.

The steps required to evaluate flexibility-influence coefficients in a multi degrees-of-freedom system are:

1. Assume unit load at point j ($j = 1$ to start with). According to the definition, the deflection at various points i ($i = 1, 2, 3, \ldots, n$) resulting from this load gives the flexibility-influence coefficients a_{ij}, $i = 1, 2, 3, \ldots, n$. Thus a_{ij} can be found by applying simple principles of statics and solid mechanics.
2. After finding a_{ij} at $j = 1$, repeat the procedure for $j = 2, 3, \ldots, n$
3. In case the stiffness matrix is known, the inverse of the same gives the flexibility-influence coefficient matrix.

To illustrate these concepts, let us analyse the system shown in Fig. 4.7

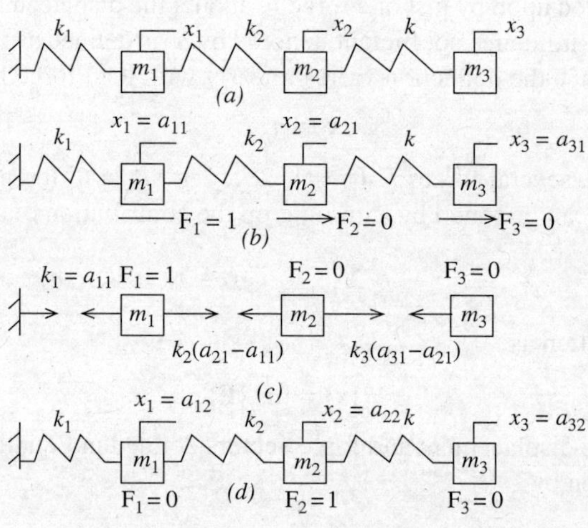

Figure 4.7(a–d) Determination of Flexibility-influence Coefficients

STEP 1

If x_1, x_2, and x_3 denote displacements of masses m_1, m_2 and m_3 [Fig. 4.7(b)], the loads are $F_1 = 1$ and $F_2 = F_3 = 0$.

Mass m_1 displaces by a_{11}
Mass m_2 displaces by a_{21}
Mass m_3 displaces by a_{31}

STEP 2

Free-body diagram for masses m_1, m_2, and m_3 [Fig. 4.7(c)],

Mass m_1: $k_1\, a_{11} = k_2(a_{21} - a_{11}) + 1$
Mass m_2: $k_2(a_{21} - a_{11}) = k_3(a_{31} - a_{21})$

Mass m_3: $k_3(a_{31} - a_{21}) = 0$

Solution of above equations gives

$a_{11} = 1/k_1$, $a_{21} = 1/k_1$, and $a_{31} = 1/k_1$

STEP 3

Apply unit force $F_2 = 1, F_2 = F_3 = 0$ [Fig. 4.7(d)]

Mass m_1 displaces by a_{12}

Mass m_1 displaces by a_{11}

Mass m_2 displaces by a_{22}

Mass m_3 displaces by a_{32}

STEP 4

Free-body diagram for masses m_1, m_2, and m_3 [Fig. 4.7(e)]

Mass m_1: $k_1 a_{12} = k_2(a_{22} - a_{12})$

Mass m_2: $k_2(a_{22} - a_{12}) = k_3(a_{32} - a_{22}) + 1$

Mass m_3: $k_3(a_{32} - a_{22}) = 0$

Solution of above equations gives

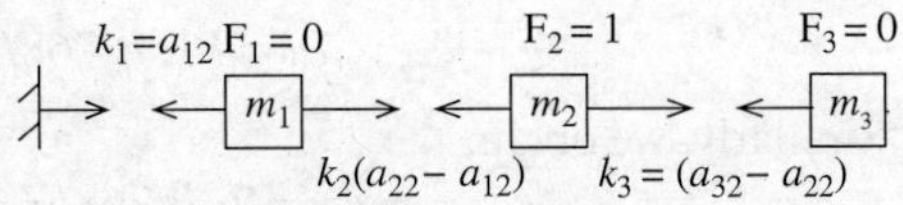

Figure 4.7(e)

$a_{12} = 1/k_1$, $a_{22} = 1/k_1 + 1/k_2$ and

$a_{32} = 1/k_1 + 1/k_2$

Similarly, we can show

$a_{13} = 1/k_1$, $a_{23} = 1/k_1 + 1/k_2$ and $a_{33} = 1/k_1 + 1/k_2 + 1/k_3$

Thus, the flexibility-influence coefficient matrix is

$$[a_{ij}] = \begin{pmatrix} 1/k_1, & 1/k_1, & 1/k_1 \\ 1/k_1, & 1/k_1 + 1/k_2, & 1/k_1 + 1/k_2 \\ 1/k_1 & 1/k_1 + 1/k_2 & 1/k_1 + 1/k_2 + 1/k_3, \end{pmatrix} = [a],$$

$$[k] = \begin{pmatrix} k_1 + k_2 & -k_2 & 0 \\ -k_2 & k_2 + k_3, & -k_3 \\ 0 & -k_3, & k_3 \end{pmatrix}$$

Thus, we can verify that $[k] = [a]^{-1}$.

As an another example, let us consider a massless rotor with uniform stiffness EI, carrying three discs with masses M_1, M_2, and M_3. The rotor is supported at both ends on bearings as shown in Fig. 4.8.

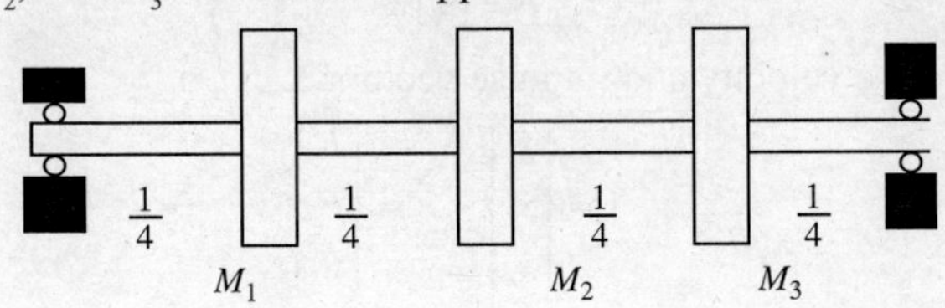

Figure 4.8 Three-disc Rotor

Let us assume the rotor shaft assembly as a pinned–pinned massless beam with equally spaced masses M_1, M_2, and M_3. Let $X_1(t)$, $X_2(t)$, and $X_3(t)$ be the displacements under the masses M_1, M_2, and M_3. In order to find the influence coefficients a_{ij} ($j = 1, 2, 3$), consider unit load at the location of mass M_1 and zero load at M_2 and M_3 locations (Fig. 4.9). Using principles of strength of materials

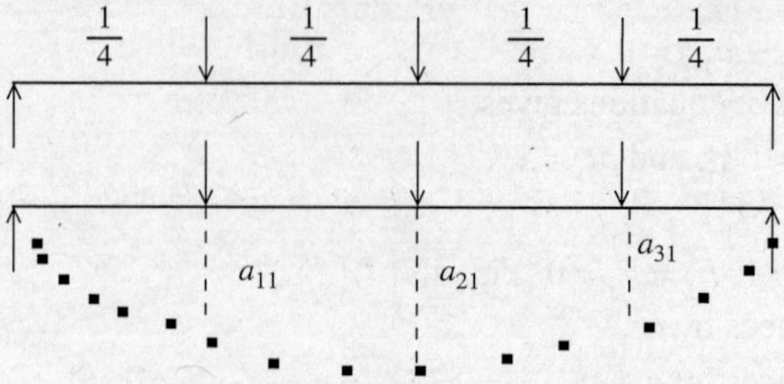

Figure 4.9 Beam Deflection

$$a_{11} = 9.\, l^3/768EI,\ a_{12} = 11.\, l^3/768EI \text{ and } a_{13} = 7.\, l^3/768EI \quad \text{(a)}$$

Similarly, we can now apply unit load at the location of M_2 and zero loads at M_1 and M_3 locations. We obtain,

$$a_{12} = a_{21} = 11.\, l^3/768EI,\ a_{22} = l^3/48EI,\ a_{32} = a_{23} = 11.\, l^3/768EI \quad \text{(b)}$$

Similarly, we obtain

$$a_{31} = a_{13} = 7.\, l^3/768EI,\ a_{32} = a_{23} = 11.\, l^3/768EI,\ a_{33} = 9.\, l^3/768EI \quad \text{(c)}$$

We thus obtain the flexibility matrix as

$$[a] = \frac{l^3}{768EI}\begin{bmatrix} 9 & 11 & 7 \\ 11 & 16 & 11 \\ 7 & 11 & 9 \end{bmatrix} \quad (\text{since } a_{22} = l^3/48EI = 16l^3/768EI)$$

Inertia-Influence Coefficients

The elements of mass matrix m_{ij} are known as inertia-influence coefficients. The inertia-influence coefficients can be derived using expression of kinetic energy of the system as we shall see later. They can also be found out using impulse-momentum principle. In this, the inertia-influence coefficients $m_{1j}, m_{2j}, \ldots, m_{nj}$ are defined as the set of impulses applied at points 1, 2, 3,…, n, respectively, to produce a unit velocity at point j and zero velocity at every other point (i.e. $\dot{x}_j = 1, \dot{x}_1 = \dot{x}_2 = \dot{x}_3 = \dot{x}_{j-1} = 0$). Thus for a multi degrees-of-freedom system, the total impulse at point i can be found by summing up the impulses causing velocities $\dot{x}_j$ ($j = 1, 2, 3, \ldots, n$) as

$$\vec{F} = \sum_{j=1}^{n} m_{ij}\dot{x}_j \quad (4.30)$$

Equation 4.30 can be written in matrix form

$$\vec{F} = [m]\dot{\vec{x}} \quad (4.31)$$

In this equation, $\dot{\vec{x}}$ and $\vec{F}$ are velocity and impulse vectors.

$$\dot{\vec{x}} = \begin{Bmatrix} \dot{x}_1 \\ \dot{x}_2 \\ \dot{x}_n \end{Bmatrix},\ \vec{F} = \begin{Bmatrix} F_1 \\ F_2 \\ F_n \end{Bmatrix} \quad (4.32)$$

And the mass matrix $[m]$ is given by

$$[m] = \begin{bmatrix} m_{11} & m_{12} & m_{13} & . & m_{1n} \\ m_{21} & . & . & . & m_{2n} \\ . & . & . & . & . \\ . & . & . & . & . \\ m_{n1} & . & . & . & m_{nn} \end{bmatrix} \quad (4.33)$$

The inertia-influence coefficients are symmetric for a linear system which means that $m_{ij} = m_{ji}$.

Procedure to derive inertia-influence coefficients is outlined as follows.

- Assume that a set of impulses are applied at various points i (= 1, 2, 3,…, n) so as to produce a unit velocity at point j ($x_j = 1$ with $j = 1$ to start with). By definition, the set of impulses ($i = 1,2,\ldots, n$) denote the inertia-influence coefficients m_{ij} ($i = 1,2,\ldots, n$).
- After completing the above step for $j = 1$, repeat the procedure for $j = 2,3,\ldots, n$.

It may be noted that if x_j denotes angular coordinate, then $\dot{x}_j$ represents angular velocity and indicates angular impulse. We shall illustrate these steps through following example.

Problem 4.1

For the system shown in Fig. 4.10, find the inertia-influence coefficients and also derive the Equation of motion of the pendulum.

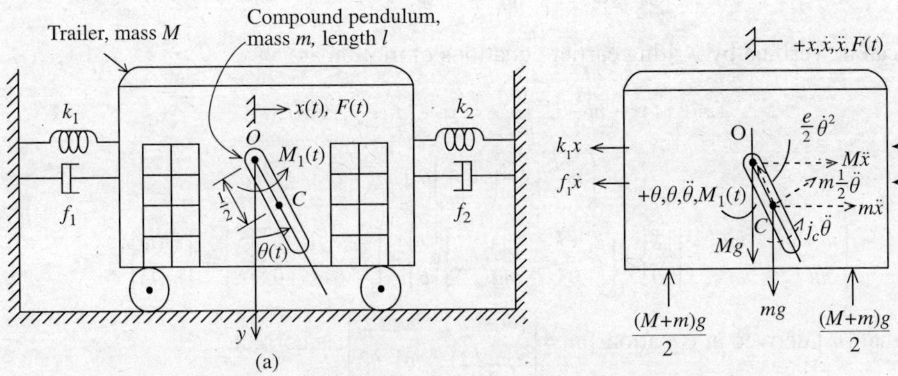

Figure 4.10 Free-body Diagram for Trailer and Compound Pendulum

Solution:

Inertia forces are: (1) $M\ddot{x}$ of trailer. (2) $m\ddot{x}$ of pendulum, (3) $m\frac{l}{2}\ddot{\theta}\cos\theta$—component of tangential acceleration (caused by angular motion of pendulum in x direction), (4) $m\frac{l}{2}\dot{\theta}^2\sin\theta$—component of centripetal force in x direction.

Spring forces: (1) k_1x, (2) k_2x.

Damping forces: (1) $f_1\dot{x}$, (2) $f_2\dot{x}$.

Thus, equation of motion (translational) is

$$\text{Applied force } F(t) = M\ddot{x} + m\ddot{x} + m\frac{l}{2}\ddot{\theta}\cos\theta - m\frac{l}{2}\dot{\theta}^2\sin\theta + k_1x + k_2x + f_1\dot{x} + f_2\dot{x} = F(t)$$

Since θ is small, cos $\theta \approx 1$, sin $\theta \sim 0$ and term $\dot{\theta}^2\sin\theta$ is small. Thus, equation of motion becomes

$$(m+M)\ddot{x} + m\frac{l}{2}\ddot{\theta} + (k_1+k_2)x + (f_1+f_2)\dot{x} = F(t).$$

Similarly, the equation of rotational motion is

$$\left(m\frac{l}{2}\ddot{\theta}\right)\frac{l}{2} + \left(\frac{ml^2}{12}\right)\ddot{\theta} + m\ddot{x}\frac{l}{2}\cos\theta = -mg\frac{l}{2}\sin\theta + M(t).$$

Since θ is small, sin $\theta \sim 0$ and cos $\theta \sim 1$.

Thus, we get,

$$\left(m\frac{l}{2}\right)\ddot{x}+\left(\frac{ml^2}{4}+\frac{ml^2}{12}\right)\ddot{\theta}+mg\frac{l}{2}\theta=M(t)=\left(m\frac{l}{2}\right)\ddot{x}+\left(\frac{ml^2}{3}\right)\ddot{\theta}+mg\frac{l}{2}\theta$$

The last two equations are the equations of motion.

To derive the inertia-influence coefficients, impulses of magnitude m_{11} and m_{21} are applied in the direction $x(t)$ and $\theta(t)$ to result in velocities $\dot{x}=1$, and $\dot{\theta}=0$. The linear impulse–linear momentum equation gives m_{11} = $(m + M)$ and linear-angular impulse and linear-angular momentum gives $m_{21} = (ml/2)$.

Next impulses of magnitudes m_{12} and m_{22} are applied along $x(t)$ and $\theta(t)$ to obtain velocities $\dot{x}=0$, and $\dot{\theta}=1$. Then the impulse-momentum relations give $m_{12} = m\frac{l}{2}$ and angular-impulse momentum equation gives $m_{22} = (ml^2/3)(l)$. Thus, the mass matrix is given by

$$[m]=\begin{bmatrix}(m+M) & ml/2\\ ml/2 & ml^2/3\end{bmatrix}$$

This is true can be verified by writing earlier equations of motion as

$$(m+M)\ddot{x}+m\frac{l}{2}\ddot{\theta}+(k_1+k_2)x+(f_1+f_2)\dot{x}=F(t)$$

$$\left(m\frac{l}{2}\right)\ddot{x}+\left(\frac{ml^2}{3}\right)\ddot{\theta}+mg\frac{l}{2}\theta\ =M(t)$$

$$\begin{bmatrix}m+M & ml/2\\ ml/2 & ml^2/3\end{bmatrix}\begin{Bmatrix}\ddot{x}\\ \ddot{\theta}\end{Bmatrix}+\begin{bmatrix}k_1+k_2 & 0\\ 0 & mgl/2\end{bmatrix}\begin{Bmatrix}x\\ \theta\end{Bmatrix}+\begin{bmatrix}f_1+f_2 & 0\\ 0 & 0\end{bmatrix}\begin{Bmatrix}x\\ \theta\end{Bmatrix}=\begin{Bmatrix}F(t)\\ m(t)\end{Bmatrix}$$

This shows that $[m]$ derived at equation $[m]=\begin{bmatrix}(m+M) & ml/2\\ (ml/2) & ml^2/3\end{bmatrix}$ is correct.

4.4 GENERALIZED COORDINATES

The equations of motion of a vibratory system can be formulated in a number of (ways) different coordinate systems. 'n' independent coordinates are required to describe the motion the system having 'n' degrees-of-freedom. Any set of 'n' independent coordinates is called the generalized coordinates usually designated by q_i (i = 1, 2, 3,..., n). These coordinates may be length, angle or any other set of numbers that defines the configuration of the system at any time uniquely. They are also independent of the coordinates of the constraints. Consider for example, a triple pendulum as shown in Fig. 4.11.

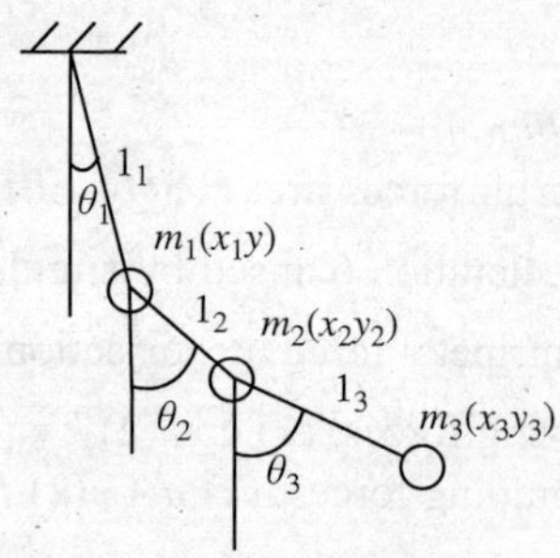

Figure 4.11 Triple Pendulum Generalized Coordination

At any instant of time t, the coordinates of masses m_1, m_2, and m_3 are (x_1, y_1), (x_2, y_2), and (x_3, y_3), respectively. Thus six coordinates are required and one may think that the system is six degrees-of-freedom system. One, however, should remember that these coordinates are not independent but are constrained by relations

$$\begin{aligned}x_1^2+y_1^2&=l_1^2\\(x_2-x_1)^2+(y_2-y_1)^2&=l_2^2\\(x_3-x_2)^2+(y_3-y_2)^2&=l_3^2\end{aligned}\tag{4.34}$$

Thus, since the coordinates (x_i, y_i) are not independent as they are related by constraint Equations 4.34, they cannot be called generalized coordinates. In fact, in absence of constraints expressed by

Equations 4.34, the masses m_1, m_2, and m_3 are free to occupy any position in the *x-y* plane. On the other hand, if angular displacement θ_i are used to specify the locations of masses m_i ($i = 1, 2, 3$) at any instant of time, there will be no constraints on θ_i. θ_i, thus form as a set of generalized coordinates q_i ($i = 1, 2, 3$) and are usually denoted as $q_j = \theta_j$ ($j = 1, 2,\ldots$).

The configuration of the system changes when external forces act on the system. The new configuration of the system can be obtained by changing the generalized coordinates q_j by δq_j ($j = 1, 2, 3,\ldots, n$) where n denotes the number of generalized coordinates (degrees-of-freedom) of the system. If U_j denotes the work done in changing the generalized coordinates q_j by δq_j ($j = 1, 2, 3,\ldots, n$), the corresponding generalized force Q_j can be defined as

$$Q_j = \frac{U_j}{\delta q_j}, j = 1,2,3,\ldots n \tag{4.35}$$

In this equation, Q_j will be force when q_j is linear and will be moment when q_j is angular displacement.

4.5 ENERGIES IN VIBRATING SYSTEMS

In the previous chapter, we discussed about the potential and kinetic energies of two degrees-of-freedom system. We also discussed about the use of Lagrange's equations to derive the equations of motion. We shall now extend these concepts to multi degrees-of-freedom systems.

Let x_i denote the displacement of mass m_i and F_i be the force applied in the direction of x_i at mass m_i in an 'n' degrees-of-freedom system similar to the one shown in Fig. 4.4. The elastic potential energy or strain energy of the ith spring is given by

$$V_i = \frac{1}{2}F_i x_i \tag{4.36}$$

The total potential/strain energy is given by

$$V = \sum_{i=1}^{n} V_i = \frac{1}{2}\sum_{i=1}^{n} F_i x_i \tag{4.37}$$

Since

$$F_i = \sum_{j=1}^{n} k_{ij} x_j \tag{4.38}$$

Equation 4.37 may be written as

$$V = \frac{1}{2}\sum_{i=1}^{n}\left(\sum_{j=1}^{n} k_{ij} x_j\right) x_i = \frac{1}{2}\sum_{i=1}^{n}\sum_{j=1}^{n} k_{ij} x_i x_j \tag{4.39}$$

Equation 4.39 can be written in matrix form as

$$V = \frac{1}{2}\bar{X}^T [K] \bar{X} \tag{4.40}$$

In the Equation 4.40, the displacement vector $\bar{X}$ is given by

$$\bar{X} = \begin{Bmatrix} x_1 \\ x_2 \\ \cdot \\ x_n \end{Bmatrix}, \bar{X}^T = (\text{transpose}.\,\bar{X}) = \{x_1 \quad x_2 \quad \cdot \quad x_n\} \tag{4.41}$$

and

$$[K] = \begin{bmatrix} k_{11} & k_{12} & . & k_{1n} \\ & . & & k_{2n} \\ k_{21} & & & \\ . & . & . & . \\ k_{n1} & . & & k_{nn} \end{bmatrix} \tag{4.42}$$

The kinetic energy associated with mass m_i is $T_i = \frac{1}{2} m_i \dot{x}_i^2$.
Thus, the total kinetic energy is given by

$$T = \sum_{i=1}^{n} T_i = \frac{1}{2} \sum m_i \dot{x}_i^2 \tag{4.43}$$

This can also be written in matrix form as

$$T = \frac{1}{2} \bar{\dot{X}}^T [M] \bar{\dot{X}} \tag{4.44}$$

In the above equation, the velocity vector and the mass matrix are given by

$$\bar{\dot{X}} = \begin{Bmatrix} \dot{x}_1 \\ . \\ \dot{x}_n \end{Bmatrix}, \bar{\dot{X}}^T = \text{transpose.} \bar{\dot{X}} = \{\dot{x}_1 \quad . \quad \dot{x}_n\}, [M] = \text{mass.matrix} = \begin{bmatrix} m_1 & .0...... & 0 \\ 0 & m_2..... & 0 \\ 0 & 0.... & m_n \end{bmatrix} \tag{4.45}$$

In the above equations $[M]$ is a diagonal matrix.

If generalized coordinates, q_i (discussed in the previous section), are used instead of the physical displacements x_i; the kinetic energy can be expressed as

$$T = \frac{1}{2} \bar{\dot{q}}^T [m] \bar{\dot{q}} \tag{4.46}$$

In the above equation

$$\bar{\dot{q}} = \begin{Bmatrix} \dot{q}_1 \\ \dot{q}_2 \\ . \\ \dot{q}_n \end{Bmatrix}, \bar{\dot{q}}^T = \{\dot{q}_1 \quad \dot{q}_2 \quad . \quad \dot{q}_n\}, [m] = \begin{bmatrix} m_{11} & m_{12} & . & m_{1n} \\ m_{21} & m_{22} & . & m_{2n} \\ . & . & . & . \\ m_{n1} & . & . & m_{nn} \end{bmatrix} \tag{4.47}$$

With $m_{ij} = m_{ji}$, the generalized mass matrix in Equation 4.47 is a full matrix whereas the mass matrix in Equation 4.45 is a diagonal matrix.

It can be seen that the potential energy is a quadratic function of displacement and kinetic energy is a quadratic function of velocities. Since the kinetic energy can never be negative and can vanish only when all velocities are zero, Equations 4.43 and 4.46 are called positive definite quadratic forms and the mass matrix is called the positive definite matrix. On the other hand, the potential energy expression (Equation 4.40) is a positive definite quadratic form, while the matrix $[K]$ is positive definite only if the system is stable. There are systems for which the potential energy is zero without displacements $x_1, x_2, \ldots, x_n$ being zero. In these cases, the potential energy will be a positive quadratic function rather than positive definite; correspondingly, the matrix $[K]$ is said to be positive. A system for which $[K]$ is positive and $[m]$ positive definite is called semi-definite system as shown in Fig. 4.12.

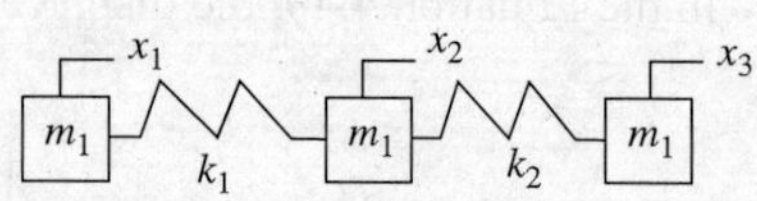

Figure 4.12 Semi-definite System

For the purpose of explanation, let us assume $m_1 = m_2 = m_3 = m$ and $k_1 = k_2 = k_3 = k$. The KE of the system is

$$T = \frac{1}{2}(m_1\dot{x}_1^2 + m_2\dot{x}_2^2 + m_3\dot{x}_3^2) = \frac{1}{2}\bar{\dot{X}}^T[M]\bar{\dot{X}}$$

In the above equation

$$\bar{\dot{X}} = \begin{Bmatrix} \dot{x}_1 \\ \dot{x}_2 \\ \dot{x}_3 \end{Bmatrix} \text{ and } [M] = \begin{bmatrix} m_1 & 0 & 0 \\ 0 & m_2 & 0 \\ 0 & 0 & m_3 \end{bmatrix}$$

The elongations of the springs k_1 and k_2 are $(x_2 - x_1)$ and $(x_3 - x_2)$, so the PE is

$$V = \frac{1}{2}\{k_1(x_2 - x_1)^2 + k_2(x_3 - x_2)^2\} = \frac{1}{2}\bar{X}^T[K]\bar{X}$$

In the above equation

$$[K] = \begin{bmatrix} k_1 & -k_1 & 0 \\ -k_1 & k_1 + k_2 & -k_2 \\ 0 & -k_2 & k_2 \end{bmatrix}$$

It can be verified that matrix $[K]$ is singular. Furthermore, if all the displacements can be same ($x_1 = x_2, = x_3 = x$ which means rigid body motion), the potential energy V can be zero.

We have already dealt with the use of Lagrange's equations for deriving the equations of motion in Chapter 3. We shall now use them for multi degrees-of-freedom systems.

4.5.1 Use of Lagrange's Equation

The Lagrange's equations referred in Chapter 3 can also be written as

$$\frac{d}{dt}\left(\frac{\partial T}{\partial \dot{q}_j}\right) - \frac{\partial T}{\partial q_j} + \frac{\partial V}{\partial q_j} = Q_j^{(n)} \tag{4.48}$$

In this equation $\dot{q}_j = \frac{\partial q_j}{\partial t}$ is generalized velocity and $Q_j^{(n)}$ is the non-conservative generalized force corresponding to the generalized coordinate q_j. The forces represented by $Q_j^{(n)}$ may be dissipative (damping) forces or other external forces that are not derivable from a potential function. For example, if F_{xk}, F_{yk}, and F_{zk} represent the external forces in the x, y, and z direction acting on the kth mass of the system, then the generalized force $Q_j^{(n)}$ can be computed as

$$Q_j^{(n)} = \sum_k \left(F_{xk}\frac{\partial x_k}{\partial q_j} + F_{yk}\frac{\partial y_k}{\partial q_j} + F_{zk}\frac{\partial z_k}{\partial q_j}\right) \tag{4.49}$$

In the above equation, x_k, y_k, and z_k are the displacements of the kth mass. For conservative systems $Q_j^{(n)} = 0$ so that Equation 4.48 becomes

$$\frac{d}{dt}\left(\frac{\partial T}{\partial \dot{q}_j}\right) - \frac{\partial T}{\partial q_j} + \frac{\partial V}{\partial q_j} = 0 \tag{4.50}$$

Equations 4.48 or 4.50 represent a system of n differential equations. Thus, the equations of motion can be derived if energy expressions are available. This is done as follows.

$$q_j = \frac{\partial q_j}{\partial t} \tag{4.51}$$

$$\frac{d}{dt}\left(\frac{\partial T}{\partial \dot{x}_i}\right) - \frac{\partial T}{\partial x_i} + \frac{\partial V}{\partial x_i} = F_i, i = 1,2,3\ldots n \tag{4.52}$$

In the Equation 4.52, F_i is the non-conservative generalized force corresponding to the ith generalized coordinate x_i, and $\dot{x}_i$ is the generalized velocity. The KE and PE of the system are given, as shown earlier by

$$T = \frac{1}{2}\bar{\dot{x}}^T[m]\bar{\dot{x}} \tag{4.53}$$

$$V = \frac{1}{2}\bar{x}^T[k]\bar{x} \tag{4.54}$$

The various quantities in this equation are

$$\bar{x} = \begin{Bmatrix} x_1 \\ x_2 \\ \cdot \\ x_n \end{Bmatrix}, \bar{\dot{x}} = \begin{Bmatrix} \dot{x}_1 \\ \dot{x}_2 \\ \cdot \\ \dot{x}_n \end{Bmatrix} \tag{4.55}$$

Since the matrix $[m]$ is symmetric, from the theory of matrices, we obtain,

$$\frac{\partial T}{\partial \dot{x}_1} = \frac{1}{2}\bar{\delta}^T[m]\bar{\dot{x}} + \frac{1}{2}\bar{\dot{x}}^T[m]\bar{\delta}^T = \bar{\delta}^T[m]\bar{\dot{x}} = \bar{m}_i^T\bar{\dot{x}}_i, \quad i = 1, 2,\ldots, n \tag{4.56}$$

where δ_{ij} is the Kronecker delta ($\delta_{ij} = 1$ if $i = j$ and zero when $i \neq j$), $\bar{\delta}$ is a column vector $\begin{Bmatrix} 1 \\ 0 \end{Bmatrix}$, that is, when $i = j$, $\delta_{ij} = 1$ and 0 when $i \neq j$, and $\bar{m}_i^T$ is a row vector $(m_1, m_2,\ldots, m_n)$, which is identical to the ith row of the matrix $[m]$. All the relations expressed in Equation 4.56 can be expressed as

$$\frac{\partial T}{\partial \dot{x}_1} = \bar{m}_i^T\bar{\dot{x}} \tag{4.57}$$

Differentiation of Equation 4.57 with respect to time gives

$$\frac{d}{dt}\left(\frac{\partial T}{\partial \dot{x}_1}\right) = \bar{m}_i^T\ddot{x} \quad i = 1, 2,\ldots, n \tag{4.58}$$

Also, KE is a function of velocity $\dot{x}$, hence $\dfrac{\partial T}{\partial x_i} = 0$ (4.59)

Similarly, we can differentiate Equation 4.54 and noting that $[k]$ is symmetric,

$$\frac{\partial V}{\partial x_i} = \frac{1}{2}\bar{\delta}^T[k]\bar{x} + \frac{1}{2}\bar{x}^T[k]\bar{\delta} = \bar{\delta}^T[k]\bar{x} = \bar{k}_i^T\bar{x} \ \ i = 1, 2,\ldots, n \tag{4.60}$$

In the Equation 4.60, $\bar{k}_i^T$ is a row vector identical to the ith row of $[k]$.
By substituting equations 4.58 to 4.60 in Equation 4.52, we get the desired equation

$$[m]\bar{\ddot{x}} + [k]\bar{x} = \bar{F} \tag{4.61}$$

The various quantities in the Equation 4.61 are $\bar{\ddot{x}} = \begin{Bmatrix} \ddot{x}_1 \\ \ddot{x}_2 \\ \cdot \\ \cdot \\ \ddot{x}_n \end{Bmatrix}, \bar{x} = \begin{Bmatrix} x_1 \\ x_2 \\ \cdot \\ \cdot \\ x_n \end{Bmatrix}$

We have thus demonstrated that Lagrange's equations can be used to derive equations of motion of multi degrees-of-freedom vibration systems.

The problem that we engineers are interested in, is to find the natural frequencies, the mode shape (deflected shape) in each of the natural frequencies and lastly, the response of the n degrees-of-freedom to various perturbation forces, acting alone or simultaneously. The first part of the problem just mentioned is the determination of natural frequencies, and the mode shapes in each of these natural frequencies is of utmost importance to vibration engineers. We shall initially concentrate upon only the undamped systems, as in many cases, the influence of damping on the natural frequency is to lower the natural frequency by an amount of the order $\sqrt{(1-\zeta^2}$ [*i.e.* $\omega_d = \omega_n\sqrt{(1-\zeta^2)}$], which in many practical problems is insignificant due to low damping. Also, we need to identify the safe zone of the frequency of perturbation force so that resonance-related failure problems are not encountered. It is not unusual to keep the natural frequency(ies) 10–20% away from the excitation frequency(ies). However, when near-resonance operation is unavoidable either during transient or steady-state operation, the effect of damping must be evaluated.

The problem of determination of the natural frequencies and the associated mode shapes is popularly known as the **eigen value problem** and is of great interest to vibration engineers.

4.6 EIGEN VALUE PROBLEM

Equation 4.61 corresponds to n degrees-of-freedom vibration system with no damping and thus is the equation for undamped free vibration of the system.

In this case, when the system is disturbed from its condition of rest by some energy say in the form of initial displacement or velocities or both, it vibrates indefinitely as there is no damping and the associated dissipation of energy.

Assume the solution of the Equation 4.61 of the form

$$x_i(t) = X_i T(t), \quad 1, 2, \ldots n \tag{4.62}$$

where X_i is a constant and T is a function of time t. Equation 4.62 shows that the amplitude ratio of the two coordinates

$$\left\{\frac{x_i(t)}{x_j(t)}\right\} = \text{constant}$$

Physically this means that all coordinates have synchronous motions. The configuration of the system does not change its shape but its amplitude does. The configuration of the system, given by the vector

$\bar{X} = \begin{Bmatrix} X_1 \\ X_2 \\ \cdot \\ \cdot \\ X_n \end{Bmatrix}$, is called the mode shape of the system. Substituting Equation 4.62 in Equation 4.61, we obtain

$$[m]\bar{X}\ddot{T}(t) + [k]\bar{X}\,T(t) = 0 \tag{4.63}$$

Equation 4.63 can be written in scalar form as n separate equations

$$\left(\sum_{j=1}^{n} m_{ij}X_j\right)\ddot{T}(t) + \left(\sum_{j=1}^{n} k_{ij}X_j\right)T(t) = 0 \tag{4.64}$$

From this equation, we obtain

$$-\frac{\ddot{T}(t)}{T(t)} = \frac{\left(\sum_{j=1}^{n} k_{ij}X_j\right)}{\left(\sum_{j=1}^{n} m_{ij}X_j\right)}, i = 1, 2 \ldots n \tag{4.65}$$

Closer examination of Equation 4.65 shows that left side of Equation 4.65 is independent of the index i and the right side is independent of t and both sides must be equal to some constant. By assuming this constant as ω^2, we can write Equation 4.65 as

$$\ddot{T}(t) + \omega^2 T(t) = 0$$

$$\sum_{j=1}^{n} (k_{ij} - \omega^2 m_{ij}) X_j = 0, \quad i = 1,2,\dots n \tag{4.66}$$

or

$$[[k] - \omega^2 [m]]\overline{X} = 0 \tag{4.67}$$

The solution of Equation 4.66 can be expressed as

$$T(t) = C_1 \cos(\omega . t + \phi) \tag{4.68}$$

In this, C_1 and Φ are constants. These are nothing but the amplitude and the phase on undamped natural vibrations. Equation 4.68 also shows that all coordinates perform harmonic motion with the same frequency ω and the same phase angle Φ. Equation 4.67 represents a set of n linear homogeneous (because the right-hand side is zero) equations, which can have a non-trivial solution ($X_i \neq 0$) only when the determinant of

$$\left|[[k] - \omega^2 [m]]\right| = 0$$

or

$$\Delta = \left|k_{ij} - \omega^2 m_{ij}\right| = 0 = \left|[k] - \omega^2 [m]\right| \tag{4.69}$$

Equation 4.67 is known as **eigen value problem or characteristic value problem.** In this, ω^2 is called **eigen value or characteristic value.** Also, the various ω_i's are the natural frequencies of the system.

The expansion of Equation 4.69 leads to the nth-order polynomial equation in ω^2. The roots of the polynomial equation give n values of ω^2. It can be shown that all n roots are real and positive when matrices $[k]$ and $[m]$ are symmetric and positive definite. If $\omega_1^2, \omega_2^2 ......\omega_n^2$ denote n roots in ascending order of magnitude, their positive square roots give n natural frequencies of the system $\omega_1 \leq \omega_2 \leq .... \leq \omega_n$. The lowest value ω_1 is called the fundamental or first natural frequency. In general, all the natural frequencies ω_i are distinct, although in some cases two natural frequencies might possess almost identical or equal values.

Equation 4.67 can also be expressed as

$$[\lambda[k] - [m]]\overline{X} = 0\text{, with } \lambda = 1/\omega^2 \tag{4.70}$$

By pre-multiplying Equation 4.70 by $[k]^{-1}$, we obtain

$$[\lambda[I] - [D]]\overline{X} = \overline{0} \tag{4.71}$$

where, $[I]$ is the identity matrix and $\quad [D] = [k]^{-1}[m]. \quad (4.72)$

$$\therefore \lambda[I]\overline{X} = [D]\overline{X} \tag{4.73}$$

Here, $[D] = [k]^{-1}[m]$ is called the dynamical matrix. The eigen value problem of Equation 4.73 is known as the standard eigen value problem. The non-trivial solution of $\overline{X}$ must satisfy the condition

$$\Delta = \left|\lambda[I] - [D]\right| = 0 \tag{4.74}$$

Expansion of Equation 4.74 yields frequency equation or the characteristic equation. If the degree-of-freedom of the system (n) is very large, the solution of this polynomial equation becomes very tedious due to which suitable numerical methods are required to be used.

We shall illustrate the procedure of finding eigen values by solving a few typical problems.

Problem 4.2

Calculate the natural frequencies of the three degrees-of-freedom system shown in Fig. 4.13.

Solution:

Figure 4.13 shows free-body diagram from which, we obtain

$$4m\ddot{x}_1 = -3kx_1 - k(x_1 - x_2)$$

$$2m\ddot{x}_2 = k(x_1 - x_2) - k(x_2 - x_3)$$

$$m\ddot{x}_3 = k(x_2 - x_3)$$

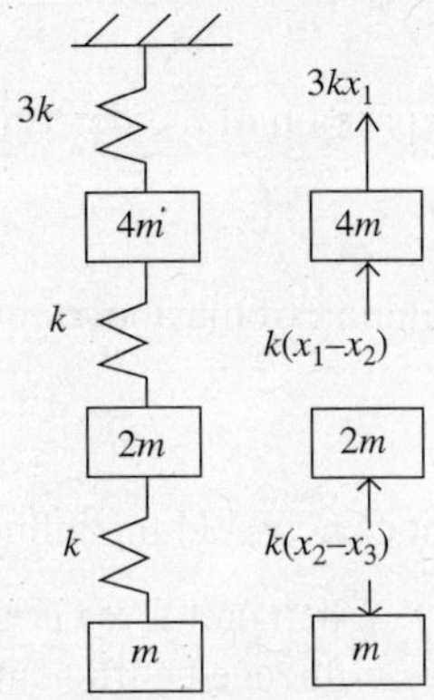

Figure 4.13

These when simplified it gives the following equations:

$$4m\ddot{x}_1 + 4kx_1 - kx_2 = 0$$

$$2m\ddot{x}_2 + 2kx_2 - kx_3 - kx_1 = 0$$

$$m\ddot{x}_3 + kx_3 - kx_2 = 0$$

The above equations can be written in the matrix form as

$$\begin{bmatrix} 4m & 0 & 0 \\ 0 & 2m & 0 \\ 0 & 0 & m \end{bmatrix} \begin{Bmatrix} \ddot{x}_1 \\ \ddot{x}_2 \\ \ddot{x}_3 \end{Bmatrix} + \begin{bmatrix} 4k & -k & 0 \\ -k & 2k & -k \\ 0 & -k & k \end{bmatrix} \begin{Bmatrix} x_1 \\ x_2 \\ x_3 \end{Bmatrix} = \begin{Bmatrix} 0 \\ 0 \\ 0 \end{Bmatrix}$$

We can use Equation 4.74

$\Delta = |\lambda[I] - [D]| = 0$ to yield the frequency equation but that requires computation of $[D] = [k]^{-1}[m]$ as given below

$$[m] = \begin{bmatrix} 4m & 0 & 0 \\ 0 & 2m & 0 \\ 0 & 0 & m \end{bmatrix}, [k] = \begin{bmatrix} 4k & -k & 0 \\ -k & 2k & -k \\ 0 & -k & k \end{bmatrix}$$

We can use a simpler method as shown here.

Assume,

$$x_1 = A\cos(\omega.t + \Psi). \therefore \ddot{x}_1 = -A\omega^2 \cos(\omega.t + \Psi)$$

$$x_2 = B\cos(\omega.t + \Psi). \therefore \ddot{x}_2 = -B\omega^2 \cos(\omega.t + \Psi)$$

$$x_3 = C\cos(\omega.t + \Psi). \therefore \ddot{x}_3 = -C\omega^2 \cos\omega.t + \Psi)$$

Substituting these in equations of equilibrium A and simplifying, we get,

$$(4k - 4m\omega^2)A - kB = 0$$

$$-kA + (2k - 2m\omega^2)B - kC = 0$$

$$-kB + (k - m\omega^2)C = 0$$

The frequency equation is obtained by equating the determinant of Equations B

$$\begin{vmatrix} (4k-4m\omega^2) & -k & 0 \\ -k & (2k-2m\omega^2) & -k \\ 0 & -k & (k-m\omega^2) \end{vmatrix} = 0$$

Expansion of the determinant gives

$$(k-m\omega^2)(8m^2\omega^4 - 16km\omega^2 + 3k^2) = 0$$

Solution of the above equation gives

$$\omega_1 = 0.46\sqrt{k/m},\ \omega_2 = \sqrt{k/m},\ \omega_3 = 1.34\sqrt{k/m}$$

Let us now find the influence coefficients.

(1) When unit force is applied to mass $4m$, the spring of stiffness $3k$ will stretch by $1/3k$, which is the influence coefficient a_{11}. Thus $a_{11} = 1/3k$. When mass $4m$ deflects $a_{11} = 1/3k$ under the action of unit force, the masses $2m$ and m will simply move by the same amount, which means $a_{21} = 1/3k$, $a_{31} = 1/3k$. But $a_{13} = a_{31}$ and $a_{12} = a_{21}$ and both equal to $1/3k$.

(2) Now apply unit load to mass $2m$. The two springs $3k$ and k are in series and their equivalent spring constant is given by $1/3k + 1/k = 1/k_{eq}$ or $k_{eq} = 3k/4$. The deflection is $F/k_{eq} = 1/3k/4 = 4/3k = a_{22}$ and as mass m hangs on mass $2m$, $a_{32} = a_{22} = a_{23} = 4/3k$.

(3) To find a_{33}, apply unit force at mass m. The three springs are in series and their equivalent stiffness is given by

$$1/3k + 1/k + 1/k = 1/k_{eq} = 7/3k \quad \text{or} \quad k_{eq} = 3k/7$$

$a_{33} = F/k_{eq} = 7/3k$. Thus, we have found all the influence coefficients from which we will now formulate the flexibility matrix.

$$a_{11} = 1/3k,\ a_{12} = 1/3k,\ a_{13} = 1/3k$$
$$a_{21} = 1/3k,\ a_{22} = 4/3k,\ a_{23} = 4/3k$$
$$a_{31} = 1/3k,\ a_{32} = 4/3k,\ a_{33} = 7/3k$$

$$[a] = \frac{1}{k}\begin{bmatrix} 1/3 & 1/3 & 1/3 \\ 1/3 & 4/3 & 4/3 \\ 1/3 & 4/3 & 7/3 \end{bmatrix},\ [m] = m\begin{bmatrix} 4 & 0 & 0 \\ 0 & 2 & 0 \\ 0 & 0 & 1 \end{bmatrix}$$

We can verify that $[a] = [k]^{-1}$. This example has been taken with an intention to show the usefulness of flexibility-coefficient matrix and the method of arriving at the influence coefficients. For the given example involving only three masses and three springs, it is not very difficult to find out the inverse of the stiffness matrix but for larger systems the task of finding inverse of the stiffness matrix involves a lot of computational effort. However, finding influence coefficients and the resulting flexibility matrix is relatively an easier task.

Problem 4.3

For the system shown in Fig. 4.14, find the frequency equation

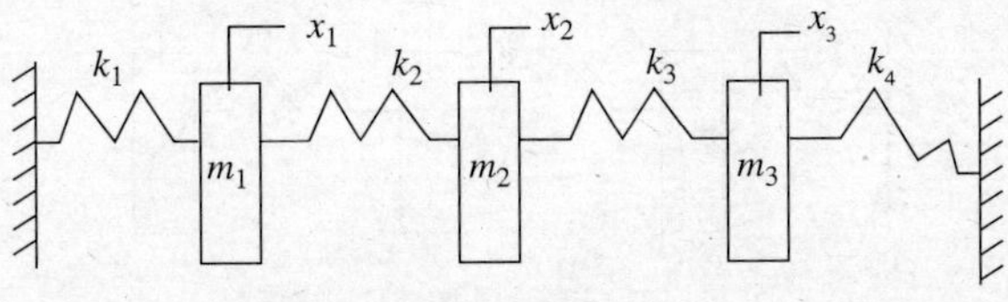

Figure 4.14

Solution:
The first step in this type of problem is to arrive at the equations of motion. We can show that the following equations are the equations of motion.

$$m_1\ddot{x}_1 = -k_1x_1 - k_2(x_1 - x_2)$$
$$m_2\ddot{x}_2 = -k_2(x_2 - x_1) - k_3(x_2 - x_3)$$
$$m_3\ddot{x}_3 = -k_3(x_3 - x_2) - k_4x_3$$

Rearranging the above equations, we get

$$m_1\ddot{x}_1 + (k_1 + k_2)x_1 - k_2x_2 = 0$$
$$m_2\ddot{x}_2 + (k_2 + k_3)x_2 - k_2x_3 = 0$$
$$m_3\ddot{x}_3 + (k_3 + k_4)x_3 - k_4x_3 = 0$$

Assume the solution as

$$x_i = A_i \sin(\omega.t + \phi), \therefore \ddot{x}_i = -A_i\omega^2 \sin(\omega.t + \phi)$$

Substituting the above relations in the rearranged equations of motion and simplifying, we get

$$(k_1 + k_2 - m_1\omega^2)\, A_1 - k_2A_2 = 0$$
$$-k_2A_1 + (k_2 + k_3 - m_2\omega^2)A_2 - k_3A_3 = 0$$
$$-k_3A_2 + (k_3 + k_4 - m_3\omega^2)A_3 = 0$$

The frequency equation is obtained by equating the determinant of the above Equation to zero

$$\begin{vmatrix} k_1 + k_2 - m_1\omega^2 & -k_2 & 0 \\ -k_2 & k_2 + k_3 - m_2\omega^2 & -k_3 \\ 0 & -k_3 & k_3 + k_4 - m_3\omega^2 \end{vmatrix} = 0$$

Expansion of this determinant gives

$$\omega^6 - \left[\frac{k_1 + k_2}{m_1} + \frac{k_2 + k_3}{m_2} + \frac{k_3 + k_4}{m_3}\right]\omega^4 + \left[\frac{k_1k_2 + k_2k_3 + k_3k_4}{m_1m_2} + \frac{k_2k_3 + k_3k_4 + k_4k_2}{m_2m_3} + \frac{(k_1 + k_2)(k_3 + k_4)}{m_3m_1}\right]\omega^2$$
$$-\left[\frac{k_1k_2k_3 + k_2k_3k_4 + k_3k_4k_1 + k_4k_1k_2}{m_1m_2m_3}\right] = 0$$

The roots of the above equation can be found using graphical method or by Newton-Raphson method.

Problem 4.4
For the system shown in Fig. 4.14(a) (without spring 4 in Fig. 4.14), find the natural frequencies and mode shapes. Assume $k_i = 2$ and $m_i = 5$.

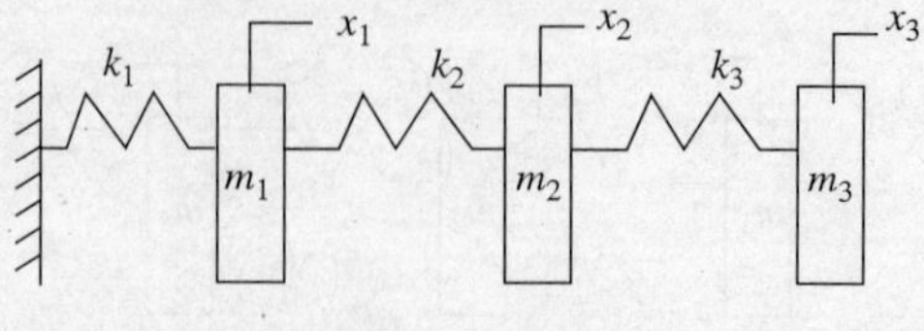

Figure 4.14(a)

Solution:

As mentioned in the earlier problems, it is lot convenient to determine the flexibility matrix using flexibility-influence coefficients rather than inverting the stiffness matrix, especially when the number of degree-of-freedom of the system is large.

The flexibility influence coefficients a_{ij} of the system can be determined in terms of spring stiffnesses k_1, k_2, and k_3 as follows:

Let x_1, x_2, and x_3 be the displacement of the masses m_1, m_2, and m_3.

STEP 1

Apply unit force at mass m_1 and no force at other masses ($F_1 = 1$ and $F_2, F_3 = 0$). The resulting deflections of the masses m_1, m_2, and m_3 are according to definition of influence coefficients a_{11}, a_{21}, and a_{31}. Equilibrium of forces gives

Mass m_1: $k_1 a_{11} = k_2(a_{21} - a_{11}) + 1$

Mass m_2: $k_2(a_{21} - a_{11}) + k_3(a_{21} - a_{31}) = 0$

Mass m_3: $k_3(a_{31} - a_{21}) = 0$

Solution of these equations gives a_{11}, a_{21}, and $a_{31} = 1/k_1$.

STEP 2

Apply unit force at mass m_2 with forces at m_1 and $m_3 = 0$.

Mass m_1: $k_1 a_{12} = k_2(a_{22} - a_{12})$

Mass m_2: $k_2(a_{22} - a_{12}) = k_3(a_{32} - a_{22}) + 1$

Mass m_3: $k_3(a_{32} - a_{22}) = 0$

Solution of above equations gives

$a_{12} = 1/k_1$. $a_{22} = 1/k_1 + 1/k_2$ and $a_{32} = 1/k_1 + 1/k_2$.

STEP 3

Apply unit force at m_3. The force equilibrium gives

Mass m_1: $k_1 a_{13} = k_2(a_{23} - a_{13})$

Mass m_2: $k_2(a_{23} - a_{13}) = k_3(a_{33} - a_{23})$

Mass m_3: $k_3(a_{33} - a_{23}) = 1$

Solution of the above Equations gives

$a_{13} = 1/k_1$, $a_{23} = 1/k_1 + 1/k_2$, and $a_{33} = 1/k_1 + 1/k_2 + 1/k_3$

Since $k_i = 2$ and $m_i = 5$, the flexibility matrix is given by

$$[a] = \frac{1}{2}\begin{bmatrix} 1 & 1 & 1 \\ 1 & 2 & 2 \\ 1 & 2 & 3 \end{bmatrix} \text{ and the mass matrix } [m] = 5\begin{bmatrix} 1 & 0 & 0 \\ 0 & 1 & 0 \\ 0 & 0 & 1 \end{bmatrix}$$

The dynamical matrix $[D] = [K]^{-1} = [a][m]$

$$\therefore [D] = \frac{5}{2}\begin{bmatrix} 1 & 1 & 1 \\ 1 & 2 & 2 \\ 1 & 2 & 3 \end{bmatrix}$$

We have

$$\Delta = |\lambda[I]-[D]| = \left|\begin{bmatrix} \lambda & 0 & 0 \\ 0 & \lambda & 0 \\ 0 & 0 & \lambda \end{bmatrix} - \frac{5}{2}\begin{bmatrix} 1 & 1 & 1 \\ 1 & 2 & 2 \\ 1 & 2 & 3 \end{bmatrix}\right| = 0$$

$$\lambda = 1/\omega^2$$

By dividing throughout by λ, and introducing $a = m/k\lambda$, we get

$$\begin{vmatrix} 1-a & -a & -a \\ -a & 1-2a & -2a \\ -a & -2a & 1-3a \end{vmatrix} = a^3 - 5a^2 + 6a - 1 = 0$$

The roots of this cubic equation are

$$a_1 = m\omega_1^2/k = 0.19806, \quad \omega_1 = 0.44\sqrt{k/m}$$

$$a_2 = m\omega_2^2/k = 1.55, \quad \omega_2 = 1.24\sqrt{k/m}$$

$$a_3 = m\omega_3^2/k = 3.24, \quad \omega_3 = 1.80\sqrt{k/m}$$

Once the natural frequencies are known, the mode shapes or eigen vectors can be calculated using $[\lambda[I]-[D]]\bar{X}^i = \bar{0}, i = 1,2,3$, where $\bar{X}^i = \begin{Bmatrix} X_1^i \\ X_2^i \\ X_3^i \end{Bmatrix}$ denotes the ith mode shape. In order to find the mode shapes, substitute values of natural frequencies one by one to obtain shapes at different modes.

First or fundamental mode: $\omega_1 = 0.44\sqrt{k/m}$ $(\lambda_1 = 5.04)$

$$\left[5.04\frac{m}{k}\begin{bmatrix} 1 & 0 & 0 \\ 0 & 1 & 0 \\ 0 & 0 & 1 \end{bmatrix} - \frac{m}{k}\begin{bmatrix} 1 & 1 & 1 \\ 1 & 2 & 2 \\ 1 & 2 & 3 \end{bmatrix}\right]\begin{Bmatrix} X_1^{(1)} \\ X_2^{(1)} \\ X_3^{(1)} \end{Bmatrix} = \begin{Bmatrix} 0 \\ 0 \\ 0 \end{Bmatrix}.$$

Simplifying this, we get,

$$\begin{bmatrix} 4.04 & -1.0 & -1.0 \\ -1.0 & 3.04 & -2.0 \\ -1.0 & -2.0 & 2.04 \end{bmatrix}\begin{Bmatrix} X_1^{(1)} \\ X_2^{(1)} \\ X_3^{(1)} \end{Bmatrix} = \begin{Bmatrix} 0 \\ 0 \\ 0 \end{Bmatrix}$$

The above equation denotes a system of homogeneous linear equations with three unknowns $X_1^{(1)}, X_2^{(1)}, X_3^{(1)}$. Any two of these unknowns can be expressed in terms of the remaining one. Let us express $X_2^{(1)}$ and $X_3^{(1)}$ in, terms of $X_1^{(1)}$.

$$X_2^{(1)} + X_3^{(1)} = 4.04X_1^{(1)}$$

$$3.04X_2^{(1)} - 2.0X_3^{(1)} = X_1^{(1)}$$

From these we obtain

$$X_2^{(1)} = 1.80X_1^{(1)}, \; X_3^{(1)} = 2.24X_1^{(1)}.$$

Therefore, the first mode shape is

$$\bar{X}^{(1)} = X_1^1 \begin{Bmatrix} 1 \\ 1.80 \\ 2.24 \end{Bmatrix}.$$

Likewise other mode shapes at ω_2, ω_3 can be evaluated. They are as shown in Fig. 4.15

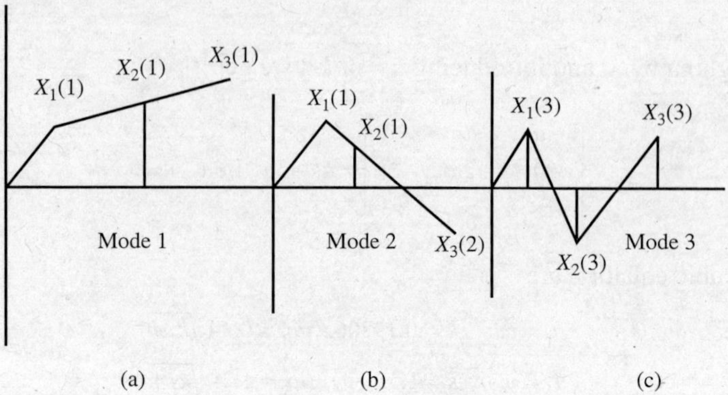

Figure 4.15 Mode Shapes

Problem 4.5

Calculate the natural frequencies of the system shown in Fig. 4.16. Assume uniform tension T in the string.

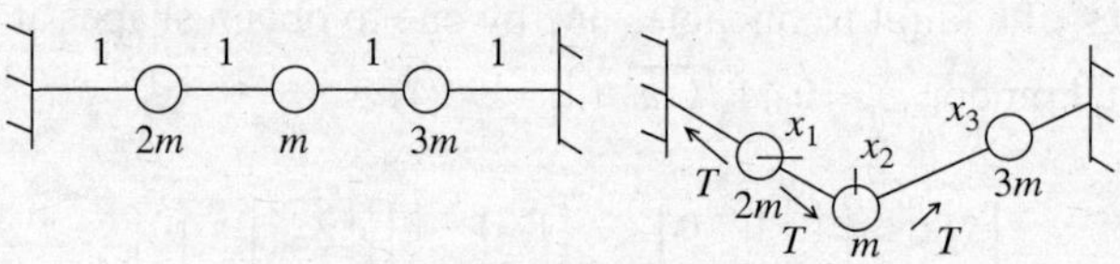

Figure 4.16 Three Mass String System

Solution:

Consider the mass $2m$ as shown in Fig. 4.16(b). Consider the force equilibrium $\sum F = ma$. This gives

$$2m\ddot{x}_1 = T[\frac{x_2 - x_1}{L}] - T[\frac{x_1}{L}]$$

$$m\ddot{x}_2 = T[\frac{x_2 - x_1}{L}] - T[\frac{x_2 - x_3}{L}] \tag{1}$$

$$3m\ddot{x}_3 = T[\frac{x_2 - x_3}{L}] - D[\frac{x_2}{L}]$$

The next step is to assume the solution as

$$x_1 = A\sin(\omega.t + \phi), \quad x_2 = B\sin(\omega.t + \phi), \quad x_3 = C\sin(\omega.t + \phi) \tag{2}$$

Substituting Equation 1 in Equation 2 and simplifying, we get

$$(2T/L - 2m\omega^2)A - (T/L)B = 0$$

$$-(T/L)A + (2T/L - m\omega^2) - (T/L)C = 0 \tag{3}$$

$$-(T/L)B + (2T/L - 3m\omega^2)C = 0$$

The frequency equation is obtained by making the determinant of the Equations 3 equal to zero

$$\begin{vmatrix} 2T/L - 2m\omega^2 & -T/L & 0 \\ -T/L & 2T/L - m\omega^2 & -T/L \\ 0 & -T/L & 2T/L - 3m\omega^2 \end{vmatrix} = 0 \tag{4}$$

Expansion of the determinant gives

$$6m^2\omega^6 - (22m^2/L)\omega^4 + (19T^2m/L^2)\omega^2 - 4T^3/L = 0.$$

The solution of this equation gives

$$\omega_1 = 0.56\sqrt{T/Lm}, \quad \omega_2 = 0.83\sqrt{T/Lm}, \quad \omega_3 = 1.59\sqrt{T/Lm} \text{ rad/s}$$

Problem 4.6

Calculate the natural frequencies of the system shown in Fig. 4.17. There is no force in the springs when all the pendulums are in vertical position.

Solution:

The best method to deal with such problems is application of Lagrange's equation. ($2a\dot{\theta}$) is the tangential velocity. The total kinetic energy of the system is given by

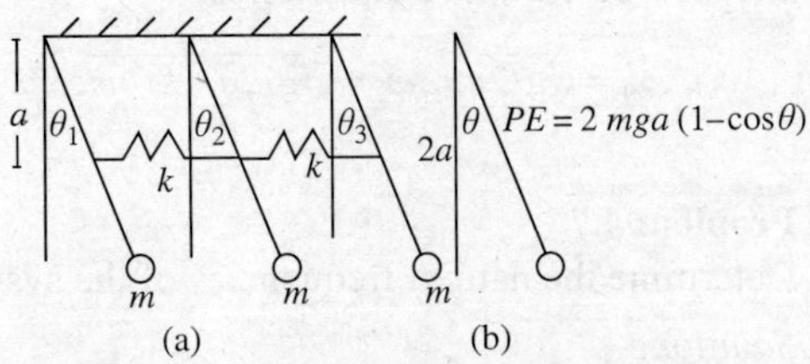

Figure 4.17 (a) Three Pendulum System (b) PE of Pendulum

$$KE = \frac{1}{2}m(2a\dot{\theta}_1)^2 + \frac{1}{2}m(2a\dot{\theta}_2)^2 + \frac{1}{2}m(2a\dot{\theta}_3)^2 \tag{1}$$

The potential energy of the system is given by

$$PE = 2mga[(1-\cos\theta_1) + (1-\cos\theta_2) + (1-\cos\theta_3)] + \frac{1}{2}k(a\theta_2 - a\theta_1)^2 + \frac{1}{2}k(a\theta_3 - a\theta_2)^2$$

Lagrange's equation is

$$\frac{d}{dt}\left(\frac{\partial KE}{\partial \dot{q}_j}\right) - \frac{\partial KE}{\partial q_j} + \frac{\partial PE}{\partial q_j} = 0$$

In the above Equation q_i are the generalized coordinates which in this case are θ_1, θ_2, and θ_3. The various components of Lagrange's equation are evaluated as given here.

$$\frac{d}{dt}\left(\frac{\partial KE}{\partial \dot{\theta}_1}\right) = 4ma^2\ddot{\theta}_1, \quad \frac{\partial KE}{\partial \theta_1} = 0, \quad \frac{\partial PE}{\partial \theta_1} = 2mga\sin\theta_1 - ka(a\theta_2 - a\theta_1)$$

Assuming θ to be small such that sin $\theta \approx \theta$, the first equation of motion (for the first mass) becomes

$$4ma^2\ddot{\theta}_1 + (2mag + ka^2)\theta_1 - ka^2\theta_2 = 0.$$

Similarly, we can obtain,

$$4ma^2\ddot{\theta}_2 + (2mag + 2ka^2)\theta_2 - ka^2\theta_1 - ka^2\theta_3 = 0$$

$$4ma^2 + (2mag + ka^2)\theta_3 - ka^2\theta_2 = 0$$

Assuming the solution of the above equations as

$$\theta_1 = A\sin(\omega.t + \phi), \quad \theta_2 = B\sin(\omega.t + \phi), \quad \theta_3 = C\sin(\omega.t + \phi),$$

substituting in above equations and simplifying, we get

$$(2mag + ka^2 - 4m\omega^2)A - ka^2B = 0$$

$$-ka^2A + (2mag + ka^2 - 4m\omega^2)B - ka^2C = 0$$

$$-ka^2B + (2mag + ka^2 - 4m\omega^2)C = 0$$

The frequency equation is obtained by equating the determinant of the above equations to zero. The result is

$$\begin{vmatrix} (2mag + ka^2 - 4ma^2\omega^2) & -ka^2 & 0 \\ -ka^2 & (2mag + 2ka^2 - 4ma^2\omega^2) & -ka^2 \\ 0 & -ka^{2^2} & (2mag + ka^2 - 4ma^2\omega^2) \end{vmatrix} = 0$$

The expansion of this determinant gives

$$(2mag + ka^2 - 4ma^2\omega^2)[(2mag + ka^2 - 4ma^2\omega^2)(2mag + 2ka^2 - 4ma^2\omega^2) - 2k^2a^4] = 0$$

Solution of the above equation is

$$\omega_1 = \sqrt{g/2a}, \quad \omega_2 = \sqrt{g/2a + k/4m}, \quad \omega_3 = \sqrt{g/2a + 3k/m} \text{ rad/s}$$

Problem 4.7

Determine the natural frequencies of the system shown in Fig. 4.18.

Solution:

For this, we write equation of equilibrium $\Sigma F = ma$

$$m\ddot{x}_1 = -kx_1 - k(x_1 - x_2) - k(x_1 - x_3) \text{ or,} \quad m\ddot{x}_1 + 3kx_1 - kx_2 - kx_3 = 0$$

$$m\ddot{x}_2 = -kx_2 - k(x_2 - x_1) - k(x_2 - x_3) \text{ or,} \quad m\ddot{x}_2 + 3kx_2 - kx_3 - kx_1 = 0$$

$$m\ddot{x}_3 = -kx_3 - k(x_3 - x_1) - k(x_3 - x_2) \text{ or,} \quad m\ddot{x}_3 + 3kx_3 - kx_1 - kx_2 = 0$$

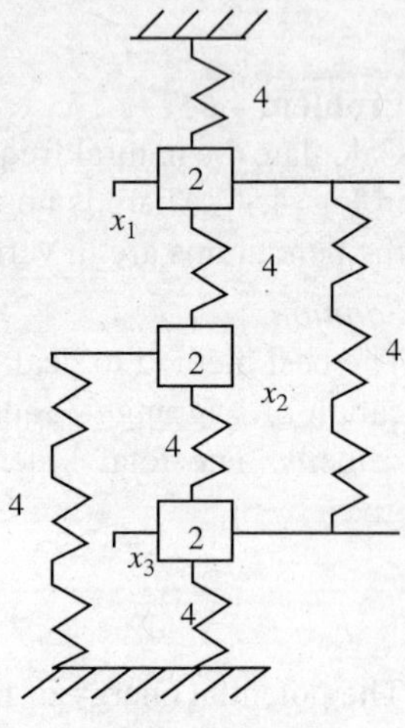

Figure 4.18

Rest of the procedure is same as followed in earlier examples. The various frequencies are

$$\omega_1 = \sqrt{k/m} = \sqrt{4/2} = 1.41, \omega_2 = \omega_3 = 2\sqrt{k/m} = 2.82 \text{ rad/s}$$

Since two of the natural frequencies of the system are equal, the system is said to be a degenerate system.

Problem 4.8

Derive the equation of motion for the triple pendulum shown in Fig. 4.19. Also find the frequencies when all the masses and lengths of the pendulums are equal.

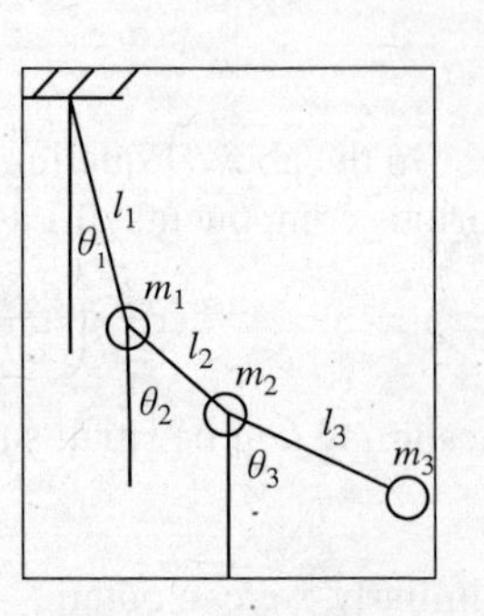

Figure 4.19 Triple Pendulum

Solution:

The absolute velocities of m_1, m_2, and m_3 are

$$\bar{v}_1 = l_1\dot{\bar{\theta}}_1, \quad \bar{v}_2 = l_2\dot{\bar{\theta}}_2 + \bar{v}_1, \quad \bar{v}_3 = l_3\dot{\bar{\theta}}_3 + \bar{v}_2$$

$$v_1^2 = l_1^2\dot{\theta}_1^2$$

$$v_2^2 = (l_1\dot{\theta}_1)^2 + 2l_1l_2\dot{\theta}_1\dot{\theta}_2\cos(\theta_2 - \theta_1) + (l_2\dot{\theta}_2)^2$$

$$v_3^2 = (l_1\dot{\theta}_1)^2 + 2l_1l_2\dot{\theta}_1\dot{\theta}_2\cos(\theta_2 - \theta_1) + (l_2\dot{\theta}_2)^2 + (l_3\dot{\theta}_3)^2$$
$$+ 2\sqrt{(l_1\dot{\theta}_1)^2 + 2l_1l_2\dot{\theta}_1\dot{\theta}_2\cos(\theta_2 - \theta_1) + (l_2\dot{\theta}_2)}.l_3\dot{\theta}_3\cos(\theta_3 - \theta_2 + \theta_1)$$

$$KE = \frac{1}{2}[m_1 v_1^2 + m_2 v_2^2 + m_3 v_3^2]$$

Keeping the configuration of the system in mind, evaluate PE. Use Lagrange's equation to find equation of motion. Rest of the procedure is as shown in previous example. The final answers are $\omega_1 = 0.65\sqrt{g/l}, \omega_2 = 1.52\sqrt{g/l}, \omega_3 = 2.5\sqrt{g/l}$. These are the results, provided all the masses are equal and all pendulums are of equal length.

Problem 4.9

Three blocks, each having a mass of 1 kg are connected by three springs of unit stiffness and supported on frictionless surface (Fig. 4.20). If the block m_3 is given a unit displacement, determine the resulting motion of the system.

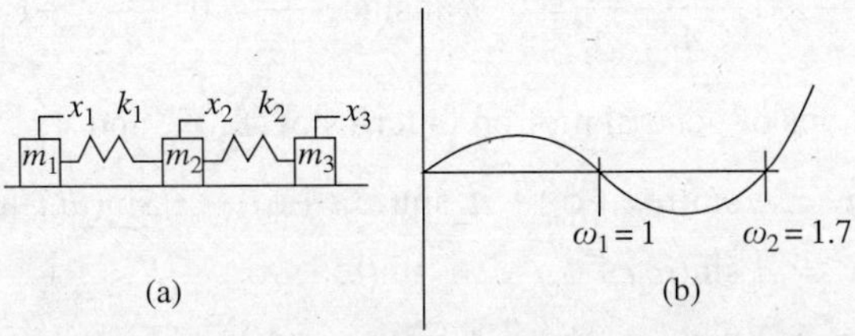

Figure 4.20

The system shown in Fig. 4.20 is a semi-definite system. The differential equations of motion can be shown to be

$$\ddot{x}_1 + x_1 - x_2 = 0$$
$$\ddot{x}_2 + 2x_2 - x_3 - x_1 = 0$$
$$\ddot{x}_3 + x_3 - x_2 = 0$$

We can assume the solution as $x_1 = A\sin(\omega.t + \phi), x_2 = B\sin(\omega.t + \phi), x_3 = C\sin(\omega.t + \phi)$

Substituting these in the equations of motion as shown above, we get

$$\begin{aligned} (1-\omega^2)A - B &= 0 \\ -A + (2-\omega^2)B - C &= 0 \\ -B + (1-\omega^2)C &= 0 \end{aligned} \tag{1}$$

Following the procedure followed in previous example, we get the frequency equation as $(1-\omega^2)(\omega^2-3)\omega^2 = 0$. The natural frequencies are $\omega_1 = 0, \omega_2 = 1, \omega_3 = \sqrt{3}$.

The general equations of motion can be expressed as

$$\begin{aligned} x_1 &= A_1\sin(\omega_1 t + \phi_1) + A_2\sin(\omega_2 t + \phi_2) + A_3\sin(\omega_3 t + \phi_3) \\ x_2 &= B_1\sin(\omega_1 t + \phi_1) + B_2\sin(\omega_2 t + \phi_2) + B_3\sin(\omega_3 t + \phi_3) \\ x_3 &= C_1\sin(\omega_1 t + \phi_1) + C_2\sin(\omega_2 t + \phi_2) + C_3\sin(\omega_3 t + \phi_3) \end{aligned} \tag{2}$$

From the first of the three equations shown at 2, we get,

$B/A = (1-\omega^2)$. Substituting the values of natural frequencies, we get,

$$B_1/A_1 = 1, B_2/A_2 = 0, B_3/A_3 = -2$$

Using first and third equations of (2), we get,

$$C/A = \frac{1-\omega}{1-\omega^2} = 1.$$

This gives $C_1/A_1 = -1, C_2/A_2 = 0/0, C_3/A_3 = 1$.

Since C_2/A_2 is indeterminate, use the second equation of (2) to obtain

$$-A_2 + (2-\omega^2)B_2 - C_2 = 0.$$

Substituting $\omega_2 = 1$, we get $-A_2 + B_2 - C_2 = 0$.

This can also be written as

$$-1 + \frac{B_2}{A_2} - \frac{C_2}{A_2} = 0. \text{ But since } \frac{B_2}{A_2} = 0, \ \frac{C_2}{A_2} = -1$$

We can now write equations of general motion in terms of A_1, A_2, and A_3.

$$x_1 = A_1 \sin(\omega_1 t + \phi_1) + A_2 \sin(\omega_2 t + \phi_2) + A_3 \sin(\omega_3 t + \phi_3)$$
$$x_2 = A_1 \sin(\omega_1 t + \phi_1) - 2A_2 \sin(\omega_3 t + \phi_3)$$
$$x_3 = A_1 \sin(\omega_1 t + \phi_1) - A_2 \sin(\omega_2 t + \phi_2) + A_3 \sin(\omega_3 t + \phi_3)$$

The six initial conditions required for evaluating the six constants are

$$\dot{x}_1(0) = 0, \dot{x}_2(0) = 0, \dot{x}_3(0) = 0, x_1(0) = 0, x_2(0) = 0, x_3(0) = 1.$$

Using these, we get,

$$A_1 \sin\phi_1 + A_2 \sin\phi_2 + A_3 \sin\phi_3 = 0 \quad (1)$$
$$A_1 \sin\phi_1 - 2A_3 \sin\phi_2 = 0 \quad (2)$$
$$\sin\phi_1 - A_2 \sin\phi_2 + A_3 \sin\phi_3 = 1 \quad (3)$$
$$\cos\phi_1 + A_2 \cos\phi_2 + \sqrt{3}A_3 \cos\phi_3 = 0 \quad (4)$$
$$\cos\phi_1 - 2\sqrt{3}A_3 \cos\phi_3 = 0 \quad (5)$$
$$A_1 \cos\phi_1 - A_2 \cos\phi_2 + \sqrt{3}A_3 \cos\phi_3 = 0 \quad (6)$$

Adding Equations 4, 5, and 6, we get $3A_1 \cos\phi_1 = 0$, *or* $\phi_1 = \Pi/2$. From Equations 5 and 6, we get $\phi_2 = \phi_3 = \Pi/2$. From Equations 1, 2, and 3, we get $A_1 = 1/3, A_2 = -1/2, A_3 = 1/6$.

Thus, the general equations of motion are

$$x_1(t) = -\frac{1}{3} - \frac{1}{2}\sin(t + \frac{\Pi}{2}) + \frac{1}{6}\sin(\sqrt{3}t + \frac{\Pi}{2})$$
$$x_2(t) = \frac{1}{3} - \frac{1}{3}\sin(\sqrt{3}t + \frac{\Pi}{2})$$
$$x_3(t) = \frac{1}{3} + \frac{1}{2}\sin(t + \frac{\Pi}{2}) + \frac{1}{6}\sin(\sqrt{3}t + \frac{\Pi}{2})$$

Problem 4.10

Figure 4.21 shows a triple pendulum of equal lengths and masses. Find the influence coefficients.

Solution:

Let us first evaluate a_{11}. For this, refer to Fig. 4.21. Apply unit force at mass m_1. Let T be the tension in the first string. Writing equation of equilibrium,

$$T \sin \theta = 1 \quad (1)$$

$$T \cos \theta = g(m_1 + m_2 + m_3) \quad (2)$$

$$\tan \theta = 1/g(m_1 + m_2 + m_3) \quad (3)$$

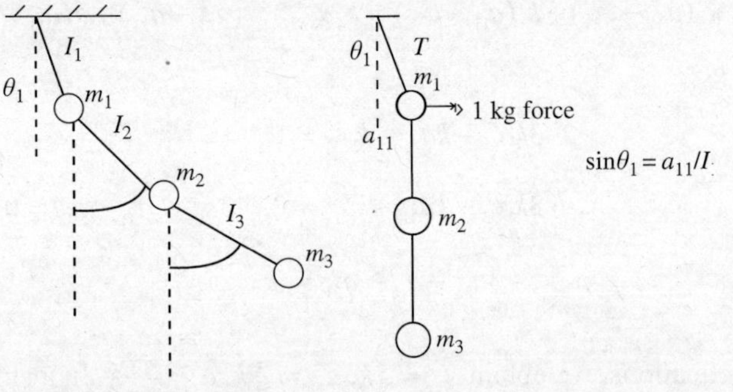

Figure 4.21

For small angles of oscillations, $\tan \theta \approx \sin \theta$ and from the configuration of the system $\sin \theta$ = displacement in the direction of the force/length $= a_{11} = l_1/g(m_1 + m_2 + m_3) = a_{21} = a_{31}$.

Now apply a unit force at mass m_2.

With this, mass m_1 will displace by a_{11}, while the masses m_2 and m_3 will displace by additional distance $l_2/g(m_1 + m_2)$. Therefore $a_{22} = a_{32} = a_{11} + l_2/g\ (m_2 + m_3)$, $a_{12} = a_{11}$.

Now apply unit force at mass m_3. With this m_1 will displace by a_{11}, m_2 will displace by $a_{22} = a_{11} + l_2/g(m_2 + m_3)$, while mass m_3 will be displaced by additional distance equal to l_3/gm_3. Thus $a_{13} = a_{11}$, $a_{23} = a_{22}$ and $a_{33} = a_{22} + l_3/gm_3$. Thus the influence coefficients are

$$a_{11} = a_{12} = a_{13} = l/3gm = a_{21} = a_{31}$$

$$a_{22} = a_{23} = a_{32} = l/3gm + l/2gm = 5l/6gm$$

$$a_{33} = l/3gm + l/2gm + l/gm = 11l/6gm.$$

Problem 4.11

For the system shown in Fig. 4.22, calculate the influence coefficients. Assume masses equal to m and all springs having stiffness equal to k. Also, write the equation of motion.

Solution:

STEP 1

Apply a unit force at mass m_1. By definition,

a_{11} = deflection at mass m_1 due to unit force at m_1

a_{21} = deflection at mass m_2 due to unit force at m_1

a_{31} = deflection at mass m_3 due to unit force at m_1

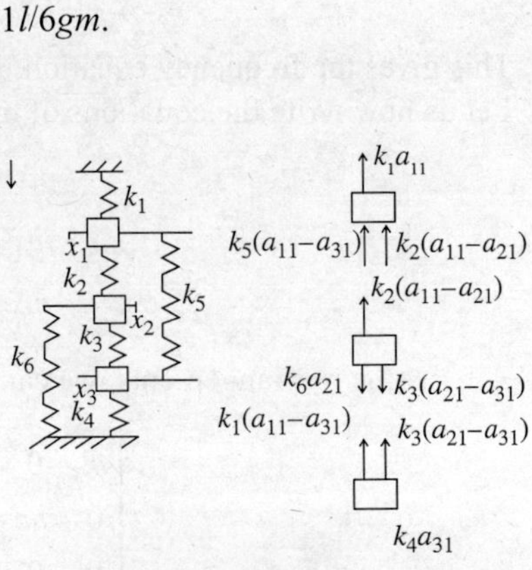

Figure 4.22

Considering the force equilibrium

$$k_1a_{11} + k_2(a_{11} - a_{21}) + k_3(a_{11} - a_{31}) = 1, \quad ka_{11} + k(a_{11} - a_{21}) + k(a_{11} - a_{31}) = 1 \tag{1}$$

For the second mass

$$k_2(a_{11} - a_{21}) = k_3(a_{21} - a_{31}) + k_6a_{21}, \quad k(a_{11} - a_{21}) = k(a_{21} - a_{31}) + ka_{21} \tag{2}$$

For the third mass

$$k_3(a_{21} - a_{31}) + k_1(a_{11} - a_{31}) = k_4a_{31}, \quad k(a_{21} - a_{31}) + k(a_{11} - a_{31}) = ka_{31} \tag{3}$$

Simplifying, we get,

$$3ka_{11} - ka_{21} - ka_{31} = 1 \tag{A}$$

$$3ka_{21} - ka_{31} - ka_{11} = 0, \quad 3a_{21} - a_{31} - a_{11} = 0 \tag{B}$$

$$3a_{31} - a_{11} - a_{21} = 0 \tag{C}$$

From these equations, we obtain $a_{11} = \frac{1}{2}k$, $a_{21} = \frac{1}{4}k$, $a_{31} = \frac{1}{4}k$. In a similar manner, we apply unit loads at other masses one by one and repeat the procedure. The final results are

$$a_{12} = \frac{1}{4}k,\ a_{22} = \frac{1}{2}k,\ a_{32} = \frac{1}{4}k,\ a_{13} = \frac{1}{4}k,\ a_{23} = \frac{1}{4}k \text{ and } a_{33} = \frac{1}{2}k$$

It must be remembered that flexibility-influence coefficient matrix $[a]$ is the inverse of stiffness matrix $[k]$ and the equation of motion of the system is given by

$$[\lambda[I] - [k]^{-1}[m]]\bar{X} = \bar{0}, \quad \text{or}$$

$$[\lambda[I] - [a][m]]\bar{X} = \bar{0}$$

The frequency equation is obtained by making the determinant of the above equation zero.

Thus,
$$\Delta = \left| \lambda \begin{bmatrix} 1 & 0 & 0 \\ 0 & 1 & 0 \\ 0 & 0 & 1 \end{bmatrix} - \begin{bmatrix} \frac{1}{2}K & \frac{1}{4}K & \frac{1}{4}K \\ \frac{1}{4}K & \frac{1}{2}K & \frac{1}{4}K \\ \frac{1}{4}K & \frac{1}{4}K & \frac{1}{2}K \end{bmatrix} . \begin{bmatrix} m & 0 & 0 \\ 0 & m & 0 \\ 0 & 0 & m \end{bmatrix} \right| = 0.$$

This gives the frequency equation.
Let us now write the equations of motion

$$m\ddot{x}_1 + k(x_1 - x_2) + k(x_1 - x_3) + k\ x_1 = 0$$

$$m\ddot{x}_2 + k(x_2 - x_1) + kx_2 + k(x_2 - x_3) = 0$$

$$m\ddot{x}_3 + kx_3 + k(x_3 - x_2) + k(x_3 - x_1) = 0$$

After some rearrangements, we can write the above equations in matrix form

$$\begin{bmatrix} m & 0 & 0 \\ 0 & m & 0 \\ 0 & 0 & m \end{bmatrix} \begin{Bmatrix} \ddot{x}_1 \\ \ddot{x}_2 \\ \ddot{x}_3 \end{Bmatrix} + \begin{bmatrix} 2k & -k & -k \\ -k & 2k & -k \\ -k & -k & 2k \end{bmatrix} \begin{Bmatrix} x \\ x_2 \\ x_3 \end{Bmatrix} = \begin{Bmatrix} 0 \\ 0 \\ 0 \end{Bmatrix}$$

4.7 ORTHOGONALITY OF NORMAL MODES

In the previous section, we considered the method of finding the n natural frequencies $\omega_i\,(i=1,2,3,\ldots,n)$ and the corresponding normal modes, also called modal vectors. We shall now see an important property of the normal modes, which is orthogonality. The natural frequency ω_i and the corresponding modal vector $\bar{X}_i$ satisfy equation 4.67 ($[[k]-\omega^2[m]]\,\bar{X}=\bar{0}$) so that

$$\omega_i^2[m]\bar{X}^i=[k]\bar{X}^i \tag{4.75}$$

If we consider another natural frequency ω_j and the corresponding modal vector $\bar{X}^j$, they also satisfy Equation 4.67 so that

$$\omega_j^2[m]\bar{X}^j=[k]\bar{X}^j \tag{4.76}$$

Pre-multiplying Equations 4.75 and 4.76 by $\bar{X}^{jT}$ and $\bar{X}^{iT}$ (transposes of the vectors) and also considering the symmetry of the matrices $[m]$ and $[k]$, we get

$$\omega_i^2\bar{X}^{JT}[m]\bar{X}^i=\bar{X}^{j^T}[k]\bar{X}^i=\bar{X}^{i^T}[k]\bar{X}^j \tag{4.77}$$

$$\omega_j^2\bar{X}^{i^T}[m]\bar{X}^j=\omega_j^2\bar{X}^j[m]\bar{X}^i=\bar{X}^{i^T}[k]\bar{X}^i \tag{4.78}$$

Subtracting Equation 4.78 from Equation 4.77, we get

$$(\omega_j^2-\omega_i^2)\bar{X}^{jT}[m]\bar{X}^i=0 \tag{4.79}$$

Since, in general $\omega_1\neq\omega_2$, Equation 4.79 leads to

$$\bar{X}^{jT}[m]\bar{X}^i=0\,,\quad i\neq j \tag{4.80}$$

We can also obtain

$$\bar{X}^{jT}[k]\bar{X}^i=0,\quad i\neq j \tag{4.81}$$

Equations 4.80 and 4.81 indicate that modal vectors $\bar{X}^i$ and $\bar{X}^j$ are orthogonal with respect to both mass and stiffness matrices. When $j=i$, the left sides of Equations 4.80 and 4.81 are not equal to zero, but they do yield the generalized mass and stiffness coefficients of the ith mode.

$$M_{ii}=\bar{X}^{iT}[m]\bar{X}^i,\quad i=1,2,3\ldots n \tag{4.82}$$

$$K_{ii}=\bar{X}^{iT}[k]\bar{X}^i,\quad i=1,2,3,\ldots n \tag{4.83}$$

Equations 4.82 and 4.83 can also be written in the matrix form as

$$\begin{bmatrix} M_{11} & 0 & 0 & 0 & 0 \\ . & M_{22} & . & . & . \\ . & . & M_{33} & . & . \\ . & . & . & . & . \\ 0 & 0 & 0 & 0 & M_{nn} \end{bmatrix}=[\text{Diagonal}.\text{matrix}.m]=[X]^T[m][X] \tag{4.84}$$

$$\begin{bmatrix} k_{11} & 0 & 0 & 0 & 0 \\ . & k_{22} & . & . & . \\ . & . & k_{33} & . & . \\ . & . & . & . & . \\ 0 & 0 & 0 & 0 & k_{nn} \end{bmatrix} = [\text{Diagonal.matrix}.k] = [X]^T[k][X] \tag{4.85}$$

In the above equations, $[X]$ is called the modal matrix in which the ith column corresponds to the ith modal vector.

$$[X] = [\,\overline{X}^1\ \overline{X}^2 \ldots \overline{X}^n\,] \tag{4.86}$$

It is customary to normalize the modal vector $\overline{X}^i$ such that

$[M]$ = identity matrix $[I]$ which means

$$\overline{X}^{i^T}[m]\overline{X}^i = 1, \quad i = 1,2,3\ldots n \tag{4.87}$$

In this case, matrix $[k]$ reduces to

$$[k] = [\omega^2] = \begin{bmatrix} \omega_1^2 & 0 & . & . & 0 \\ 0 & \omega_2^2 & . & . & . \\ . & . & . & 0 & 0 \\ 0 & . & . & . & 0 \\ 0 & 0 & 0 & 0 & \omega_n^2 \end{bmatrix} \tag{4.88}$$

Since the eigen vectors are orthogonal and linearly independent in the n-dimensional space, any vector in the n-dimensional space can be expressed as a linear combination of the n independent vectors. If $\overline{x}$ is an arbitrary vector in n-dimensional space, it can be expressed as

$$\overline{x} = \sum_{i=1}^{n} C_i \overline{X}^i \tag{4.89}$$

C_i (i = 1, 2, …, n) are constants. By pre-multiplying Equation 4.89 throughout by $\overline{X}^{i^T}$, the values of the constants C_i can be determined from

$$C_i = \frac{\overline{X}^{i^T}[m]\overline{x}}{\overline{x}^{i^T}[m]\overline{X}^i} = \frac{\overline{X}^i[m]\overline{x}}{M_{ii}}, \quad i = 1,2,3\ldots n \tag{4.90}$$

In the Equation 4.90, M_{ii} is the generalized mass in the ith mode. If the modal vector $\overline{x}^i$ are normalized, C_i are given by

$$C_i = \overline{X}^{i^T}[m]\overline{x}, \quad i = 1,2,3\ldots n \tag{4.91}$$

The above expression is called eigen vector expansion theorem. It is a very important tool for finding out response of the multi degrees-of-freedom systems subjected to arbitrary forcing conditions according to a procedure called modal analysis.

4.8 MODAL ANALYSIS

Modal analysis is an important tool in the analysis and solution of vibration problems. A multi degrees (n) of freedom system, when excited by a force or system of forces (periodic, non-periodic, transient, etc.), undergoes vibrations. For such a system, the governing equations of motion are a set of n coupled equations of second order. For large number of degree-of-freedom, the solution of these equations is extremely complex and sometimes almost impossible unless some gross simplifications are done. For such problems, modal analysis is a very convenient method to find the solution. In this method, the expansion theorem, explained in previous section is used. In modal analysis, the displacements of the n degrees-of-freedom system are expressed as linear combinations of the normal modes of the system. This linear transformation enables uncoupling of the equations of motion, resulting in modified equations of motion, which are n uncoupled equations of motion that can be solved easily.

Modal analysis techniques are also used to experimentally determine the natural frequencies and the mode shapes associated with each frequency. In this method, the vibration system is disturbed from its position of rest by giving a blow (bang) to system by an instrumented hammer. Both the disturbing forces as well as the resulting response are measured using appropriate transducer and the data are analyzed using appropriate two channel FFT analyzer. We shall deal with these techniques in later chapter.

The equations of motion of a multi degrees-of-freedom system under external forces F are given by

$$[m]\ddot{\overline{x}} + [k]\overline{x} = \overline{F} \tag{4.92}$$

The first step in modal analysis is the solution of eigen value problem

$$\omega^2 [m]\overline{X} = [k]\overline{X} \tag{4.93}$$

and to find the various natural frequencies ω_n and the corresponding normal modes $\overline{X}^{(n)}, (n = 1, 2, 3...)$ According to the expansion theorem, the solution vector of equation 4.92 can be expressed as a linear combination of the normal modes

$$\overline{x}(t) = q_1(t)\overline{X}^{(1)} + q_2\overline{X}^{(2)} + \ldots + q_n\overline{X}^{(n)} \tag{4.94}$$

where $q_1(t)$, $q_2(t)$, $q_3(t)$,..., $q_n(t)$ are time-dependent generalized coordinates (these are also known as principal coordinates or modal coefficients). By defining a modal matrix $[X]$ in which the jth column is the vector $\overline{X}^{(j)}$ i.e.

$$[X] = [\overline{X}^{(1)} \quad \overline{X}^{(2)} \quad . \quad . \quad \overline{X}^{(n)}] \tag{4.95}$$

Equation 4.94 can be rewritten as

$$\overline{x}(t) = [X]\overline{q}(t) \tag{4.96}$$

where

$$\overline{q}(t) = \begin{Bmatrix} q_1(t) \\ q_2(t) \\ . \\ . \\ q_n(t) \end{Bmatrix} \tag{4.97}$$

Since $[X]$ is not a function of time, we obtain from equation 4.96

$$\ddot{x}(t) = [X]\ddot{q}(t) \tag{4.98}$$

The equation of motion can therefore be written as

$$[m][x]\bar{\ddot{q}} + [k][X]\bar{q} = \bar{F} \tag{4.99}$$

Pre-multiplying the above equation throughout by $[X]^T$, we get

$$[X]^T[m][X]\bar{\ddot{q}} + [X]^T[k][X]\bar{q} = [X]^T\bar{F} \tag{4.100}$$

If the normal modes are normalized according to Equations 4.90 and 4.91, we obtain

$$[X]^T[m][X] = [I] \tag{4.101}$$

$$[X]^T[k][X] = [\text{diagonal } \omega^2 \text{ matrix}] \tag{4.102}$$

Let us define a new vector of generalized forces associated with generalized coordinates

$$\bar{Q}(t) = [X]\bar{F}(t) \tag{4.103}$$

Using Equations 4.101, 4.102, and 4.103, Equation 4.100 can be rewritten as

$$\bar{\ddot{q}}(t) + [\text{diagonal } \omega^2 \text{matrix}]\bar{q}(t) = \bar{Q}(t) \tag{4.104}$$

$$\text{Note that } [\text{diagonal } \omega^2 \text{ matrix}] = \begin{bmatrix} \omega_1^2 & 0 & . & . & 0 \\ 0 & \omega_2^2 & . & & . \\ . & . & . & 0 & 0 \\ 0 & & & . & 0 \\ 0 & 0 & 0 & 0 & \omega_n^2 \end{bmatrix}$$

Equations 4.104 denote a set of uncoupled differential equations of second order whose solution can be expressed as

$$q_i(t) = q_i(0)\cos\omega_i t + (\frac{\dot{q}(0)}{\omega_i})\sin\omega_i t + \frac{1}{\omega_i}\int_0^t Q_i(\tau)\sin\omega_i(t-\tau)d\tau, \quad i = 1,2,3...n \tag{4.105}$$

The initialized generalized displacements and velocities $\{q_i(0), \dot{q}_i(0)\}$ can be obtained from the initial values of the physical displacement and physical velocities $x_i(0), \dot{x}_i(0)$ as

$$\bar{q}_i(0) = [X]^T\bar{x}_i(0) \tag{4.106}$$

$$\bar{\dot{q}}_i(0) = [X]^T\bar{\dot{x}}_i(0) \tag{4.107}$$

$$\bar{q}(0) = \begin{Bmatrix} q_1(0) \\ q_2(0) \\ . \\ . \\ q_n(0) \end{Bmatrix}, \bar{\dot{q}} = \begin{Bmatrix} \dot{q}_1(0) \\ \dot{q}_2(0) \\ . \\ . \\ \dot{q}_n(0) \end{Bmatrix}, \bar{x}(0) = \begin{Bmatrix} x_1(0) \\ x_2(0) \\ . \\ . \\ x_n(0) \end{Bmatrix}, \bar{\dot{x}}(0) = \begin{Bmatrix} \dot{x}_1(0) \\ \dot{x}_2(0) \\ . \\ . \\ \dot{x}_n(0) \end{Bmatrix}$$

Once the generalized displacements q_i are obtained from Equations 4.105 to 4.107, the physical displacements $x_i(t)$ can be found from Equation 4.98.

4.9 DETERMINATION OF NATURAL FREQUENCIES AND MODE SHAPES

The equations of motion of vibratory system of n degrees-of-freedom are represented by n-differential equations of motion. These equations can be obtained using the Newton's second law of motion or by Lagrange's equations or by influence-coefficient method. Since the equations of motion are not entirely independent, a simultaneous solution of these equations requires the complete evaluation of determinant of the nth order. This will yield all the natural frequencies of the system. At times, this is extremely tedious. The other methods of solution are Stodola method, Holzer method, and the method of matrix iterations. These are more direct numerical approaches and computer programmable for complete analysis and hence frequently employed for the analysis of complex systems.

4.9.1 Method of Matrices and Matrix Iteration

The use of matrices in vibration analysis not only simplifies the work involved but also helps us understand the procedure of solution. As discussed earlier, the equations of motion are given by

$$\begin{aligned}
m_{11}\ddot{q}_1 + m_{12}\ddot{q}_2 + .. + m_{1n}\ddot{q}_n + k_{11}q_1 + k_{12}q_2 + ... + k_{1n}q_n &= 0 \\
m_{21}\ddot{q}_1 + m_{22}\ddot{q}_2 + .. + m_{2n}\ddot{q}_n + k_{21}q_1 + k_{22}q_2 + ... + k_{2n}q_n &= 0 \\
\cdots\cdots\cdots\cdots\cdots\cdots\cdots\cdots\cdots\cdots\cdots\cdots \\
m_{n1}\ddot{q}_1 + m_{2n}\ddot{q}_2 + .. + m_{nn}\ddot{q}_n + k_{n1}q_1 + k_{n2}q_2 + ... + k_{nn}q_n &= 0
\end{aligned} \tag{4.108}$$

The above equations are for free-undamped vibrations. They can also be written in matrix form as given below.

$$\begin{bmatrix} m_{11} & m_{12} & . & . & m_{1n} \\ m_{21} & m_{22} & . & . & m_{2n} \\ . & . & . & . & . \\ . & . & . & . & . \\ m_{n1} & m_{n2} & . & . & m_{nn} \end{bmatrix} \begin{Bmatrix} \ddot{q}_1 \\ \ddot{q}_2 \\ . \\ , \\ \ddot{q}_n \end{Bmatrix} + \begin{bmatrix} k_{11} & k_{12} & . & . & k_{1n} \\ k_{21} & k_{22} & . & . & k_{2n} \\ . & . & . & . & . \\ . & . & . & . & . \\ k_{n1} & k_{n2} & . & . & k_{nn} \end{bmatrix} \begin{Bmatrix} q_1 \\ q_2 \\ . \\ . \\ q_n \end{Bmatrix} = \begin{Bmatrix} 0 \\ 0 \\ . \\ . \\ 0 \end{Bmatrix} \tag{4.109}$$

The above equations can be written in a concise form as

$$[M\{\ddot{q}\} + [K]\{q\} = \{0\}.$$

This equation can also be written as

$$\{\ddot{q}\} + [M]^{-1}[K]\{q\} = \{0\}.$$

As mentioned earlier, we write the $[M]^{-1}[K] = [C]$ so that the above equation becomes

$$\{\ddot{q}\} + [C]\{q\} = \{0\} \tag{4.110}$$

$[C]$ is called the dynamic matrix. As mentioned earlier, the natural frequencies are obtained from the characteristic equation $|[I] - [C]| = 0$ from λ-matrix theory where $[I]$ is a diagonal unitary matrix.

If p_i are the principal coordinates, then

$$\begin{aligned}
q_1 &= a_{11}p_1 + a_{12}p_2 + \cdots\cdots + a_{1n}p_n \\
q_2 &= a_{21}p_1 + a_{22}p_2 + \cdots\cdots + a_{2n}p_n \\
&\cdots\cdots\cdots\cdots\cdots\cdots \\
&\cdots\cdots\cdots\cdots\cdots\cdots \\
q_n &= a_{n1}p_1 + a_{n2}p_2 + \cdots\cdots + a_{nn}p_n
\end{aligned}$$

In matrix notation, the above equations can be written as

$$\{q\} = [a]\{p\} \quad \text{and} \quad \{p\} = [a]^{-1}\{q\} \tag{4.111}$$

In these Equations $[a]^{-1}$ is called the transformation matrix.

The notation q is used since it can represent translational as well as angular coordinates and corresponding to these, the principal coordinates (generalized coordinates) are given the notation p. Having revisited the concept of matrices as applied to analyse the vibration problem, we now discuss the matrix iteration.

Matrix iteration is an iteration procedure that helps us determine the natural frequencies and the associated principal modes. In this method, displacements of the masses are estimated/assumed. From these, the matrix equation of the system is written. The influence coefficients of the system are substituted into the matrix equation which is then expanded. The normalization of the displacements and expansion of the matrix is repeated. The process is continued until the first mode repeats itself to the desired degree of accuracy.

For the higher modes and natural frequencies, the orthogonality principle is used to obtain a new matrix equation that is free from any lower modes. The iterative procedure is repeated.

Let us now solve a few typical problems to understand the concept of matrices and the matrix-iteration method.

Problem 4.12

Figure 4.23 shows a vibration system. Assume small displacements. Calculate the inertia matrix, stiffness matrix, dynamic matrix, natural frequencies, and the principal coordinates.

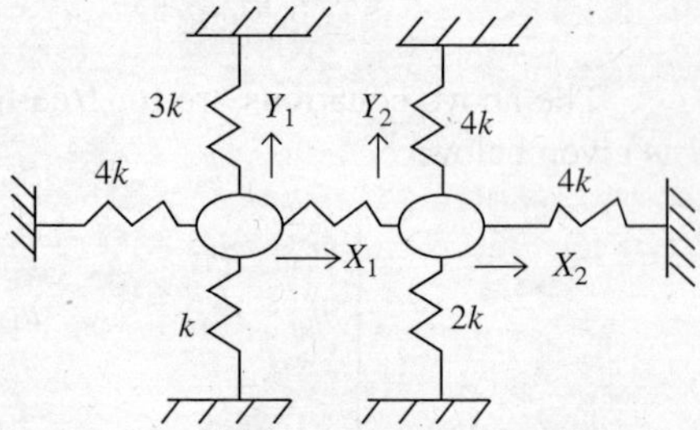

Figure 4.23

Solution:

Let x_1, y_1 and x_2, y_2 be the displacements of masses m_1 and m_2 as shown.

The equations of motion are written as

X-direction:
$$m\ddot{x}_1 + 4kx_1 + k(x_1 - x_2) = 0 = m\ddot{x}_1 + 5kx_1 - kx_2$$
$$m\ddot{x}_2 + 4kx_2 + k(x_2 - x_1) = 0 = m\ddot{x}_2 + 5kx_2 - kx_1$$

Y-direction:
$$m\ddot{y}_1 + (3k + k)y_1 = 0 = m\ddot{y}_1 + 4ky_1$$
$$m\ddot{y}_2 + (4k + 2k)y_2 = 0 = m\ddot{y}_2 + 6ky_2$$

Using notation q and p as discussed earlier, we can write the above equations as

$$\begin{bmatrix} m & 0 & 0 & 0 \\ 0 & m & 0 & 0 \\ 0 & 0 & m & 0 \\ 0 & 0 & 0 & m \end{bmatrix}\begin{Bmatrix} \ddot{q}_1 \\ \ddot{q}_2 \\ \ddot{q}_3 \\ \ddot{q}_4 \end{Bmatrix} + \begin{bmatrix} 5k & -k & 0 & 0 \\ -k & 5k & 0 & 0 \\ 0 & 0 & 4k & 0 \\ 0 & 0 & 0 & 6k \end{bmatrix}\begin{Bmatrix} q_1 \\ q_2 \\ q_3 \\ q_4 \end{Bmatrix} = \begin{Bmatrix} 0 \\ 0 \\ 0 \\ 0 \end{Bmatrix}$$

Thus, the inertia matrix
$$[M] = m\begin{bmatrix} 1 & 0 & 0 & 0 \\ 0 & 1 & 0 & 0 \\ 0 & 0 & 1 & 0 \\ 0 & 0 & 0 & 1 \end{bmatrix} = m[I]$$

The stiffness matrix $$[K] = k\begin{bmatrix} 5 & -1 & 0 & 0 \\ -1 & 5 & 0 & 0 \\ 0 & 0 & 4 & 0 \\ 0 & 0 & 0 & .6 \end{bmatrix}$$

The dynamic matrix $[C] = [M]^{-1}[K]$.

In this, $$[M]^{-1} = \frac{1}{m}\begin{bmatrix} 1 & 0 & 0 & 0 \\ 0 & 1 & 0 & 0 \\ 0 & 0 & 1 & 0 \\ 0 & 0 & 0 & 1 \end{bmatrix}$$

since $[I]$ is a unitary diagonal matrix

Therefore, $$[C] = \frac{k}{m}\begin{bmatrix} 1 & 0 & 0 & 0 \\ 0 & 1 & 0 & 0 \\ 0 & 0 & 1 & 0 \\ 0 & 0 & 0 & 1 \end{bmatrix}\begin{bmatrix} 5 & -1 & 0 & 0 \\ -1 & 5 & 0 & 0 \\ 0 & 0 & 4 & 0 \\ 0 & 0 & 0 & 6 \end{bmatrix} = \frac{k}{m}\begin{bmatrix} 5 & -1 & 0 & 0 \\ -1 & 5 & 0 & 0 \\ 0 & 0 & 4 & 0 \\ 0 & 0 & 0 & 6 \end{bmatrix}$$

Using λ-matrix theory, we obtain,

$$f(\lambda) = [\lambda[I] - [C]] = \begin{bmatrix} (\lambda - 5k/m) & k/m & 0 & 0 \\ k/m & (\lambda - 5k/m) & 0 & 0 \\ 0 & 0 & (\lambda - 4k/m) & 0 \\ 0 & 0 & 0 & (\lambda - 6k/m) \end{bmatrix}$$

and the characteristic equation is

$$\Delta(f) = |[\lambda I] - [C]| = 0 = \begin{vmatrix} (\lambda - 5k/m) & k/m & 0 & 0 \\ k/m & (\lambda - 5k/m) & 0 & 0 \\ 0 & 0 & (\lambda - 4k/m) & 0 \\ 0 & 0 & 0 & (\lambda - 6k/m) \end{vmatrix} = 0$$

The expansion of the determinant gives

$$\omega_1 = \sqrt{4k/m}, \omega_2 = \sqrt{6k/m}, \omega_3 = \sqrt{4k/m}, \omega_4 = \sqrt{6k/m}$$

Principal coordinates: We know $\{p\} = [a]^{-1}\{q\}$. Thus, the transformation matrix is formed by all principal modes of vibrations of the system. We thus need to determine the mode shapes. A closer examination of the equations of motion reveals that the Y-direction equations

$$\begin{aligned} m\ddot{y}_1 + 4ky_1 &= 0 \\ m\ddot{y}_2 + 6ky_2 &= 0 \end{aligned}$$ are uncoupled.

Also, the natural frequencies are $\omega_3 = \sqrt{4k/m}$ and $\omega_4 = \sqrt{6k/m}$. Hence coordinates are independent of each other.

The principal coordinates in Y-direction are $y_1/2$ and $y_2/2$.

We can add the two equations of motion in X-direction to obtain

$$m(\ddot{x}_1 + \ddot{x}_2) + 4k(x_1 + x_2) = 0$$

Defining $x_1 + x_2 = 2\ x_1'$, we can write this equation as

$$2m\ddot{x}_1' + 8kx_1' = 0 \tag{4.112}$$

The above equation is an uncoupled equation. Similarly, subtracting the equations of motion in X-direction, we get,

$$m(\ddot{x}_1 - \ddot{x}_2) + 6k(x_1 - x_2) = 0$$

Defining $x_1 - x_2 = 2x_2'$ and substituting in the above equations, we get,

$$2m\ddot{x}_2' + 12kx_2' = 0 \tag{4.113}$$

The above equation is also an uncoupled equation.
Obviously, x_1' and x_2' are the principal coordinates. This in other words, means that

$$x_1' = \frac{x_1}{2} + \frac{x_2}{2} \text{ and } x_2' = \frac{x_1}{2}\frac{x_2}{2} \text{ are the principal coordinates.}$$

Alternative method:

The X-direction equations are

$$m\ddot{x}_1 + 5kx_1 - kx_2 = 0 \tag{4.114}$$

$$m\ddot{x}_2 + 5kx_2 - kx_1 = 0 \tag{4.115}$$

Assume the solution as $x_1 = A\sin(\omega.t + \phi), x_2 = B\sin(\omega.t + \phi)$ (4.116)
Substituting Equations 4.116 into Equations 4.114 and 4.115, rearranging and simplifying, we get,

$$(5k - m\omega^2)A - kB = 0$$

$$(5k - m\omega^2)B - kA = 0$$

The above equations will have a non-trivial solution only if the determinant of these equations is zero. This means

$$\begin{vmatrix} (5k - m\omega^2) & -k \\ -k & (5k - \omega^2) \end{vmatrix} = 0$$

Expanding the above determinant, we obtain the frequency equation. Omitting the details, we get the two roots as $\omega_1 = \sqrt{4k/m}$ and $\omega_2 = \sqrt{6k/m}$ as found earlier.
The general solution of the equations of motion in X-direction can be written as

$$x_1 = A_1\sin(\omega_1 t + \phi_1) + A_2\sin(\omega_2 t + \phi_2) \tag{4.117}$$

$$x_2 = B_1\sin(\omega_1 t + \phi_1) + B_2\sin(\omega_2 t + \phi_2) \tag{4.118}$$

where A_1 and B_1 in the above equations are arbitrary constants. The amplitude ratios are

$$\frac{A_1}{B_1} = \frac{1}{\lambda_1} = \frac{k}{5k - m\omega_1^2} = \frac{5k - m\omega_1^2}{k}.$$

Substituting $\omega_1 = \sqrt{4k/m}$, we get,

$$\frac{A_1}{B_1} = \frac{1}{\lambda_1} = \frac{5k - m.4k/m}{k} = 1$$

Similarly, we shall get,

$$\frac{A_2}{B_2} = \frac{1}{\lambda_2} = \frac{5k - m\omega_2^2}{k} = \frac{5k - m.6k/m}{k} = -1$$

Thus $\lambda_1 = 1$ and $\lambda_2 = 1$. Equations 6 and 7 can now be written as

$$x_1(t) = A_1 \sin(\sqrt{4k/m}t + \phi_1) + A_2 \sin(\sqrt{6k/m}t + \phi_2) \tag{4.119}$$

$$x_2(t) = A_1 \sin(\sqrt{4k/m}t + \phi_1) - A_2 \sin(\sqrt{6k/m}t + \phi_2) \tag{4.120}$$

Let $x_1' = A_1\sin(\sqrt{4k/m}\,t + \phi_1)$ and $x_2' = A_2\sin(\sqrt{6k/m}t + \phi_2)$, which are pure sine waves. Then $x_1' + x_2' = x_1$ and $x_1' - x_2' = x_2$. These relations give the principal coordinates as

$$x_1' = \frac{x_1}{2} + \frac{x_2}{2}$$

$$x_2' = \frac{x_1}{2} - \frac{x_2}{2}$$

Problem 4.13

Find, for the system shown in Fig. 4.24, the natural frequencies.

Solution:

Let us demonstrate the matrix-iteration method. If $[a]$ denotes the flexibility-influence-coefficient matrix, the equations of motion can be very conveniently written using the influence coefficients as shown below.

$$-x_1 = a_{11} 4m\ddot{x}_1 + a_{12} 2m\ddot{x}_2 + a_{13} m\ddot{x}_3 \tag{1}$$

$$-x_2 = a_{21} 4m\ddot{x}_1 + a_{22} 2m\ddot{x}_2 + a_{23} m\ddot{x}_3 \tag{2}$$

$$-x_3 = a_{31} 4m\ddot{x}_1 + a_{23} 2m\ddot{x}_2 + a_{33} m\ddot{x}_3 \tag{3}$$

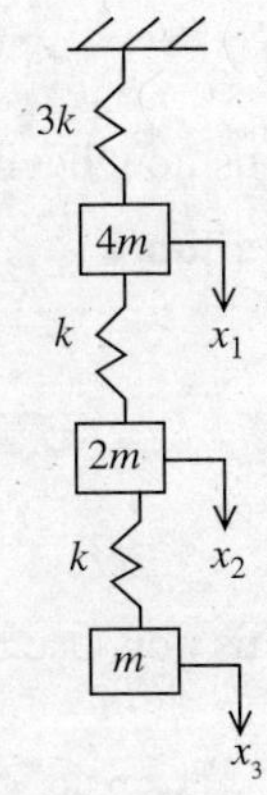

Figure 4.24

Let $x_i = X_i \sin(\omega.t + \phi)$, $\quad \therefore \ddot{x}_i = -\omega^2 X_i \sin(\omega.t + \phi)$

Substituting the above equations in Equations 1, 2, and 3, we get

$$X_1 = a_{11} 4mX_1\omega^2 + a_{12} 2mX_2\omega^2 + a_{13} mX_3\omega^2 \tag{4}$$

$$X_2 = a_{21} 4mX_1\omega^2 + a_{22} 2mX_2\omega^2 + a_{23} mX_3\omega^2 \tag{5}$$

$$X_3 = a_{31} 4mX_1\omega^2 + a_{32} 2mX_2\omega^2 + a_{33} mX_3\omega^2 \tag{6}$$

Equations 4, 5, and 6 can be written in matrix form as

$$\begin{Bmatrix} X_1 \\ X_2 \\ X_3 \end{Bmatrix} = m\omega^2 \begin{bmatrix} 4a_{11} & 2a_{12} & a_{13} \\ 4a_{21} & 2a_{22} & a_{23} \\ 4a_{31} & 2a_{32} & a_{33} \end{bmatrix} \begin{Bmatrix} X_1 \\ X_2 \\ X_3 \end{Bmatrix} \tag{7}$$

Let us now find the influence coefficients.

(a) First apply unit force at mass $4m$. With this, spring of stiffness $3k$ will deform by $1/3k$, which by definition is a_{11}. Therefore, $a_{11} = 1/3k$.

When the mass $4m$ deflects by $a_{11} = 1/3k$ under the action of unit force the masses $2m$ and m will simply move bodily by the same amount. This means that $a_{21} = a_{31} = 1/3k$. Additionally, since $a_{ij} = a_{ji}$, $a_{12} = a_{13} = 1/3k$.

(b) Now apply unit force at $2m$ location. The springs $3k$ and k are in series and their equivalent spring constant is given by $1/k_{eq} = 1/3k + 1/k, \therefore k_{eq} = 3k/4$. The deflection is therefore $1/3k/4 = 4/3k = a_{22}$ and as mass m hangs on mass $2m$, $a_{32} = a_{22} = a_{23} = 4/3k$.

(c) Now apply unit force at mass m. Since the springs $3k$, k and k are in series, the equivalent spring stiffness is given by $1/k_{eq} = 1/3k + 1/k + 1/k, \therefore k_{eq} = 7k/4$ and $a_{33} = 1/3k/7 = 7/3k$.

Thus, the influence-coefficient matrix becomes

$$[a] = \begin{bmatrix} \frac{1}{3}k & \frac{1}{3}k & \frac{1}{3}k \\ \frac{1}{3}k & \frac{4}{3}k & \frac{4}{3}k \\ \frac{1}{3}k & \frac{4}{3}k & \frac{7}{3}k \end{bmatrix}$$

Substituting the above equation in the matrix Equation 7, we get,

$$\begin{Bmatrix} X_1 \\ X_2 \\ X_3 \end{Bmatrix} = (\omega^2 m/3k) \begin{bmatrix} 4 & 2 & 1 \\ 4 & 8 & 4 \\ 4 & 8 & 7 \end{bmatrix} \begin{Bmatrix} X_1 \\ X_2 \\ X_3 \end{Bmatrix} \tag{8}$$

Let us now begin the iteration process. To start with, let us assume $X_1 = 1, X_2 = 2, X_3 = 4$

Iteration 1

$$\begin{Bmatrix} 1 \\ 2 \\ 4 \end{Bmatrix} = \frac{\omega^2 m}{3k} \begin{bmatrix} 4 & 2 & 1 \\ 4 & 8 & 4 \\ 4 & 8 & 7 \end{bmatrix} \begin{Bmatrix} 1 \\ 2 \\ 4 \end{Bmatrix} = \frac{\omega^2 m}{3k} \begin{Bmatrix} 12 \\ 36 \\ 48 \end{Bmatrix} = \frac{\omega^2 m}{3k}(12) \begin{Bmatrix} 1 \\ 3 \\ 4 \end{Bmatrix}$$

Let us now use new vector $\begin{Bmatrix} 1 \\ 3 \\ 4 \end{Bmatrix}$

Iteration 2

$$\begin{Bmatrix} 1 \\ 3 \\ 4 \end{Bmatrix} = \frac{\omega^2 m}{3k} \begin{bmatrix} 4 & 2 & 1 \\ 4 & 8 & 4 \\ 4 & 8 & 7 \end{bmatrix} \begin{Bmatrix} 1 \\ 3 \\ 4 \end{Bmatrix} = \frac{\omega^2 m}{3k} \begin{Bmatrix} 14 \\ 44 \\ 56 \end{Bmatrix} = \frac{\omega^2 m}{3k}(14) \begin{Bmatrix} 1 \\ 3.2 \\ 4 \end{Bmatrix}$$

Let us now use the new vector $\begin{Bmatrix} 1 \\ 3.2 \\ 4 \end{Bmatrix}$ for the third iteration.

Iteration 3

$$\begin{Bmatrix} 1 \\ 3.2 \\ 4 \end{Bmatrix} = \frac{\omega^2 m}{3k} \begin{bmatrix} 4 & 2 & 1 \\ 4 & 8 & 4 \\ 4 & 8 & 7 \end{bmatrix} \begin{Bmatrix} 1 \\ 3.2 \\ 4 \end{Bmatrix} = \frac{\omega^2 m}{3k} \begin{Bmatrix} 14.4 \\ 45.6 \\ 57.6 \end{Bmatrix} = \frac{\omega^2 m}{3k}(14.4) \begin{Bmatrix} 1 \\ 3.18 \\ 4 \end{Bmatrix}$$

We can now note that $\begin{Bmatrix} 1 \\ 3.2 \\ 4 \end{Bmatrix} \cong \begin{Bmatrix} 1 \\ 3.18 \\ 4 \end{Bmatrix}$. We can therefore conclude that the vector $\begin{Bmatrix} 1 \\ 3.2 \\ 4 \end{Bmatrix}$ is almost the correct vector. We can further refine over this but for all practical purposes, we have found the correct vector as $\begin{Bmatrix} 1 \\ 3.2 \\ 4 \end{Bmatrix}$. Substituting this in Equation 8, we get,

$$\begin{Bmatrix} 1 \\ 3.2 \\ 4 \end{Bmatrix} = \frac{14.4m\omega^2}{3k} \begin{Bmatrix} 1 \\ 3.18 \\ 4 \end{Bmatrix}, \quad \therefore 1 = \frac{14.4m\omega^2}{3k}, \quad \therefore \omega_1 = 0.46\sqrt{k/m} \text{ rad/s}$$

To obtain the second and third natural frequency, we make use of orthogonality principle.
For the first and second mode, we have,

$$m_1 A_1 A_2 + m_2 B_1 B_2 + m_3 C_1 C_2 = 0 \tag{9}$$

In this equation, $A_1 = 1$, $B_1 = 3.2$, $C_1 = 4$.
Substituting these in Equation 9, we get,

$$A_2 = -1.6B_2 - C_2,\ B_2 = B_2,\ C_2 = C_2.$$

This equation can be written as

$$\begin{Bmatrix} A_2 \\ B_2 \\ C_2 \end{Bmatrix} = \begin{bmatrix} 0 & -1.6 & -1.0 \\ 0 & 1 & 0 \\ 0 & 0 & 1 \end{bmatrix} \begin{Bmatrix} A_2 \\ B_2 \\ C_2 \end{Bmatrix} \tag{10}$$

When Equation 10 is combined with matrix equation of first mode, it will converge to second mode

$$\begin{Bmatrix} X_1 \\ X_2 \\ X_3 \end{Bmatrix} = \frac{\omega^2 M}{3k} \begin{bmatrix} 4 & 2 & 1 \\ 4 & 8 & 4 \\ 4 & 8 & 7 \end{bmatrix} \begin{bmatrix} 0 & -1.6 & -1.0 \\ 0 & 1 & 0 \\ 0 & 0 & 1 \end{bmatrix} \begin{Bmatrix} X_1 \\ X_2 \\ X_3 \end{Bmatrix} = \frac{\omega^2 m}{3k} \begin{bmatrix} 0 & -4.4 & -3 \\ 0 & 1.6 & 0 \\ 0 & 1.6 & 3 \end{bmatrix} \begin{Bmatrix} X_{1_1} \\ X_2 \\ X_3 \end{Bmatrix} \tag{11}$$

Let us assume second mode $\begin{Bmatrix} 1 \\ 0 \\ -1 \end{Bmatrix}$. When this is used to start the iteration process, we have

$$\begin{Bmatrix} 1 \\ 0 \\ -1 \end{Bmatrix} = \frac{\omega^2 m}{3k} \begin{bmatrix} 0 & -4.4 & -3 \\ 0 & 1.6 & 0 \\ 0 & 1.6 & 3 \end{bmatrix} \begin{Bmatrix} 1 \\ 0 \\ -1 \end{Bmatrix} = \frac{\omega^2 m}{3k}(3) \begin{Bmatrix} 1 \\ 0 \\ -1 \end{Bmatrix}.$$

Thus, the assumed second mode $\begin{Bmatrix} 1 \\ 0 \\ -1 \end{Bmatrix}$ is correct as the equation repeats.

Thus,

$$\omega^2 m/k = 1 \quad \text{or} \quad \omega_2 = \sqrt{k/m}.$$

It is important to clearly understand the steps at Equations 9–11. Equation 9 is the equation of Orthogonality of modes. By writing equation 9 and finding expression for A_2, B_2, and C_2, we have suppressed first mode, thereby enabling us obtain the second mode. This process of successive suppression of modes must be well understood and remembered.

To obtain the third mode, use the orthogonality equations

$$m_1 A_2 A_3 + m_2 B_2 B_3 + m_3 C_2 C_3 = 0$$
$$m_1 A_1 A_3 + m_2 B_1 B_3 + m_3 C_1 C_3 = 0$$

We have $A_1 = 1, B_1 = 3.2, C_1 = 4, A_2 = 1, B_2 = 0, C_2 = -1$.
From this we obtain,

$$A_3 = 0.25C_3, B_3 = -0.781C_3.$$

This gives

$$\begin{Bmatrix} A_3 \\ B_3 \\ C_3 \end{Bmatrix} = \begin{bmatrix} 0 & 0 & 0.25 \\ 0 & 0 & -0.78 \\ 0 & 0 & 1 \end{bmatrix} \begin{Bmatrix} A_3 \\ B_3 \\ C_3 \end{Bmatrix}$$

When the above equation is combined with matrix Equation 7, it will yield the third mode

$$\begin{Bmatrix} X_1 \\ X_2 \\ X_3 \end{Bmatrix} = \frac{\omega^2 m}{3k} \begin{bmatrix} 0 & -4.4 & -3 \\ 0 & 1.6 & 0 \\ 0 & 1.6 & 3 \end{bmatrix} \begin{bmatrix} 0 & 0 & 0.25 \\ 0 & 0 & -0.78 \\ 0 & 0 & 1 \end{bmatrix} \begin{Bmatrix} X_1 \\ X_2 \\ X_3 \end{Bmatrix} = \frac{\omega^2 m}{3k} \begin{bmatrix} 0 & 0 & 0.43 \\ 0 & 0 & -1.25 \\ 0 & 0 & 1.0 \end{bmatrix} \begin{Bmatrix} X_1 \\ X_2 \\ X_3 \end{Bmatrix}$$

We can show that the vector $\{ 0.25, \ -0.72, \ 1 \}^T$ represents the third mode and $\omega_3 = 1.32\sqrt{k/m}$.
Discussion: The matrix-iteration method assumes that the natural frequencies ($\omega_1 < \omega_2 < \omega_3 < .... < \omega_n$) are distinct. The iteration is started by selecting a trial vector $\bar{X}_1$ (we assumed $\bar{X}_1 = \begin{Bmatrix} 1 \\ 2 \\ 4 \end{Bmatrix}$). This trial vector is pre-multiplied by the matrix [*a*] in this example or by dynamical matrix [*D*]. The resulting column vector $\begin{Bmatrix} 12 \\ 36 \\ 48 \end{Bmatrix}$ is then normalized by making its lowest component unity, that is, $\begin{Bmatrix} 1 \\ 3 \\ 4 \end{Bmatrix}$ in this example. This normalized vector is pre-multiplied by matrix [*a*] in this example or by [*D*] to obtain a third-column vector $\begin{Bmatrix} 14 \\ 36 \\ 48 \end{Bmatrix}$ in this example. This vector is once again normalized to $\begin{Bmatrix} 1 \\ 3.2 \\ 4 \end{Bmatrix}$, which again becomes a trial vector. The process is continued until the successive normalized column vectors converge to a common vector [$\begin{Bmatrix} 1 \\ 3.2 \\ 4 \end{Bmatrix}$ is almost equal to $\begin{Bmatrix} 1 \\ 3.18 \\ 4 \end{Bmatrix}$]. The iterations are then stopped since the first eigen vector has been found out. The normalizing factor gives $\lambda = 1/\omega^2$, that is, the fundamental frequency.

In the example illustrated, we dealt with flexibility-influence coefficient matrix and wrote the frequency Equation 7 in the example.

In general, the equation of motion is (as discussed earlier)

$$[[k] - \omega^2 [m]]\bar{X} = \bar{0}$$

or

$$[\lambda[I]-[D]]\overline{X}=\overline{0} \qquad [4.121(a)]$$

or

$$[D]\overline{X}^{(i)}=\lambda_i[I]\overline{X}^{(i)}$$

In the above equations, $[D]$ = $[k]^{-1}[m]$ or = $[a][m]$. $[k]$ is the stiffness matrix and $[m]$ is the mass matrix. We illustrated the interactive process by an example. Let us know understand the theory behind it.

According to expansion theorem, any arbitrary vector $\overline{X}_1$ can be expressed as a linear combination of n orthogonal eigen vectors of the system $\overline{X}^{(i)}, i=1,2,3...n$ as given here.

$$\overline{X}_1=C_1\overline{X}^{(1)}+C_2\overline{X}^{(2)}+\cdots+C_n\overline{X}^{(n)} \qquad [4.121(b)]$$

C_1, C_2, …, C_n are constants. In the iteration method, the trial vector $\overline{X}_1$ is selected arbitrarily and is therefore, a known vector. The unknown modal vectors $\overline{X}^{(i)}$ are constant vectors because they depend upon only the properties of the system. The constants C_i are unknown numbers to be determined. As discussed while explaining the steps in the illustrated example, we pre-multiply $\overline{X}_1$ by the matrix $[D]$. In view of Equation 4.121, this gives

$$D\overline{X}_1=C_1[D]\overline{X}^{(1)}+C_2[D]\overline{X}^{(2)}+.....+C_n[D]\overline{X}^{(n)} \qquad (4.122)$$

But, as per Equation 4.121a

$$[D]\overline{X}^{(i)}=\lambda_i[I]\overline{X}^{(i)}=\frac{1}{\omega^2}\overline{X}^{(i)}, \quad i=1,2,3\cdots n \qquad (4.123)$$

Substituting Equation 4.123 into 4.122, we get

$$[D]\overline{X}_1=\overline{X}_2=\frac{C_1}{\omega_1^2}\overline{X}^{(1)}+\frac{C_2}{\omega_2^2}\overline{X}^{(2)}+......+\frac{C_n}{\omega_n^2}\overline{X}^{(n)} \qquad (4.124)$$

where $\overline{X}_2$ is now the new trial vector. We now repeat the process and pre-multiply $\overline{X}_2$ by $[D]$ to obtain

$$[D]\overline{X}_2=\overline{X}_3=\frac{C_1}{\omega_1^4}\overline{X}^{(1)}+\frac{C_2}{\omega_2^4}\overline{X}^{(2)}+......+\frac{C_n}{\omega_n^4}\overline{X}^{(n)} \qquad (4.125)$$

By repeating the process, we obtain after rth iteration

$$[D]\overline{X}_r=\overline{X}_{r+1}=\frac{C_1}{\omega_1^{2r}}\overline{X}^{(1)}+\frac{C_2}{\omega_2^{2r}}\overline{X}^{(2)}+......+\frac{C_n}{\omega_n^{2r}}\overline{X}^{(n)} \qquad (4.126)$$

Since $\omega_1<\omega_2<\omega_3<....<\omega_n$, a sufficiently large value of r yields

$$\frac{1}{\omega_1^{2r}}>>\frac{1}{\omega_2^{2r}}>>\frac{1}{\omega_3^{2r}}.......>>\frac{1}{\omega_n^{2r}} \qquad (4.127)$$

Thus the first term on the right hand side of Equation 4.126 becomes the only significant one. Hence, we have

$$\overline{X}_{r+1}=\frac{C_1}{\omega_1^{2r}}\overline{X}^{(1)} \qquad (4.128)$$

This means that the $(r+1)$th trial vector becomes identical to the fundamental modal vector to within a multiplication constant. Since

$$\overline{X}_r = \frac{C_1}{\omega_1^{2(r-1)}} \overline{X}^{(1)} \tag{4.129}$$

the fundamental natural frequency can be found by taking the ratio of any two corresponding components in the vector $\overline{X}_r$ and $\overline{X}_{r+1}$

$$\omega_1^2 \cong \frac{X_{i,r}}{X_{i,r+1}} \quad \text{for any } i = 1, 2, \ldots, n, \tag{4.130}$$

where $X_{i,r}$ and $X_{i,r+1}$ are the elements of vectors $\overline{X}_r$ and $\overline{X}_{r+1}$, respectively.

Highest mode: The eigen value equation is

$\lambda[I]\overline{X} = [D]\overline{X}, \lambda = 1/\omega^2$ can also be written as

$$[D]^{-1}\overline{X} = \omega^2[I]\overline{X} = \omega^2\overline{X} \tag{4.131}$$

where $[D]^{-1} = [m]^{-1}[k]$. We can now select any arbitrary trial vector and pre-multiply it by $[D]^{-1}$ to obtain an improved trial vector $\overline{X}_2$. The sequence of trial vectors $\overline{X}_{i+1}, i = 1,2,3$ obtained by pre-multiplying by $[D]^{-1}$ converges to highest normal mode $\overline{X}^{(n)}$. The procedure is similar to the one already described. The constant of proportionality in this case is ω^2 instead of $1/\omega^2$.

Intermediate modes: It should be remembered that any arbitrary vector when pre-multiplied by $[D]^{-1}$ would again lead to the largest eigen value. It is thus necessary to remove the largest eigen value from matrix D. The method to do so was explained in the problem we solved where we used Orthogonality principle of normal modes. The succeeding eigen values and eigen vectors can be obtained by eliminating the root λ_i from the characteristic equation

$$\left|[D] - \lambda[I]\right| = 0 \tag{4.132}$$

A procedure known as matrix deflation can also be used for this purpose. To find the eigen vector $\overline{X}^{(i)}$ by this procedure, the previous eigen vector $\overline{X}^{(i-1)}$is normalized with respect to the mass matrix such that

$$\overline{X}^{(i-1)}[m]\overline{X}^{(i-1)} = \overline{1} \tag{4.133}$$

The deflated matrix $[D_i]$ is then constructed as

$$[D_i] = [D_{(i-1)}] - \lambda_{(i-1)}\overline{X}^{(i-1)}\overline{X}^{(i-1)T}[m], i = 1,2,3\ldots n \tag{4.134}$$

Once $[D]^{-1}$ is constructed, the iterative scheme

$$\overline{X}_{r+1} = [D_i]\overline{X}_r \tag{4.135}$$

is used where $\overline{X}_1$ is an arbitrary trial eigen vector.

It can thus be seen that the procedure for obtaining natural frequencies and mode shapes is lot easier when we use equations of motion in terms of influence coefficients. Although the procedure to be adopted when we use [D] matrix may appear to be somewhat tedious, it is very well suited for computer programming.

Problem 4.14

Find the natural frequencies of the triple pendulum shown in Fig. 4.25.

Solution:

We write the equations of equilibrium in terms of influence coefficients as

$$-x_1 = a_{11}m_1\ddot{x}_1 + a_{12}m_2\ddot{x}_2 + a_{13}m_3\ddot{x}_3$$
$$-x_2 = a_{21}m_1\ddot{x}_1 + a_{22}m_2\ddot{x}_2 + a_{23}m_3\ddot{x}_3$$
$$-x_3 = a_{31}m_1\ddot{x}_1 + a_{23}m_2\ddot{x}_2 + a_{33}m_3\ddot{x}_3$$

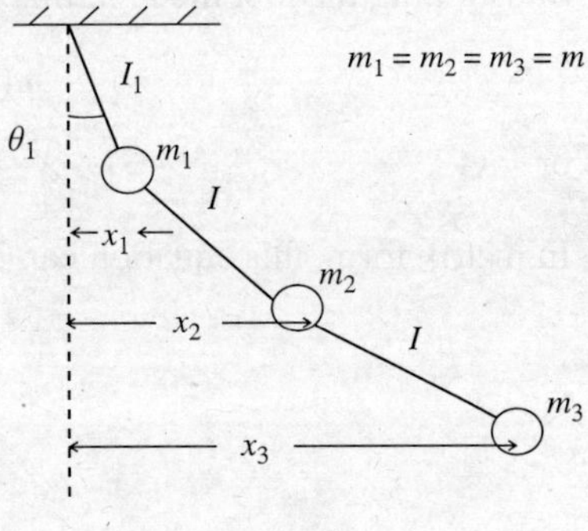

Figure 4.25

Let $x_i = X_i \sin(\omega.t + \phi), \therefore \ddot{x}_i = -\omega^2 X_i \sin(\omega.t + \phi)$. By substituting in Equations above,

$$X_1 = a_{11}m_1X_1\omega^2 + a_{12}m_2X_2\omega^2 + a_{13}m_3X_3\omega^2$$
$$X_2 = a_{21}m_1X_1\omega^2 + a_{22}m_2X_2\omega^2 + a_{23}m_3X_3\omega^2$$
$$X_3 = a_{31}m_1X_1\omega^2 + a_{32}\ m_2X_2\omega^2 + a_{33}m_3X_3\omega^2$$

Refer to the solution of problem 4.10, where we have evaluated the influence coefficients.

$$a_{11} = a_{12} = a_{13} = l/3gm = a_{21} = a_{31}$$
$$a_{22} = a_{23} = a_{32} = l/3gm + l/2gm = 5l/6gm$$

$a_{33} = l/3gm + l/2gm + l/gm = 11l/6gm$. In matrix notation, the Equations of motion become

$$\begin{Bmatrix} X_1 \\ X_2 \\ X_3 \end{Bmatrix} = \frac{l\omega^2}{6g}\begin{bmatrix} 2 & 2 & 2 \\ 2 & 5 & 5 \\ 2 & 5 & 11 \end{bmatrix}\begin{Bmatrix} X_1 \\ X_2 \\ X_3 \end{Bmatrix} \qquad (1)$$

Let us assume trial vector as $\begin{Bmatrix} 0.2 \\ 0.6 \\ 1 \end{Bmatrix}$

Iteration 1

$$\begin{Bmatrix} 0.2 \\ 0.6 \\ 1 \end{Bmatrix} = \frac{l\omega^2}{6g}\begin{bmatrix} 2 & 2 & 2 \\ 2 & 5 & 5 \\ 2 & 5 & 11 \end{bmatrix}\begin{Bmatrix} 0.2 \\ 0.6 \\ 1 \end{Bmatrix} = \frac{l\omega^2}{6g}\begin{Bmatrix} 3.6 \\ 8.4 \\ 14.4 \end{Bmatrix} = \frac{l\omega^2}{6g}(14.4)\begin{Bmatrix} 0.25 \\ 0.58 \\ 1 \end{Bmatrix}$$

Iteration 2

$$\begin{Bmatrix} 0.25 \\ 0.58 \\ 1 \end{Bmatrix} = \frac{l\omega^2}{6g}\begin{bmatrix} 2 & 2 & 2 \\ 2 & 5 & 5 \\ 2 & 5 & 11 \end{bmatrix}\begin{Bmatrix} 0.25 \\ 0.58 \\ 1 \end{Bmatrix} = \frac{l\omega^2}{6g}\begin{bmatrix} 3.62 \\ 8.42 \\ 14.4 \end{bmatrix} = \frac{l\omega^2}{6g}(14.4)\begin{Bmatrix} 0.25 \\ 0.58 \\ 1 \end{Bmatrix}$$

Since the solution repeats itself, the iteration process is complete. To find the natural frequency, we write $\frac{l\omega^2}{6g}(14.4) = 1$. From this we get, $\omega_1 = 0.65\sqrt{g/l}$. The first mode is $\begin{Bmatrix} 0.25 \\ 0.58 \\ 1 \end{Bmatrix}$

To obtain the second mode, the first mode has to be suppressed during the iteration process. This is achieved using Orthogonality principle ($m_1A_1A_2 + m_2B_1B_2 + m_3C_1C_2) = 0$.

Substituting the first mode in this equation, we get,

$$m(0.25)X_1 + m90.58)X_2 + m(1)X_3 = 0$$

or

$$X_1 = -2.32X_2 - 4X_3, X_2 = X_2, X_3 = X_3.$$

In matrix form, this equation can be written as

$$\begin{Bmatrix} X_1 \\ X_2 \\ X_3 \end{Bmatrix} = \begin{bmatrix} 0 & -2.32 & -4 \\ 0 & 1 & 0 \\ 0 & 0 & 1 \end{bmatrix} \begin{Bmatrix} X_1 \\ X_2 \\ X_3 \end{Bmatrix}$$

Now combine this with the fundamental matrix of Equation 1 that has no first mode present.

$$\begin{Bmatrix} X_1 \\ X_2 \\ X_3 \end{Bmatrix} = \frac{l\omega^2}{6g} \begin{bmatrix} 2 & 2 & 2 \\ 2 & 5 & 5 \\ 2 & 5 & 11 \end{bmatrix} \begin{bmatrix} 0 & -2.32 & -4 \\ 0 & 1 & 0 \\ 0 & 0 & 1 \end{bmatrix} \begin{Bmatrix} X_1 \\ X_2 \\ X_3 \end{Bmatrix} = \frac{l\omega^2}{6g} \begin{bmatrix} 0 & -2.6 & -6 \\ 0 & 0.4 & -3 \\ 0 & 0.4 & 3 \end{bmatrix} \begin{Bmatrix} X_1 \\ X_2 \\ X_3 \end{Bmatrix}$$

Now let us begin iterations for second mode. Let us assume $\{-1 \quad -1 \quad -1\}^T$ as starting vector.

First iteration

$$\begin{Bmatrix} -1 \\ -1 \\ -1 \end{Bmatrix} = \frac{l\omega^2}{6g} \begin{bmatrix} 0 & -2.6 & -6 \\ 0 & 0.4 & -3 \\ 0 & 0.4 & 3 \end{bmatrix} \begin{Bmatrix} -1 \\ -1 \\ -1 \end{Bmatrix} = \frac{l\omega^2}{6g}(2.6) \begin{Bmatrix} -1.3 \\ -1.3 \\ 1 \end{Bmatrix}$$

Second iteration

$$\begin{Bmatrix} -1.3 \\ -1.3 \\ 1 \end{Bmatrix} = \frac{l\omega^2}{6g} \begin{bmatrix} 0 & -2.6 & -6 \\ 0 & 0.4 & -3 \\ 0 & 0.4 & 3 \end{bmatrix} \begin{Bmatrix} -1.3 \\ -1.3 \\ 1 \end{Bmatrix} = \frac{l\omega^2}{6g} \begin{Bmatrix} -2.6 \\ -3.5 \\ 2.5 \end{Bmatrix} = \frac{l\omega^2}{6g}(2.5) \begin{Bmatrix} -1.05 \\ -1.4 \\ 1 \end{Bmatrix}$$

Third iteration

$$\begin{Bmatrix} -1.04 \\ -1.4 \\ 1 \end{Bmatrix} = \frac{l\omega^2}{6g} \begin{bmatrix} 0 & -2.6 & -6 \\ 0 & 0.4 & -3 \\ 0 & 0.4 & 3 \end{bmatrix} \begin{Bmatrix} -1.04 \\ -1.4 \\ 1 \end{Bmatrix} = \frac{l\omega^2}{6g} \begin{Bmatrix} -2.4 \\ -3.5 \\ 2.5 \end{Bmatrix} \cong \frac{l\omega^2}{6g}(2.5) \begin{Bmatrix} -1 \\ -1.4 \\ 1 \end{Bmatrix}$$

The third iteration appears to be satisfactory. The second mode shape and the frequency are given by

$$\begin{Bmatrix} -1 \\ -1.4 \\ 1 \end{Bmatrix}, \ 1 = \frac{l\omega^2}{6g}(2.50, \therefore \omega_2 = 1.52\sqrt{g/l}$$

In order to obtain the third mode and third natural frequency of the system, both first and second modes must be suppressed so that they do not appear in the iteration process. This is again done by using the Orthogonality principle as expressed by

$$m_1 A_1 A_3 + m_2 B_1 B_3 + m_3 C_1 C_3 = 0.$$

Thus, for the first and third mode, we get

$$m(0.25)X_1 + m(0.58)X_2 + m(1)X_3 = 0 \tag{2}$$

For the second and third modes, we have,

$$m_1A_2A_3 + m_2B_2B_3 + m_3C_2C_3 = 0$$

$$m(-1)X_1 + m(-1.4)X_2 + m(1)X_3 = 0 \tag{3}$$

Solving Equations 2 and 3, we get,

$$X_1 = 8X_3, X_2 = -5X_3, X_3 = X_3.$$

We can write the solution in matrix form

$$\begin{Bmatrix} X_1 \\ X_2 \\ X_3 \end{Bmatrix} = \begin{bmatrix} 0 & 0 & 8 \\ 0 & 0 & -5 \\ 0 & 0 & 1 \end{bmatrix} \begin{Bmatrix} X_1 \\ X_2 \\ X_3 \end{Bmatrix}$$

When this is combined with matrix Equation of second mode, we obtain,

$$\begin{Bmatrix} X_1 \\ X_2 \\ X_3 \end{Bmatrix} = \frac{l\omega^2}{6g} \begin{bmatrix} 0 & -2.6 & -6 \\ 0 & 0.4 & -3 \\ 0 & 0.4 & 3 \end{bmatrix} \begin{bmatrix} 0 & 0 & 8 \\ 0 & 0 & -5 \\ 0 & 0 & 1 \end{bmatrix} \begin{Bmatrix} X_1 \\ X_2 \\ X_3 \end{Bmatrix} = \frac{l\omega^2}{6g} \begin{bmatrix} 0 & 0 & 7 \\ 0 & 0 & -5 \\ 0 & 0 & 1 \end{bmatrix} \begin{Bmatrix} X_1 \\ X_2 \\ X_3 \end{Bmatrix}$$

Assume any convenient vector for the iteration. It will be found that the mode shape $\begin{Bmatrix} 7 \\ -5 \\ 1 \end{Bmatrix}$ is obtained repeatedly. This means that $\begin{Bmatrix} 7 \\ -5 \\ 1 \end{Bmatrix}$ is the third mode. The third-mode natural frequency is given by $\frac{l\omega^2}{6g} = 1, \therefore \omega_3 = 2.45\sqrt{g/l}.$

Besides the matrix iteration method, there are many other methods available for evaluating the natural frequencies of the multi degrees-of-freedom systems. Some of the important methods are: (1) Holzer method, which is extensively used in the analysis of torsional vibrations, (2) Stodola method, (3) Dunkerley method, and (4) Rayleigh's method. We shall discuss them one by one.

4.9.2 Holzer Method

Holzer method is a popular method for carrying out vibration analysis of torsional systems. We shall deal with this method in the analysis of various natural frequencies of multi-disc-rotor systems. This method is equally applicable to translational vibration systems. Holzer method is a tabular method used for the determination of natural frequencies of free systems; it is also applicable for the analysis of forced vibrations with or without damping. The method essentially is based upon successive assumptions of the natural frequency of the system, each followed by the calculation of the configuration governed by that assumed frequency. The method can be used for computing all the natural frequencies of a system and each calculation is completely independent of the others.

Figure 4.26 shows a spring–mass system free at both ends (semi-definite system). The equations of motion are

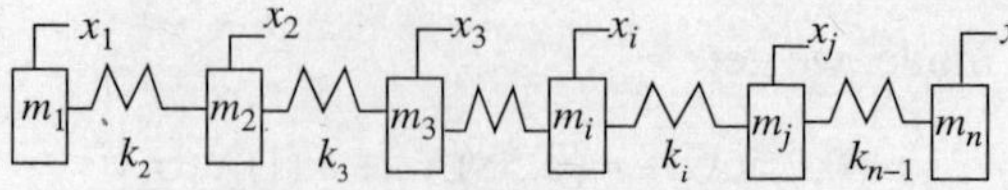

Figure 4.26 Holzer Method

$$m_1\ddot{x}_1 + k_1(x_1 - x_2) = 0$$
$$m_2\bar{x}_2 + k_1(x_2 - x_1) + k_2(x_2 - x_3) = 0 \quad (4.136)$$
$$\cdots\cdots$$
$$m_n\ddot{x}_n + k_{n-1}(x_n - x_{n-1}) = 0$$

For harmonic oscillations, let $x_i = X_i \sin\omega.t, X_i$ = amplitude of mass m_i. Using this, Equations 4.136 can be written as

$$\omega^2 m_1 X_1 = k_1(X_1 - X_2)$$
$$\omega^2 m_2 X_2 = k_1(X_2 - X_1) + k_2(X_2 - X_3) \quad (4.137)$$
$$\cdots\cdots$$
$$\omega^2 m_n X_n = k_{n-1}(X_n - X_{n-1})$$

In Holzer method, we start with a trial natural frequency and amplitude of mass m_1 as $X_1 = 1$. With this, Equation 4.137 can be used to obtain the amplitudes of masses m_2, m_3 .... m_n. This means

$$X_2 = X_1 - \frac{\omega^2 m_1}{k_1} X_1 \quad (4.138)$$

$$X_3 = X_2 - \frac{\omega^2}{k_2}(m_1 X_1 + m_2 X_2) \quad (4.139)$$

$$\cdots\cdots$$

$$X_i = X_{i-1} - \frac{\omega^2}{k_{i-1}} \sum_{k=1}^{i-1} (m_k X_k), i = 1,2,3,...n \quad (4.140)$$

The resultant of all inertia forces can be computed as

$$F = \sum_{i=1}^{n} \omega^2 m_i X_i \quad (4.141)$$

For the correct assumed frequency ω, $F = 0$ when $\omega = \omega_n$

The calculations are repeated with several other frequencies. The natural frequencies are identified as those values of ω that give $F = 0$ for the semi-definite systems. The best way to do this is to plot a graph between F and ω and identify the zero-crossing points. These points are nothing but the natural frequencies.

Let us demonstrate the procedure by solving Problem 4.15 [Fig. 4.26(a)].

Problem 4.15

For the system shown in Fig. 4.26(a), find the natural frequencies. Assume all masses equal to 1 and all springs having same stiffness $k = 1$.

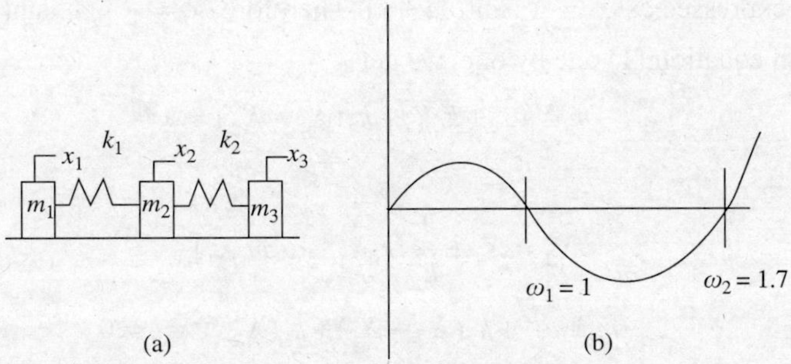

Figure 4.26 (a) Problem 4.15, (b) Identification of Natural Frequencies

Solution:
Look into the table below.

Position	m_i	$m_i\omega^2$	x_i	$m_i x_i \omega^2$	$m_i x_i \omega^2$	k_{ij}	$\Sigma m_i x_i \omega^2/k_{ij}$
Assumed frequency $\omega = 0.5$							
1	1	0.25	1	0.25	0.25	1	0.25
2	1	0.25	0.75	0.19	0.44	1	0.44
3	1	0.25	0.31	0.07	0.51		
Assumed frequency $\omega = 0.75$							
1	1	0.56	1	0.56	0.56	1	0.56
2	1	0.56	0.44	0.24	0.8	1	0.8
3	1	0.56	–0.36	–0.2	0.6		

In this manner, for various assumed frequencies the last column ($F = m_i x_i \omega^2/k_{ij}$) is evaluated and plotted as shown in Fig. 4.26(b). The zero-cross-over points give the natural frequencies ($\omega_1 = 0$, $\omega_2 = 1$ and $\omega_3 = 1.7$).

Problem 4.16
Find the generalized equation for the fixed–free system shown in Fig. 4.27.

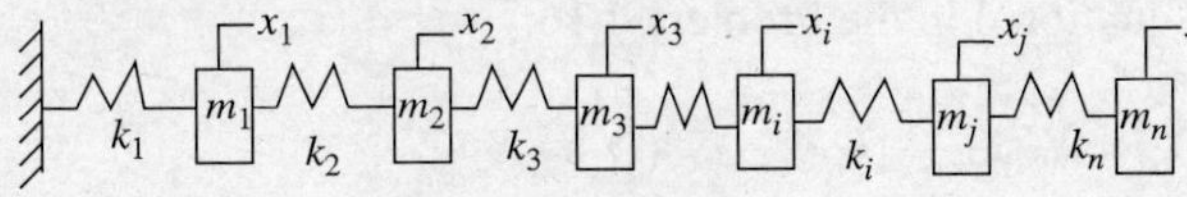

Figure 4.27

Solution:
The equations of motion for the system can be written as

$$
\begin{aligned}
m_1\ddot{x}_1 + k_1x_1 + k_2(x_1 - x_2) &= 0 \\
m_2\ddot{x}_2 + k_2(x_2 - x_1) + k_3(x_2 - x_3) &= 0 \\
m_3\ddot{x}_3 + k_3(x_3 - x_2) + k_4(x_3 - x_4) &= 0 \qquad (1) \\
&\cdots\cdots \\
m_n\ddot{x}_n + k_n(x_n - x_{n-1}) &= 0
\end{aligned}
$$

Let the solution be expressed as $x_i = X_i \sin(\omega.t + \phi)$, Therefore, $\ddot{x}_i = -X_i\omega^2 \sin(\omega.t + \phi)$

Substituting these in equation (1) one by one, we get

$$-m_1 X_1 \omega^2 + k_1 X_1 + k_2(X_1 - X_2) = 0 \tag{2}$$

This gives

$$X_2 = X_1 + \frac{1}{k_2}[k_1 X_1 - \omega^2 m_1 X_1] \tag{3}$$

$$-m_2 \ddot{x}_2 \omega^2 + k_2(X_2 - X_1) + k_3(X_2 - X_3) = 0 \tag{4}$$

Adding Equations a and c, we get

$$-(m_1 X_1 + m_2 X_2)\omega^2 + k_1 X_1 + k_3(X_2 - X_3) = 0$$

This gives

$$k_3 X_3 = k_1 X_1 + k_3 X_2 - (m_1 X_1 + m_2 X_2)\omega^2$$

$$X_3 = \frac{k_1}{k_3} X_1 + X_2 - \frac{(m_1 X_1 + m_2 X_2)\omega^2}{k_3}$$

or

$$X_3 = X_2 + \frac{1}{k_3}[k_1 X_1 - (m_1 X_1 + m_2 X_2)\omega^2] \tag{5}$$

We can now generalize

$$X_i = X_{i-1} + \frac{1}{k_i}[k_1 X_1 - \omega^2 (\sum_{i=1}^{i-1} m_i X_i)] \tag{6}$$

In a similar fashion, we can derive the generalized equation for X_i for a fixed–fixed system. The thing to remember is that whereas for fixed–free system the total inertia force is zero at free end, the amplitude of vibration at fixed end must be zero.

Problem 4.17

Figure 4.28 shows a fixed–free system. Use Holzer method to determine the natural frequencies. Assume $m = k = 1$.

Solution:

Although the tabular method is normally used, we shall describe the procedure using the generalized Equation 6 in the Problem 4.16.

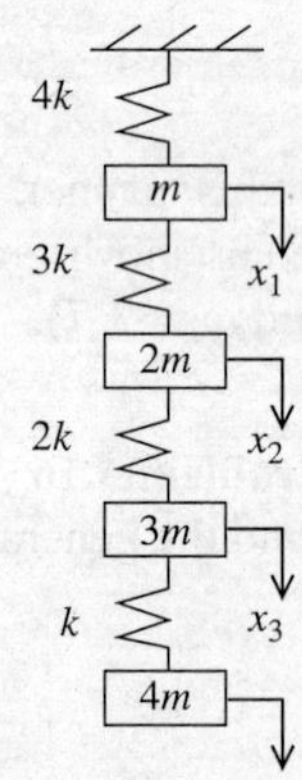

Figure 4.28

$$X_i = X_{i-1} + \frac{1}{k_i}[k_1 X_1 - \omega^2 (\sum_{i=1}^{i-1} m_i X_i)]$$

Let us first assume $\omega = 0.2$ and $X_1 = 1$

$$X_2 = X_1 + \frac{1}{k_1}(k_1 X_1 - \omega^2 m_1 X_1) = 1 + \frac{1}{3}(4 * 1 - 0.2^2 * 1 * 1) = 2.32$$

$$X_3 = X_2 + \frac{1}{k_3}[k_1 X_1 - (m_1 X_1 + m_2 X_2)\omega^2] = 2.32 + \frac{1}{2}[4 * 1 - (1 + 4.64) * 0.04] = 2.32 + 1.887 = 4.2$$

$$X_4 = X_3 + \frac{1}{k_4}[k_1 X_1 - (m_1 X_1 + m_2 X_2 + m_3 X_3)\omega^2] = 4.2 + \frac{1}{1}[4 * 1 - (1 + 2 * 2.32 + 3 * 4.2)0.04] = 7.473$$

Repeat now the procedure assuming $\omega = 1$ and $X_1 = 1$. We shall obtain $X_2 = 2$, $X_3 = 1.5$, $X_3 = -4$. We may now choose $\omega = 1.5$ and $X_1 = 1$ and go through the same procedure. For each of these (starting from $\omega = 0$ to 1.5 and beyond, if necessary), calculate $k_1X_1 - \sum^{i} m_iX_i\omega^2$ and plot this against ω. The zero cross-over points indicate the natural frequencies.

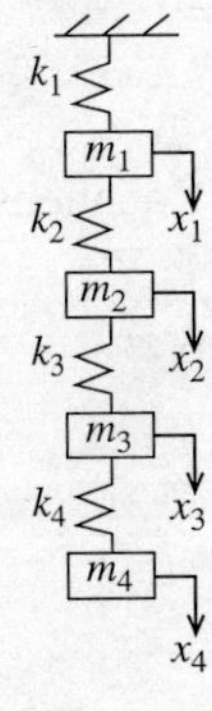

Figure 4.29 Four DOF System

Proof of the above: Let us consider as an illustration, four degrees-of-freedom system shown in Fig. 4.29 with one end fixed and another end free. We write the equations of motion as below.

$$\begin{aligned} m_1\ddot{x}_1 + k_1x_1 + k_2(x_1 - x_2) &= 0 \\ m_2\ddot{x}_2 + k_2(x_2 - x_1) + k_3(x_2 - x_3) &= 0 \\ m_3\ddot{x}_3 + k_3(x_3 - x_2) + k_4(x_3 - x_4) &= 0 \\ m_4\ddot{x}_4 + k_4(x_4 - x_3) &= 0 \end{aligned} \tag{4.142}$$

Adding all these equations, we get,

$$m_1\ddot{x}_1 + m_2\ddot{x}_2 + m_3\ddot{x}_3 + m_4\ddot{x}_4 + k_1x_1 = 0 \tag{4.143}$$

Let $x_i = X_i \sin(\omega.t + \phi), \therefore \ddot{x}_i = -X_i\omega^2 \sin(\omega.t + \phi)$.

Putting this in Equation 4.145, we get,

$$-(m_1X_1\omega^2 + m_2X_2\omega^2 + m_3X_3\omega^2 + m_4X_4\omega^2) + k_1X_1 = 0 \tag{4.144}$$

Generalizing, for n degrees of fixed free system, we will have $k_1X_1 - \sum_1^{i} m_iX_i\omega^2 = 0$ whenever $\omega = \omega_n$. This explains the rational behind plotting the curve $k_1X_1 - \sum_1^{i} m_iX_i\omega^2$ for various values of ω versus ω and identifying the zero cross-over points as the natural frequencies as explained.

Fixed–Fixed System

Similar analysis can be done for fixed–fixed system (Fig. 4.30). Similar to the previous case we write the equations of motion as

$$\begin{aligned} m_1\ddot{x}_1 + k_1x_1 + k_2(x_1 - x_2) &= 0 \\ m_2\ddot{x}_2 + k_2(x_2 - x_1) + k_3(x_2 - x_3) &= 0 \\ m_3\ddot{x}_3 + k_3(x_3 - x_2) + k_4(x_3 - x_4) &= 0 \\ m_4\ddot{x}_4 + k_4(x_4 - x_3) + k_5x_4 &= 0 \end{aligned} \tag{4.145}$$

Let $x_i = X_i \sin(\omega.t + \phi), \therefore \ddot{x}_i = -X_i\omega^2 \sin(\omega.t + \phi)$. Putting this in Equation 4.145 and adding all the four equations, we get

$$-\omega^2 \sum_{i=1}^{4} m_iX_i + k_1X_1 + k_5X_4 = 0 \tag{4.146}$$

Let us now find relation for X_i.

$x_i = X_i \sin(\omega.t + \phi), \therefore \ddot{x}_i = -X_i\omega^2 \sin(\omega.t + \phi)$, then

$$-m_1\,\omega^2X_1 + (k_1 + k_2)X_1 - k_2X_2 = 0 \tag{4.146(a)}$$

$$\therefore X_2 = \frac{-m_1\omega^2X_1 + k_1X_1 + k_2X_1}{k_2} = X_1 + \frac{1}{k_2}[k_1X_1 - m_1\omega^2X_1] \tag{4.146(b)}$$

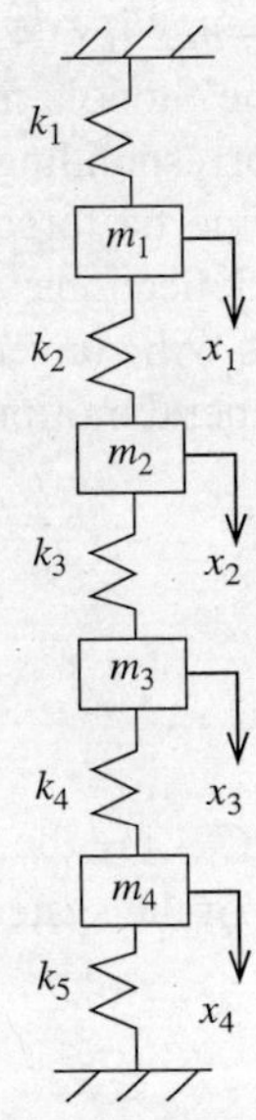

Figure 4.30 Fixed–Fixed System

$$-m_2\omega^2 X_2 + k_2(X_2 - X_1) + k_3(X_2 - X_3) = 0. \qquad [4.146(c)]$$

Adding Equations a and b, we get

$$-(m_1 X_1\omega^2 + m_2 X_2\omega^2) + k_1 X_1 + k_3(X_2 - X_3) = 0$$

$$\therefore k_3 X_3 = -(m_1 X_1\omega^2 + m_2 X_2\omega^2) + k_1 X_1 + k_3 X_2$$

This gives

$$X_3 = X_2 + \frac{1}{k_3}[k_1 X_1 - (m_1 X_1 + m_2 X_2)\omega^2 \qquad [4.146(d)]$$

We may now generalize as follows

$$X_i = X_{i-1} + \frac{1}{k_i}[k_1 X_1 - \sum_{1}^{i-1} m_i X_i \omega^2] \qquad (4.147)$$

The criterion for correct natural frequency is

$$k_1 X_1 + k_{i+1} X_i - \omega^2 \sum_{1}^{i} m_i X_i = 0 \qquad (4.148)$$

We shall now discuss Stodola method.

4.9.3 Stodola Method

The Stodola method is an iterative procedure used for calculation of the principal modes and the natural frequencies of multi degrees-of-freedom vibration systems. It is purely a physical approach and does not require derivation of differential equations of motion.

In general, the inertia force is a maximum and is in the direction opposite to that of deflection. In other words, the inertia force can be interpreted as the dynamic loading. When the vibration system vibrates at one of its principal modes with natural frequency ω, it is acted upon by inertia forces $-m_i\ddot{x}_i, (x_i = X_i \sin\omega.t, \therefore -m_i\ddot{x}_i = \omega^2 m_i X_i)$.

The Stodola method begins with assumed deflections for the fundamental mode of the system. The corresponding inertia forces due to these assumed deflections are calculated. Compared with actual inertia forces and deflections of the system, the inertia forces just found will produce a new set of deflections which are used to start the next iteration. The process is repeated. Eventually, this process will converge to the fundamental mode.

The general motion of an n degrees-of-freedom system is given by

$$\begin{aligned} x_1 &= A_1\sin(\omega_1 t + \phi_1) + A_2\sin(\omega_2 t + \phi_2) + \dots\dots + A_n\sin(\omega_n t + \phi_n) \\ x_2 &= B_1\sin(\omega_1 t + \phi_1) + B_2\sin(\omega_2 t + \phi_2) + \dots\dots + B_n\sin(\omega_n t + \phi_n) \\ x_3 &= C_1\sin(\omega_1 t + \phi_1) + C_2\sin(\omega_2 t + \phi_2) + \dots\dots + C_n\sin(\omega_n t + \phi_n) \\ &\dots\dots\dots\dots\dots\dots \end{aligned} \qquad (4.149)$$

$A_i, B_i, C_i ...$ are the modal components. Let the assumed deflection be an arbitrary superposition of all modes of the system with constants $a_1, a_2, a_3 .., a_n$. This means

$$\begin{aligned} x_1 &= a_1 A_1 + a_2 A_2 + \dots\dots + a_n A_n \\ x_2 &= a_1 B_1 + a_2 B_2 + \dots\dots + a_n B_n \\ x_3 &= a_1 C_1 + a_2 C_2 + \dots\dots + a_n C_n \end{aligned} \qquad (4.150)$$

The corresponding inertia forces are

$$\begin{aligned} &m_1(a_1A_1 + a_2A_2 + ... + a_nA_n)\omega^2 \\ &m_2(a_1B_1 + a_2B_2 + ... + a_nB_n)\omega^2 \\ &m_3(a_1C_1 + a_2C_2 + ... + a_nC_n)\omega^2 \end{aligned} \tag{4.151}$$

Now, if the system is vibrating with all principal modes, the inertia forces and the corresponding deflections are

$$\begin{aligned} &m_1(A_1\omega_1^2 + A_2\omega_2^2 + ... + A_n\omega_n^2), (A_1 + A_2 + ... + A_n) \\ &m_2(B_1\omega_1^2 + B_2\omega_2^2 + ... + \ \omega_n^2), (B_1 + B_2 + ... + B_n) \\ &m_3(C_1\omega_1^2 + C_2\omega_2^2 + ... + C_n\omega_n^2), (C_1 + C_2 + ... + C_n) \end{aligned} \tag{4.152}$$

Hence the inertia forces in Equation 4.152 will produce a new set of deflections

$$\begin{aligned} &\omega^2(a_1A_1 / \omega_1^2 + a_2A_2 / \omega_2^2 + ... + a_nA_n / \omega_n^2 \\ &\omega^2(a_1B_1 / \omega_1^2 + a_2B_2 / \omega_2^2 + ... + a_nB_n / \omega_n^2 \\ &\omega^2(a_1C_1 / \omega_1^2 + a_2C_2 / \omega_2^2 + ... + a_nC_n / \omega_n^2 \end{aligned} \tag{4.153}$$

Now,

$$\begin{aligned} x_1 &= \omega^2(a_1A_1 / \omega_1^2 + a_2A_2 / \omega_2^2 + ... + a_nA_n / \omega_n^2 \\ x_2 &= \omega^2(a_1B_1 / \omega_1^2 + a_2B_2 / \omega_2^2 + ... + a_nB_n / \omega_n^2 \\ x_3 &= \omega^2(a_1C_1 / \omega_1^2 + a_2C_2 / \omega_2^2 + ... + a_nC_n / \omega_n^2 \end{aligned} \tag{4.154}$$

Using the deflections in Equations 4.154 as the assumed deflections and carrying out exactly the steps in last iteration, we have

$$\begin{aligned} x_1 &= \omega^4(a_1A_1 / \omega_1^4 + a_2A_2 / \omega_2^4 + ... + a_nA_n / \omega_n^4 \\ x_2 &= \omega^4(a_1B_1 / \omega_1^4 + a_2B_2 / \omega_2^4 + ... + a_nB_n / \omega_n^4 \\ x_3 &= \omega^4(a_1C_1 / \omega_1^4 + a_2C_2 / \omega_2^4 + ... + a_nC_n / \omega_n^4 \end{aligned} \tag{4.155}$$

After r iterations, the assumed deflections take the following general form

$$\begin{aligned} x_1 &= \omega^{2r}(a_1A_1 / \omega_1^{2r} + a_2A_2 / \omega_2^{2r} + ... + a_nA_n / \omega_n^{2r} \\ x_2 &= \omega^{2r}(a_1B_1 / \omega_1^{2r} + a_2B_2 / \omega_2^{2r} + ... + a_nB_n / \omega_n^{2r} \\ x_3 &= \omega^{2r}(a_1C_1 / \omega_1^{2r} + a_2C_2 / \omega_2^{2r} + ... + a_nC_n / \omega_n^{2r} \end{aligned} \tag{4.156}$$

or

$$\begin{aligned} x_1 &= (a_1\omega^{2r} / \omega_1^{2r})(A_1 + a_2A_2\omega_1^{2r} / a_1\omega_2^{2r} + ... + a_nA_n\omega_1^{2r} / a_1\omega_n^{2r}) \\ x_2 &= (a_1\omega^{2r} / \omega_1^{2r})(B_1 + a_2B_2\omega_1^{2r} / a_1\omega_2^{2r} + ... + a_nB_n\omega_1^{2r} / a_1\omega_n^{2r}) \\ x_3 &= (a_1\omega^{2r} / \omega_1^{2r})(C_1 + a_2C_2\omega_1^{2r} / a_1\omega_2^{2r} + ... + a_nC_n\omega_1^{2r} / a_1\omega_n^{2r}) \\ &\ldots\ldots\ldots\ldots\ldots\ldots\ldots\ldots\ldots\ldots \end{aligned} \tag{4.157}$$

As ω_1 is the fundamental frequency, in general $\omega_1 < \omega_2 < \omega_3 \ldots < \omega_n$, and as the number of iterations is sufficiently large or r assumes a large number, the ratios of natural frequencies become very small. Thus for a sufficiently large number of iterations, the calculated deflections in Equations 4.157 become

$$x_1 = (a_1 A_1 \omega^{2r} / \omega_1^{2r})$$
$$x_2 = (a_1 B_1 \omega^{2r} / \omega_1^{2r})$$
$$x_3 = (a_1 C_1 \omega^{2r} / \omega_1^{2r})$$
$$\ldots\ldots\ldots\ldots\ldots\ldots$$

or

$$\begin{Bmatrix} x_1 \\ x_2 \\ x_3 \end{Bmatrix} = \frac{a_1 \omega^{2r}}{\omega_1^{2r}} \begin{Bmatrix} A_1 \\ B_1 \\ C_1 \end{Bmatrix} \text{ approaches the first mode.}$$

Thus Stodola method converges to the fundamental mode of vibration for an n degrees-of-freedom system. We shall now illustrate by an example.

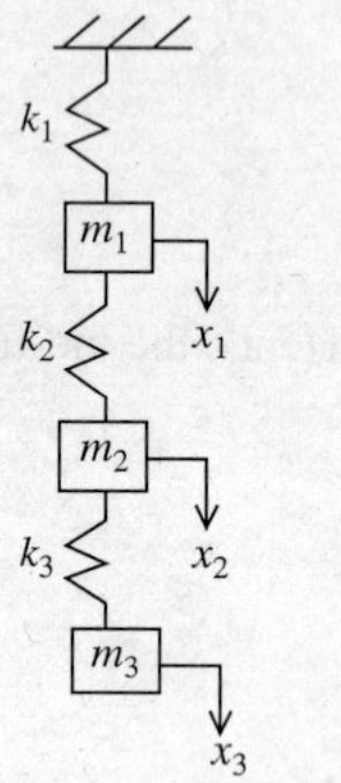

Figure 4.31

Problem 4.18

Figure 4.31 shows a three degrees-of-freedom system. Use Stodola method to find the first mode and the natural frequency. Consider all springs having $k = 1$ and all masses equal, each being unity.

Solution:

The iteration procedure is tabulated as given here.

	K_1	M_1	K_2	M_2	K_3	M_3	Remark
Assumed deflection		**1**		**1**		**1**	**Initial vector**
Inertia force		ω^2		ω^2		ω^2	
Spring force	$3\omega^2$		$2\omega^2$		ω^2		
Spring deflection	$3\omega^2$		$2\omega^2$		ω^2		
Calculated deflection		$3\omega^2$		$5\omega^2$		$6\omega^2$	
(A)		**1**		**1.67**		**2**	**Initial vector**
Assumed deflection		1		1.67		2	
Inertia force		ω^2		$1.67\ \omega^2$		$2\omega^2$	
Spring force	$4.67\omega^2$		$3.67\omega^2$		$2\omega^2$		
Spring deflection	$4.67\omega^2$		$3.67\omega^2$		$2\omega^2$		
Calculated deflection		$4.67\omega^2$		$8.34\omega^2$		$10.34\omega^2$	
(B)		**1**		**1.79**		**2.21**	**Initial vector**
Assumed deflection		1		1.79		2.21	
Inertia force			ω^2	$1.79\omega^2$		$2.21\omega^2$	
Spring force	$5\omega^2$		$4\omega^2$		$2.21\omega^2$		
Spring deflection	$5\omega^2$		$4\omega^2$		$2.21\omega^2$		
Calculated deflection		$5\omega^2$		$9\omega^2$		$11.21\omega^2$	
(C)		**1**		**1.8**		**2.24**	

Comparing results B and C, assumed deflection $\begin{Bmatrix} 1 \\ 1.79 \\ 2.21 \end{Bmatrix}$ is very close to calculated $\begin{Bmatrix} 1 \\ 1.8 \\ 2.24 \end{Bmatrix}$. Hence the principal mode is $\begin{Bmatrix} 1 \\ 1.8 \\ 2.24 \end{Bmatrix}$ and the natural frequency is $(5+9+11.21)\,\omega^2 = (1+1.8+2.34)$ or $\omega = 0.44$ rad/s

Explanation of the process

- All masses are equal (i.e., $m_1 = m_2 = m_3 = 1$)
- All springs have equal stiffnesses (i.e., $K_1 = K_2 = K_3 = 1$)

Step 1

Assumed deflection at m_1, m_2 and m_3 are 1, 1, 1.

Step 2

Inertia forces on masses are (for ω)
Mass 1 Inertia force $= \omega^2$
Mass 2 Inertia force $= \omega^2$
Mass 3 Inertia force $= \omega^2$

Step 3

Spring force on 1st spring $3\omega^2$
Spring force on 2nd spring $2\omega^2$
Spring force on 3rd spring ω^2
Since stiffness $k = 1$ for all spring

Step 4

Spring deflection on 1st spring $3\omega^2$
Spring deflection on 2nd spring $2\omega^2$
Spring deflection on 3rd spring ω^2

Step 5

Deflection at mass 1 in $3\omega^2$
Deflection at mass 2 in $3\omega^2 + 2\omega^2 = 5\omega^2$
Deflection at mass 3 in $3\omega^2 + 2\omega^2 = 6\omega^2$

Step 6

We normalize the above by dividing the deflection by $3\omega^2$.

- Thus the new assumed deflection are at mass m_1 1, at mass m_2 1.67, at mass m_3 2.
- The Process is repeated.
- The deflections are normalized and compared with initial assumed values.
- The process is continued. One can see that the values at **B** and **C** almost agree. The deflections arrived at are modal deflections. The calculated deflection in row 5 are found by adding the deflections due to springs, with the mass near the fixed end having least deflection or so, that is, mass 1.

Problem 4.19

Find the natural frequency of the system shown in Fig. 4.32.

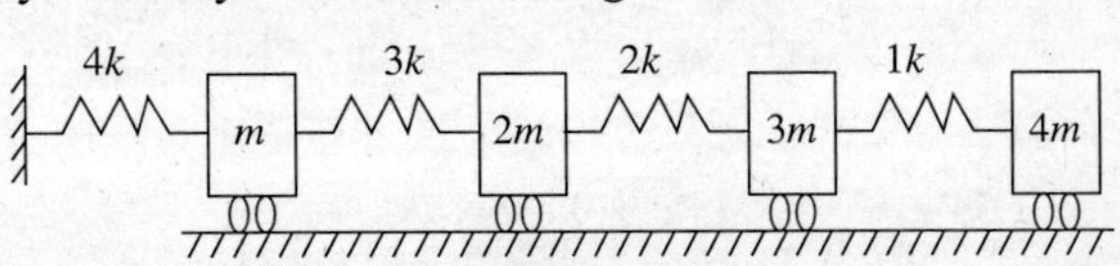

Figure 4.32

Solution:

We have already discussed the procedure. Following the same, we arrive at various iterations as given in the table below.

	$K_1 = 4k$	$M_1 = m$	$K_2 = 3k$	$M_2 = 2m$	$K_3 = 2k$	$M_3 = 3m$	$K_4 = k$	$M_4 = 4m$
Assumed deflection		**4**		**3**		**2**		**1**
Inertia force		$4\omega^2$		$6\omega^2$		$6\omega^2$		$4\omega^2$
Spring force	$20\omega^2$		$16\omega^2$		$10\omega^2$		$4\omega^2$	
Spring deflection	$5\omega^2$		$5.33\omega^2$		$5\omega^2$		$4\omega^2$	
Calculated deflection		$5\omega^2$		$10.33\omega^2$		$15.33\omega^2$		$19.33\omega^2$
Assumed deflection		**1**		**2**		**3**		**4**
Inertia force		ω^2		$4\omega^2$		$9\omega^2$		$16\omega^2$
Spring force	$30\omega^2$		$29\omega^2$		$25\omega^2$		$16\omega^2$	
Spring deflection	$7.5\omega^2$		$9.7\omega^2$		$12.5\omega^2$		$16\omega^2$	
Calculated deflection		$7.5\omega^2$		$17.7\omega^2$		$29.7\omega^2$		$45.7\omega^2$
Assumed deflection		**1**		**2**		**4**		**6**
Inertia force		ω^2		$4\omega^2$		$12\omega^2$		$24\omega^2$
Spring force	$41\omega^2$		$40\omega^2$		$36\omega^2$		$24\omega^2$	
Spring deflection	$10.25\omega^2$		$13.3\omega^2$		$18\omega^2$		$24\omega^2$	
Calculated		$10.25\omega^2$		$23.5\omega^2$		$41.5\omega^2$		$65.5\omega^2$
		1		2.2		4		6.4

Omitting the details, final mode shape is $\begin{Bmatrix} 1 \\ 2.3 \\ 4.05 \\ 6.42 \end{Bmatrix}$ and natural frequency is $\omega = 0.306$ rad/s. Readers may verify this.

Problem 4.20

Find the lowest natural frequency for the system shown in Fig. 4.33

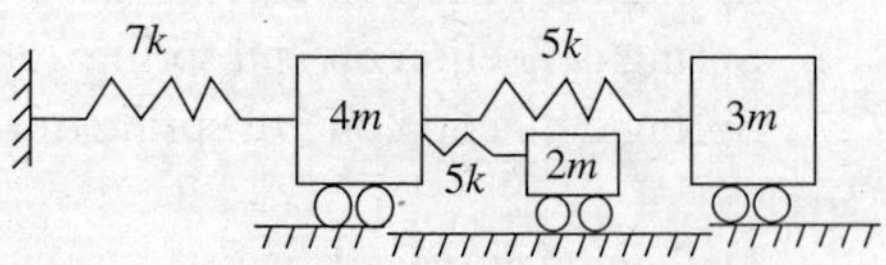

Figure 4.33

Solution:

The system shown in Fig. 4.33 is a complex system and requires correct understanding of the force distribution.

One may note that the spring force on mass m_1 is $9\omega^2$, on mass m_2 it is $3\omega^2$ and mass m_3 it is $2\omega^2$. This is because m_2 and m_3 are not coupled. Hence, while doing these iterations, the physical connections between masses and springs must be correctly taken into account.

The following table shows two iterations.

	$K_1 = 7k$	$M_1 = 4m$	$K_2 = 5k$	$M_2 = 3m$	$K_3 = 5k$	$M_3 = 2m$
Assumed deflection		1		1		1
Inertia force		$4\omega^2$		$3\omega^2$		$2\omega^2$
Spring force	$9\omega^2$		$3\omega^2$		$2\omega^2$	
Spring deflection	$1.3\omega^2$		$0.6\omega^2$		$0.4\omega^2$	$1.7\omega^2$
Calculated deflection		$1.3\omega^2$		$1.9\omega^2$		
		1		1.46		1.31
Assumed deflection		1		1.4		1.3
Inertia force		$4\omega^2$		$4.2\omega^2$		$2.6\omega^2$
Spring force	$10.8\omega^2$		$4.2\omega^2$		$2.6\omega^2$	
Spring deflection	1.54		$0.84\omega^2$		$0.52\omega^2$	
Calculated deflection		$1.54\omega^2$		$2.38\omega^2$		$2.06\omega^2$
		1.0		1.54		1.34

4.9.4 Dunkerley's Method

Dunkerley's method gives the approximate value of the fundamental frequency of a multi degrees-of-freedom or composite systems in terms of natural frequencies of its component parts. It is derived by making use of the fact that the higher natural frequencies of most of the complex vibratory systems are large compared to their fundamental frequency.

We know that the frequency equation for an n degrees-of-freedom system is given by the solution of the equation

$$\left|-[k]+\omega^2[m]\right| = 0, or, \left|-\frac{1}{\omega^2}[I]+[a][m]\right| = 0 \tag{4.158}$$

where $1/\omega^2 = \lambda$, $[a]$ = flexibility matrix = $[k]^{-1}$ and [m] = mass matrix.
Equation 4.158, for lumped-mass system becomes

$$\left| -\frac{1}{\omega^2}\begin{bmatrix} 1 & 0 & 0 & 0 & 0 \\ 0 & 1 & 0 & . & 0 \\ . & . & 1 & . & 0 \\ 0 & . & . & . & 0 \\ . & 0 & . & 0 & 1 \end{bmatrix} + \begin{bmatrix} a_{11} & a_{12} & \ldots\ldots & . & a_{1n} \\ a_{21} & a_{22} & \ldots\ldots & .. & a_{2n} \\ .. & .. & .. & . & .. \\ .. & .. & .. & .. & .. \\ a_{n1} & a_{n2} & .. & .. & a_{nn} \end{bmatrix} \begin{bmatrix} m_1 & 0 & .. & .. & 0 \\ 0 & m_2 & .. & .. & 0 \\ 0 & 0 & .. & .. & .. \\ .. & \ldots & .. & .. & 0 \\ 0 & .. & .. & 0 & m_n \end{bmatrix} \right| = 0$$

Expanding the above, we get,

$$\begin{vmatrix} (-1/\omega^2 + a_{11}m_1 & a_{12}m_2 & \ldots\ldots & a_{1n}m_n \\ a_{21}m_1 & (-1/\omega^2 + a_{22}m_2 & \ldots\ldots & a_{2n}m_n \\ \ldots & \ldots & \ldots\ldots & \ldots\ldots \\ a_{n1}m_1 & a_{n2}m_2 & \ldots\ldots & (-1/\omega^2 + a_{nn}m_n \end{vmatrix} = 0 \tag{4.159}$$

The expansion of Equation 4.159 results into

$$(\frac{1}{\omega^2})^n - (a_{11}m_1 + a_{22}m_2 + \ldots\ldots + a_{nn}m_n)(\frac{1}{\omega^2})^{n-1} + (a_{11}a_{22}m_1m_2 + a_{22}a_{33}m_2m_3 + \ldots$$
$$\ldots + a_{(n-1),(n-1)}a_{nn}m_{n-1}m_n - a_{12}a_{21}m_1m_2 - \ldots\ldots a_{n-1,n}a_{n,n+1}m_{n-1}m_n)(\frac{1}{\omega^2})^{n-1} - \ldots = 0 \tag{4.160}$$

Equation 4.160 is a polynomial of nth degrees in $1/\omega^2$. The roots of this equation are, say,

$$1/\omega_1^{\ 2}, 1/\omega_2^{\ 2}, \ldots \quad \text{and} \quad 1/\omega_n^{\ 2}.$$

Thus,

$$(1/\omega^2 - 1/\omega_1^{\ 2})(1/\omega^2 - 1/\omega_2^{\ 2}) \ldots (1/\omega^2 - 1/\omega_n^{\ 2}) = (1/\omega^2)^n - (1/\omega_1^{\ 2} + 1/\omega_2^{\ 2}$$
$$+ \ldots + 1/\omega_n^{\ 2})(1/\omega^2)^{n-1} = 0 \tag{4.161}$$

Equating coefficient of $(1/\omega^2)^{n-1}$ in Equations 4.160 and 4.161, we get

$$\frac{1}{\omega_1^2} + \frac{1}{\omega_2^2} + \ldots\ldots + \frac{1}{\omega_n^2} = a_{11}m_1 + a_{22}m_2 + \ldots\ldots + a_{nn}m_n \tag{4.162}$$

In most cases, $\omega_2, \omega_3, \omega_4 \ldots\ldots \omega_n$ are much larger than the fundamental frequency ω_1 and thus,

$$\frac{1}{\omega_1^2} \cong a_{11}m_1 + a_{22}m_2 + \ldots + a_{nn}m_n \tag{4.163}$$

Equation 4.163 is known as Dunkerley's formula. The fundamental frequency given by Equation 4.156 will always be smaller than the exact value. Equation 4.163 can also be written as

$$\frac{1}{\omega_1^2} = \frac{1}{\omega_{1n}^2} + \frac{1}{\omega_{2n}^2} + \ldots\ldots = (k_{ii} / m_i)^{1/2} \tag{4.164}$$

One may remember that $\omega_{in} = (\frac{1}{a_{ii} m_i})^{1/2}$ denotes natural frequency of a single degree-of-freedom system.

Problem 4.21

Use Dunkerley's formula to find the lowest natural frequency of the system shown in Fig. 4.34.

Solution:

Figure 4.34(a) shows the components of the system shown in Fig. 4.34. The natural frequencies of the individual systems are

$$\omega_1 = \sqrt{3k/4m}, \omega_2 = \sqrt{k/2m}, \omega_3 = \sqrt{k/m}$$

$$\frac{1}{\omega^2} = \frac{1}{\omega_1^2} + \frac{1}{\omega_2^2} + \frac{1}{\omega_3^2} = \frac{13m}{3k}, \omega = 0.475\sqrt{k/m}$$

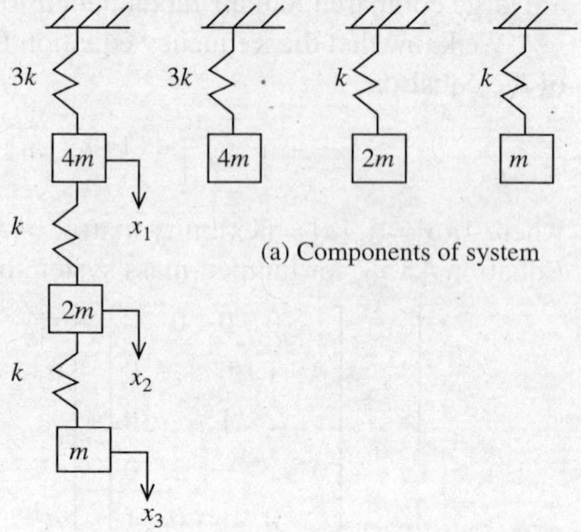

Figure 4.34

4.9.5 Rayleigh's Method

We have already discussed this method in the earlier chapter for single degree-of-freedom system. The principle applied there holds good for n degrees-of-freedom systems also. The kinetic and potential energies of n degrees-of-freedom discrete system can be expressed as

$$T = \frac{1}{2}\bar{\dot{x}}^T [m]\bar{\dot{x}} \tag{4.165}$$

$$V = \frac{1}{2}\bar{x}^T [k]\bar{x} \tag{4.166}$$

Assume,

$$\bar{x} = \bar{X} \cos \omega.t \tag{4.167}$$

$\bar{X}$ denotes vector of amplitudes (mode shape) and ω represents the natural frequency. Since there is no dissipation of energy (no damping) $T_{\max} = V_{\max}$ and we find

$$T_{\max} = \frac{1}{2}\bar{X}^T [m]\bar{X}$$

$$V_{\max} = \frac{1}{2}\bar{X}^T [k]\bar{X}$$

Thus we obtain

$$\frac{V_{\max}}{T_{\max}} = \frac{\bar{X}^T [k]\bar{X}}{\bar{X}^T [m]\bar{X}} \tag{4.168}$$

We also know that $[k]\bar{X} = \omega^2 [m]\bar{X}$, hence Equation 4.168 becomes

$$\frac{V_{\max}}{T_{\max}} = \frac{\bar{X}^Y \omega^2 [m]\bar{X}}{\bar{X}^T [m]\bar{X}} = \omega^2 \tag{4.169}$$

The quotient $\frac{V_{max}}{T_{max}} = \frac{\bar{X}^T[k]\bar{X}}{\bar{X}^T[m]\bar{X}} = R(\bar{X})$ is called Rayleigh's quotient. This quotient has the following important properties.

- Rayleigh's quotient has a stationary value. This means that the plot of $\bar{X}$ versus $R(\bar{X})$ has a horizontal tangent (slope = 0), when any arbitrary vector $\bar{X}$ is in the near neighborhood of any eigen vector.
- Rayleigh's quotient is never lower than first eigen vector.
- Rayleigh's quotient is never higher than the highest eigen vector.

Equation 4.162 can be used to find approximate value of the first natural frequency ω of the system. For this, we select a trial vector $\bar{X}$ to represent the first natural mode $\bar{X}^{(1)}$ and substitute in R.H.S. of Equation 4.162. This yields approximate value of ω_1^2. Since Rayleigh's quotient is stationary, remarkably good estimate of ω_1^2 can be obtained even if trial vector deviates significantly from the true natural mode $\bar{X}^{(1)}$. Naturally, the estimated value of the fundamental frequency, ω_1, is more accurate if the trial vector is close to true natural mode. We shall illustrate through a worked example.

Problem 4.22

For the system shown in Fig. 4.35, find the fundamental frequency using Rayleigh's method.

Solution:

The stiffness and mass matrices are

$$[k] = \begin{bmatrix} 2k & -k & 0 \\ -k & 2k & -k \\ 0 & -k & -k \end{bmatrix}, [m] = \begin{bmatrix} m & 0 & 0 \\ 0 & m & 0 \\ 0 & 0 & m \end{bmatrix}.$$

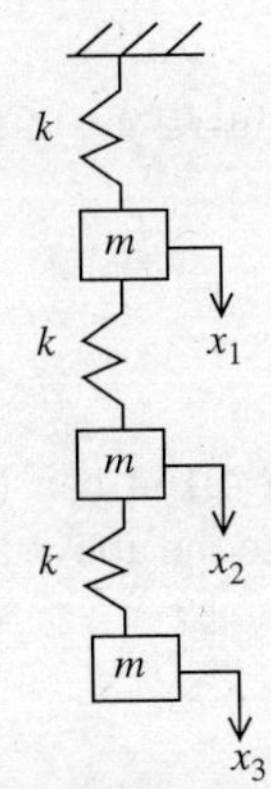

Figure 4.35

Let us take a trial vector $\bar{X} = \begin{Bmatrix} 1 \\ 2 \\ 3 \end{Bmatrix}$

The Rayleigh's quotient is given by

$$R(\bar{X}) = \omega^2 = \frac{\bar{X}^T[k]\bar{X}}{\bar{X}^T[m]\bar{X}} = \frac{\{1 \;\; 2 \;\; 3\} k \begin{bmatrix} 2 & -1 & 0 \\ -1 & 2 & -1 \\ 0 & -1 & -1 \end{bmatrix} \begin{Bmatrix} 1 \\ 2 \\ 3 \end{Bmatrix}}{\{1 \;\; 2 \;\; 3\} m \begin{bmatrix} 1 & 0 & 0 \\ 0 & 1 & 0 \\ 0 & 0 & 1 \end{bmatrix} \begin{Bmatrix} 1 \\ 2 \\ 3 \end{Bmatrix}} = \frac{3k}{14m}$$

From this we get $\omega_1 = 0.4629\sqrt{k/m}$.

Although the procedure explained before is applicable to all discrete systems, it is equally applicable for lateral vibrations of beams or rotors carrying masses such as pulleys, flywheel, gears, bladed discs etc. For these, a simpler equation can be derived for the fundamental frequency. In this case, the trial vector is static-deflection curve, which is used as an approximation to the dynamic-deflection curve.

Figure 4.36 shows a rotor carrying several discs on shaft with negligible mass compared to disc masses.

Let w_1, w_2, and w_3 be the static deflections under the masses m_1, m_2, and m_3 (we are considering only three disc masses but actual rotor may have many more. The methodology, we shall discuss is applicable to rotors having large number of discs).

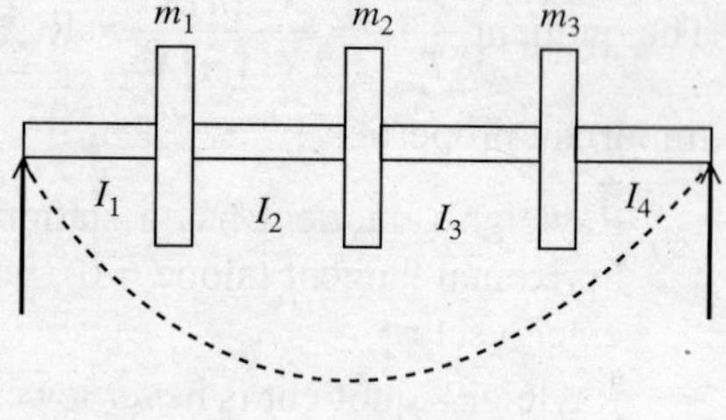

Figure 4.36 Multidisc Rotor

The potential energy of the system is the strain energy of the deflected shaft, which is equal to work done by the static loads. Thus,

$$V_{\max} = \frac{1}{2} g[m_1 w_1 + m_2 w_2 + m_3 w_3] \tag{4.170}$$

In the above Equation, gm_i is the static load due to mass m_i and w_i is the static deflection under the mass m_i. For harmonic oscillations (free), the maximum kinetic energy due to the masses is

$$T_{\max} = \frac{\omega^2}{2}(m_1 w_1^2 + m_2 w_2^2 + m_3 w_3^2)$$

The natural frequency is ω. Since the maximum potential energy is equal to maximum kinetic energy, we get

$$\frac{V_{\max}}{T_{\max}} = \frac{1}{\omega^2}\left\{\frac{m_1 w_1^2 + m_2 w_2^2 + m_3 w_3^2}{g(m_1 w_1 + m_2 w_2 + m_3 w_3)}\right\} = 1 \tag{4.171}$$

Equation 4.164 can be used to evaluate the fundamental natural frequency of the system. We can generalize the above approach and write the solution for massless rotor having n discs as given below.

$$\omega = \left\{\frac{g(m_1 w_1 + m_2 w_2 + \dots\dots\dots\dots + m_n w_n)}{m_1 w_1^2 + m_2 w_2^2 + \dots\dots + m_n w_n^2}\right\}^{1/2} \tag{4.172}$$

Let us illustrate the procedure through solution of a typical problem.

Problem 4.23

Estimate the fundamental frequency of lateral vibrations of the system shown in Fig. 4.37. The three masses are 20, 50, and 40 kg, respectively. The various spans are: $l_1 = m$, $l_2 = 3m$, $l_3 = 4m$ and $l_4 = 2m$. $E = 2.07 \times 10^{11}\text{N/m}^2$, $I = 4.90 \times 10^{-6}\text{ m}^4$.

Solution:

The deflections w_1, w_2, w_3 can be obtained considering effect of each load acting alone at each location and then adding. This means that we first consider only m_1 acting at l_1 and evaluate w_1', w_2', w_3' followed by considering only m_2 at $(l_1 + l_2)$ from LH and evaluate the deflections as before. In the end the effect of third mass is evaluated in a similar fashion.

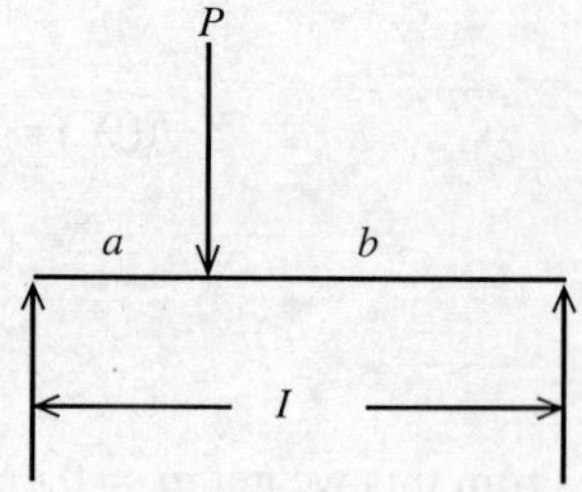

Figure 4.37

Thus, we obtain

$$w_1 = w_1' + w_1'' + w_1''', w_2 = w_2' + w_2'' + w_2''', w_3 = w_3' + w_3'' + w_3'''$$

For evaluating the deflection, we consider the formula

$$w(x) = \frac{Pbx}{6EIL}(L^2 - b^2 - x^2), 0 \le x \le a$$
$$w(x) = -Pa(L - x)[a^2 + x^2 - 2Lx], a \le x \le L$$

See Fig. 4.37.
Considering load $m_1 = 20$ kg acting alone at location 1

$$w_1' = 529.74 / EI$$
$$w_2' = 1236 / EI$$
$$w_3' = 621 / EI$$

Considering load $m_2 = 50$ kg acting alone at location 2

$$w_1'' = 3090 / EI$$
$$w_2'' = 9417 / EI$$
$$w_3'' = 5232 / EI$$

Considering load $m_3 = 40$ kg acting alone at location 3

$$w_1''' = 1242.6 / EI$$
$$w_2''' = 4185 / EI$$
$$w_3''' = 3348 / EI$$

Total deflection of masses m_1, m_2, and m_3 when all of them act together

$$w_1 = w_1' + w_1'' + w_1''' = 4862.5 / EI$$
$$w_2 = w_2' + w_2'' + w_2''' = 14840 / EI$$
$$w_3 = w_3' + w_3'' + w_3''' = 9201 / EI$$

Since $\omega = \left\{\frac{g(m_1 w_1 + m_2 w_2 + \ldots\ldots\ldots\ldots + m_n w_n)}{m_1 w_1^2 + m_2 w_2^2 + \ldots\ldots + m_n w_n^2}\right\}^{1/2}$

Substituting the values from the given data and above calculations, we get

$\omega = 28.4$ rad/s (calculation details omitted as the main aim is to demonstrate the method).

Lastly, we shall discuss the mechanical-impedance method, which is a very important method in vibration analysis.

4.9.6 Mechanical-Impedance Method

Certain vibration problems are very conveniently solved by mechanical-impedance method. The impedances of mass, spring, and dash pot are given by $-m\omega^2$, k and if ω ($i = \sqrt{-1}$). This method yields the steady-state response for forced vibrations and also leads to the frequency equation of the system for free vibrations. The following steps are followed while using mechanical-impedance method.

- Multiply the amplitude of each connecting point or junction of the system by the impedances of the elements connected to it.
- Subtract from this quantity the 'slippage terms' which are defined as the products of the impedances of the elements attached to the junction and the amplitudes of their opposite ends.
- Set this quantity equal to zero for free vibrations and equal to maximum value of the sinusoidal force for forced vibrations. If more than one force is applied to the junction, proper account must be taken of their phase relationship.
- Solve the equation for amplitude of vibrations. The expression for the amplitude of each junction can be put in the form $F/(A + iB)$. The numerical value of the amplitude is $F/\sqrt{(A^2 + B^2)}$ and the motion lags behind the force by $\tan^{-1}(B/A)$.

We shall demonstrate this method through a few solved examples.

Problem 4.24

Figure 4.38 shows a three degrees-of-freedom system. Find the natural frequency equation by mechanical-impedance method.

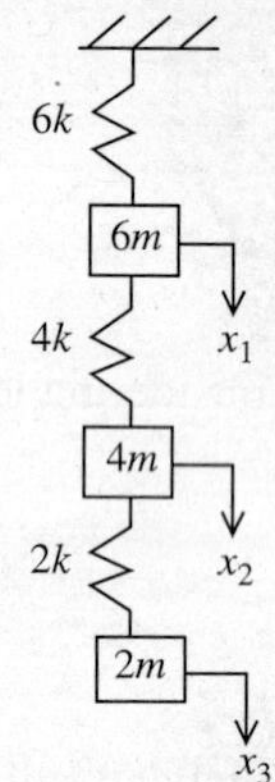

Figure 4.38

Solution:

The mechanical impedance of spring and mass are k and $-m\omega^2$.

For junction 1, this becomes

$$(6k + 4k - 6m\omega^2)\, x_1.$$

The slippage term is $4kx_2$. Since there is no external force, we obtain the equation

$$(6k + 4k - 6m\omega^2)\, x_1 - 4kx_2 = 0$$

For junction 2, the equation is

$$(4k + 2k - 4m\omega^2)\, x_2 - 2kx_3 = 0 \text{ } (2kx_3 \text{ is the slippage term})$$

For junction 3, the Equation is

$$(2k - 2m\omega^2)\, x_3 - 2kx_2 = 0 \text{ } (2kx_2 \text{ is the slippage term})$$

Rearrange the above equations to get

$$(10k - 6m\omega^2)x_1 - 4kx_2 = 0$$

$$-4kx_1 + (6k - 4m\omega^2)x_2 - 2kx_3 = 0$$

$$-2kx_2 + (2k - 2m\omega^2)x_3 = 0$$

The frequency equation is thus

$$\begin{vmatrix} (10k - 6m\omega^2) & -4k & 0 \\ -4k & (6k - 4m\omega^2) & -2k \\ 0 & -2k & (2k - 2m\omega^2) \end{vmatrix} = 0$$

The above equation can be solved for evaluating the three natural frequencies.

Problem 4.25

Determine the steady state vibration response of the system shown in Fig. 4.39 using the mechanical impedance method. Assume $k_i = f_i = m_i = 1$.

Solution:

Junction 1: Displacement is x_1.

The impedance is

$$(k_1 + if_1\omega - m_1\omega^2 + k_2 + if_2\omega)x_1$$

The slippage terms for this junction are $k_2 x_2 + if_2\omega x_2$.

Therefore first equation is

$$(k_1 + k_2 + if_1\omega + if_2\omega - m_1\omega^2)x_1 - k_2x_2 - if_2\omega x_2 = F_0.$$

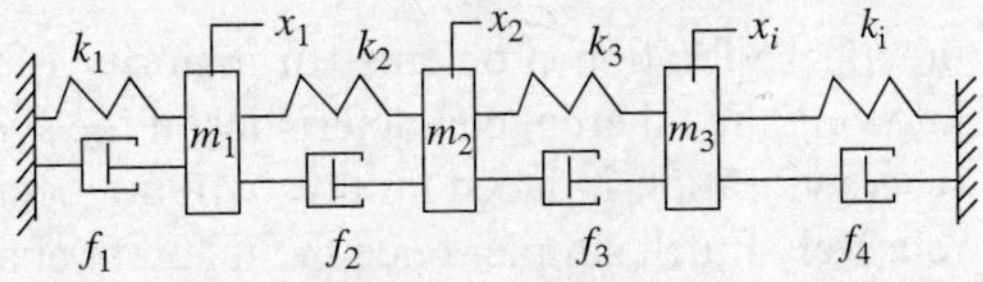

Figure 4.39

Similarly for junctions 2 and 3

$$(k_2 + k_1 + if_2\omega + if_3\omega - m_2\omega^2)x_2 - k_2x_1 - if_2\omega x_1 - k_3x_3 - if_3\omega x_3 = 0$$

$$(k_3 + k_4 + if_3\omega + if_4\omega - m_3\omega^2)x_3 - k_3x_2 - if_3\omega x_2 = 0$$

Substituting the given values, equations of motion become

$$(i + 2i)x_1 - (1+i)x_2 = F_0$$

$$-(1+i)x_1 + (1+2i)x_2 - (1+i)x_3 = 0$$

$$-(1+i)x_2 + (1+2i)x_3 = 0$$

Solving these equations using Cramer's rule, we get

$$x_1 = \frac{\begin{vmatrix} F_0 & -(1+i) & 0 \\ 0 & (1+2i) & -(1+i) \\ 0 & -(1+i) & (1+2i) \end{vmatrix}}{\begin{vmatrix} (1+2i) & -(1+i) & 0 \\ -(1+i) & (1+2i) & -(1+i) \\ 0 & -(1+i) & (1+2i) \end{vmatrix}} = \frac{F_0(1+2i)(1+2i) - F_0(1+i)^2}{(1+2i)^3 - (1+2i)(1+i)^2 - (1+2i)(1+i)^2}$$

$$x_2 = \frac{\begin{vmatrix} (1+2i) & F_0 & 0 \\ -(1+i) & 0 & -(1+i) \\ 0 & 0 & (1+2i) \end{vmatrix}}{\begin{vmatrix} (1+2i) & -(1+i) & 0 \\ -(1+i) & (1+2i) & -(1+i) \\ 0 & -(1+i) & (1+2i) \end{vmatrix}} = \frac{F_0(1+i)(1+2i)}{(1+2i)^3 - (1+2i)(1+i)^2 - (1+2i)(1+i)^2}$$

$$x_3 = \frac{\begin{vmatrix} (1+2i) & -(1+i) & F_0 \\ -(1+i) & (1+2i) & 0 \\ 0 & -(1+i) & 0 \end{vmatrix}}{\begin{vmatrix} (1+2i) & -(1+i) & 0 \\ -(1+i) & (1+2i) & -(1+i) \\ 0 & -(1+i) & (1+2i) \end{vmatrix}} = \frac{F_0(1+i)^2}{(1+2i)^3 - (1+2i)(1+i)^2 - (1+2i)(1+i)^2}$$

The equation for x_1 simplifies to $x_1 = 13\ F_0/(24i-3)$.

This gives amplitude of $x_1 = 0.54\ F_0$ and $\Phi = -82.9°$. Similarly for x_2, the amplitude is $0.47\ F_0$ and phase is –45° and for x_3 the amplitude is $0.29\ F_0$ and the phase is –26°.

We in this chapter discussed more about finding the natural frequencies and mode shapes. This is so because knowing the natural frequencies and the mode shape is the most important step in solving practical vibration problems. In majority of the cases, the vibration engineer is more interested in knowing the inherent characteristics of the system to solve the vibration problem or design equipment where vibration-induced failures will not occur. The determination of the forced vibration response is relatively much simpler once the inherent characteristics of the system.

CONCLUSION

In this chapter, we have discussed various methods of analysis of multi degrees-of-freedom vibration systems. We have considered discrete systems comprising of various springs, masses, and damper elements.

In the beginning of this chapter, we have shown how a continuous system can be modelled as a system comprising of masses and springs. We have seen that the accuracy of modelling improves as we increase the number of masses and springs. We have discussed various methods of arriving at equations of motion of multi degrees-of-freedom systems (Section 4.3). We discussed the method of using Newton's laws of motion and derived them in matrix form. The various matrices such as mass matrix, spring or stiffness matrix and damping matrix have been defined for a general n degrees-of-freedom vibration system.

We also discussed the concept of influence coefficients and method of writing the Equations of motion in terms of influence coefficients. The various influence coefficients such as stiffness influence coefficients, flexibility-influence coefficients, and inertia influence coefficients have been defined and methods to evaluate them have been presented. These concepts have been applied to a few numerical examples. It is shown that the influence coefficient method is very powerful tool in the analysis of vibrations of multi degrees-of-freedom systems

In Section 4.4, we have discussed generalized coordinates and their importance in analysis of vibration problems. In Section 4.5, we have discussed about energies in vibratory systems. We have shown that potential energy is a quadratic function of displacements and the kinetic energy is quadratic function of velocities. Based upon this, we explained the concept of positive definite quadratic form and positive definite matrices as applied in vibration analysis. We also discussed Lagrange's equations and explained their application in derivation of equations of motion of vibratory system.

Based upon the equations of motion $[m]\bar{\ddot{x}} + [k]\bar{x} = 0$, we have defined and discussed the eigen value problem in section 4.6 and shown that solution of $[[k] - \omega^2\{m]]\bar{X} = 0$ is possible only when $\left|[k] - \omega^2[m]\right| = 0$. This gives us the frequency equation. Defining $[D]$ = dynamical matrix = $[k]^{-1}[m]$, we have defined the standard eigen value problem as $\lambda[I]\bar{X} = [D]\bar{X}$, which must satisfy the condition that $\Delta = \left|\lambda[I] - [D]\right| = 0$

Various numerical problems have solved to clarify the concepts of solution of eigen value problem and definitions of eigen frequencies (natural frequencies) and eigen vectors (mode shapes).

Section 4.7 deals with Orthogonality of principal or normal modes. This concept has tremendous applications in mode-suppression technique, once the fundamental mode is established. Another important concept of eigen value expansion theorem has been presented in Section 4.7.

Section 4.8 deals with modal analysis, which is a very powerful tool in vibration analysis. Section 4.9 deals with various methods of determination of natural frequencies and mode shapes.

The n Equations of motion, which can be obtained using either Newton's laws of motion or Lagrange's Equation or by influence-coefficient method, are not entirely independent. Their simultaneous solution, therefore, is very complex. The other methods of solution such as Stodola method, Holzer method, matrix-iteration method, Rayleigh's method, Dunkerley's method are discussed and illustrated by a few solved problems.

The obvious follow-up of what has been described in this chapter is the Finite Element Method of Vibration analysis. This by itself is a separate subject and several books have been published. The author would recommend book by O.C. Zienkiewiez entitled 'The Finite Element Method', McGraw–Hill, London (1987) and book by S.S. Rao entitled 'The Finite Element Method in Engineering', Butterworth-Heinemann, Boston (1999).

EXERCISES

4.1 Find the stiffness matrix of the frame shown here.

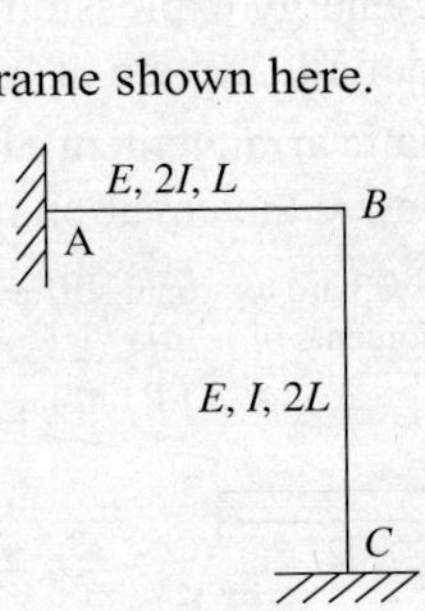

Problem 4.1

Hint: Segments AB and BC of the frame can be considered as beams. We can therefore use beam force deflection formulae to generate the stiffness matrix of the frame. For this, we need to find the forces necessary to cause a deflection along one coordinate while maintaining zero displacement along other coordinates of the beam. For example, the ends A and C are fixed and hence the joint B will have three possible displacements—x, y, and θ.

Answer: $[k] = \frac{EI}{l^3}\begin{bmatrix} 3/2 & 0 & 3l/2 \\ 0 & 24 & -12l \\ 3l/2 & -12l & 10l^2 \end{bmatrix}$

4.2 Evaluate the stiffness matrix for the system shown here.

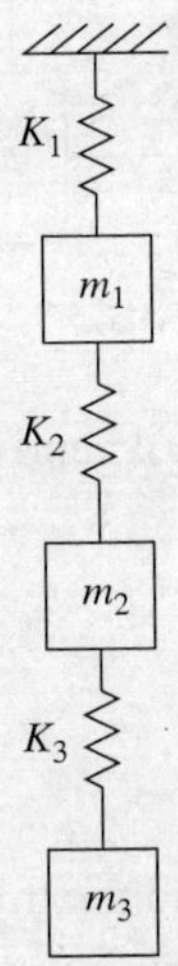

Problem 4.2

Hint: Use the method of stiffness-influence coefficients.

Answer: $k_{11} = k_1 + k_2$, $k_{12} = -k_2 = k_{21}$, $k_{13} = 0 = k_{31}$, $k_{22} = k_2 + k_3$, $k_{23} = k_{32} = -k_3$, $k_{33} = -k_3$

4.3 For a system comprising of three masses as shown in the figure of Problem 4.2, explain how you will evaluate the flexibility-influence coefficients. Assume that all the springs have unit stiffness and all the masses have unit mass. Also show that the product of stiffness matrix and flexibility matrix is a unit matrix.

Answer: $a_{11} = 1$, $a_{12} = 1$, $a_{13} = 1$, $a_{21} = 1$, $a_{22} = 2$, $a_{23} = 2$, $a_{31} = 1$, $a_{32} = 2$, $a_{33} = 3$

4.4 Derive the potential- and kinetic-energy expression for a multi degrees-of-freedom system. Based upon this, explain what a semi-definite system is.

4.5 The figure here shows the schematic arrangement of a two-cylinder turbine drive for a turbo-generator. Using Lagrange's method, derive the equation of motion for the system.

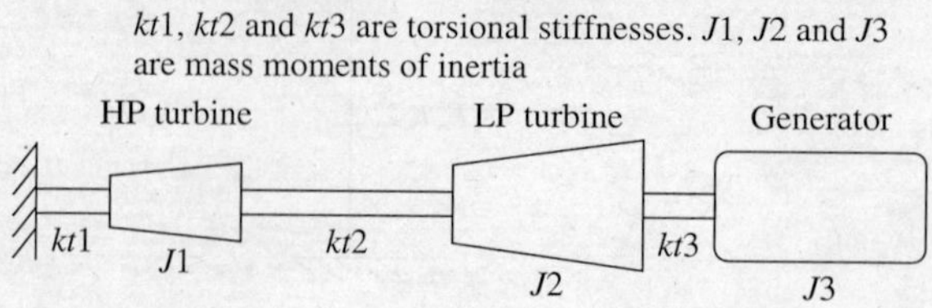

Problem 4.5

Answer:
$$\begin{bmatrix} J1 & 0 & 0 \\ 0 & J2 & 0 \\ 0 & 0 & J3 \end{bmatrix}\begin{pmatrix} \ddot{\theta}_1 \\ \ddot{\theta}_2 \\ \ddot{\theta}_3 \end{pmatrix} + \begin{bmatrix} (kt1+kt2) & -kt2 & 0 \\ -kt2 & (kt2+kt3) & -kt3 \\ 0 & -kt3 & kt3 \end{bmatrix}\begin{pmatrix} \theta_1 \\ \theta_2 \\ \theta_3 \end{pmatrix} = \begin{pmatrix} 0 \\ 0 \\ 0 \end{pmatrix}$$

4.6 Explain the concept of eigen value problem. Show how the eigen value problem is solved.

4.7 For the system shown in the figure, assuming that all masses as well as springs have unit value, find the natural frequencies and the mode shapes.

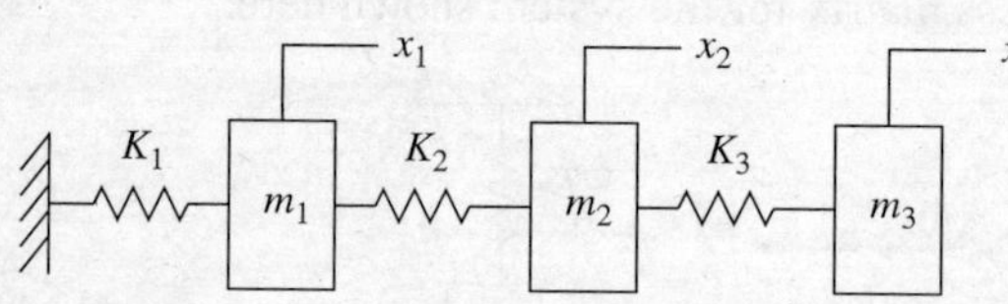

Problem 4.7

Answer: $\omega_1 = 0.445\ (k/m)^{1/2}$, $\omega_2 = 1.247\ (k/m)^{1/2}$ and $\omega_3 = 1.8\ (k/m)^{1/2}$

$$\bar{X}^{(1)} = X_1^1 \begin{Bmatrix} 1 \\ 1.80 \\ 2.24 \end{Bmatrix} \text{ is the first mode shape.}$$

$$\vec{X}^{(2)} = X_1^2 \begin{bmatrix} \begin{Bmatrix} 1.0 \\ 0.445 \\ -0.8 \end{Bmatrix} \end{bmatrix} \text{ is the second mode and the third mode is } \vec{X}^{(3)} = X_1^3 \begin{Bmatrix} 1 \\ -1.247 \\ 0.55 \end{Bmatrix}$$

4.8 What is modal analysis and how can it be used for finding the forced vibration response of typical three degrees-of-freedom (undamped) vibration system?

Answer: See the text.

STEP1

Form the equations of motion. $[m]\{\ddot{x}\}+[k]\{x\}=\{F\}$

STEP 2

Solve the eigen value problem $\omega^2[m]\{x\}=[k]\{x\}$ and find ω_1, ω_2, and ω_3

STEP 3

Find the normal modes $\vec{X}^{(1)}, \vec{X}^{(2)}, \vec{X}^{(3)}$

STEP 4

Find the solution vector expressed as linear combinations of the normal modes.

$$\{x(t)\} = q_1(t)\vec{X}^{(1)} + q_2(t)\vec{X}^{(2)} + q_3(t)\vec{X}^{(3)}$$

In the above equations, q_i are the generalized coordinates which are time-dependent. These are also called the principal coordinates or more popularly as modal-participation coefficients.

STEP 5

This involves defining a modal matrix [X] in which the jth column is the vector $\vec{X}^{(j)}$. This in other words means $[X] = [\vec{X}^{(1)} \vec{X}^{(2)} \vec{X}^{(3)}]$, since we are dealing with a three degrees-of-freedom system. We can rewrite the equation in step 4 as $\{x(t)\} = [X]\{\vec{q}(t)\}$ where $q_1(t)$, $q_2(t)$, and $q_3(t)$ are components of the $\vec{q}(t)$ vector. This is a very important step since we can now write the earlier equation simply as $\vec{x}(t) = [X]\vec{q}(t)$ and $\ddot{\vec{x}}(t) = [X]\ddot{\vec{q}}$. In these equations [X] is not function of time. This will enable us to rewrite the equations of the motion as

$$[m][X]\ddot{\vec{q}} + [k][X]\vec{q} = \vec{F} \text{ and also as } [X]^T [m][X]\ddot{\vec{q}} + [X]^T [k]\vec{q} = [X]^T \vec{F}.$$

The first term in the above equation is a unit matrix [I] while the second term is the matrix with ω^2 term-diagonal matrix.

STEP 6

Results in formation of a set of three uncoupled equations of second order as

$$\ddot{q}_i(t) + \omega_i^2 q_i(t) = [X]^T \vec{F}(t) = \vec{Q}_i(t).$$

The above equations are three equations, which can be solved in same way as we solve equation for a single degree-of-freedom system.

4.9 Derive the equation of motion for the system shown here.

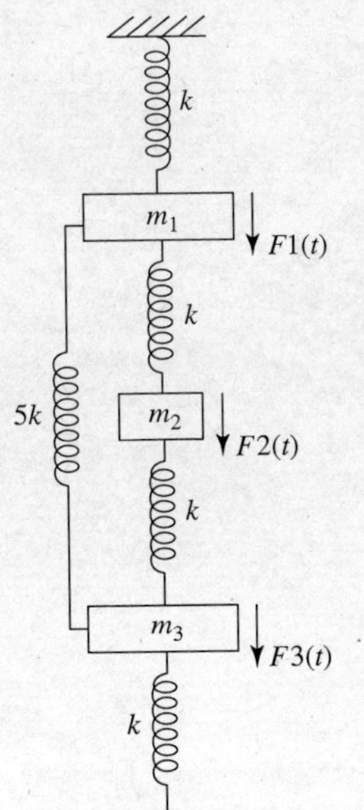

Problem 4.9

Answer:
$$\begin{bmatrix} m_1 & 0 & 0 \\ 0 & m_2 & 0 \\ 0 & 0 & m_3 \end{bmatrix} \begin{Bmatrix} \ddot{y}_1 \\ \ddot{y}_2 \\ \ddot{y}_3 \end{Bmatrix} + k \begin{bmatrix} 7 & -1 & -5 \\ -1 & 2 & -1 \\ -5 & -1 & 7 \end{bmatrix} \begin{Bmatrix} y_1 \\ y_2 \\ y_3 \end{Bmatrix} = \begin{bmatrix} F1(t) \\ F2(t) \\ F3(t) \end{bmatrix}$$

4.10 Three freight cars are coupled by two springs having equal stiffness. All the cars have same mass. Show that the stiffness matrix for the total assembly is singular. Find the natural frequencies and the mode shapes.

Hint: Use energy expression as described in the text. The eigen value problem for this system will give $[[k] - \omega^2[m]]\ \vec{X} = \vec{0}$
Since $[k]$ is singular, its inversion I not possible. Set the determinant of the coefficient matrix of $\vec{X}$ equal to zero. This will give the frequency expression. Rest of the procedure is obvious. The first mode will be a rigid-body motion of all the masses together ($\omega_1 = 0$). $\omega_2 = (k/m)^{1/2}$ and $\omega_3 = (3k/m)^{1/2}$.

4.11 Find the modal matrix for the system shown here.

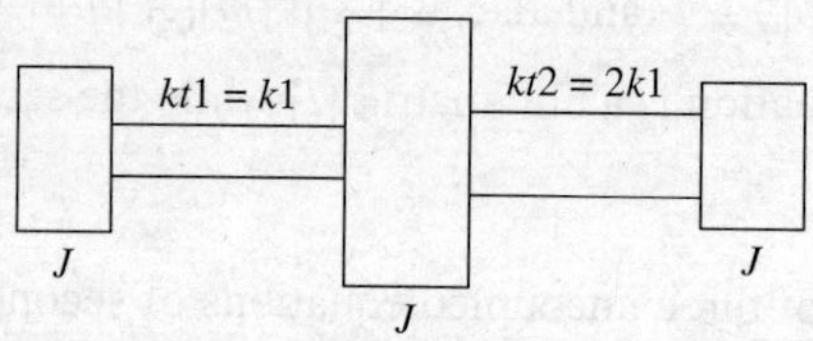

Problem 4.11

5 Torsional Vibrations

5.1 INTRODUCTION

The analysis of torsional vibrations assumes a very important role in the design of rotating systems such as multi-span rotors of power plants, compressors, pumps, etc., since these equipments are often subject to fluctuating torques, which at times may lead to their failures. This becomes much more critical because, unlike transverse vibrations where it is possible to measure and analyse the shaft as well as bearing vibrations using suitable transducers. However, it is a complex affair to measure and monitor the torsional vibrations, and hence in most of the rotating machinery, torsional vibrations are neither measured nor possible to measure. The torsional vibrations induce torsional-shear stresses, which can combine with bending stresses (which also fluctuate), and thus equipment is subjected to pulsating stresses, which in many instances, lead to fatigue failures. In absence of any experimental support for the analysis of torsional stresses, one has to totally depend upon analytical methods to compute them and provide sufficient design margins so that the dynamic/vibratory torsional stresses remain well within limits.

Torsional vibration is a periodic angular motion of elastic shafts, which carry bladed discs, impellors or rotor windings in electric motors/generators.

There is a close resemblance and similarity between rectilinear and torsional vibrations. Thus, the theory and analysis discussed in the previous chapters can be applied equally well for torsional analysis. Table 5.1 gives the analogy between rectilinear and torsional vibrations.

5.2 TORSIONAL VIBRATION SYSTEMS

We shall now deal with torsional system with various degrees-of-freedom starting from single degree-of-freedom system.

5.2.1 Single Degree-of-Freedom System

Figure 5.1 shows a typical single degree-of-freedom system, comprising a massless shaft with stiffness K (= angular displacement ϑ per unit torque) and having a damper with damping constant η (angular velocity $\dot{\vartheta}$ per unit torque and a disc having polar mass moment of inertia is the angular displacement).

Table 5.1

	Rectilinear Vibrations		Torsional Vibrations	
Time	T	s	t	s
Displacement	X	mm, cm, m	ϑ	radian
Velocity	$\dot{x}$	m/s, cm/s	$\dot{\vartheta}$	rad/s
Acceleration	$\ddot{x}$	m/s²	$\ddot{\vartheta}$	rad/s²
Spring constant	K	kg/cm	K	m kg/rad
Damping coefficient	f	kg s/cm	η	m kg s/rad
Mass	m	kg s²/m	J	mass moment of inertia
Force	$F = m\ddot{x}$	kg	$t = J\ddot{\vartheta}$	torque
Momentum	$m\dot{x}$	kg s	$J\dot{\vartheta}$	m kg s
Impulse	Ft	kg s	Tt	m kg s
Kinetic energy	$\frac{1}{2}m\dot{x}^2$	kg m	$\frac{1}{2}J\dot{\vartheta}^2$	kg m
Potential energy	$\frac{1}{2}k\dot{x}^2$	kg m	$\frac{1}{2}K\dot{\vartheta}^2$	kg m
Work done	$\int F dx$		$\int T d\vartheta$	
Natural frequency ω_η	$\sqrt{k/m}$		$\sqrt{k/J}$	
Equation of motion, SDOF	$m\ddot{x} + f\dot{x} + kx = F_0 \sin \omega t$		$J\ddot{\vartheta} + \eta\dot{\vartheta} + k\vartheta = T_0 \sin \omega t$	
Initial condition	$x(0) = x_0, \dot{x}(0) = \dot{x}_0$		$\vartheta(0) = \vartheta_0, \dot{\vartheta}(0) = \dot{\vartheta}_0$	
Transient response	$Ae^{-\xi\omega\eta t}\sin(\omega dt + Ø)$ $\omega_d = \omega_n\sqrt{(1-\xi^2)}$		$Ae^{-\xi\omega\eta t}\sin(\omega dt + Ø)$ $\omega_d = \omega_n\sqrt{(1-\xi^2)}$	
Steady response	$x_p = X\sin(\omega t - \Phi)$ $X = \frac{F_o}{\sqrt{(k-m\omega^2)^2 + (f\omega)^2}}$		$\vartheta_p = Ø\sin(\omega t - \Phi)$ $Ø = \frac{T_0}{\sqrt{(K-J\omega^2)^2 + (\eta\omega)^2}}$	

If $\vartheta = \vartheta(t)$, the inertia torque $= J\ddot{\vartheta}$, spring torque $= k\vartheta$, damping torque $= \eta\dot{\vartheta}$

Applied torque $T = T_0 \sin\omega t$.

The equation of motion is $J\ddot{\vartheta} + \eta\dot{\vartheta} + k\vartheta = T_0 \sin \omega t$ (5.1)

The solution of Equation 5.1 is ϑ = C.F. + P.I (5.2)

As explained in Chapter 2, we can obtain the solution along the same lines with m replaced by J, x (or y) replaced by ϑ, f replaced by η, etc. The salient results for free vibrations are

$$\omega_\eta = \sqrt{\frac{K}{J}}, \tag{5.3}$$

$$\xi = \frac{\eta}{\eta_{cri}} = \frac{\eta}{2J\omega_\eta} \tag{5.4}$$

$$= \frac{\eta}{2\sqrt{KJ}} \tag{5.5}$$

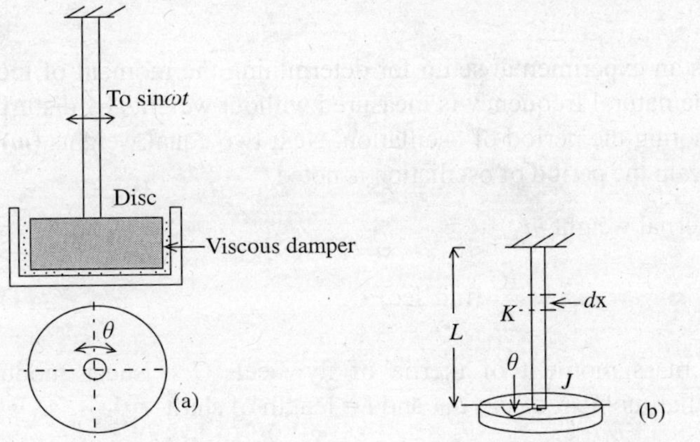

Figure 5.1 (a) SDOF Torsional Vibration System (b) Problem 5.1

The C.F. represents the transient response, $\vartheta_c = Ae^{-\xi\omega\eta t}\sin(\omega_d t + \text{Ø})$ (5.6)

where $\omega_d = \sqrt{1-\xi^2}\,\omega_\eta$

The constant A is determined by using initial condition $\vartheta\ (t = 0) = \vartheta_0$ and $\dot{\vartheta}\ (t = 0) = \dot{\vartheta}_0$.

The steady-state response is $\vartheta = \theta_P \sin(\omega t - \Phi)$ (5.7)

where $\theta_P = \dfrac{T_0}{\sqrt{(K - J\omega^2)^2 + (\eta\omega)^2}}$. (5.8)

Problem 5.1

(a) A circular disc of moment of inertia J is attached to the lower end of a massless elastic shaft having torsional stiffness. Using energy method, derive the equation of motion.

(b) Consider the shaft-mass moment of inertia J_s and derive equation of motion.

(a) The KE of the system $= \frac{1}{2}J\dot{\vartheta}^2$,

The PE of the system $= \frac{1}{2}K\dot{\vartheta}^2$

$\dfrac{d}{dt}(KE + PE) = 0$ or $J\ddot{\vartheta}\dot{\vartheta} + K\vartheta\dot{\vartheta} = 0$, since $\dot{\vartheta}$ is not always zero.

$J\ddot{\vartheta} + K\vartheta = 0$ or $\omega_\eta = \sqrt{K/J}$

(b) Let $\dot{\vartheta}$ angular velocity at free end (disc location) see Figure 5.1(b). Consider a section at a distance x from the fixed end. If J_s is the M.I. of shaft, the KE of

$$dx = \frac{1}{2}\frac{dx}{x}.J_s\left(\frac{x}{l}\omega\right)^2 = \frac{1}{2}J_s\frac{\omega^2}{l^3}x^2dx$$

Total $KE = \dfrac{1}{2}J_s\dfrac{\omega^2}{l^3}\int x^2 dx = \dfrac{1}{3}.\dfrac{1}{2}J_s\omega^2$

Hence, KE of shaft is equivalent to that of flywheel of mass movement of inertia $\dfrac{J_s}{3}$.

Therefore, Equivalent moment of inertia (MI) of flywheel $= J + \dfrac{J_s}{3}$

Thus, the equation of motion is $\left(J + \dfrac{J_s}{3}\right)\ddot{\vartheta} + K\vartheta = 0.$

Problem 5.2

Figure 5.2 shows an experimental setup for determining the moment of inertia of a flywheel. The natural frequency is measured without weights by disturbing the system and noting the period of oscillation. Next two equal weights (m) are placed at r and again the period of oscillation is noted.

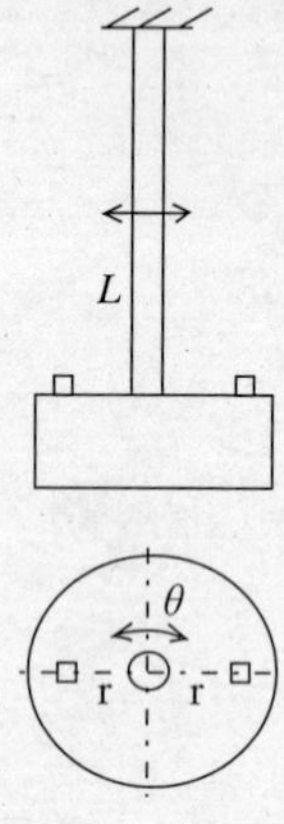

Figure 5.2

(a) Without external weight m

$$\omega_1^2 = \frac{JG}{IL} (\text{rad/sec})^2 \quad (1)$$

where I = mass moment of inertia of flywheel, G = shear modulus, J = rectangular moment of inertia, and l = length of shaft

(b) When equal weights are put

$$\omega_2^2 = \frac{JG}{(I + 2mr^2)L} \quad (2)$$

$$\omega_1^2/\omega_2^2 = \left(\frac{JG}{IL}\right) \div \frac{JG}{(I+2mr^2)L} = \text{I} + \frac{2mr^2}{I}$$

or
$$\text{I} = \frac{2mr^2\omega_2^2}{\omega_1^2 - \omega_2^2}$$

Stepped Shaft

In many cases, torsional system comprises of shafts of various diameters, that is, stepped shaft. For example, Fig. 5.3 shows a stepped shaft having diameters d_1 and d_2 in lengths L_1 and L_2, respectively. For the convenience of analysis, it is necessary to define an equivalent shaft of uniform diameter say d_1 having a length L_1.

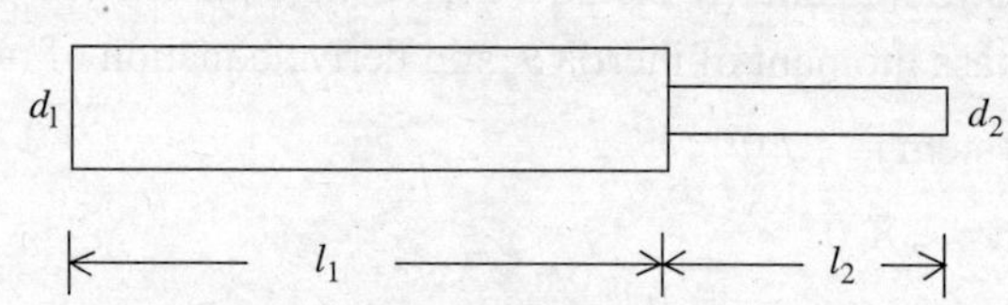

Figure 5.3 Stepped Shaft

The equivalent shaft is defined as a shaft of length L_{eq} and constant diameter d having the same torsional stiffness K as the given shaft.

Consider the action of an applied torque T to one end of a given shaft. The torque will be transmitted from the end of shaft d_1 through the connecting point of two shaft portions, to the other end of shaft d_2. The total angle of twist of the entire shaft is given by

$\vartheta = \vartheta_1$ (twist of shaft of diameter d_1) + ϑ_2 (twist of shaft of diameter d_2)

$$= \frac{32TL_1}{\pi d_1^4 G} + \frac{32TL_2}{\pi d_2^4 G} \quad (5.9)$$

where T is the applied torque and G = shear modulus.

The total angle of twist of the entire shaft can then be expressed as

$$\vartheta = \frac{32T}{\pi G d_1^4}\left(L_1 + \frac{d_1^4}{d_2^4} L_2\right)$$

or

$$\frac{32T}{\pi G d_2^4}\left(L_2 + \frac{d_2^4}{d_1^4} L_1\right) \quad (5.10)$$

Therefore, the equivalent shaft is given by a shaft of constant diameter d_1 and length $\left(L_1 + \frac{d_1^4}{d_2^4} L_2\right)$ or $\left(L_2 + \frac{d_2^4}{d_1^4} L_1\right)$.

5.3 TWO DEGREES-OF-FREEDOM TORSIONAL SYSTEMS (FREE UNCLAMPED)

Consider a two-rotor system as shown in Fig. 5.4.

We assume that mass of the shaft is much lesser than the mass of discs (J_1 and J_2) and shaft has torsional stiffness K. When equal and opposite torques are applied at two ends and suddenly removed, the system will undergo free vibration. The two discs (rotor) move in opposite direction.

Let ϑ_1 and ϑ_2 be the angle of twist (rotations) of rotors A and B, respectively. The equation of motion is then given by

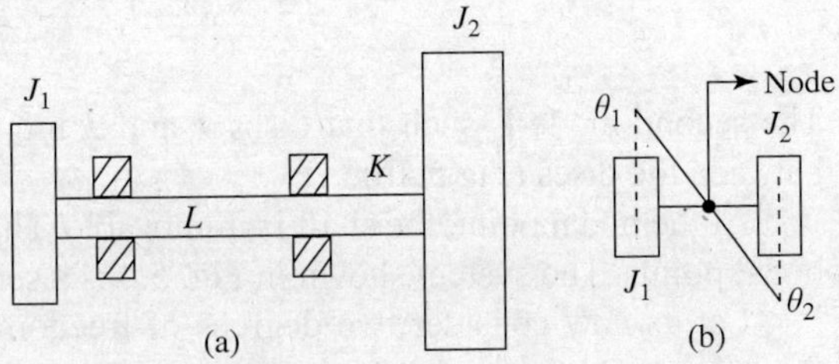

Figure 5.4 (a) Two-rotor System (b) Mode Shape

$$J_1 \ddot{\vartheta}_1 + K(\vartheta_1 + \vartheta_2) = 0 \quad (5.11)$$

$$J_2 \ddot{\vartheta}_2 + K(\vartheta_1 + \vartheta_2) = 0 \quad (5.12)$$

where K is the torsional stiffness, and J_1 and J_2 are the mass momentum of inertias of discs (rotors) A and B.

The solution of Equations 5.11 and 5.12 can be assumed as

$$\vartheta_1 = \mathrm{A}\sin(\omega t + \text{Ø}) \quad (5.13)$$

$$\vartheta_2 = \mathrm{B}\sin(\omega t + \text{Ø}) \quad (5.14)$$

Thus,

$$\ddot{\vartheta}_1 = -\mathrm{A}\omega^2 \sin(\omega t + \text{Ø}) \quad (5.15)$$

$$\ddot{\vartheta}_2 = -\mathrm{B}\omega^2 \sin(\omega t + \text{Ø}) \quad (5.16)$$

Using the relations in the Equations 5.11 and 5.12, we get,

$$(K - J_1\omega^2)\,\mathrm{A} + K\mathrm{B} = 0 \quad (5.17)$$

$$K\mathrm{A} + (K - J_2\omega^2)\mathrm{B} = 0 \quad (5.18)$$

Equations 5.17 and 5.18 homogeneous linear-algebraic equation in A and B and will have a solution different from the trivial solution ($A = 0$, $B = 0$) only when the determinant is zero, that is,

$$\begin{vmatrix} (K - J_1\omega^2) & K \\ K & (K - J_1\omega^2) \end{vmatrix} = 0 \quad (5.19)$$

Expanding the determinant, we get,

$$\omega^2[\omega^2 J_1 J_2 - k(J_1 + J_2)] = 0 \tag{5.20}$$

The solutions are

$$\omega_1 = 0 \quad \text{and} \quad \omega_2 = \sqrt{\frac{k(J_1 + J_2)}{J_1 J_2}} \tag{5.21}$$

The amplitude ratios are obtained from the algebraic equations

$$\frac{A}{B} = \frac{-K}{K - J_1\omega^2} = \frac{-(K - J_2\omega^2)}{K}$$

$$\frac{A_2}{B_2} = \frac{-K}{K - J_1 \dfrac{K(J_1 + J_2)}{J_1 J_2}} = \frac{-KJ_1J_2}{-KJ_1^2} = \frac{J_2}{J_1} \tag{5.22}$$

The second mode is such that discs J_1 and J_2 move in opposite direction. Thus there is a node point between the discs (Fig 5.4b).

At the nodal point, the shaft is immovable. Thus, end rotors move in opposite directions about the nodal points. The system shown in Fig. 5.4 is a semi-definite system.

Let us now consider two degrees-of-freedom system (Fig. 5.5) with an end condition different from the one shown in Fig. 5.4.

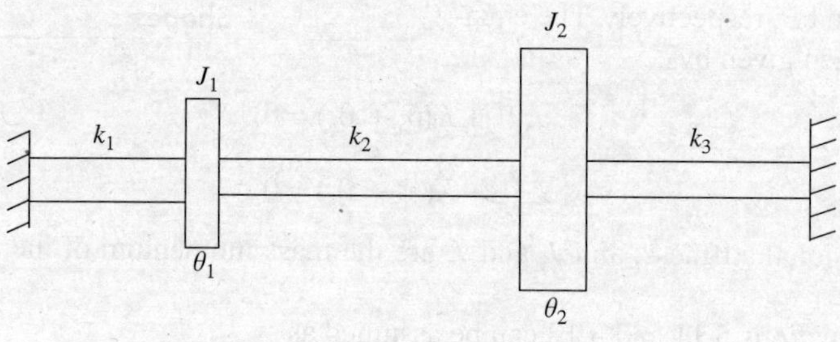

Figure 5.5 Two Degrees-of-Freedom Torsional System

The equations of motion are given by

$$J_1 \ddot{\vartheta} + K_1 \vartheta_1 + K_2(\vartheta_1 - \vartheta_2) = 0 \tag{5.23}$$

or

$$J_1 \ddot{\vartheta}_1 + (K_1 + K_2)\vartheta_1 - K_2\vartheta_2 = 0 \tag{5.24}$$

and

$$J_2 \ddot{\vartheta}_2 + K_2(\vartheta_2 - \vartheta_1) + K_3\vartheta_2 = 0 \tag{5.25}$$

or

$$J_2 \ddot{\vartheta}_2 + (K_2 + K_3)\vartheta_2 - K_2\vartheta_1 = 0 \tag{5.26}$$

Adding Equations 5.24 and 5.26, we get,

$$J_1 \ddot{\vartheta}_1 + J_2 \ddot{\vartheta}_2 + K_1\vartheta_1 + K_3\vartheta_2 = 0$$

or we can have a general relation

$$\sum_i J_i \ddot{\vartheta}_i + K_1\vartheta_1 + K_{i+1}\vartheta_i = 0$$

for a system consisting of 'i' rotors and $i + 1$ shafts.

This type of equation will be useful in the analysis of multi-degrees-of-freedom system. In the special case where $K_1 = K_3 = 0$ (semi-definite system),

$$\sum_1^2 J_i \ddot{\vartheta}_i = 0.$$

We now turn our attention to Equations 5.24 and 5.26. Assuming the solution as

$$\ddot{\vartheta}_1 = \text{A} \sin(\omega t + \text{Ø}),$$

$$\ddot{\vartheta}_2 = \text{B} \sin(\omega t + \text{Ø})$$

Using these equations in Equations 5.24 and 5.26 become

$$(K_1 + K_2 \;\; J_1\omega^2)A - K_2 B = 0 \tag{5.27}$$

$$-K_2 A + (K_1 + K_2 - J_2\omega^2)B = 0 \tag{5.28}$$

Equations 5.27 and 5.28 are homogeneous algebraic equations, which will have other than non-trivial solution ($A = 0$, $B = 0$), only when determinant is zero, and hence

$$\begin{vmatrix} (K_1 + K_2 - J_1\omega^2) & -K_2 \\ -K_2 & K_1 + K_2 - J_2\omega^2 \end{vmatrix} = 0 \tag{5.29}$$

This gives

$$\omega^4 - \left[\frac{K_1 + K_2}{J_1} + \frac{K_2 + K_3}{J_2}\right]\omega^2 + \frac{K_1K_2 + K_2K_3 + K_3K_1}{J_2J_1} = 0 \tag{5.30}$$

This is the frequency equation. The general solution of this is composed of two harmonic motions of natural frequencies ω_1 and ω_2 as

$$\vartheta_1(t) = A_1 \sin(\omega_1 t + \phi_1) + A_2 \sin(\omega_2 t + \phi_2) \tag{5.31}$$

$$\vartheta_2(t) = B_1 \sin(\omega_1 t + \phi_1) + B_2 \sin(\omega_2 t + \phi_2) \tag{5.32}$$

where A_i, B_i, and ϕ_i ($i = 1, 2$) are arbitrary constants. But the amplitude ratios are determined from algebraic equations as

$$\frac{A_1}{B_1} = \frac{K_2}{K_1 + K_2 - J_1\omega_1^2} = \frac{K_2 + K_3 - J_2\omega_1^2}{K_2} = \frac{1}{\lambda_1} \tag{5.33}$$

$$\frac{A_2}{B_2} = \frac{K_2}{K_1 + K_2 - J_1\omega_2^2} = \frac{K_2 + K_3 - J_2\omega_2^2}{K_2} = \frac{1}{\lambda_2} \tag{5.34}$$

Hence, the general motion finally becomes

$$\vartheta_1(t) = A_1 \sin(\omega_1 t + \phi_1) + A_2 \sin(\omega_2 t + \phi_2) \tag{5.35}$$

$$\vartheta_2(t) = \lambda_1 A \sin(\omega_1 t + \phi_1) + \lambda_2 A_2 \sin(\omega_2 t + \phi_2) \tag{5.36}$$

The constants A_1, A_2, ϕ_1, and ϕ_2 are evaluated by the four initial conditions $\vartheta_1(0)$, $\dot{\vartheta}_1(0)$, $\vartheta_2(0)$, and $\dot{\vartheta}_2(0)$.

5.4 GEARED SYSTEMS

A large number of rotating systems employ geared drives. As a result, torsional vibrations of the geared system are of considerable interest. Consider the system shown in Fig. 5.6.

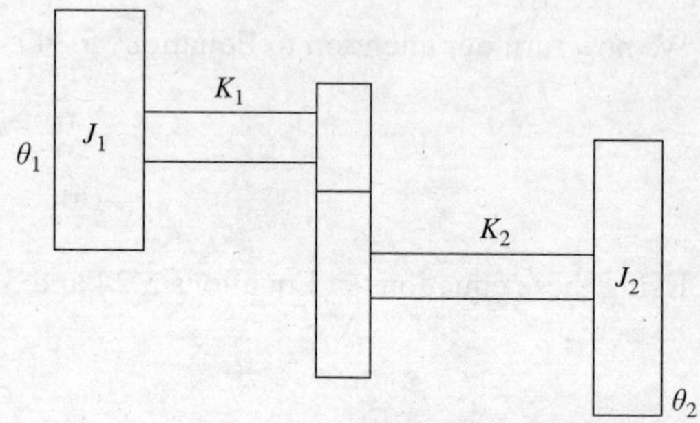

Figure 5.6 Geared System

In this case, the inertias of gears are neglected. Let η be the speed ratio. Let ϑ_1 and ϑ_2 be the angular displacements of the rotors J_1 and J_2. Because of the gears $\vartheta_2 = \eta\,\vartheta_1$, the total energy of the system which consists of both kinetic energy and potential energy; remains constant

$$KE = \tfrac{1}{2} J_1 \dot{\vartheta}_1^2 + \tfrac{1}{2} J_2 \dot{\vartheta}_2^2$$

$$PE = \tfrac{1}{2} K_1 \vartheta_1^2 + \tfrac{1}{2} K_2 \vartheta_2^2$$

Or

$$KE = \tfrac{1}{2} J_1 \dot{\vartheta}_1^2 + \tfrac{1}{2} J_2 (\eta\,\dot{\vartheta}_1)^2$$

$$PE = \tfrac{1}{2} K_1 \vartheta_1^2 + \tfrac{1}{2} K_2 (\eta\,\vartheta_1)^2$$

The total energy will be the same when the gear system is replaced by its equivalent system. This is done by replacing J_2 with $J_2' = \eta^2 J_2$ and K_2 with $K_2 = \eta^2 K$ (Fig. 5.7).

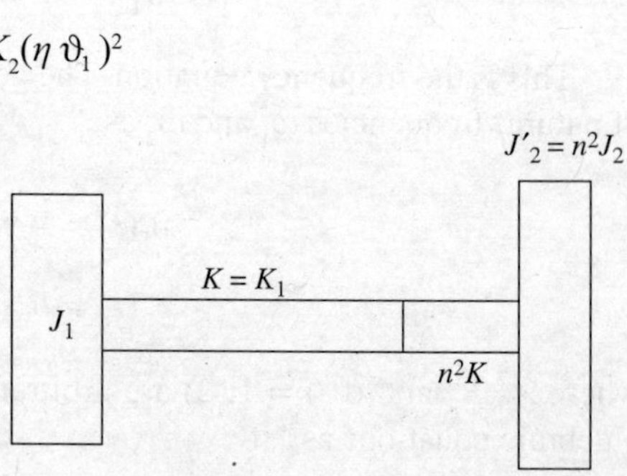

Figure 5.7 Equivalent of Geared System

The natural frequency of the equivalent system is given by

$$\omega_\eta = \sqrt{K_{eq}(J_1 + J_2')/(J_1 J_2')} \quad \text{where} \quad J_2' = \eta^2 J_2 \tag{5.37}$$

$$K_{eq} = \frac{\eta^2 K^2}{K + \eta^2 K} \tag{5.38}$$

5.5 MULTI DEGREES-OF-FREEDOM SYSTEMS

5.5.1 Semi-Definite Systems

A large number of torsional systems can be represented as a series of rotors connected as shown in the Fig. 5.8. For example, in case of large turbine generators, we can have HP rotor, IP rotor, LP rotor, and generator rotor coupled together, and there may be very small shaft extensions at first and last rotor to accommodate supporting bearings and smaller auxilliaries. Modelling them as semi-definite systems is reasonably accurate.

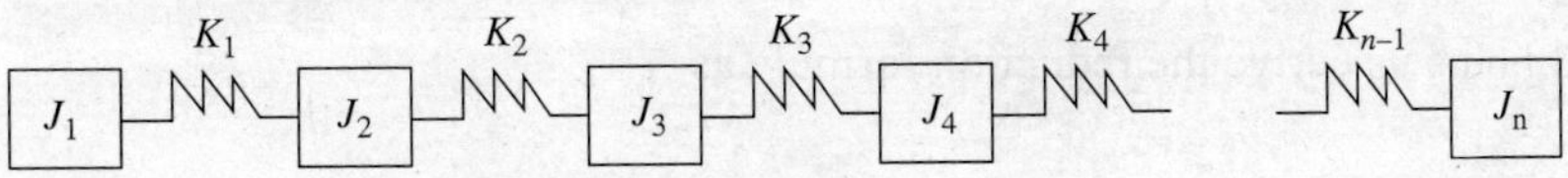

Figure 5.8 Multi-rotor System

Let $\vartheta_1, \vartheta_2 ..., \vartheta_\eta$ be the angular displacements at rotors $J_1, J_2, J_3,...,J_\eta$. The damping in the system is usually due to internal strains and fluids being handled. No external dampers are used. Therefore, it is, quite reasonable to assume zero damping. The equations of motion can then be written as

$$J_1\ddot{\vartheta}_1 + K_1(\vartheta_1 - \vartheta_2) = 0. \tag{5.39}$$

or

$$J_1\ddot{\vartheta}_1 + K_1\vartheta_1 = K_1\vartheta_2$$

or

$$\vartheta_2 = \vartheta_1 + \frac{1}{K_1} J_1\ddot{\vartheta}_1 \tag{5.40}$$

$$J_2\ddot{\vartheta}_2 + K_1(\vartheta_2 - \vartheta_1) + K_2(\vartheta_2 - \vartheta_3) = 0. \tag{5.41}$$

$$K_2\vartheta_3 = (K_1 + K_2)\vartheta_2 - K_1\vartheta_1 + J_2\ddot{\vartheta}_2$$

$$\vartheta_3 = \left(\frac{K_1 + K_2}{K_2}\right)\vartheta_2 - \frac{K_1\vartheta_1}{K_2} + \frac{J_2\ddot{\vartheta}_2}{K_2}$$

$$\vartheta_3 = \left(\frac{K_1 + K_2}{K_2}\right)\left(\vartheta_1 + \frac{1}{K_1}J_1\ddot{\vartheta}_1\right) - \frac{K_1\vartheta_1}{K_2} + \frac{J\ddot{\vartheta}_2}{K_2}$$

$$= \left(\frac{K_1}{K_2} + 1\right)\left(\vartheta_1 + \frac{1}{K_1}J_1\ddot{\vartheta}_1\right) - \frac{K_1\vartheta_1}{K_2} + \frac{J\ddot{\vartheta}_2}{K_2}$$

$$= \left(\frac{K_1}{K_2}\vartheta_1\right) + \frac{J_1\ddot{\vartheta}_1}{K_2} + \vartheta_1 + \left(\frac{1}{K_1}J_1\ddot{\vartheta}_1\right) - \frac{K_1\vartheta_1}{K_2} + \frac{J\ddot{\vartheta}_2}{K_2}$$

$$= \left(\vartheta_1 + \frac{J_1\ddot{\vartheta}_1}{K_1}\right) + \left(\frac{J_1\ddot{\vartheta}_1 + J_2\ddot{\vartheta}_2}{K_2}\right)$$

$$= \vartheta_2 + \frac{1}{K_2}\sum_{i=1}^{2} J_i\ddot{\vartheta}_i$$

If we have yet another disc, J_4, in the semi-definite system, we can write the equation for the third disc as taking into account the shaft having stiffness K_4 as

$$J_3\ddot{\vartheta}_3 + K_3(\vartheta_3 - \vartheta_2) + K_4(\vartheta_3 - \vartheta_4) = 0 \text{ etc.} \tag{5.42}$$

In general show, we can then show

$$\vartheta_i = \vartheta_{i-1} + \frac{1}{K_{i-1}}\sum_1^{i-1} J_i\ddot{\vartheta}_i \tag{5.43}$$

Equation 5.43 helps us derive the reduction formula as

$$\vartheta_i = \vartheta_{i-1} + \frac{1}{K_{i-1}}\sum_1^{i} J_i\ddot{\vartheta}_i$$

writing

$$\vartheta_i = \vartheta_i \sin(\omega t + \phi)$$

$$\ddot{\vartheta}_i = -\omega^2\vartheta_i \sin(\omega t + \phi)$$

Thus,

$$\vartheta_i \sin(\omega t + \phi) = \vartheta_{i-1}\sin(\omega t + \phi) - \frac{1}{K_{i-1}}\omega^2 \sum_{i=1}^{i-1} J_i\vartheta_i$$

or,

$$\vartheta_i = \vartheta_{i-1} - \frac{\omega^2}{K_{i-1}}\sum J_i\vartheta_i \tag{5.44}$$

Adding Equations 5.41, 5.42 and 5.43, etc. for all discs in the semi-definite system, we shall get,

$$\sum J_i\ddot{\vartheta}_i = 0, \quad \sum \omega^2 J_i\vartheta_i = 0 \tag{5.45}$$

Thus, the sum of all inertia torques is equal to zero.
Equations 5.44 and 5.45 are used in estimating the natural frequencies by Holzer method.

5.5.2 One End Fixed, Other End Free and Both Ends Fixed

The analysis is similar to the one discussed in Section 4.9.2 of Chapter 4. Replace x by ϑ m by J in Equation 4.134 and 4.140/4.141 to arrive at the relation to be used.

$$\vartheta_i = \vartheta_{i-1} + \frac{1}{K_i}\left(k_1\vartheta_1 - \omega^2\sum_1^{i-1} J_i\vartheta_i\right) \tag{5.46}$$

and the criterion for correct natural frequency is $K_1\vartheta_1 - \omega^2\sum J_i\vartheta_i = 0$

The natural frequencies can be evaluated using Holzer method or Stodola method. Let us now solve a few typical problems.

Problem 5.3

For the system shown in Fig. 5.9, find natural-frequencies using Holzer's method.

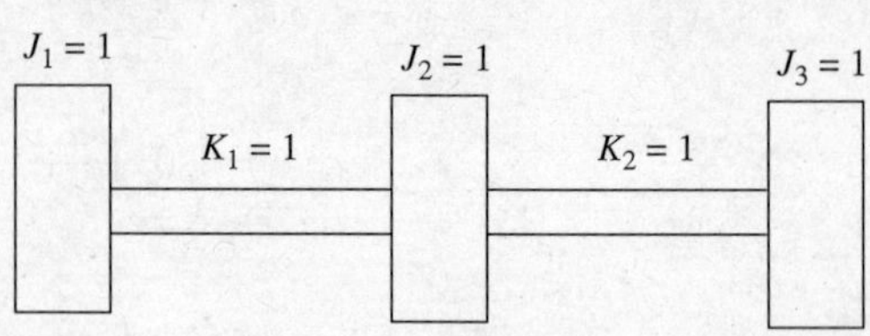

Figure 5.9

This is a semi-definite system and equations to be used for Holzer analysis are

$$\sum \omega^2 J_i \vartheta_i = 0 \quad \text{and} \quad \vartheta_i = \vartheta_{i-1} - \frac{\omega^2}{K_i} \sum J_i \vartheta$$

Item	J_i	$J_i\omega^2$	ϑ_i	$\omega^2 J_i \vartheta$	$\sum \omega^2 J_i \vartheta_i$	K_i	$\sum \omega^2 J_i \vartheta_i / K_i$
Assumed frequency ω = 0.5							
1	1	0.25	1	0.25	0.25	1	0.25
2	1	0.25	0.75	0.19	0.44	1	0.44
3	1	0.25	0.31	0.07	0.51		
Assumed frequency = 0.75							
1	1	0.56	1	0.56	0.56	1	0.56
2	1	0.56	0.44	0.24	0.80	1	0.80
3	1	0.56	−0.36	−0.2	0.60		
Assumed frequency = 1.0							
1	1	1	1	1	1	1	1
2	1	1	0	0	1	1	1
3	1	1	−1	−1	0		
Assumed frequency = 1.5							
1	1	2.25	1.0	2.25	2.25	1	2.25
2	1	2.25	−1.25	−2.82	−0.57	1	−0.57
3	1	2.25	−0.68	−1.53	−2.10		
Assumed frequency = 1.79							
1	1	3.21	1	3.21	3.21	1	3.21
2	1	3.21	−2.21	−7.08	−3.87	1	−3.87
3	1	3.21					

we now plot the results as Fig 5.10 graph.
Thus, the natural frequencies are: $\omega_1 = 0$, $\omega_2 = 1$, $\omega_3 = 1.7$ rad/s.

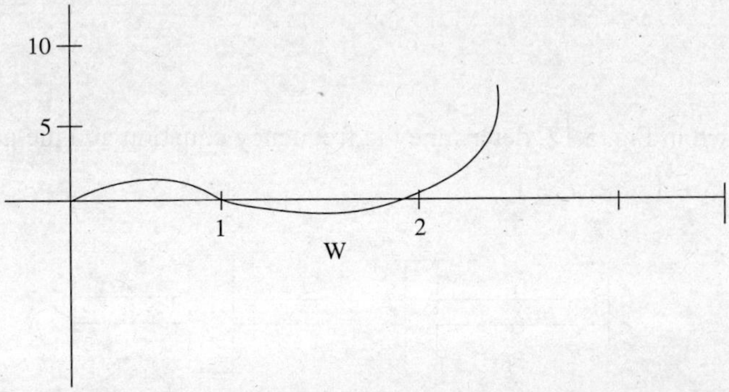

Figure 5.10 Graphical Solution of Holzer Method

Problem 5.4

Use Holzer method to determine the natural frequency of the system shown in Fig. 5.11.

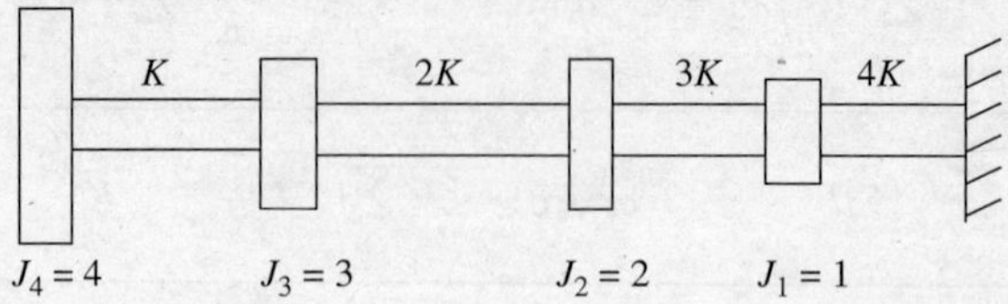

Figure 5.11

Use the relation,

$$\vartheta_i = \vartheta_{i-1} + \frac{1}{K_i}\left(k_1\vartheta_1 - \omega^2 \sum_1^{i-1} J_i\vartheta_i\right) \tag{1}$$

and criterion

$$K_1\vartheta_1 - \omega^2 \sum J_i\vartheta_i = 0 \tag{2}$$

Steps

Assume various values of ω, $\vartheta_1 = 1$ using relation (1), evaluate ϑ_1, ϑ_2, ϑ_3, ϑ_4 and then evaluate relation $K_1\vartheta_1 - \omega^2 \sum J_i\vartheta_i$ for all frequencies. Plot ω versus $K_1\vartheta_1 - \omega^2 \sum J_i\vartheta_i$ and note the zero-cross-over points. They are torsional critical speeds.

Solution:

$$\omega_1 = 0.3\sqrt{\frac{K}{J}}$$

$$\omega_2 = 0.81\sqrt{\frac{K}{J}}$$

$$\omega_3 = 1.45\sqrt{\frac{K}{J}}$$

$$\omega_4 = 2.83\sqrt{\frac{K}{J}}$$

Problem 5.5

For the system shown in Fig. 5.12, determine the frequency equation and the general motion.

Assume $K_1 = K_2 = K_3 = 1$, $J_1 = J_2 = 1$.

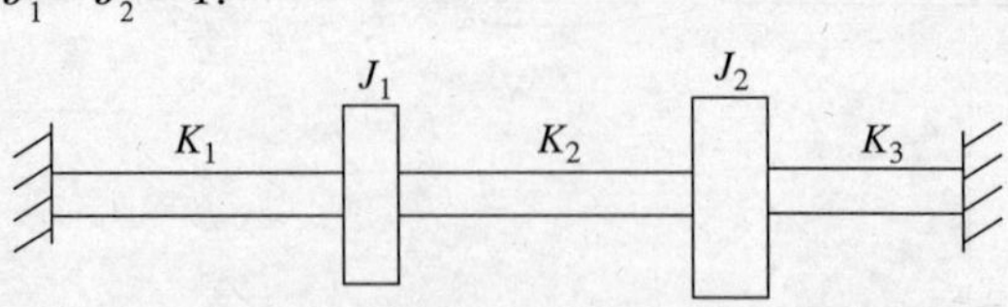

Figure 5.12

The equation motion is

$$J_1 \ddot{\vartheta}_1 + K_1 \vartheta_1 + K_2(\vartheta_1 - \vartheta_2) = 0$$

$$J_2 \ddot{\vartheta}_2 + K_2(\vartheta_2 - \vartheta_1) + K_3 \vartheta_2 = 0$$

or,

$$J_1 \ddot{\vartheta}_1 + (K_1 + K_2)\vartheta_1 - K_2 \vartheta_2 = 0$$

$$J_2 \ddot{\vartheta}_2 + (K_2 + K_3)\vartheta_2 - K_2 \vartheta_1 = 0$$

Let

$$\vartheta_1 = \text{A} \cos(\omega t + \text{Ø}), \quad \ddot{\vartheta}_1 = -\omega^2 \text{A} \cos(\omega t + \text{Ø})$$

$$\vartheta_2 = \text{B} \cos(\omega t + \text{Ø}), \quad \ddot{\vartheta}_2 = -\omega^2 \text{B} \cos(\omega t + \text{Ø})$$

Substituting in the equations of motion, we get,

$$(K_1 + K_2 - J_1\omega^2)\text{A} - K_2\text{B} = 0 \qquad (1)$$

$$-K_2\text{A} + (K_2 + K_3 - J_2\omega^2)\text{B} = 0 \qquad (2)$$

For the Equations 1 and 2 to have a non-trivial solution, the determinant must be zero or,

$$\begin{vmatrix} K_1 + K_2 - J_1\omega^2 & -K_2 \\ -K_2 & K_2 + K_3 - J_2\omega^2 \end{vmatrix} = 0$$

Expression of this gives

$$\omega^4 - \left[\frac{K_1 + K_2}{J_1} + \frac{K_2 + K_3}{J_2}\right]\omega^2 + \frac{K_1K_2 + K_2K_3 + K_3K_1}{J_1J_2} = 0 \text{ is the frequency equation.}$$

$K_1 = K_2 = K_3 = 1, J_1 = J_2 = 1$ putting these in frequency equation, we get,

$$\omega^4 - (2+2)\omega^2 + 3 = 0,$$

or,

$$\omega^4 - 4\omega^2 + 3 = 0$$

$$(\omega^2 - 3)(\omega^2 - 1) = 0$$

or,

$$\omega = 1 \quad \text{and} \quad \omega = \sqrt{3}$$

The general solution is

$$\vartheta_1(t) = A_1 \cos(t + \phi_1) + A_2 \cos(\sqrt{3}t + \phi_2)$$

$$\vartheta_2(t) = B_1 \cos(t + \phi_1) + B_2 \cos(\sqrt{3}t + \phi_2)$$

Equations 1 and 2 gives,

$$\frac{A_1}{B_1} = \frac{K_2}{K_1 + K_2 - J_1\omega_1^2} = \frac{K_2 + K_3 - J_2\omega_1^2}{K_2}$$

$$\frac{A_1}{B_1}=\frac{1}{2-\omega_1^{\,2}}=\frac{2-\omega_1^{\,2}}{1}=\frac{1}{2-1}=\frac{2-1}{1}=1=\frac{1}{\lambda_1}\quad \text{or, } \lambda_1=1$$

$$\frac{A_2}{B_2}=\frac{K_2}{K_1+K_2-J_1\omega_2^{\,2}}=\frac{K_2+K_3-J_2\omega_2^{\,2}}{K_2}=\frac{1}{2-\omega_1^{\,2}}=\frac{2-\omega_1^{\,2}}{1}=\frac{1}{2-3}=\frac{2-3}{1}=-1=\lambda_2$$

Hence, the general solution is

$$\vartheta_1(t)=A_1\cos(t+\phi_1)+A_2\cos(\sqrt{3}t+\phi_2)$$

$$\vartheta_2(t)=A_1\cos(t+\phi_1)-A_2\cos(\sqrt{3}t+\phi_2)$$

Problem 5.6

A harmonic torque $T_0 \sin\omega t$ is applied to the first rotor of the torsional system shown in the Fig. 5.13. Find the steady-state response.

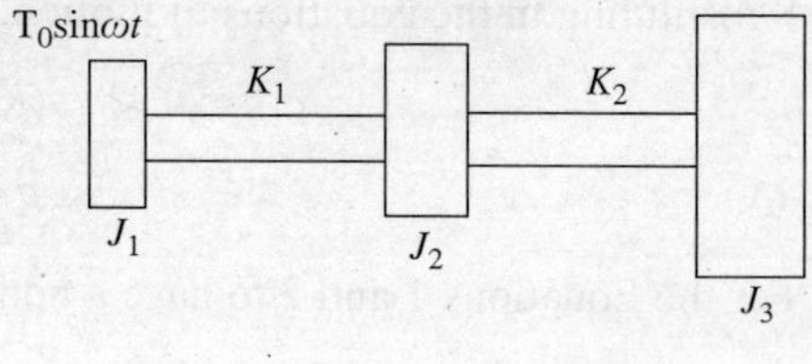

Figure 5.13

$J_1 = 5, J_2 = 10, J_3 = 15$ in lb s²/rad, $K_1 = 10 \times 10^6$, $K_2 = 20 \times 10^6$ in lb/rad, $\omega = 1000$ rad/s

Equations of motion are

$$J_1\ddot{\vartheta}_1+K_1(\vartheta_1-\vartheta_2)=T_0\sin\omega t$$

$$J_2\ddot{\vartheta}_2+K_2(\vartheta_2-\vartheta_1)+K_3(\vartheta_2-\vartheta_1)=0$$

$$J_2\ddot{\vartheta}_3+K_2(\vartheta_3-\vartheta_2)=0$$

Let,

$$\vartheta_1=\text{A}\sin\omega t, \text{ and } \ddot{\vartheta}_1=-\text{A}\omega^2\text{A}\sin\omega t$$

$$\vartheta_2=\text{B}\sin\omega t, \text{ and } \ddot{\vartheta}_2=-\text{B}\omega^2\sin\omega t$$

$$\vartheta_3=\text{C}\sin\omega t,\ \ddot{\vartheta}_3=-\text{C}\omega^2\sin\omega t$$

Substituting these in equations of motion and simplifying,

$$(K_1-J_1\,\omega^2)\text{A}-K_1\text{B}=T_0$$

$$-K_1\text{A}+(K_1+K_2-J_2\,\omega^2)\text{B}-K_2\text{C}=0$$

$$-K_2\text{B}+(K_2-J_3\,\omega^2)=0$$

Solving these equations by Cramer's rule, we obtain,

$$\text{A}=1.5\times10^{-7}\,T J_3\ddot{\vartheta}_3+K_3(\vartheta_3-\vartheta_2)+K_4(\vartheta_3-\vartheta_4)=0. \tag{1}$$

$$\text{B}=0.25\times10^{-7}\,T_0,\ \text{C}=-1\times10^{-7}T_0.$$

Hence, steady-state vibrations are

$$\vartheta_1(t) = 1.5 \times 10^{-7}\, T_0 \sin\omega t$$

$$\vartheta_2(t) = 0.25 \times 10^{-7}\, T_0 \sin\omega t$$

$$\vartheta_3(t) = -1 \times 10^{-7}\, T_0 \sin\omega t$$

CONCLUSION

Although torsional vibration analysis can be done on principles described in the previous chapter, considering practical applications, torsional vibrations are separately considered in this chapter.

After discussing SDOF system, we discussed about multi-degrees-of-freedom system, starting with two-degrees-of-freedom system (Section 5.3). In this, we considered both semi-definite as well as fixed–fixed two rotors system. We also considered geared system (Section 5.4) considering numerous applications of geared system in practice.

We have given a detailed analysis of multi-degrees-of-freedom system in section 5.5. In this section, we considered three conditions at the beginning and the end of the rotor (first and last rotors). These conditions are semi-definite system, one end fixed and the other end free, and fixed–fixed system. Formulae have been derived to find angular displacements at various stations using iterative procedure (Holzer method), and also the criterion for convergence of the solution has been presented for all these cases. Stodola method described in the previous chapter can also be applied for torsional systems; however, Holzer's method gives results faster. In the next chapter, we will consider the continuous system.

EXERCISES

5.1 Explain why it is necessary to analytically evaluate the torsional vibration response of a multi-rotor system.

5.2 For the system shown below, derive the equations of motion in torsional vibrations. Find the natural frequencies and the mode shape. Rotor 1 is the impeller of a blower while Rotor 3 is rotor of a reciprocating engine. In case the natural frequency of the system is in close vicinity of the engine speed, what corrective action would you suggest?

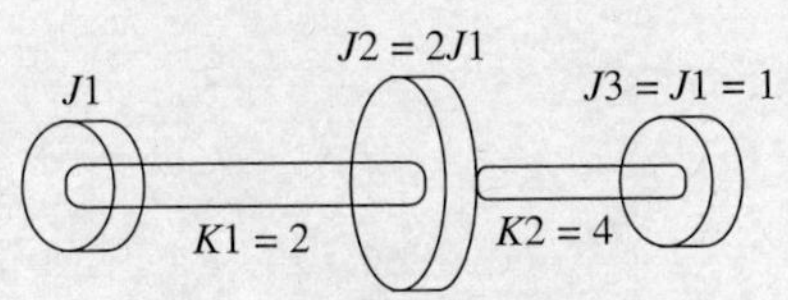

Problem 5.2

5.3 Derive the equation of motion of a five-rotor system fixed at both ends. Assume that all the rotors have equal mass moment of inertia and the connected shafts have equal torsional stiffness. Extend the analysis to six-rotor system fixed at both ends by arriving at a generalized relation that you may develop while analysing a five rotor system.

5.4 For the system described in Problem 5.3, find out the torsional natural frequency using Holzer's method.

5.5 For the system shown in the figure, find the natural frequency of the system.

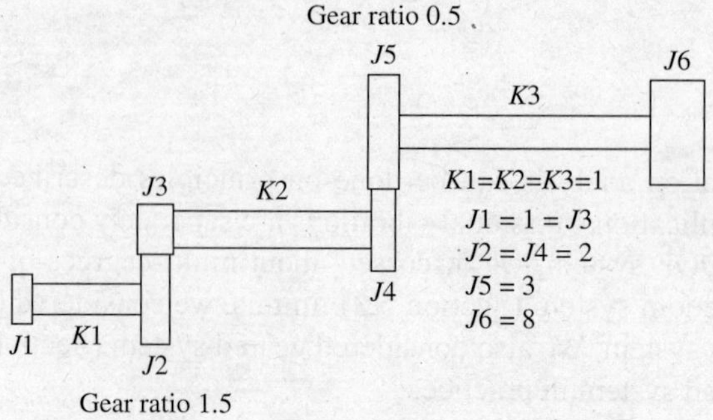

Problem 5.5

Hint: The system can be modelled as a four-rotor semi-definite system. *J*2 and *J*3 can be combined as Rotor 2 (taking into account gear ratio), *J*4 and *J*5 as another rotor and *J*6 is the fourth rotor. *J*1 is the first rotor. Use Holzer's method.

6 Transverse Vibrations

6.1 INTRODUCTION

We have until now dealt with discrete systems where it was assumed that the mass (inertia), damping (dissipation), and elasticity (springs) are present at certain discrete points in the system. For example, as explained earlier, a three-storey building [Fig. 6.1(a)] is idealized as a three-mass-and-three-spring system [Fig. 6.1(b)].

Under the idealization concept, the inertia of the system is assumed to be concentrated as a three-point mass located at the floor levels and the elasticities of supporting columns are idealized as massless springs. This sort of idealization, many a times, proves reasonably acceptable, since the results of computed natural frequencies and the mode shapes do not greatly differ from those obtained from more complex analyses, which take into account the elasticity of floors and the masses of columns. In reality, the system shown in Fig. 6.1(a) consists of the assemblage of continuous systems comprising of three slabs and eight columns.

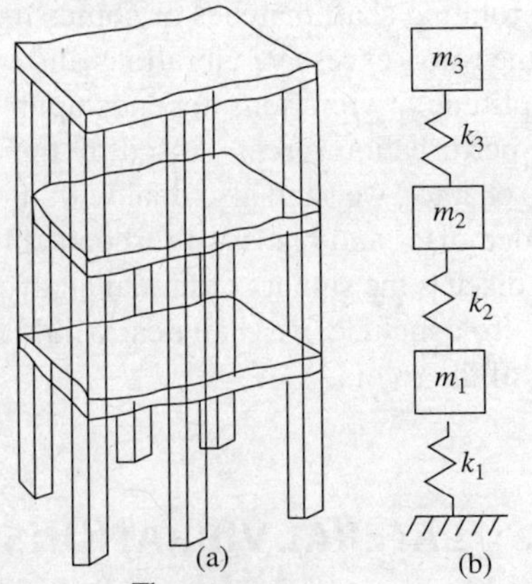

(a) Three-storey building (b) Lumped spring–mass model

Figure 6.1 (a) Three-storey Building, (b) Idealization

Another popular method of approximating a continuous system as a multi degrees-of-freedom system involves replacing the geometry of the system by an assemblage of large number of small elements. By assuming a simple solution within each element, the principles of compatibility and equilibrium are used to find an approximate solution to the original system. This method, known as finite element method, usually is (wrongly) regarded as most accurate/exact, although in reality, it is also an approximate method. Of course, the finite-element approximation is much closer to continuous system compared to idealization by discrete spring–mass–damper system.

As stated earlier in this book, most engineering systems are continuous and have an infinite degrees-of-freedom. In-order to arrive at system equations, we still use equations of equilibrium. The continuous system equations are partial differential equations that are against the ordinary linear differential equations derived for the discrete system. The solution of partial differential equations is quite complex and at times, quite difficult. In fact, the analytical or closed-form solutions do not exist for many partial differential equations. On the other hand, the analysis of a multi degrees-of-freedom (discrete) system requires a solution of set of ordinary differential equations, which is relatively much simpler. Hence, for simplicity of analysis, continuous systems are often approximated as multi degrees-of-freedom systems. The choice between the exact method and approximate method must be made very carefully with due consideration to the factors such as the purpose of analysis, the influence of analysis on design, the computational to ease, and time available.

In this chapter, we shall consider the vibrations of simple continuous systems, such as beams. For the other advanced and complex systems, readers may refer to *Vibration Problems in Engineering* (4th ed.) by S. Timoshenko, D. H. Young and W. Weaver or *Dynamics of Continuous Elements* by S. K. Clark, Prentice Hall, Englewood, Cliffs NJ.

In general, the frequency equation of a continuous system is a transcendental equation that yields an infinite number of natural frequencies and mode shapes. We need to apply boundary conditions to find the natural frequencies and mode shapes of a continuous system. The question of boundary conditions does not arise in the case of discrete system except in an indirect way and that is because the influence coefficients depend upon the manner in which the system is supported.

One of the most important aspects of the vibrations of rotating machinery is the lateral or transverse vibration of the rotors, also called the whirling of rotors. Additionally, these turbomachinery rotors have bladed (vaned) wheels/discs. The vibration of the blades (vanes) as well as discs is also a subject of great interest, since their vibration characteristics greatly influence the designs of the turbomachinery. In this chapter, we will see that when the frequency of transverse (lateral) vibration of the rotating shaft matches or comes in the near neighbourhood of the natural frequency of transverse of the rotor, excessive vibrations and vibratory stresses develop in the rotor. We shall also see that the amplitude of vibrations depends upon the state of balance of the rotor. One of the means of reducing the perturbation forces caused by the unbalance of the rotor is appropriately balancing the rotor. In this chapter, we shall also deal with the subject of balancing of the rotors. Analysis of the vibration of bladed discs and blades of turbomachinery is subject by itself. We shall however deal with this when we discuss the subject of modal analysis in the next chapter.

To conclude, we shall deal with transverse-vibration beams/shafts, whirling of rotors, and balancing of the rotors.

6.2 LATERAL VIBRATIONS OF BEAMS

Earlier, we dealt with methods to derive the equation of fundamental frequency of lateral vibration of beams or shafts carrying several masses such as pulleys, bladed discs/impellers or flywheel. In these cases, the static-deflection curve obtained from the strength of material approach was used as an approximation to dynamic-deflection curve. We had assumed that the beam/shaft is massless (negligible mass compared to the masses on the beam/shaft) and beam/shaft was considered to be having only elasticity. In many situations, this provides a reasonable accuracy and hence is quite a popular method.

We shall now consider both the inertia and elasticity of the beam/shaft.

Consider the free-body diagram of an element of beam/shaft [Fig. 6.2(a)] and consider the various forces/moments acting [Fig. 6.2(b)].

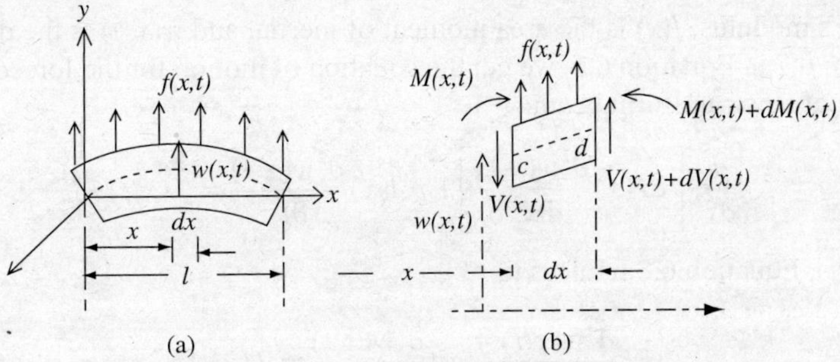

Figure 6.2 Element of Vibrating Beam

$M(x, t)$ is the bending moment (BM) at x, $V(x, t)$ is the shear force (SF) at x, and $f(x, t)$ is the external force per unit length of the beam. Over the length dx, the (BM) and (SF) are $M + dM$ and $V + dV$, respectively. The dynamic deflection at x and at time t is $w(x, t)$. This means that in all moments, shear forces as well as disturbing forces are time-dependent. The inertia force acting on the element of beam is

$$\rho A dx \frac{\partial^2 w(x.t)}{\partial t^2} \tag{6.1}$$

where ρ is the density, $A(x)$ is area of cross-section, dx is the length of element and $w(x, t)$ is the dynamic deflection.

The force-equilibrium equation in the Z-direction is given by

$$-(V + dV) + f(x,t)dx + V = \rho A dx \frac{\partial^2 w(x,t)}{\partial t^2} \tag{6.2}$$

The moment equation about the y-axis passing through point O in Fig. 6.2 leads to

$$(M + dM) - (V + dV)dx + f(x,t)dx\frac{dx}{2} - M = 0 \tag{6.3}$$

Let $dV = \frac{\partial V}{\partial x}dx$ and $dM = \frac{\partial M}{\partial x}dx$ and let us neglect the terms involving powers of dx, since dx is infinitesimal. Then Equations 6.2 and 6.3 become

$$-\frac{\partial V(x,t)}{\partial x} + f(x,t) = \rho A \frac{\partial^2 w(x,t)}{\partial t^2} \tag{6.4}$$

and

$$\frac{\partial M(x,t)}{\partial x} - V(x,t) = 0 \tag{6.5}$$

Since $V = \frac{\partial M}{\partial x}$ from Equation 6.5, Equation 6.4 becomes

$$-\frac{\partial^2 M(x,t)}{\partial x^2} + f(x,t) = \rho A \frac{\partial^2 w(x,t)}{\partial t^2} \tag{6.6}$$

From the elementary theory of beams (also known as the Euler–Bernoulli or thin-beam theory), the relationship between bending moment and the deflection of the beam can be expressed as

$$M(x,t) = EI \frac{\partial^2 w(x,t)}{\partial x^2} \tag{6.7}$$

where E is Young's modulus, $I(x)$ is the area moment of inertia, and $w(x, t)$ is the deflection of beam. Inserting Equation 6.7 in Equation 6.6, we get the equation of motion for the forced-lateral vibrations of the beam/shaft of a non-uniform beam.

$$\frac{\partial^2}{\partial x^2}\left[EI(x)\frac{\partial^2 w(x,t)}{\partial x^2}\right]+\rho A(x)\frac{\partial^2 w(x,t)}{\partial t^2}=f(x,t) \tag{6.8}$$

For uniform beams, Equation 6.8 reduces to

$$EI\frac{\partial^4 w(x,t)}{\partial x^4}+\rho A\frac{\partial^2 w(x,t)}{\partial t^2}=f(x,t) \tag{6.9}$$

For free vibrations, $f(x, t) = 0$ and thus the equation of motion becomes

$$c^2\frac{\partial^4 w(x,t)}{\partial x^4}+\frac{\partial^2 w(x,t)}{\partial t^2}=0 \tag{6.10}$$

where
$$c^2=\frac{EI}{\rho A} \tag{6.11}$$

Study of Equations of motion 6.9 and 6.10 shows that they involve second-order derivative with respect to time and fourth order derivative with respect to x. Thus, to solve these equations, we need two initial conditions and four boundary conditions. Usually, the values of lateral displacements and velocities are specified $w_0(x), \dot{w}_0(x)$ at $t = 0$ and the initial conditions become

$$w(x,t=0)=w_0(x)$$

$$\frac{\partial w(x,t=0)}{\partial t}=\dot{w}_0(x) \tag{6.12}$$

6.2.1 Free Vibrations

The free-vibration equation is

$$c^2\frac{\partial^4 w(x,t)}{\partial x^4}+\frac{\partial^2 w(x,t)}{\partial t^2}=0$$

where
$$c^2=\frac{EI}{\rho A}.$$

We assume $w(x, t) = W(x)\, T(t)$, according to the method of separation of variables. Substituting this in Equation 6.10, we get,

$$\frac{c^2}{W(x)}\frac{d^4W(x)}{dx^4}=-\frac{1}{T(t)}\frac{d^2T(t)}{dt^2}=a=\omega^2 \tag{6.13}$$

In the Equation 6.13, $a = \omega^2$ is a positive constant. Equation 6.13 can also be written as

$$\frac{d^4W(x)}{dx^4}-\beta^4W(x)=0 \tag{6.14}$$

and
$$\frac{d^2T(t)}{dt^2}+\omega^2T(t)=0 \tag{6.15}$$

In the Equations 6.14 and 6.15,

$$\beta^4 = \frac{\omega^2}{c^2} = \frac{\rho A \omega^2}{EI} \tag{6.16}$$

The solution of Equation 6.16 is expressed as

$$T(t) = \text{A} \cos \omega t + \text{B} \sin \omega t \tag{6.17}$$

where A and B are constants, which can be found from the initial conditions. For the Equation 6.14, we assume

$$W(x) = \text{C}e^{sx} \tag{6.18}$$

where s and C are constants.

The auxilliary equation is $$x^4 - \beta^4 = 0 \tag{6.19}$$

Roots of the Equation 6.19 are

$$s_{1,2} = \pm\beta, \quad s_{3,4} = \pm i\beta \tag{6.20}$$

Hence, solution of Equation 6.14 becomes

$$W(x) = c_1 e^{\beta x} + c_2 e^{-\beta x} + c_3 e^{i\beta x} + c_4 e^{-i\beta x} \tag{6.21}$$

where c_1, c_2, c_3, and c_4 are constants. Equation 6.21 can also be expressed as

$$W(x) = c_1(\cos\beta x + \cosh\beta x) + c_2(\cos\beta x - \cosh\beta x) + c_3(\sin\beta x + \sinh\beta x) + c_4(\sin\beta x - \sinh\beta x) \tag{6.22}$$

where c_1, c_2, c_3, and c_4 are different constants and can be found from the boundary conditions. The natural frequencies of the beam/shaft are computed from Equation 6.16 as

$$\omega = \beta^2 \sqrt{\frac{EI}{\rho A}} = (\beta l)^2 \sqrt{\frac{EI}{\rho A l^4}}, \tag{6.23}$$

where l = length of the beam.

The function $W(x)$ is known as the normal mode or characteristic function of the beam and ω is known as the natural frequency. For any beam, there will be an infinite number of normal modes with one natural frequency associated with each normal mode. The unknowns constants c_1, c_2, c_3, and c_4 in Equations 6.21 and 6.22 and the value of β in Equation 6.23 can be obtained from the boundary conditions of the beam as follows:

(1) Free end $$BM = EI\frac{\partial^2 W}{\partial x^2} = 0, \quad SF = \frac{\partial}{\partial x}\left(EI\frac{\partial^2 W}{\partial x^2}\right) = 0 \tag{6.24}$$

(2) Simply-supported (pinned) beam-end $$\textit{Deflection}\ W = 0, \quad BM = EI\frac{\partial^2 W}{\partial x^2} = 0 \tag{6.25}$$

(3) Fixed end $$\textit{Deflection}\ W = 0, \quad \textit{Slope} = \frac{\partial W}{\partial x} = 0 \tag{6.26}$$

The frequency equations, the mode shapes (normal functions), and the natural frequencies for beams with common boundary conditions are given here in Table 6.1.

Table 6.1 Transverse vibration of beams

End Conditions	Frequency Equation	Mode Shape	Value of
Pinned–pinned	$\sin \beta_n l = 0$	$W_n(x) = c_n[\sin \beta_n x]$	$\beta_1 l = \pi$ $\beta_2 l = 2\pi$ $\beta_3 l = 3\pi$ $\beta_4 l = 4\pi$
Free–free	$\cos \beta_n l - \cosh \beta_n l = 1$	$W_n(x) = c_n[\sin \beta_n x + \sinh \beta_n l + a_n(\cosh \beta_n x + \cos \beta_n x)]$, where $a_n = \left(\dfrac{\sinh \beta_n l - \sin \beta_n l}{\cos \beta_n l - \cosh \beta_n l}\right)$	$\beta_1 l = 4.730$ $\beta_2 l = 7.853$ $\beta_3 l = 10.995$ $\beta_4 l = 14.137$ $\beta l = 0$ for rigid body mode
Fixed–fixed	$\cos \beta_n l.\cosh \beta_n l = 1$	$W_n(x) = c_n[\sinh \beta_n x - \sin \beta_n l + a_n(\cosh \beta_n x - \cos \beta_n x)]$, where $a_n = \left(\dfrac{\sinh \beta_n l - \sin \beta_n l}{\cos \beta_n l - \cosh \beta_n l}\right)$	$\beta_1 l = 4.730$ $\beta_2 l = 7.850$ $\beta_3 l = 10.995$ $\beta_4 l = 14.137$
Fixed–free	$\cos \beta_n l.\cosh \beta_n l = -1$	$W_n(x) = c_n[\sin \beta_n x - \sinh \beta_n l - a_n(\cos \beta_n x - \cosh \beta_n x)]$	$\beta_1 l = 1.875$ $\beta_2 l = 4.694$ $\beta_3 l = 7.854$ $\beta_4 l = 10.995$
Fixed–pinned	$\tan \beta_n l - \tanh \beta_n l = 0$	$W_n(x) = c_n[\sin \beta_n x - \sinh \beta_n l + a_n(\cosh \beta_n x - \cos \beta_n x)]$, $a_n = \left(\dfrac{\sinh \beta_n l - \sin \beta_n l}{\cos \beta_n l - \cosh \beta_n l}\right)$	$\beta_1 l = 3.926$ $\beta_2 l = 7.0685$ $\beta_3 l = 10.21$ $\beta_4 l = 13.351$

A configuration comprising of ends connected to a linear spring, damper, and mass (Fig. 6.3) is of particular and practical importance. When the end of a beam undergoes transverse vibration displacement w and slope $\partial w/\partial x$ with velocity $\partial w/\partial t$ and acceleration $\partial^2 w/\partial t^2$, the resisting forces due to spring, damper, and mass are proportional to w, $\partial w/\partial t$, and $\partial^2 w/\partial t^2$, respectively. The shear force at the ends balances this resulting force. Thus,

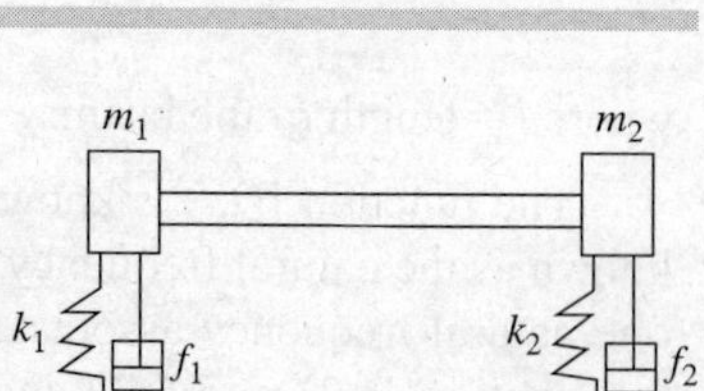

Figure 6.3 Beam on Spring/Damper Support

$$\partial/\partial x\left(\mathrm{EI}\frac{\partial^2 w}{\partial x^2}\right) = a\left[kw + f\frac{\partial w}{\partial t} + m\frac{\partial^2 w}{\partial t^2}\right] \tag{6.27}$$

where $a = -1$ for left end and $+1$ for right end of the beam. In addition, the bending moment must be zero; hence,

$$EI\frac{\partial^2 w}{\partial x^2} = 0 \tag{6.28}$$

Next, we consider a configuration consisting of ends connected to a torsional spring, torsional damper, and a rotating inertia (Fig. 6.4).
In this, the boundary conditions are

$$EI\frac{\partial^2 w}{\partial x^2} = a\left[k_t\frac{\partial w}{\partial x} + \frac{\partial^2 w}{\partial x \partial t} + I_0\frac{\partial^3 w}{\partial x \partial t^2}\right] \qquad (6.29)$$

where a = −1 for left and +1 for right end

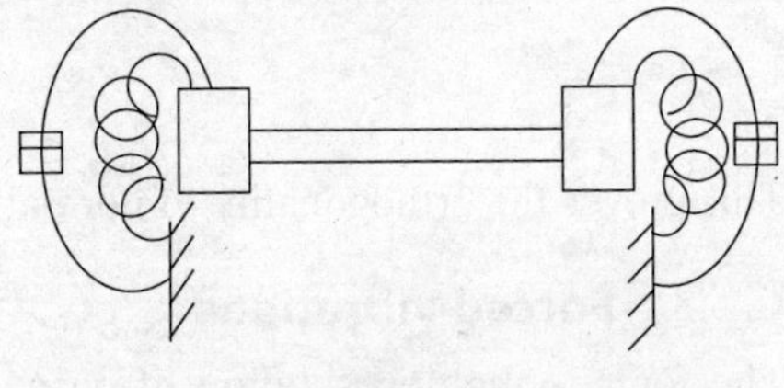

Figure 6.4 Beam-ends Connected to Torsional Spring/Damper

and
$$\frac{\partial}{\partial x}\left[EI\frac{\partial^2 w}{\partial x^2}\right] = 0 \qquad (6.30)$$

6.2.2 Orthogonality of Normal Functions

As in the case of discrete systems, the principle of orthogonality of modes holds good for beam vibrations. Refer to Equation 6.13

$$c^2\frac{d^4W(x)}{dx^4} - \omega^2 W(x) = 0 \qquad (6.31)$$

Let $W_i(x)$ and $W_j(x)$ be the normal functions corresponding to the natural frequencies ω_i and $\omega_j (i \neq j)$ so that

$$c^2\frac{d^4W_i(x)}{dx^4} - \omega_i^2 W(x) = 0 \qquad (6.32)$$

and
$$c^2\frac{d^4W_j(x)}{dx^4} - \omega_j^2 W(x) = 0 \qquad (6.33)$$

Multiplying Equation 6.32 by W_j and Equation 6.33 by W_i, subtracting the resulting equations one from the other and integrating from 0 to l gives

$$\int_0^l\left[c^2\frac{d^4W_i}{dx^4}W_j - \omega_i^2 W_i W_j\right]dx - \int_0^l\left[c^2\frac{d^4W_j}{dx^4}W_i - \omega_j^2 W_j W_i\right]dx = 0$$

or

$$\int_0^l W_i W_j dx = -\frac{c^2}{\omega_i^2 - \omega_j^2}\int_0^l (W_i'''' W_j - W_i W_j'''')\,dx \qquad (6.34)$$

(Primes indicate differentiation with respect to x)
The right-hand side of Equation 6.34 can be evaluated using integration by parts to obtain

$$\int_0^l W_i W_j dx = -\frac{c^2}{\omega_i^2 - \omega_j^2}\left[W_i W_j''' - W_j W_i''' + W_j' W_i'' - W_i' W_j''\right]_0^l \qquad (6.35)$$

The right-hand side of Equation 6.35 can be shown to be zero for any combination of free, fixed or simply-supported end conditions. For example, at free end BM = SF = 0 or w″ = w‴ = 0, etc. Since each term on the RHS of equation is zero at $x = 0$ or $x = l$ for any combination of the boundary conditions, we obtain,

$$\int W_i W_j dx = 0 \tag{6.36}$$

This proves the orthogonality of normal modes.

6.2.3 Forced Vibrations

The forced-vibration solution of beam can be determined using the principle of mode superposition. (See the section on modal analysis in the previous chapter where this concept was explained for discrete systems.) For the continuous beams, the deflection of the beam is assumed as

$$w(x,t) = \sum_{n=1}^{\infty} W_n(x) q_n(t) \tag{6.37}$$

W_n in Equation 6.37 is the normal mode or characteristic function satisfying the differential equation

$$EI \frac{d^4 W(x)}{dx^4} - \omega^2 W(x) = 0 \tag{6.37(a)}$$

Thus,

$$EI \frac{d^4 W_n(x)}{dx^4} - \omega_n^2 \rho A W_n(x) = 0, \quad n = 1, 2, \ldots .n, \tag{6.38}$$

and q_n is the generalized coordinate in the nth mode. Substituting Equation 6.37 into Equation 6.9, that is,

$$EI \frac{\partial^4 w(x,t)}{\partial x^4} + \rho A \frac{\partial^2 w(x,t)}{\partial t^2} = f(x,t)$$

$$EI \sum_{n=1}^{\infty} \frac{d^4 W_n(x)}{dx^4} q_n(t) + \rho A \sum_{n=1}^{\infty} W_n(x) \frac{d^2 q_n(t)}{dt^2} = f(x,t) \tag{6.39}$$

In view of Equations 6.37(a), Equation 6.39 can be written as

$$\sum_{n=1}^{\infty} \omega_n^2 W_n(x) q_n(t) + \sum_{n=1}^{\infty} W_n(x) \frac{d^2 q_n(t)}{dt^2} = \frac{1}{\rho A} f(x,t) \tag{6.40}$$

Multiplying Equation 6.40 throughout by $W_m(x)$, integrating from 0 to l (length of the beam) and using orthogonality condition $\left(\int_0^l W_i W_j = 0 \right)$, we get

$$\frac{d^2 q_n(t)}{dt^2} + \omega_n^2 q_n(t) = \frac{1}{\rho A b} Q_n(t) \tag{6.41}$$

$Q_n(t)$, in the above equation, is called the generalized force corresponding to $q_n(t)$ and is given by

$$Q_n(t) = \int_0^l f(x,t) W_n(x) dx \tag{6.42}$$

and

$$b = \int_0^l W_n^2(x) dx \tag{6.43}$$

Equation 6.41 is similar to the equation of motion of an undamped single degree-of-freedom system. The solution of this equation is

$$q_n(t) = A_n \cos\omega_n t + B_n \sin\omega_n t + \frac{1}{\rho A b \omega_n}\int_0^t Q_n(t)\sin\omega_n(t-\tau)d\tau \tag{6.44}$$

The first two terms on RHS of Equation 6.44 represents the transient or free vibrations (resulting from the initial conditions) and the last term denotes the steady-state vibration resulting from forcing function.

Problem 6.1

Find the steady-state response of a pinned–pinned beam subjected to harmonic force (such as unbalance force) $f_0 \sin \omega t$.

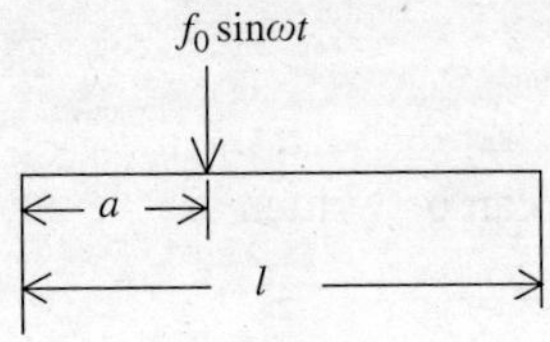

Figure 6.5 Forced Vibration of Beam

Solution:

We shall use mode superposition method. The normal mode functions for a pinned–pinned beam (Table 6.1) is

$$W_n(x) = \sin\beta_n x = \sin\frac{n\pi x}{l}$$

$$\beta_n l = n\pi$$

The generalized force $Q_n = \int_0^l f(x,t)\sin\beta_n x dx = f_0(\sin n\pi a/l)\sin\omega t$

The steady-state response

$$q_n(t) = \frac{1}{\rho A b \omega_n}\int_0^l Q_n(\tau)\sin\omega_n(t-\tau)d\tau$$

where $$b = \int_0^l W_n^2(x)dx = \int_0^l \sin^2\beta_n x dx = l/2$$

Thus, the solution is

$$q_n(t) = \frac{2f_0}{\rho A l}\frac{\sin n\pi a/l}{\omega_n^2 - \omega^2}\sin\omega t$$

$$w(x,t) = \sum_1^\infty W_n(x) q_n(t)$$

Therefore, the response of the beam is given by

$$w(x,t) = \frac{2f_0}{\rho A l}\sum_1^\infty \frac{1}{\omega_n^2-\omega^2}\sin\frac{n\pi a}{l}\sin\frac{n\pi x}{l}\sin\omega t$$

Problem 6.2

Determine the normal modes of transverse vibrations of a simply supported beam of length l and uniform cross-section as shown in Fig. 6.6.

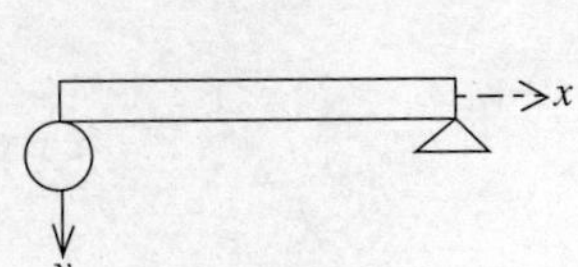

Figure 6.6

Solution:

Let $y(x, t)$ be the dynamic deflection at any point x. The governing equation is

$$\frac{\partial^2 y}{\partial t^2} + c^2\frac{\partial^4 y}{\partial x^4} = 0, \quad c^2 = \frac{EI}{\rho A} \tag{1}$$

Assume the solution is composed of two parts—one being function of x, say $X(x)$ and other being function of time, say $T(t)$. Thus,

$$y = X(x)\,T(t).$$

Substituting in the governing Equation 1, we get,

$$X\ddot{T} + c^2 X'''' T = 0, \quad X'''' = \frac{d^4 X}{dx^4} \tag{2}$$

This can be written as

$$X'''' / X = -\ddot{T} / c^2 T \tag{3}$$

The LHS of Equation 3 is a function of x alone and RHS is a function of t alone. This is possible only when side is equal to a constant. Let this constant be P^2/c^2. This leads to two ordinary differential equations

$$(1)\ X'''' - (P^2/c^2)X = 0 \quad \text{and} \quad (2)\ \ddot{T} + P^2 T = 0$$

The solution of second equation is well-known

$$T(t) = \text{A} \cos Pt + \text{B} \sin Pt \tag{4}$$

The solution of Equation 1 is of the form $X = e^{\lambda x}$, where λ is given by

$$\lambda = k, -k, ik, \quad \text{and} \quad -ik\ (k^4 = P^2/c^2)$$

The general solution can be written in the form

$$X(x) = c_1 \cos kx + c_2 \sin kx + c_3 \cosh kx + c_4 \sinh kx$$

or

$$X(x) = c_1(\cos kx + \cosh kx) + c_2(\sin kx + \sinh kx) + c_3(\cos kx - \cosh kx) + c_4(\sin kx - \sinh kx) \tag{5}$$

c_1, c_2, c_3, c_4 are the new constants.

In the case of simply supported beam, the displacements and bending moments are both equal to zero at each end of the beam. In other words, it means that we have the following four-boundary conditions.

$$(X)_{x=0} = 0, \quad X_{x=1} = 0, \quad \left(\frac{d^2 X}{dx^2}\right)_{x=0} = 0, \quad \left(\frac{d^2 X}{dx^2}\right)_{x=l} = 0$$

$$(X)_{x=0} = c_1 + c_3 = 0, \quad \left(\frac{d^2 X}{dx^2}\right)_{x=0} = -c_1 + c_3 = 0 \tag{6}$$

From this we obtain $c_1 = c_3 = 0$.

$$(X)_l = c_2 \sin kl + c_4 \sinh kl = 0 \tag{7}$$

$$\left(\frac{d_2 X}{dx^2}\right)_l = -k^2 c_2 \sin kl + c_4 k^2 \sin kl = 0 \tag{8}$$

$$\therefore c_4 = 0, \quad \sin kl = 0 \tag{9}$$

Therefore, we get,

$$kl = i\Pi \quad (i = 1, 2, 3, \ldots).$$

Again, we get,

$$\sqrt{P/Cl} = i\pi$$

Thus, the natural frequencies are $P_i = i\pi^2 c^2/l^2$ and the normal function is X_i = sin $i\Pi x/l$. The mode shape are shown in Fig. 6.7.

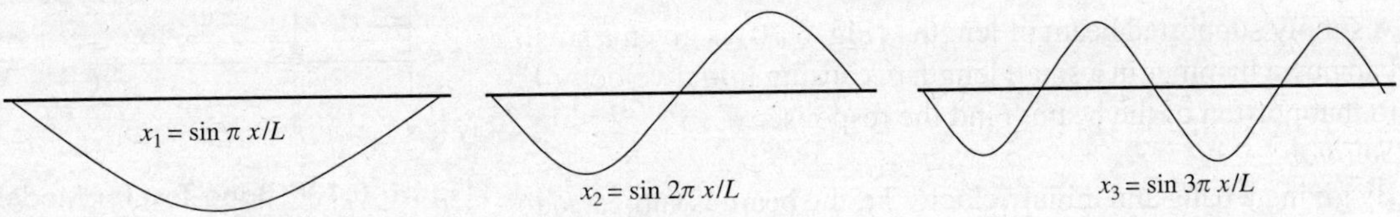

Figure 6.7 Mode Shapes of the Beam

In many cases as explained in Section 6.2.1, the boundary conditions can be different. For example, let us consider two different boundary conditions for the cantilever [Fig. 6.8(a) and (b)].

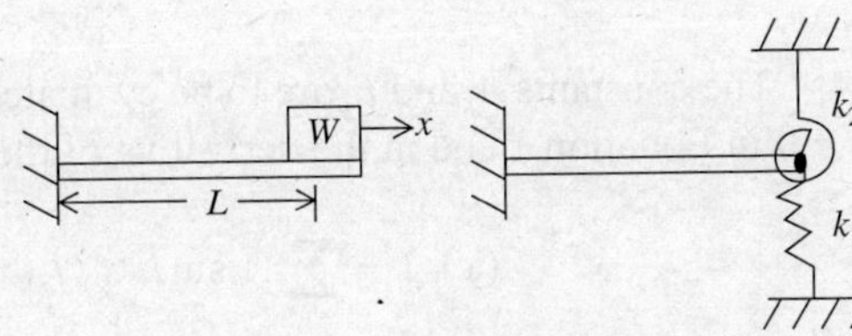

Figure 6.8 Boundary Conditions

For the boundary condition shown in Fig. 6.8(a), the displacement and the slope are equal to zero at built-up end, whereas at free end, the inertia force is equal to the shearing force. This means

$$(y)_{x=0} = 0, \quad \left(\frac{\partial y}{\partial x}\right)_{x=0} = 0$$

$$\left[EI\frac{\partial^3 y}{\partial x^3}\right]_{x=l} = \left[-\frac{W}{g}\frac{d^2 y}{dt^2}\right]_{x=l} = \left[EI\frac{d^2 y}{dx^2}\right]_{x=l} = 0$$

For the boundary condition shown in Fig. 6.8(b), displacement and the slope are zero at the built-up end whereas spring forces balance the shearing force and the bending moment at other end. This means

$$(y)_{x=0} = 0, \quad \left(\frac{\partial y}{\partial x}\right)_{x=0} = 0$$

$$\left[EI\frac{\partial^2 y}{\partial x^2}\right]_{x=l} = \left[-k_2\frac{dy}{dx}\right]_{x=l}, \quad \left[EI\frac{\partial^3 y}{\partial x^3}\right]_{x=l} = k_1(y)_{x=l}$$

Remember that k_2 = BM/slope.

Problem 6.3

Find the frequency equation for a beam fixed (Fig. 6.9) at both ends.

Solution:

Figure 6.9

As explained earlier, the solution of equation $\frac{\partial^2 y}{\partial t^2} + c^2\frac{\partial^4 y}{\partial x^4} = 0, c^2 = \frac{EI}{\rho A}$ can be written as $X'''' / X = -\ddot{T}/c^2 T$, which leads to two independent ordinary differential equations

$$X'''' - (P^2/c^2)X = 0, \text{ and } \ddot{T} + P^2T = 0$$

In this the boundary conditions are

$$(y)_{x=0} = 0, \quad \left(\frac{dy}{dx}\right)_{x=0} = 0, \quad (y)_{x=l} = 0, \quad \left(\frac{dy}{dx}\right)_{x=l} = 0$$

Rest of the procedure is as explained in previous problems.

Problem 6.4

A simply supported beam of length l (Fig. 6.10) is given a small bang by a hammer in a small length δ, causing initial velocity V_0 to that portion of the beam. Find the response.

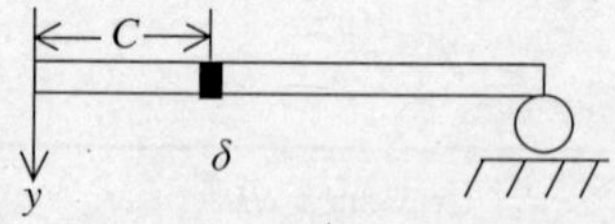

Figure 6.10 Bang Test for Model Analysis of Beam

Solution:

By giving a bang and initial velocity V_0, the beam is subjected to free vibrations given by

$$y(x,t) = \sum_{i=1,2}^{\infty} \sin\frac{i\pi x}{l}(A_i \cos p_i t + B_i \sin p_i t) \tag{1}$$

The constants A_i and B_i are to be evaluated using initial conditions and $p_i = c_i^2\pi^2/l^2$. Substituting $t = 0$ in Equation 1 and in the derivative of the above expression with respect to t, we obtain

$$(y)_{t=0} = \sum_{i=1,2}^{\infty} A_i \sin i\pi x/l, (\dot{y})_{t=0} = V_0, \quad (C - \delta/2) \le x \le (C + \delta/2)$$

$$\therefore A_i = 0, [V_0]_{C-\delta/2}^{C+\delta/2} = \sum_{i}^{\infty} p_i B_i \sin\frac{i\pi x}{l}$$

$$\therefore \int_0^l p_i B_i \sin\frac{i\pi x}{l} \sin\frac{i\pi x}{l} dx = \int_{C-\delta/2}^{C+\delta/2} V_0 \sin\frac{i\pi x}{l} dx$$

and

$$Bi = \frac{2V_0 l}{i\pi l p_i} \sin\frac{i\pi C}{l}\left(2\sin\frac{i\pi\delta}{2}\right)$$

$$\therefore y(x,t) = \frac{4V_0}{i\pi}\sum\frac{1}{p_i}\sin\frac{i\pi C}{l}\sin\frac{i\pi x}{l}\sin p_i t \sin\frac{i\pi\delta}{2l}$$

The response when analysed in a spectrum analyser will show all natural frequencies of the beam. This forms the basis for the bang test in extracting the information about natural frequencies of the structures and is a widely used method. Determination of natural frequencies is one of the most important exercises in experimental vibration testing.

We have until now considered rigorous methods of finding the natural frequencies and the mode shapes for various configurations of beams/shafts. The solutions invariably consist of infinite series and infinite natural frequencies. Thus, these methods are quite tedious.

Also, in actual practice, we are interested in knowing only a few of the natural frequencies and verify as to whether or not any of the operating frequencies, such as rotational speed of the shaft and its harmonics, are equal to or in the near neighbourhood of any of the natural frequencies. This is to ensure that large amplification of forces/stresses takes place. If such amplification is taking place, suitable corrective actions required to be taken.

Also, in majority of the cases, knowing the lowest or fundamental natural frequency is most important. Also, since in practical vibration problems, we strive to keep the forcing frequency much away from the lowest natural frequency, knowing the natural frequency range is enough.

We may recall that in the previous chapter on multi degrees-of-freedom discrete-vibration systems, we discussed various methods such as Dunkerley method, Stodola method, and Rayleigh's method. Amongst these, Rayleigh's method can be applied to find the fundamental frequency of continuous systems. This method is much simpler than the exact analysis especially with varying distribution of masses and stiffness. Rayleigh's method is applicable to all continuous systems; however, we shall consider its application to beam/shaft-transverse vibrations.

6.3 RAYLEIGH'S METHOD

Consider the beam shown in Fig. 6.2. In order to apply Rayleigh's method, we need to derive expressions for maximum kinetic and potential energies and Rayleigh's quotient. The kinetic energy of the beam is

$$T = \frac{1}{2}\int_0^l \dot{w}^2 dm = \frac{1}{2}\int_0^l \dot{w}^2 \rho A(x) dx \tag{6.45}$$

Assume the transverse vibration (deflection) $w(x, t)$ as harmonic variation, that is,

$$w(x, t) = W(x) \cos \omega t.$$

$$\therefore T_{\max} = \frac{\omega^2}{2}\int_0^l \rho A(x) W^2(x) dx \tag{6.46}$$

The potential energy of the beam is work done in deforming the beam. Neglecting the work done by shear forces, we have

$$V = \frac{1}{2}\int_0^l M d\theta \tag{6.47}$$

where M is the bending moment given by

$$M = M(x,t) = EI(x)\frac{\partial^2 w(x,t)}{\partial x^2} \tag{6.48}$$

θ is the slope of the deformed beam given by $\theta = \dfrac{\partial w}{\partial x}$.

Thus, Equation 6.47 becomes Equation 6.48.

Since the maximum value (amplitude) of w is $W(x)$, the maximum value of V is given by

$$V_{\max} = \frac{1}{2}\int EI(x)\left(\frac{d^2 W(x)}{dx^2}\right)^2 dx \tag{6.49}$$

By equating KE and PE, we obtain Rayleigh's quotient as

$$R(\omega) = \omega^2 \frac{\int_0^l EI\left(\dfrac{d^2 W(x)}{dx^2}\right)^2 dx}{\int_0^l \rho A (W(x))^2 dx} \tag{6.50}$$

Thus, the natural frequency of the beam can be found once the deflection $W(x)$ is known. In general, $W(x)$ is not known and must therefore be assumed. Generally, the static-equilibrium shape is assumed for $W(x)$ to obtain the fundamental frequency. It must always be remembered that the assumed shape $W(x)$ unintentionally introduces a constraint on the system in the form of additional stiffness to the system. Due to this, the frequency given by Equation 6.50 is higher than the exact frequency.

For a stepped beam/shaft

$$R(\omega) = \omega^2 = \frac{E_1 I_1 \int_0^{l_1} \frac{d^2W}{dx^2} dx + E_2 I_2 \int^{l_2} \frac{d^2W}{dx^2} dx + \cdots\cdots}{\rho A_1 \int_0^{l_1} W^2 dx + \rho A_2 \int_0^{l_2} W^2 dx + \cdots\cdots} \tag{6.51}$$

Problem 6.5

For a tapered shaft (Fig. 6.11), find the natural frequency.

Solution:

For the shaft, we have

$$A(x) = \frac{hx}{l} \times 1, \quad I(x) = \frac{1}{2}\left(\frac{hx}{l}\right)^3, \quad W_{assumed}(x) = (l - x/l)^2$$

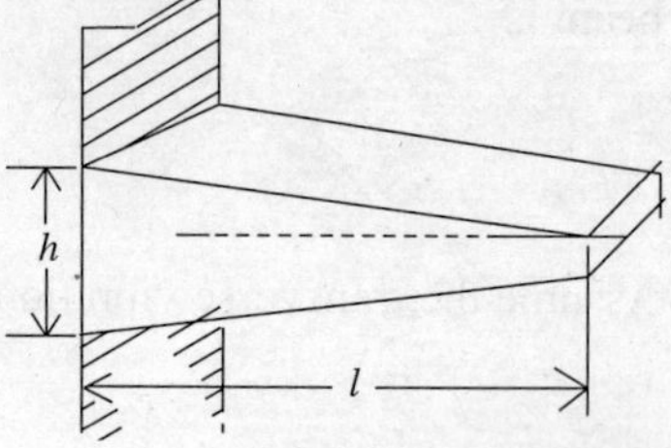

Figure 6.11 Natural Frequency of Tapered Beam

The Rayleigh's quotient

$$R(\omega) = \omega^2 = \frac{\int_0^L E\left(\frac{h^3x^3}{12l^3}\right)(2/l^2)dx}{\int_0^l \rho(hx/l)(1 - x/l)^4 dx} = 2.5Eh^2/\rho l^4$$

$$\omega = 1.5811\sqrt{(Eh^2/\rho l^4)}$$

The exact value, from literature is

$$\omega = 1.5343(Eh^2/\rho l^4)^{1/2}$$

This shows that frequency obtained from Rayleigh's method is about 3% higher than the exact value. This is accepted in practice.

One of the biggest disadvantages of Rayleigh's method is that it yields the information about only the lowest or the first natural frequency of the beam/shaft, and information about higher natural frequencies and mode shapes remains undetermined. This difficulty is overcome in Rayleigh–Ritz method.

6.4 RAYLEIGH–RITZ METHOD

Rayleigh–Ritz (R–R) method can be considered as an extension of Rayleigh's method. It is based upon an assumption (hypothesis) that a closer approximation to the exact natural mode can be obtained by superposing a number of assumed functions instead of using a single assumed function (static-deflection curve) as in the Rayleigh's method. If the assumed functions are appropriately chosen, this method provides not only the approximate value of the first natural frequency, but also the values of higher modes and the mode shapes. An arbitrary number of functions can be used and the number of frequencies that can be obtained is equal to the number of functions, although it involves more computational effort. Of course, it leads to results that are more accurate.

For the analysis of lateral vibrations of the beam, if n functions are chosen for approximating the deflection curve, $W(x)$, we have

$$W(x) = c_1 w_1(x) + c_2 w_2(x) + c_3 w_3(x) + \cdots\cdots + c_n w_n(x) \tag{6.52}$$

where w_n ($n = 1, 2, 3,\ldots, n$) are known linearly independent functions of the coordinate x that satisfy all the boundary conditions of the problem and c_n ($n = 1, 2, 3, \ldots$) are coefficients that need to be evaluated. The coefficients c_i ($i = 1, 2, 3\ldots, n$) are to be determined so that assumed functions $w_i(x)$ provides the best possible approximation to the natural modes. To obtain such approximations, the coefficients c_i are adjusted, and the natural frequency is made stationary at the natural modes.

For this, we substitute Equation 6.52 in Rayleigh's quotient

$$R(\omega) = \frac{\int_0^l EI\left(\frac{d^2 w(x)}{dx^2}\right)^2 dx}{\int_0^l \rho A(W(x)^2)dx}$$

and the resulting expression is partially differentiated with respect to each of the coefficients c_i. To make the natural frequency stationary, we set each of the partial derivative equal to zero and obtain

$$\frac{\partial}{\partial c_i}(\omega^2) = 0, \qquad i = 1,2,3\ldots n \tag{6.53}$$

Equations 6.53 are a set of linear algebraic equations in coefficients c_1, c_2, c_3, … c_n containing unknown quantity ω^2. This defines an algebraic eigen value problem similar to the ones that arise in discrete multi degrees-of-freedom system. The solution of this eigen value problem gives n natural frequencies $\omega^2(i)$ ($i = 1,2, \ldots, n$) and n eigen vectors each containing a set of numbers for $c_1, c_2, c_3, \ldots, c_n$. The ith eigen vector corresponding to ω_i may be expressed as

$$\bar{C}^i = \begin{Bmatrix} c_1^i \\ c_2^i \\ \cdot \\ c_n^i \end{Bmatrix} \tag{6.54}$$

When the above eigen vector $\bar{C}^i = \begin{Bmatrix} c_1^i \\ c_2^i \\ \cdot \\ c_n^i \end{Bmatrix}$ is substituted in Equation 6.53, we obtain the best possible approximation to the ith mode of the beam.

We shall illustrate this by solving the Problem 6.5 (Fig. 6.11). In this problem, the beam is tapered and its natural frequency was evaluated using Rayleigh's method. We now assume deflection functions as

$$w_1(x) = (1 - x/l)^2 \tag{6.55}$$

$$w_2(x) = \frac{x}{l}(1 - x/l)^2 \tag{6.56}$$

$$w_3(x) = \frac{x^2}{l^2}(1 - x/l)^2 \tag{6.57}$$

If we use the one-term approximation $w(x) = c_1(1 - x/l)^2$, the fundamental frequency will be same as

$$R(\omega) = \omega^2 = \frac{\int_0^L E\left(\frac{h^3 x^3}{12 l^3}\right)(2/l^2)^2\, dx}{\int_0^l \rho(hx/l)(1 - x/l)^4\, dx} = 2.5Eh^2/\rho l^4$$

$$\omega = 1.5811\sqrt{(Eh^2/\rho l^4)}$$

We now use the two-term approximation

$$W(x) = c_1(1 - x/l)^2 + c_2 x/l(1 - x/l)^2 \tag{6.58}$$

The Rayleigh's quotient is then

$$R(\omega) = \omega^2 = \frac{\int_0^l EI(x)\left(\frac{d^2W(x)}{dx^2}\right)^2 dx}{\int_0^l \rho A(x)[W(x)]^2\, dx} = \frac{X}{Y} \tag{6.59}$$

If Equation 6.58 is substituted in Equation 6.59, Equation 6.59 becomes function of c_1 and c_2. The conditions that make ω^2 or $R(\omega^2)$ stationary are

$$\frac{\partial(\omega^2)}{\partial c_1} = \frac{Y\frac{\partial X}{\partial c_1} - X\frac{\partial Y}{\partial c_1}}{Y^2} = 0$$

$$\frac{\partial(\omega^2)}{\partial c_2} = \frac{Y\frac{\partial X}{\partial c_2} - X\frac{\partial Y}{\partial c_2}}{Y^2} = 0$$

These equations can also be written as

$$\frac{\partial X}{\partial c_1} - \frac{X}{Y}\frac{\partial Y}{\partial c_1} = \frac{\partial X}{\partial c_1} - \omega^2\frac{\partial Y}{\partial c_1} = 0 \tag{6.60}$$

$$\frac{\partial X}{\partial c_2} - \frac{X}{Y}\frac{\partial Y}{\partial c_2} = \frac{\partial X}{\partial c_2} - \omega^2\frac{\partial Y}{\partial c_2} = 0 \tag{6.61}$$

But $$W(x) = c_1(1 - x/l)^2 + c_2 x/l(1 - x/l)^2$$

$$X = \int_0^l EI(x)\left(\frac{d^2W(x)}{dx^2}\right)^2 dx$$

$$Y = \int_0^l \rho A(x)[W(x)]^2\, dx$$

We can obtain

$$X = \frac{Eh^3}{3l^3}(c_1^2/4 + c_2^2/10 + c_1c_2/5)$$

$$Y = \rho hl(c_1^2/30 + c_2^2/280 + 2c_1c_2/105)$$

Using these in Equations 6.60 and 6.61, we obtain

$$\begin{bmatrix} (1/2-\omega^{*2}/15) & (1/5-2\omega^{*2}/105) \\ (1/5-2\omega^{*2}/105) & (1/5-\omega^{*2}/140) \end{bmatrix} \begin{Bmatrix} c_1 \\ c_2 \end{Bmatrix} = \begin{Bmatrix} 0 \\ \end{Bmatrix} \tag{6.62}$$

where

$$\omega^{*2} = 3\omega^2 \rho l^4 / Eh^2 \tag{6.63}$$

Equation 6.62 gives frequency equation

$$\frac{1}{8820}\omega^{*4} - \frac{13}{1400}\omega^{*2} + \frac{3}{50} = 0$$

This equation gives

$$\omega_1{}^* = 2.6500, \quad \omega_2{}^* = 8.6492,$$

Using Equation 6.63, we get

$$\omega_1 = 1.5367(Eh^2/\rho l^4)^{1/2}$$

$$\omega_2 = 4.9936(Eh^2/\rho l^4)^{1/2}$$

We can note that the exact solution gives the fundamental frequency as $\omega = 1.5343(Eh^2/\rho l^4)^{1/2}$. Thus see a close matching of result of R–R method with result from the exact solution.

The choice for the functions $w_1(x)$, $w_2(x)$, …, $w_n(x)$ should be made looking at the geometry of the beam and the end conditions, remembering that at fixed ends, the deflections and slopes of deflection curve are zero and the displacements at simple supports are zero. Also, one should visualize all possible modes or deflected shapes of the beams at various natural frequencies. The use of polynomials for the dynamic deflection curve generally yields very good results.

The concept of Rayleigh–Ritz method is really not empirical if we recollect the fact that any dynamic-deflected curve can be shown to be a suitable superposition of mode shapes at various natural frequencies.

Generally for all practical purposes, it is sufficient to use the static-deflection curve to evaluate the fundamental natural frequency. Thus, simpler method such as Dunkerley method can be very effectively used. In the previous chapter, we have discussed the Dunkerley method with illustration.

6.5 WHIRLING OF ROTATING SHAFTS

In many practical engineering applications such as turbines, compressors, electrical motors, pumps, heavy rotors (bladed disc/impeller, rotor windings, etc) are mounted on a shaft that undergoes dynamic deflections. These deflections are caused by forces due to factors such as unbalance of the rotors, misalignment of rotors, and variety of forces caused by mechanical (constructional) deficiencies. Depending upon the deflected shape of the rotor–shaft assembly, the system can be categorized as a rigid system or flexible system. When the shaft mounted with rotor elements rotates at a speed less than the first natural frequency it is considered as rigid rotor while when it rotates at a speed higher than the first natural frequency it is said to be flexible. In either case, the shaft carrying the rotor element runs in a bent condition. At certain speeds, called critical speeds, this bending may be large or small depending upon the magnitude of the forces that cause the bending, damping in the system, gyroscopic effect, etc. These speeds are also called the whirling speeds. In general, whirling is defined as the rotation of the plane made by the line of centres of the bearings and the bent shaft. We shall now consider the aspects of modelling the rotor system, critical speeds, response, and stability of the system.

6.5.1 Equations of Motion

Consider a shaft carrying a rotor or disc of mass m supported by two bearings as shown in Fig. 6.12.

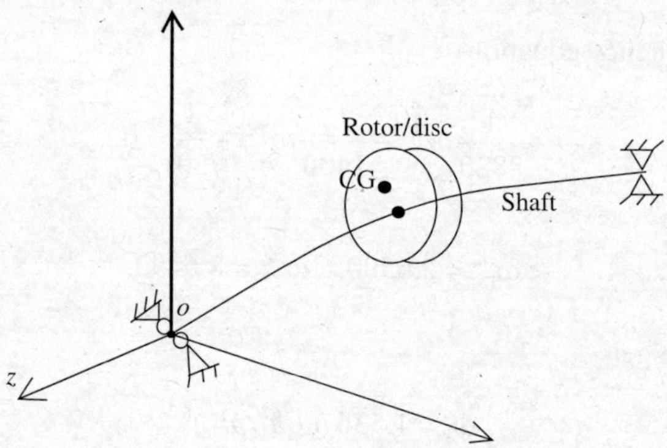

Figure 6.12 Whirling of Shafts

We shall assume that the rotor is subjected to a steady-state excitation force due to mass unbalance, which means that the CG of the rotor do not coincide with the centre of rotation C and is eccentric by an amount e. Due to this eccentricity, during the rotation of shaft–rotor assembly, the system is subjected to an unbalance force of magnitude $me\omega^2$, where ω is rotational speed of the rotor. The forces acting on the rotor are the inertia force due to the acceleration of mass centre ($me\omega^2$), the spring force due to elasticity of the shaft and the external/internal damping forces (fluid friction if the rotor/disc handles fluid, strain-induced, etc.)

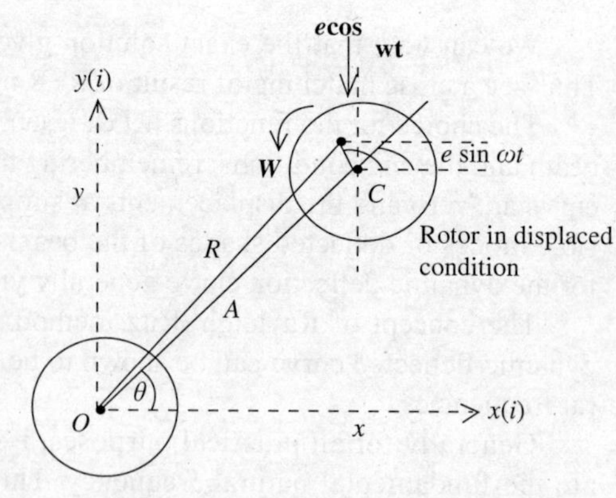

Figure 6.13 Rotor with Eccentricity

Refer to Fig. 6.13 where O denotes the equilibrium position of the shaft when perfectly balanced (which means CG of the rotor matches with the centre of rotation). The shaft line (CG) is assumed to rotate with a constant angular velocity ω. During rotation, the rotor deflects radially by a distance A = OC (in steady state). This type of displacement occurs since the shaft is supported on bearings (hydrodynamic or hydrostatic or anti-friction bearings) in which the bearing clearance (required for film of lubrication) allows the shaft to ride up in the direction of rotation more so in case shaft is supported on the hydrodynamic bearings. The rotor/disc is assumed to have an eccentricity e (which means the centre of gravity of the disc/rotor is at a distance e from the centre of rotation; the rotor is unbalanced). We use a fixed coordinate system (X and Y) with O as the origin for describing the motion of the system. The angular velocity of the line OC $\dot{\theta} = d\theta/dr$ is known as the whirling speed and in general, not equal to ω.

The equation of motion of the rotor (mass m) can be written as

$$\text{Inertia force } (\overline{F}_i) = \text{Elastic force } (\overline{F}_e) + \text{Internal damping force } (\overline{F}_{id}) + \text{External damping force } (\overline{F}_{ed}) \tag{6.64}$$

Inertia force
$$\overline{F}_i = m\ddot{\overline{R}} \tag{6.65}$$

Where $\overline{R}$ denotes the radius vector of the mass centre G given by

$$\overline{R} = (x + e\cos\omega.t)\overline{i} + (y + e\sin\omega.t)\overline{j} \tag{6.66}$$

where i and j are unit vectors, x and y are coordinates of geometric centre (centre of rotation).

Equations 6.65 and 6.66 lead to

$$\overline{F} = m((\ddot{x} - e\omega^2 \cos\omega.t)\overline{i} + (\ddot{y} - e\omega^2)\overline{j}) \tag{6.67}$$

Elastic force $\overline{F}_e = -k(x\overline{i} + y\overline{j})$ where k = stiffness of the shaft (6.68)

Internal damping force $\overline{F}_{di} = -f_i\{(\dot{x} + \omega y)\overline{i} + (\dot{y} + \omega x)\overline{j}$ where f_i is the internal damping coefficient. (6.69)

External damping force $\overline{F}_{de} = -f\{\dot{x}\overline{i} + \dot{y}\overline{j}\}$ where f is the external damping coefficient. (6.70)

Using these equations in Equation 6.69, we get,

$$m\ddot{x} + (f_i + f)\dot{x} + kx - f_i\omega y = em\omega^2 \cos\omega.t \tag{6.71}$$

$$m\ddot{y} + (f_i + f)\dot{y} + kx - f_i\omega x = em\omega^2 \sin\omega.t \tag{6.72}$$

These equations of motion which describe the lateral (transverse) vibrations of the rotor are coupled and are dependent on the speed of the state rotation of the shaft, ω. By defining a complex quantity P as

$$P = x + iy \tag{6.73}$$

where
$$i = \sqrt{-1}$$

Adding Equations 6.71 and 6.72, we get,

$$m\ddot{P} + (f_i + f)\dot{P} + kP - i\omega f_i P = em\omega^2 e^{j\omega.t} \tag{6.74}$$

6.5.2 Critical Speeds

A critical speed will exist when the frequency of rotation of the shaft (rotational speed ω) equals the natural frequency of the shaft. The undamped natural frequency of the rotor system can be obtained by solving Equations 6.71, 6.72 or 6.74 by making RHS of the equation zero and f_i, $f = 0$. The resulting homogeneous equation gives

$$\omega_n = \sqrt{k/m} \tag{6.75}$$

When the rotational speed ω matches with this critical speed ω_n, the rotor undergoes large deflections and the force transmitted to the bearings will be so enormous that they may fail. It is for this reason that when normal operating speed of the shaft is above the critical speed, during the speed-up of the shaft, when it comes in the near neighbourhood of critical speed, is done very rapidly/fast so that large shaft deformations will be experienced for a very short duration. After this, the shaft bearing system can approach safe-operating speed.

Response of the System

The rotor–shaft bearing system is excited by a harmonic force due to the unbalance of the rotor. For determining the response, let us assume that the internal damping f_i is negligible. Then we can solve Equations 6.71 and 6.72 or Equation 6.74, so that the equation of motion is

$$m\ddot{P} + f\dot{P} + kP = em\omega^2 e^{j\omega.t} \tag{6.75}$$

The solution of the above equation comprises of complimentary function, which represents the decaying transient vibrations, and particular solution, which represents the steady-state response. The steady-state part of the solution is assumed as

$$P = Ae^{j(\omega.t-\phi)} \tag{6.76(a)}$$

Substituting Equation 6.76 in Equation 6.77, we get

$$A = \frac{em\omega^2}{\left[(k-m\omega^2)^2 + \omega^2 f^2\right]^{1/2}} = \frac{er^2}{\sqrt{(1-r^2)^2 + (2\xi.r)^2}} \tag{6.77}$$

$$\phi = \tan^{-1}\left(\frac{f\omega}{k - m\omega^2}\right) = \tan^{-1}\left(\frac{2\xi.r}{1-r^2}\right) \tag{6.78}$$

In these equations, $r = \omega/\omega_n, \omega_n = \sqrt{k/m},\ \xi = f/2\sqrt{km}$. By differentiating Equation 6.77 and setting it equal to zero, we get rotational speed ω at which the whirl amplitude attains a maximum value.

$$\omega = \frac{\omega_n}{\left\{1 - \frac{1}{2}(f/\omega_n)^1\right\}^{1/2}} \tag{6.79}$$

Hence, if we neglect the damping in the system ($f = 0$), we get,

$$\omega_{\text{critical}} = \omega_n, \quad A \to \infty \tag{6.80}$$

With damping present in the system, the critical speed is given by

$$\omega_{\text{critical}} = \frac{\omega_n}{\left\{1 - \frac{1}{2}(f/\omega_n)^2\right\}^{1/2}} \tag{6.81}$$

Thus, damping in the system increases the critical speed. Also, the response of the system does not assume infinite value. A plot of Equations 6.77 and 6.78 is shown in Fig. 6.14(a) and (b).

Since the forcing function is proportional to ω^2, we normally expect that amplitude rises with increase in speed. However, with damping in the system, the amplitude of vibration actually decreases. Study of Fig. 6.14(a) suggests that the amplitude of vibration at low speeds is dictated by k (spring stiffness), and beyond critical speeds, it is dictated by mass. Around resonance, shown by the band of ω, the damping controls the maximum amplitude. Also, the study of Fig. 6.14(b) shows that at resonance, the phase angle is 90°, while at low ω as well as at high ω the phase is 0° and 180°. This fact must always be remembered, since the speed at which phase angle changes by 90° are the true critical speed. In vibration testing, where one is interested in locating the critical speed, this information is extremely useful. We shall show applications of this concept in the forthcoming chapter on vibration diagnosis and control.

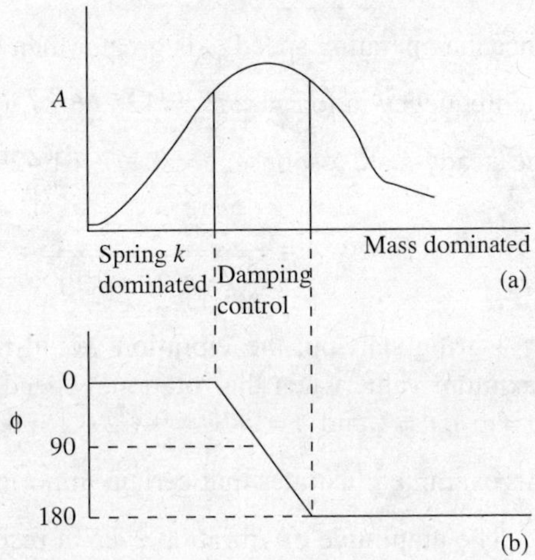

Figure 6.14 Plots of Equations 6.77 and 6.78

In Fig. 6.14(b), we can see that as the speed increases beyond ω_n, the response is dominated by the mass term $m^2\omega^4$ in Equation 6.69. Since this term is 180° out of phase with the unbalance force, the shaft moves in a direction opposite to that of the unbalance force (called backward whirl) and hence the response of shaft will be limited.

The solution of Equation 6.67, deleting the transient term was assumed as $Ae^{i(\omega t - \Phi)}$, which implicitly assumes a condition of forward synchronous whirl (under steady state), that is, $\dot{\theta} = \omega$. However as a general rule, if the steady solution of equation is assumed as $\omega(t) = Ae^{i(\gamma t - \phi)}$, the solution can be obtained as $\gamma = \pm\omega$, with $\gamma = +\omega$ representing the forward synchronous whirl and $\gamma = -\omega$ denoting backward synchronous whirl.

For determining the bearing reactions, we first find the deflection of mass centre of the disc/rotor from the bearing axis R in Fig. 6.13 as

$$R^2 = A^2 + e^2 + 2Ae\cos\Phi \quad (A \text{ as designated in Fig. 6.13}).$$

In view of Equations 6.77 and 6.78

$$R = e\left[\frac{1 + (2\xi r)^2}{(1 - r^2) + (2\xi r)^2}\right]^{1/2} \tag{6.82}$$

The bearing reaction can then be determined from the centrifugal force $mR\omega^2$.

Problem 6.6

A shaft carrying a rotor weighing 100 lbs having eccentricity of 0.1 in. rotates at 1200 rpm. Determine the steady-state whirl amplitude and maximum whirl amplitude during start up. Assume $k = 2 \times 10^5$ lb/in. and the external damping ratio 0.1. Also, find whether the shaft is rigid or flexible.

Solution:

Since the shaft is rotating at 1200 rpm, the forcing frequency is $\omega = \dfrac{2\pi N}{60} = 2\pi.1200/60 =$ 125.66 rad/sec, $m = 100/32$. The natural frequency $\omega_n = \sqrt{k/m} = \sqrt{2 \times 10^5/100/32.12} = 87.9$ rad/s.

Since the operating speed ω is greater than ω_n, the shaft is flexible.

The frequency ratio $r = \omega/\omega_n = 125.66/87.9 = 1.429$.

The steady-state response: $e = 0.1$, $r = 1.429$ and $\zeta = 0.1$.

$$A = \frac{er^2}{\sqrt{(1-r^2)+(2\xi r)^2}} = \frac{0.1.1.429^2}{\sqrt{(1-1.429^2+(2.0.1.1.429)^2}} = 0.19\,\text{in.}$$

During start-up, the vibration would rise as the speed increases. The vibrations will reach its maximum value when the rotational speed matches with the natural frequency. Thus, at resonance ($\omega = \omega_n$), $r = 1$ and $A = e/2\zeta = 0.1/2.0.1 = 0.5$ in.

This example illustrates that certain important facts that include the following.

- The amplitude of vibration even at resonance condition reduces if e is reduced. This means that the vibrations can be controlled by balancing so that the mass centre (CG) is brought as near as possible with the centre of rotation. Therefore, the balancing is one of the most-efficient technique of vibration control.
- The vibration also reduces considerably even at resonance if we increase the damping in the system.

It may be remembered that critical speeds are nothing but the natural frequencies of transverse vibrations of the shaft considered as suitably-supported/fixed/free end conditions. The fundamental frequency or the first critical speed can be very easily determined using approximate methods such as Dunkerley method, Rayleigh's method. The use of Rayleigh–Ritz method yields, apart from fundamental frequency, the higher modes. Once frequencies are known, mode shapes can also be determined using principles of orthogonality of modes.

As mentioned here, one of the methods of reducing the whirling response of the rotor and hence controlling vibration levels within permissible limits is reducing the eccentricity e of the rotor. This in other words, means balancing of rotors. We shall now deal with this aspect.

6.5.3 Balancing

Balancing is a method of reduction of vibrations at the source. The imbalance of components takes place both in rotating and reciprocating machines such as turbines, pumps, compressor etc. We shall first deal with balancing of rotating machines.

Balancing of Rotating Machines

The presence of an eccentric or unbalance mass in a rotating disc/rotor causes vibrations, which may be acceptable up to a certain level. It may also be appreciated that machining errors (which beyond a certain point cannot be avoided), presence of voids in the disc/rotor material, attachments such as blades in bladed discs/rotors, etc. contribute towards shifting of CG of the rotor from the centre of rotation (thus creating eccentricity of the mass) and resulting into unbalance. The resulting unbalance force when the shaft rotates is $me\omega^2$ [Fig. 6.15(a) and (b)].

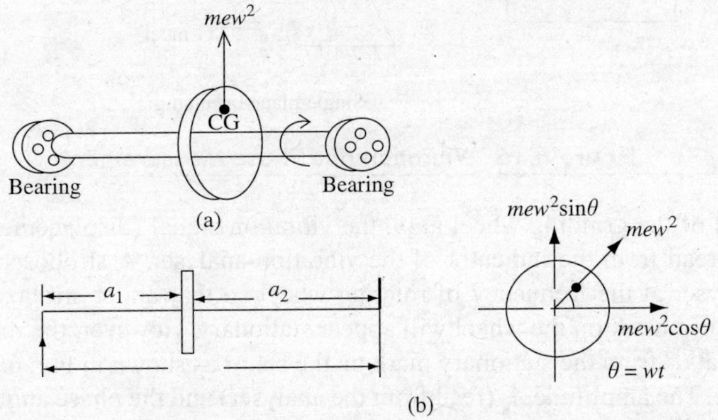

Figure 6.15 Unbalance of Rotor

This force has two components namely $me\omega^2 \cos\theta$ (= $me\,\omega^2 \cos\omega t$) and $me\omega^2 \sin\theta$ (= $me\omega^2 \sin\omega t$). When the shaft is stationary and then is displaced, it will always assume a position such that the CG is vertically downwards. Mark the lowest point of the disc/rotor with chalk. Again displace the disc/rotor arbitrarily. We can notice that the disc/rotor in stationary condition will always have chalk mark vertically downwards. The unbalance detected by this procedure is called static unbalance. The static unbalance can be corrected by removing (drilling) material at chalk mark or by adding a weight 180° from the chalk mark. Since the magnitude of the unbalance is not known, the amount of material to be removed and its position or weight to be added and its position must be determined by trial/error. The procedure described pertains to a single plane and hence is called a single-plane balancing.

The amount of unbalance can be found by rotating the shaft–disc/rotor assembly at known speed ω and measuring the reactions R_1 and R_2 since

$$R_1 = \frac{a_2}{l} me\omega^2,\quad R_2 = \frac{a_1}{l} me\omega^2$$

Since the unbalance force $me\omega^2$ rotates at frequency ω, the forces transmitted to the bearings $me\omega^2 \cos\theta$ (= $me\omega^2 \cos\omega t$) and $me\omega^2 \sin\theta$ (= $me\omega^2 \sin\omega t$ $\theta = \omega t$) are pulsating forces, which cause vibrations at the bearings at a frequency ω, that is, rotating speed. The measurement of vibrations therefore yields the information necessary for balancing. This is illustrated by the following example.

Figure 6.16 shows a set up for balancing of the grinder. Before starting balancing procedure, reference marks, also known as phase marks, are made on grinding wheel and stationary part as shown in Fig. 6.17(a). A vibration pick-up is firmly mounted on the bearing and a vibration analyser is set to a frequency corresponding to the angular velocity of the grinding wheel.

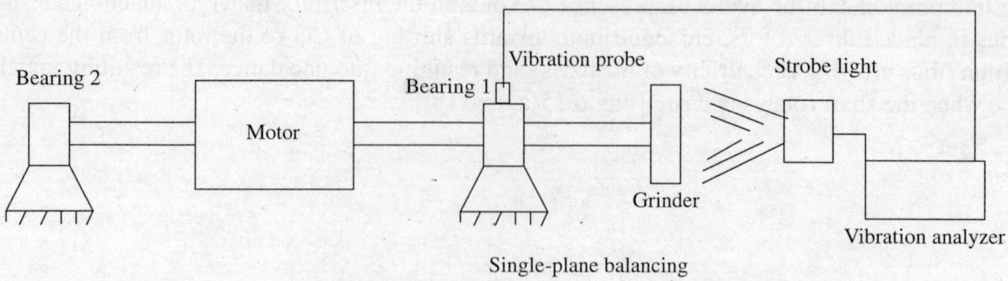

Figure 6.16 Vibration and Phase Measurement

Upon rotation of the grinding wheel at ω, the vibration signal (displacement) produced by the unbalance can be read from the indicator of the vibration analyser. A stroboscopic light is fired by the vibration analyser at the frequency of rotating wheel so that under stroboscopic light the disc and also the reference mark on the wheel will appear stationary. However, the reference mark on the wheel will appear at $\theta°$ from the stationary mark on the stator as shown in Fig. 6.17(b), due to phase-lag of the response. The amplitude A_u (read from the analyser) and the phase angle θ caused by initial unbalance is noted. The rotor is then stopped and a known trial mass W is attached to the rotor as shown in Fig 6.17b. The rotor is then once again rotated to its speed ω. The new angular position Φ and vibration amplitude A_{u+W} caused by combined unbalance of rotor and trial mass W are noted Fig. 6.17(c).

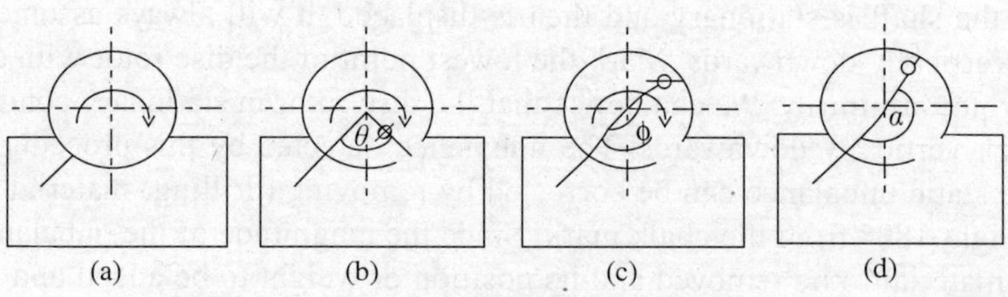

Figure 6.17 Determination of Phase Angle

We now construct a vector diagram to find the magnitude and location of correction weight (mass) required for balancing the wheel. The original unbalance vector A_u, Fig. 6.18 is drawn in an arbitrary direction with its length equal to A_u. Then the combined unbalance vector is drawn as A_{u+W} at an angle $\Phi - \theta$ from the direction of A_u with a length of A_{u+W}. The difference vector A_W ($A_W = A_{u+W} - A_u$) in Fig. 6.18 represents the unbalance vector due to the trial weight W. The magnitude of A_W can be computed using the law of cosines

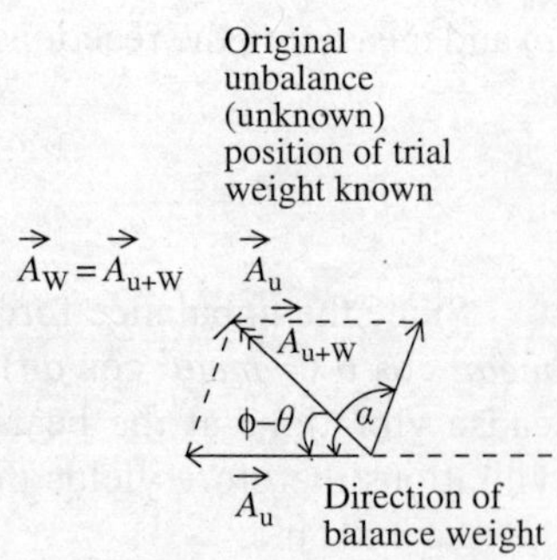

Figure 6.18 Vector Diagram

$$A_W = \{A_u^2 + A_{u+W}^2 - 2A_u A_{u+W}\cos(\phi - \theta)\}^{1/2} \tag{6.83}$$

Since the magnitude of trial weight W and its direction relative to the original unbalance (a in Fig. 6.17) are known, the original unbalance itself must be at an angle a away from the position of trial weight and is shown in Fig. 6.17(d). The angle a can be obtained from the law of cosines

$$\alpha = \cos^{-1}\left[\frac{A_u^2 + A_W^2 - A_{u+W}^2}{2A_u A_W}\right] \tag{6.84}$$

The magnitude of original unbalance is $W_0 = \left(\frac{A_u}{A_W}\right)\cdot W$ located at the same radial distance from axis of rotation of the rotor as the weight W. Once the location and magnitude of the original unbalance are known, correction weight can be added to balance the wheel properly.

Although the single-plane-balancing procedure can be used for balancing in one plane, that is, for rotors of the rigid disc type, there are situations where this brings down vibration levels in longer rotors also. In case balancing analyzer as well as vibration pick-up is not available, as was the case during earlier times, the balancing still can be done by trial-error method with shaft riders like a pencil just touching the shaft, enabling qualitative feeling of vibration levels. One can start with a trial weight and shift the balancing mass by 180°. Next trial run will be with a weight shifted by 90° and again 180° from its new position. This enables identification of quadrant where trial mass will give least vibration. Though not scientific in nature, in most cases, 5 to 6 trials will indicate the balancing position. In case the rotor is very long, the unbalance can be one of the three types namely: (*a*) static unbalance, (*b*) quasi-static unbalance, and (*c*) dynamic unbalance.

In case of static unbalance, the axis of unbalance is parallel to the axis of rotation as shown in Fig. 6.19(a). Figure 6.19(b) shows that in case of quasi-static unbalance, the axis of unbalance is skew with the axis of rotation. Figure 6.19 shows that in couple unbalance or dynamic unbalance, the unbalance produces a turning moment making shaft wobble about an axis perpendicular to the axis of rotation. The procedure for identifying the nature of unbalance will be described in the forthcoming chapter on vibration diagnosis and control.

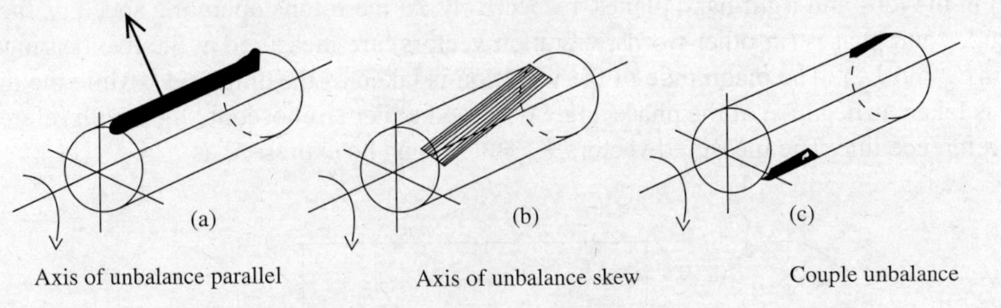

Figure 6.19 Types of Unbalance

Let us consider a very simple case of a long rotor having unbalanced mass m at a distance l/3 from the right end (Fig. 6.20).

When the rotor rotates at a speed of ω, the force due to unbalance will be $mR\omega^2$. The unbalance mass can be replaced by two masses m_1 and m_2 located at the ends of the rotor [Fig. 6.20(b)]. There are three forces now, namely $F_1 = m_1R\omega^2$, $F_2 = m_2R\omega^2$ and $F = mR\omega^2$. Obviously, we can arrange masses m_1 and m_2 such that

$$m_1R\omega^2 + m_2R\omega^2 = mR\omega^2 \tag{6.85}$$

Take moments about right end. This gives

$$m_1R\omega^2 l = mR\omega^2 l/3 \quad \text{or} \quad m = 3m_1 \quad \text{or} \quad m_1 = m/3,\ m_2 = 2m/3 \tag{6.86}$$

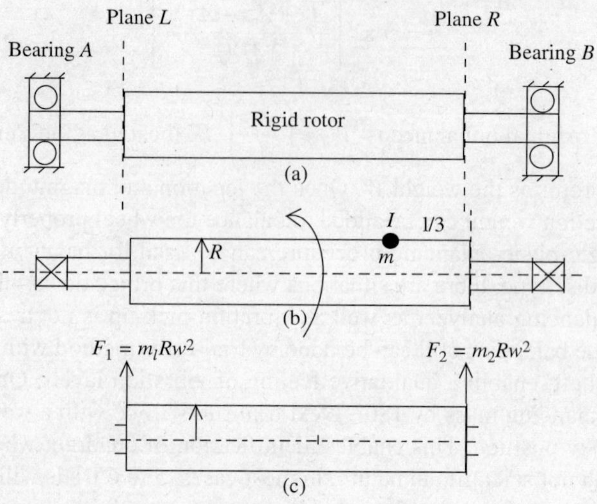

Figure 6.20 Unbalance—Two Equivalent Unbalanced Masses

Thus, if we precisely know the unbalance mass m, we can put two masses $m_1 = m/3$ and $m_2 = 2m/3$ opposite to the location of m and thus the rotor will get balanced. This is called two-plane balancing. What we have described is a process when the unbalance and its position are known *a priori*. This seldom happens. In general, we neither know the amount of unbalance nor its position. For balancing, we therefore must make measurements of vibration vectors. This is explained as follows.

In Fig. 6.21, the total unbalance in the rotor is replaced by two unbalanced weights (to be determined) in the left- and right-hand planes, respectively. At the rotor's operating speed ω, the vibration amplitudes and phases (in other words, vibration vectors) are measured at the two bearings A and B (vectors $\overline{V}_A$ and $\overline{V}_B$). The magnitude of the vibration is taken as the amplitude, while the direction of vector is taken as negative of the phase angle observed under stroboscopic light with reference to the stator-reference line. The measured vectors $\overline{V}_A$ and $\overline{V}_B$ can be expressed as

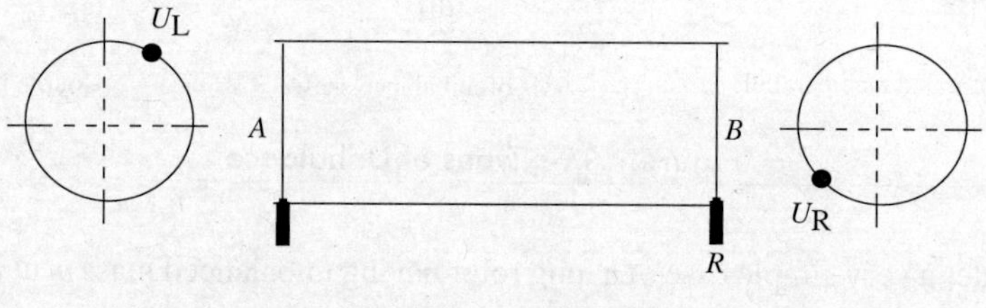

Figure 6.21 Two-Plane Balancing

$$\overline{V}_A = \overline{A}_{AL}\overline{U}_L + \overline{A}_{AR}\overline{U}_R \tag{6.87}$$

$$\overline{V}_B = \overline{A}_{BL}\overline{U}_L + \overline{A}_{BR}\overline{U}_R \tag{6.88}$$

where $\overline{A}_{ij}$ can be considered as a vector, reflecting the effect of the unbalance in the plane j (j = L, R) on the vibration at bearing i (i = A, B). It may be noted that $\overline{U}_L, \overline{U}_R$, and all the vectors $\overline{A}_{ij}$ are unknown in Equations 6.87 and 6.88. Then, the following steps are followed.

- As in the case of single-plane balancing, we add known trial weight (mass) and take measurements to obtain information about the unbalanced masses. Initially, we add a known weight $\bar{W}_L$ in the left plane at a known angular position and measure the displacement and phase of the vibration at two bearings when the rotor rotates at ω. We denote these measured vibration vectors as

$$\bar{V}_{A'} = \bar{A}_{AL}(\bar{U}_L + \bar{W}_L) + \bar{A}_{AR}\bar{U}_R \tag{6.89}$$

$$\bar{V}_{B'} = \bar{A}_{BL}(\bar{U}_L + \bar{W}_L) + \bar{A}_{BR}\bar{U}_R \tag{6.90}$$

By subtracting Equations 6.87 and 6.88 from Equations 6.89 and 6.90 and solving, we get,

$$\bar{A}_{AL} = \frac{\bar{V}_{A'} - \bar{V}_A}{W_L} \tag{6.91}$$

$$\bar{A}_{BL} = \frac{\bar{V}_{B'} - \bar{V}_B}{W_L} \tag{6.92}$$

- Remove $\bar{W}_L$ and add a known weight $\bar{W}_R$ in the right plane at a known angular position and measure the resulting vibrations when the rotor rotates at ω. The measured vibrations can be denoted as

$$\bar{V}_{A''} = \bar{A}_{AR}(\bar{U}_R + \bar{W}_R) + \bar{A}_{AL}\bar{U}_L \tag{6.93}$$

$$\bar{V}_{B''} = \bar{A}_{BR}(\bar{U}_R + \bar{W}_R) + \bar{A}_{BL}\bar{U}_L \tag{6.94}$$

As before, subtract Equations 6.87 and 6.88 from the above equations to obtain

$$\bar{A}_{AR} = \frac{\bar{V}_{A''} - \bar{V}_A}{W_R} \tag{6.95}$$

$$\bar{A}_{BR} = \frac{\bar{V}_{B''} - \bar{V}_B}{W_R} \tag{6.96}$$

Once the vector operators $\bar{A}_{ij}$ are known, Equations 6.87 and 6.88 can be solved to find the unbalance vectors $\bar{U}_L, \bar{U}_R$

$$\bar{U}_L = \frac{\bar{A}_{BR}\bar{V}_A - \bar{A}_{AR}\bar{V}_B}{A_{BR}A_{AL} - A_{AR}A_{BL}} \tag{6.97}$$

$$\bar{U}_R = \frac{\bar{A}_{BL}\bar{V}_A - \bar{A}_{AL}\bar{V}_B}{A_{BL}\bar{A}_{AR} - \bar{A}_{AL}\bar{A}_{BR}} \tag{6.98}$$

Now, the rotor can now be balanced by adding equal and opposite weights in each plane. The balancing weights in the left and right planes can be vectorially denoted as

$$\bar{B}_L = -\bar{U}, \bar{B}_R = -\bar{U}_R.$$

Thus, it can be seen that two-plane-balancing procedure is a mere straightforward extension of a single-plane-balancing procedure.

It can be seen that complex subtraction, division, multiplication are often used in the computation of balancing weights. If $A = a \angle \theta_A$ and $B = b \angle \theta_B$. Thus, we can write

$$\overline{A} = a_1 + ia_2, \quad \overline{B} = b_1 + ib_2$$
$$a_1 = a\cos\theta_A, \quad a_2 = a\sin\theta_a, \quad b_1 = b\cos\theta_B, \quad b_2 = b\sin\theta_B$$

Then

$$\overline{A} - \overline{B} = (a_1 - b_1) + i(a_2 - b_2) \tag{6.99}$$

$$\frac{\overline{A}}{\overline{B}} = \frac{(a_1b_1 + a_2b_2) + i(a_2b_1 - a_1b_2)}{b_1^2 + b_2^2} \tag{6.100}$$

$$\overline{A}.\overline{B} = (a_1b_1 - a_2b_2) + i(a_2b_1 + a_1b_2) \tag{6.101}$$

We shall illustrate this procedure by solving a problem.

Problem 6.7

While balancing (two plane) a turbine rotor, following data has been obtained through vibration measurements

Condition	Vibration Displacement		Phase Angle	
	At Brg A	At Brg B	Brg A	Brg B
Original	8.5	6.5	60°	205°
WL = 10.0 oz 270° from the reference mark	6.0	4.5	125°	230°
WR = 12 oz, 180° from ref.	6.0	10.5	35°	160°

Find the balancing weights and their positions.

Solution:

$$\overline{V}_A = 8.5\angle 60 = 4.25 + i7.3612$$

$$\overline{V}_B = 6.5\angle 205 = -5.8916 - i2.7470$$

$$\overline{V}_{A'} = 6.0\angle 125 = -3.4415 + i4.9149$$

$$\overline{V}_{B'} = 4.5\angle 230 = -2.8926 - i3.4472$$

$$\overline{V}_{A''} = 6.0\angle 35 = 4.9149 - i3.4472$$

$$\overline{V}_{B''} = 10.5\angle 160 = -9.8668 + i3.5912$$

$$\overline{W}_L = 10\angle 270 = 0.000 - i10.00$$

$$\overline{W}_R = 12\angle 180 = -12 + i0.00$$

We now calculate the various 'A' vectors.

$$\bar{A}_{AL} = \frac{\bar{V}_{A'} - \bar{V}_A}{\bar{W}_L} = \frac{-7.6915 - i2.4463}{0.000 - i10.00} = 0.2446 - i0.7691$$

$$\bar{A}_{BL} = \frac{\bar{V}_{B'} - \bar{V}_B}{\bar{W}_L} = \frac{2.9985 - i0.7002}{0.000 - i10.00} = 0.0700 + i0.2998$$

$$\bar{A}_{AR} = \frac{\bar{V}_{A''} - \bar{V}_A}{\bar{W}_R} = \frac{0.6649 - i3.9198}{-12 + i0.00} = -0.554 + i0.3266$$

$$\bar{A}_{BR} = \frac{\bar{V}_{B''} - \bar{V}_B}{\bar{W}_R} = \frac{-3.9758 + i6.3382}{-12.00 + i0.00} = 0.3313 - i0.5282$$

We now use Equations 6.97 and 6.98 to obtain the 'U' vectors

$$\bar{U}_L = \frac{(5.2962 + i0.1941) - (1.2237 - i1.7721)}{(-0.3252 - i0.3840) - (-0.1018 + i0.0063)} = -8.2930 + i5.6879$$

Similary

$$\bar{U}_R = -2.1773 - i5.4592 \quad \text{(Omitting details)}$$

The required balance weights are

$$\bar{B}_L = -\bar{U}_L = (8.2930 - i5.6879) = 10.05\angle 146°$$

$$\bar{B}_R = -\bar{U}_R = (2.1773\_i5.4592) = 5.8774\angle 248°$$

This shows that addition of 10.0561 oz at 146° in the left plane and 5.8774 oz at 246° in the right plane from reference position will balance the rotor. The balance weights are added at the same radial distance as the trial weights. If a balancing weight is to be located at a different radial position, the required balance weight should be modified in inverse proportion to the radial distances from the axis.

Balancing of Reciprocating Machines

In reciprocating machines, the linear oscillating motion of the piston is converted into continuous rotary motion of the crankshaft. This is achieved through a mechanism consisting of piston in the cylinder; connecting rod and crank. Vibrations in reciprocating machines arise due to (1) periodic variation of gas pressure in the cylinder and (2) inertia forces associated with the moving parts.

Figure 6.22 shows schematic diagram of a cylinder of reciprocating machine, say engine. The work done in the engine is due to expansion of gases. The expanding gases exert on the piston a pressure force F that is transmitted to the crankshaft through the connecting rod. The reaction to the force F can be resolved into two components: one of magnitude $F/\cos \Phi$ acting on the connecting rod and the other of magnitude $F \tan \Phi$ acting in a horizontal direction. The force $F/\cos \Phi$ induces a torque M_T, which tends to rotate the crankshaft [Fig. 6.22(b)]. M_T acts about an axis perpendicular to the plane of the paper and passes through the point Q.

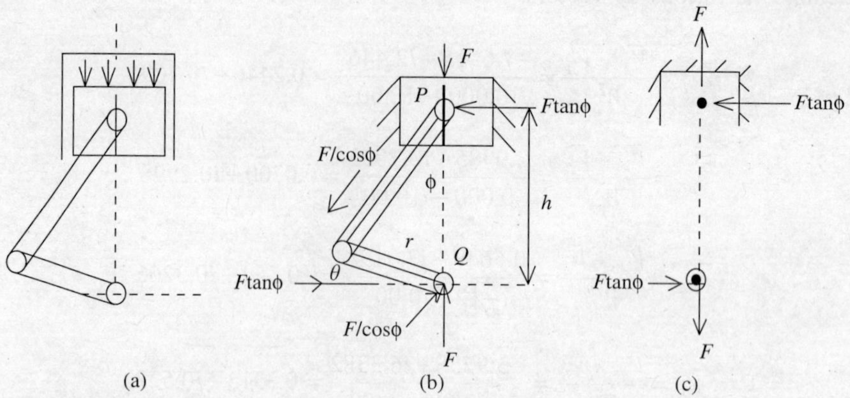

Figure 6.22 Forces in Reciprocating Engine

$$M_T = \left(\frac{F}{\cos\phi}\right) r\cos\theta \tag{6.102}$$

For force equilibrium of the overall system, the forces at the bearings of the crankshaft will be F in vertical direction and $F \tan \Phi$ in the horizontal direction.

The forces transmitted to the stationary parts of the engine are [Fig. 6.22(c)]

- Force F acting upward at the cylinder head
- Force $F \tan \Phi$ acting towards the right at the cylinder head
- Force F acting downward at the crankshaft bearing Q
- Force $F \tan \Phi$ acting toward the left of the crankshaft bearing

whereas the total resulting force is zero, there is a resultant torque $M_Q = Fh \tan \Phi$ on the body of the engine, where h can be found from the geometry of the system

$$h = \frac{r\cos\theta}{\sin\phi} \tag{6.103}$$

Thus,

$$M_Q = \frac{Fr\cos\theta}{\cos\phi} \tag{6.104}$$

We can see that M_T and M_Q, given by Equations 6.102 and 6.104, are identical. This indicates that the torque induced on the crankshaft due to gas pressure on the piston is felt at the supports of the engine. It may be noted that magnitude of the gas pressure varies as piston moves (with time) and thus it reaches maximum and minimum at a frequency governed by many factors such as number of cylinders, type of operating cycle, and rotating speed of the engine. Same thing is true for reciprocating compressors and pumps. Let us now analyse acceleration of piston and crankpin which contribute towards the inertia forces.

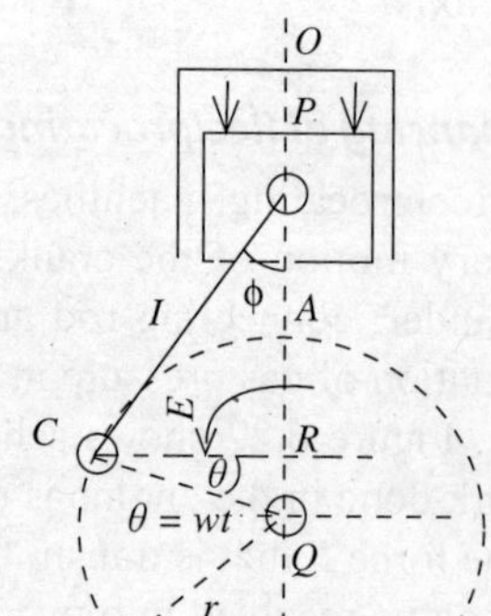

Figure 6.23 Motion of Piston, Connecting Rod and Crank

Figure 6.23 shows the crank (length r), connecting rod (length l), and piston of a reciprocating engine (which may be multi-cylinder engine, inline or V or radial). Consider O as the origin for the analysis of acceleration. It is assumed that the angular speed of the crank is constant. Point O is the uppermost position of the piston. The displacement of piston P corresponding to angular position of crank θ ($= \omega t$ at any time t) is given by

$$x_p = r + l - r\cos\theta - l\sqrt{1-\sin^2\phi} \tag{6.105}$$

But

$$l\sin\phi = r\cos\theta = r\sin\omega.t \tag{6.106}$$

Therefore,
$$\cos\phi = \sqrt{1-\frac{r^2}{l^2}\sin^2\omega t} \tag{6.107}$$

Using this in Equation 6.105, we obtain,

$$x_P = r + l - r\cos\omega t - l\left(\sqrt{1-\frac{r^2}{l^2}\sin^2\omega t}\right) \tag{6.108}$$

In general, $r/l < ¼$, and hence the term under square-root using expansion theorem can be written as

$$\sqrt{\left(1-\frac{r^2}{l^2}\sin^2\omega t\right)} = 1-\frac{r^2}{2l^2}\sin^2\omega t$$

Using this, Equation 6.108 becomes

$$x_P = r + l - r\cos\omega t - l + \frac{r^2}{2l}\sin\omega t = r(1-\cos\omega t) + \frac{r^2}{2l}\sin\omega t \tag{6.109}$$

After a few simplifications, we obtain,

$$x_P = r\left(1+\frac{r}{2l}\right) - r(\cos\omega t + \frac{r}{4l}\cos 2\omega t) \tag{6.110}$$

The velocity
$$\frac{dx}{dt} = \dot{x}_P = r\omega(\sin\omega t + \frac{r}{2l}\sin 2\omega t) \tag{6.111}$$

The acceleration
$$\frac{d^2x}{dt^2} = \ddot{x}_P = r\omega^2(\cos\omega t + \frac{r}{l}\cos 2\omega t) \tag{6.112}$$

Equation 6.112 represents the acceleration of the piston. We shall now find the same quantities for crankpin. With respect to the X−Y coordinates shown in Fig. 6.23, the vertical and horizontal displacements of the crankpin C are given by

$$x_C = OA + AB = l + r(1-\cos\omega t) \tag{6.113}$$

$$y_C = CB = r\sin\omega t \tag{6.114}$$

From the above equations we compute the velocity and accelerations of the crankpin as follows

$$\dot{x}_C = r\omega\sin\omega t \tag{6.115}$$

$$\ddot{x}_C = -r\omega^2\cos\omega t \tag{6.116}$$

$$\dot{y}_C = r\omega\cos\omega t \tag{6.117}$$

$$\ddot{y}_C = -r\omega^2 \cos\omega t \tag{6.118}$$

The connecting rod also has mass; however, for the ease of analysis it is generally idealized as a massless link with two masses concentrated at its ends—the piston end and the crank pin end (usually about one-third of the connecting rod is regarded as reciprocating, the remainder two-thirds, including big-end bearing considered as a revolving mass).

If m_P and m_C denote the total masses (including concentrated mass of connecting rod) of piston and the crankpin, the vertical component of inertia force F_X for one cylinder is

$$F_X = m_P \ddot{x}_P + m_C \ddot{x}_C \tag{6.119}$$

Substituting Equations 6.112 and 6.115

$$F_X = (m_P + m_C) r\omega^2 \cos\omega t + m_P \frac{r^2\omega^2}{l} \cos 2\omega t \tag{6.120}$$

In the above equation, the coefficient of cos ωt (i.e. $[m_P$ and $m_C]$ $r\omega^2$) is called primary (part) force at frequency ω and the coefficient $(m_P r^2\omega^2/l)$ is called secondary (part) force at a frequency 2ω (two times the rotational frequency).

Similarly, the horizontal components of inertia for *a cylinder* can be obtained as

$$F_Y = m_P \ddot{y}_P + m_C \ddot{y}_C \tag{6.121}$$

But $\ddot{y}_P = 0$ (assuming no lateral slap of the piston) and $\ddot{y}_C = -\omega^2 \sin\omega y$, then

$$F_Y = -m_C r\omega^2 \sin\omega t \tag{6.122}$$

Thus, the horizontal component of inertia force has only primary force at a frequency ω and no secondary force.

Balancing The unbalanced or inertia forces on a single-cylinder-reciprocating engine are given by Equations 6.120 and 6.122. In these equations, m_P and m_C represent equivalent reciprocating and rotating masses, respectively. The mass m_P is always positive but m_C can be made zero by counterbalancing the crank. It is therefore possible to reduce the horizontal inertia force F_Y to zero, but the vertical component F_X (see Fig. 6.23 for the coordinate system) always remains as it is not possible to balance secondary unbalance force occurring at two times the running speed *in a single-cylinder engine/compressor/pump.* Thus, the single-cylinder-reciprocating machine is inherently unbalanced.

In a multi-cylinder engine, on the other hand, it is possible to balance some or all inertia forces and torques by proper arrangement of cranks. This is done as follows:

Figure 6.24 shows the general arrangement of *N*-cylinders engine (only six cranks are shown for the purpose of illustration in Fig. 6.24(a)).

The lengths of all cranks and connecting rods are equal, which means that for all cylinders *r* and *l* are same. It is assumed that angular speed ω of the crankshaft is constant. The axial displacement and angular orientation of the *i*th cylinder are assumed to be l_i and a_i (i = 1, 2, 3, …, *N*). For force balance, the total inertia force in the X-direction and Y-direction must be zero, that is,

$$(F_X)_{total} = \sum_{1}^{N} (F_X)_i = 0 \tag{6.123}$$

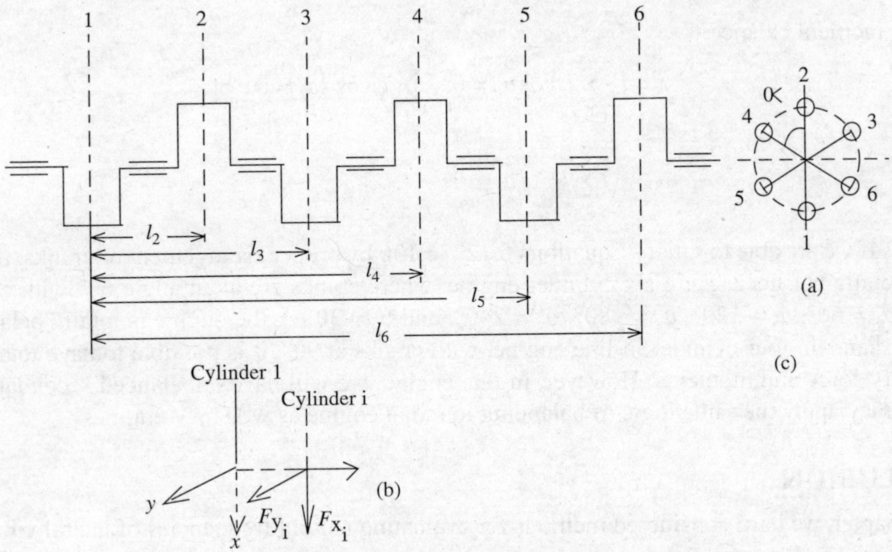

Figure 6.24 Crank Arrangement for Six-cylinder Engine

$$(F_Y)_{total} = \sum_1^N (F_Y)_i = 0 \tag{6.124}$$

$$(F_X)_i = (m_P + m_C)_i r\omega^2 \cos(\omega.t + a_i) + (m_P)_i \frac{r^2\omega^2}{l} \cos(2\omega.t + 2a_i) \tag{6.125}$$

$$(F_Y)_i = (-m_C) r\omega^2 \sin(\omega.t + a_i) \tag{6.126}$$

Usually all pistons, connecting rods and cranks have the same mass, that is,

$$(m_P)_i = m_P, (m_C)_i = m_C, 1 = 1, 2, 3..N$$

Thus Equations 6.123 and 6.124 give (for total force balance)

$$\sum \cos a_i = 0, \quad \sum \cos 2a_i = 0 \tag{6.127}$$

$$\sum \sin a_i = 0 \tag{6.128}$$

The inertia forces $(F_X)_i$ and $(F_Y)_i$ of the ith cylinder [Fig. 6.24(b)] induce moments about Y- and X-axis. These are

$$M_Z = \sum_1^N (F_X)_i l_i = 0 \tag{6.129}$$

$$M_X = \sum_1^N (F_Y)_i l_i = 0 \tag{6.130}$$

Thus for moment balance

$$\sum_{1}^{N} l_i \cos a_i = 0, \quad \sum_{1}^{N} l_i \cos 2a_i = 0 \tag{6.131}$$

$$\sum_{1}^{N} l_i \sin a_i = 0 \tag{6.132}$$

Thus if we are able to satisfy Equations 6.127–6.130 by proper arrangement of cranks, the engine will be totally balanced. For a six cylinder engine, where cranks are at an angle 60° with each other ($a_1 = 0$, $a_2 = 60°$, $a_3 = 120°$, $a_4 = 180°$, $a_5 = 240°$, and $a_6 = 300°$), the engine is totally balanced. On the other hand, in four cylinder in-line engine, with cranks at 90°, it is possible to have total balance of primary force and moments. However, in this engine, we will have unbalanced secondary forces. Readers may apply the same logic of balancing to radial engine as well as V-engines.

CONCLUSION

In this chapter, we have considered methods for evaluating natural frequencies of lateral vibrations in beams. We have discussed the rigorous mathematical method as well as reasonably accurate methods such as Rayleigh's method, Rayleigh–Ritz method, Dunkerley method, etc. For all practical purposes, it is sufficient to determine the static deflection curve. We discussed the important aspects of whirling of rotating shafts and the critical speeds. We have shown that these critical speeds are merely the natural frequencies of lateral/transverse vibrations of the rotor–shaft assembly considered as beams with appropriate end conditions.

Due to the dynamic forces, which cause the bending and hence the transverse vibrations of the rotor, one of the best methods to control these vibrations is to balance the rotor. Both single-plane and two-plane balancing method have been discussed. We also discussed important aspects of balancing of reciprocating machines such as engines, compressors, pumps, etc. We have shown that it is impossible to totally balance a single-cylinder reciprocating machine since, although, the rotating masses can be completely balanced, the reciprocating masses remain unbalanced. We have also discussed how to achieve balancing of multi-cylinder reciprocating engines by adopting suitable crank arrangements.

References

1. S.K. Clark, *Dynamics of Continuous Elements*, Prentice Hall, Englewood Cliffs. N.J. 1972.
2. S. Timoshenko, D.H. Young and R. Weaver, Jr., *Vibration Problems in Engineering* (4th ed.), Wiley, New York, 1974.
3. Singiresu S. Rao, *Mechanical Vibrations* (4th ed.), Pearson Education, 2004.
4. R.W. Fitzgerald, *Mechanics of Materials* (2nd ed.) Addison-Wesley, Reading, Mass, 1982.
5. J.R. Hutchinson, *Transverse Vibrations of Beams: Exact Versus Approximate Solutions*, Journal of *Applied Mechanics*, vol. 48, 1981. pp. 923–928.

EXERCISES

6.1 Determine the natural frequencies of vibrations of a uniform beam fixed at one end and simply supported at the other end.

Answer: $\omega_n = (\beta_n l)^2 \left(\dfrac{EI}{\rho A l^4} \right)^{1/4}$

6.2 A fixed–fixed beam of length 2 m carries at its mid-span a 100-kg-weight motor running at 3000 rpm. The motor has a rotational unbalance amounting to 0.5 kg-m. Determine the steady-state response of the beam. The beam has a cross-section 10 cm × 10 cm and the material of the beam is steel ($E = 1.6 \times 10^{11}$ N/m^2). Neglect the weight of the beam.

Hint: The normal function of a fixed–fixed beam can be chosen from the table given in this chapter. Find the generalized force using Equation 6.42. Use Equation 6.44. Compare the result of this analysis with the result you obtain using Dunkerley's method.

6.3 Find the fundamental frequency of transverse vibration of a non-uniform cantilever beam whose cross-section varies as $A(x) = hx/l$ and moment of inertia varies as $I(x) = (1/12)(hx/l)^3$. You may choose any suitable static-deflection curve and use Rayleigh's quotient.

Answer: $1.5811(Eh^2/\rho l^4)^{1/2}$

6.4 Assuming suitable functions for deflection such as $w_1(x) = (1 - x/l)^2$, $w_2(x) = (x/l)(1 - x/l)^2$ and $w_3 = (x^2/l^2)(1 - x/l)^2$, find the natural frequencies using Rayleigh–Ritz method for the beam in Problem 6.3.

Hint: See the methodology as explained in the text.

6.5 A rotating shaft, which can be considered as a simply supported beam of length $4l$ carrying three rotors/masses each of weight w, equally located on the beam. Find out the natural frequency (1) using assumed static-deflection curve, (2) using Dunkerley's method.

Hint: We can assume that each mass produces deflection as per the strength of material formula.

$$w(x) = \frac{Pbx}{6EIL}(L^2 - b^2 - x^2),\ 0 \le x \le a$$

$$w(x) = \frac{Pa(L-x)}{6EIL}[a^2 + x^2 - 2Lx],\ a \le x \le L$$

Calculate the deflection under each load by the action of only one load. This means that we shall calculate the deflection at the location of the first load, at the location of second load, and at the location of third load when only one load is acting as shown in the figure below. In other words, find w_1 at the location of first rotor/mass, w_2 at the location of second rotor/mass due to load at the location of first mass, and w_3 at the location of the third mass but the load being at first location, that is, P_1. Now repeat the process by considering the load at location 2, that is, P_2 and finally at location 3. Now sum them up to obtain total deflections w_1, w_2, and w_3. The natural frequency is then given by

$$\omega = \left\{ \frac{g(m_1 w_1 + m_2 w_2 + m_3 w_3)}{(m_1 w_1^2 + m_2 w_2^2 + m_3 w_3^2)} \right\}$$

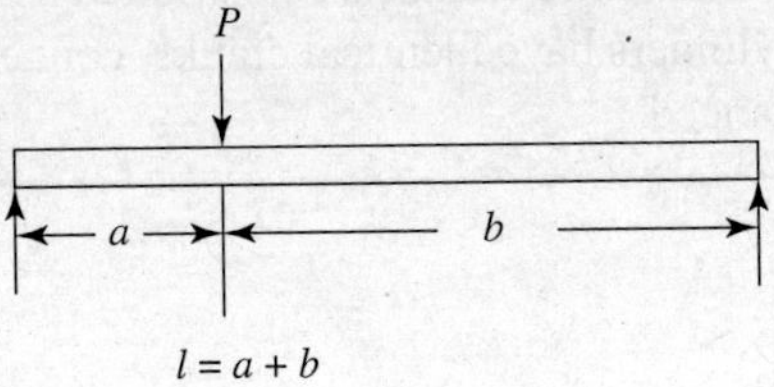

Problem 6.5

Now use Dunkerley's method to find the natural frequency and compare it with the value obtained by the above procedure.

6.6 A uniform fixed–fixed beam of length $2l$ is simply supported at the middle point. Derive the frequency equation for the transverse vibration of the beam.

Hint: See the derivation of Equations 6.10 and 6.11. In this particular case, the middle point being simply supported, the boundary conditions are required to be specified. At the middle, the deflection w is zero but the slope dw/dt is not zero and the support reaction balances the shear force.

Answer: $\cos \beta l \cosh \beta l = 1$, $\tan \beta l = \tanh \beta l$

6.7 A uniform beam, simply supported at ends, is found to vibrate in its first mode with an amplitude 10 mm at its centre. The area of cross-section of the beam is 120 mm^2, moment of inertia = 1000 mm^4, $E = 20.5 \times 10^{11}$, the density = 7.83×10^3 kg/m^3, and the span is 1 m. Determine the maximum bending moment.

6.8 What is whirling of rotating shafts? Derive the equations of motion of the lateral vibration of the rotor having an eccentricity 'a' and rotating at N rpm. Using this, explain the concept of critical speed.

6.9 A flywheel weighing 100 lb and having eccentricity 0f 0.5 in. is mounted at the centre of a steel shaft of 1 in diameter. If the length of the shaft between the bearings is 30 in. and the system rotates at 1200 rpm, find (a) critical speed, (b) the amplitude of vibration of the rotor, and (c) the bearing reactions.

Answer: 20.23 N-m

6.10 A steel shaft of diameter 2.5 cm and length 1 m is supported at ends in bearings. It carries a turbine disc of mass 20 kg and eccentricity 0.005 m at the middle and operates at 6000 rpm. The damping in the system is given by $\zeta = 0.01$. Determine the whirl amplitude at (a) rotating speed, (b) critical speed, and (c) at 1.5 times the critical speed.

Answer: 0.005124 m, 0.0608, and 0.00846.

6.11 The spacing of cylinders in a four-cylinder inline engine is 12 in. The cranks have equal lengths (4 in.) and their angular positions are 0^0, 180^0, 180^0 and 0^0. The length of the connecting rods is 10 in. and the reciprocating weight is 2 lb for each cylinder. Find the unbalanced forces and moments when the engine speed is 3000 rpm. Use the centre line through cylinder 1 as the reference plane.

Hint: See the analysis presented in the text at Paragraph 6.5.3.2. Equations 6.117, 6.118, 6.121 and 6.122 can be used considering $\alpha = 90^0$.

Answer: (1) Primary forces balanced, (2) Secondary unbalance force = 3270 lb, (3) Both couples balanced.

6.12 Consider a crank arrangement for a six-cylinder in-line engine. The cylinders are separated by a distance a in axial direction and the cranks are positioned as $\alpha_1 = \alpha_6 = 0^0$, $\alpha_2 = \alpha_5 = 120^0$, $\alpha_3 = \alpha_4 = 240^0$. Assuming that all cylinders have identical cranks, connecting rod and pistons, show that the entire engine is totally balanced.

7 Vibration Diagnosis and Control

7.1 INTRODUCTION

In the previous chapters, we dealt with the analysis of vibration of single/multi degrees-of-freedom systems including the continuous system. We also discussed the principles of analysis of whirling of shafts and also the torsional vibrations of complex systems. In all the cases, we either analyzed the response of free vibrations and/or forced vibrations when the bodies/structures/components are subjected to a known disturbing force(s). These types of problems are called direct problems. While knowing the analysis of such problems is very important, the indirect problems, wherein we are required to pinpoint the source of vibrations and suggest suitable remedial measures, are much more complex to solve. The practising engineers, more often than not, come across indirect problems of vibrations/noise which require certain diagnostic capabilities. The aim of this chapter is to bring out certain principles which are very useful in solving practical vibrations/noise problems.

The vibrations experienced by machinery are caused by perturbation forces which could be mechanical, electrical, or hydraulic/aerodynamic in nature. Many a times, these perturbation forces exist together, thereby making their identification quite a difficult task. In order to identify the nature of the perturbation forces, it is necessary to use high-quality equipment that is capable of recording the relevant variables or quantities with least possible errors. The quantities are (1) bearing and shaft vibrations along with phase information, (2) vibrations of the individual elements of the unit and deformation and displacements of components, (3) pressure pulsations/variations at different sections of flow passage, in case vibrations/noise problem is associated with flow, and (4) noise.

The data obtained through the (appropriate) sources provide the information about vibrations, pressure pulsations, noise, etc. in time domain. This time-domain information, in general, does not provide any sufficient data regarding the likely reasons for the vibration problem unless the signals are properly processed and converted to frequency domain. The conversion of time-domain signal to frequency domain facilitates identification of distress in the machine as each of the frequencies in the frequency domain is identifiable to a characteristic malfunction in the system. This is the basis of diagnosis of vibration problems which in principle is similar to a cardiologist, who after measuring the pulse rate, carries out an electrocardiogram (ECG) to judge the nature of the distress in the heart.

One of the methods of solving the vibration problem is to eliminate or reduce the disturbing force. For example, if the vibration problem is diagnosed as being due to imbalance, we can balance the rotor; or, if the vibrations are due to misalignment of the connected rotors, we can eliminate or reduce the same significantly. But there are cases, where it is not possible to entirely eliminate or reduce the perturbation forces. In such cases, the only alternative is to alter the response of the system by carrying out structural modifications such as stiffening. The evaluation of such a modification can be done only when we know the natural frequency and mode shapes of the components. These quantities, in practice, are determined through an experimental procedure known as modal analysis. We shall also briefly deal with principles of experimental modal analysis in this chapter. In order to understand the principles of techniques that are employed, it is worthwhile to review, in brief, some of the fundamental concepts of the analysis of periodic signals such as vibrational characteristics of the periodic vibration (Fig. 7.1) and spectrum analysis.

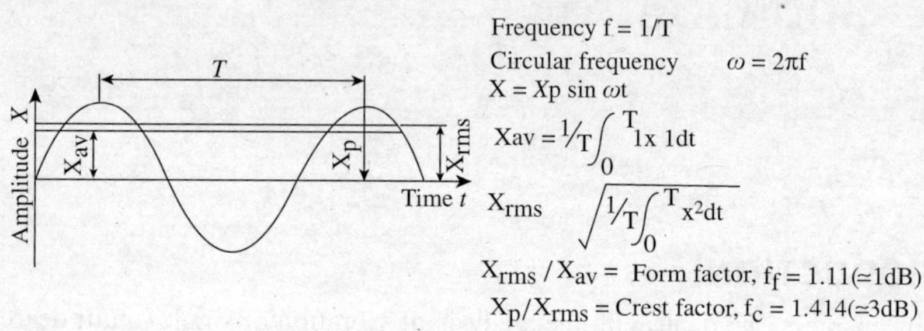

Figure 7.1 Sinusoidal Vibration Signal

The periodic vibration is an oscillating nature of a particle or a body about a reference position such that the motion repeats exactly after a certain time. The simplest of this form of vibration is harmonic motion as shown in the Fig. 7.1 and it can be represented as

$$X = X_P \,\text{Sin}\omega t,\ \dot{X}_i = \omega X_p \,\text{Cos}\omega t \quad \text{or} \quad \ddot{X} = -\omega^2 X_P \,\text{Sin}\omega t. \qquad (7.1)$$

Frequency f = 1 / T, Circular frequency $\omega = 2\pi f$

$$X = X_P \,\text{Sin}\omega t, X_{av} \int_0^T |X|\, dt \quad X_{rms} = \sqrt{1/T} \int_0^T x^2 dt$$

$$X_{rms} / X_{av} = \text{form factor } f_f = 1.11\ (\approx \text{dB})$$

$$X_p / X_{rms} = \text{crest factor } f_c = 1.414\ (\approx 3\text{dB})$$

Where X is the displacement, $\dot{X}$ is the velocity, and $\ddot{X}$ is the acceleration of vibration, respectively. It is apparent that the form and the period of vibration remain the same regardless of whether the vibration is specified in displacement, velocity, or acceleration mode. Of course, the phases of these quantities differ, that is, velocity leads displacement by 90°. It is, therefore, enough to characterize the vibration by peak values of displacement (X_p), velocity V_p ($= \omega X_p$) or acceleration a_p ($= \omega^2 X_p$), and frequency, ω. The various other quantities such as average absolute value, root-mean-square (rms) value, and so on, are shown in Fig. 7.1. The figure also shows the parameters such as form factor and crest factor, which are used for the indication of wave shape of vibrations.

Most of the vibrations experienced in daily life are not pure harmonic motion as shown in Fig. 7.1, even though many of them may be periodic. Typical non-harmonic periodic motions are shown in Fig. 7.2.

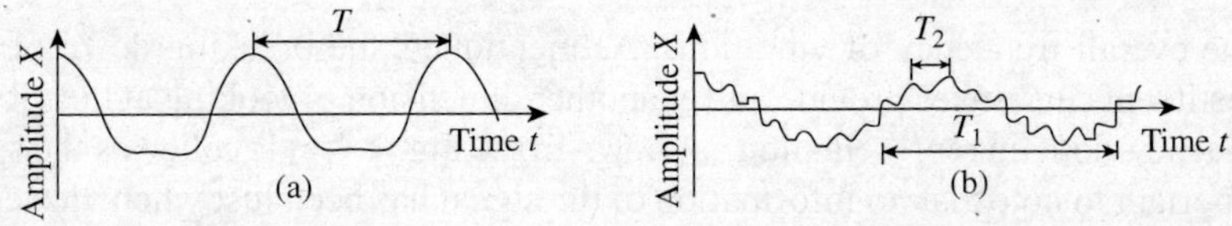

Figure 7.2 Non-Harmonic Periodic Motion

For the periodic motions shown in Fig. 7.2, we can obtain information regarding peak amplitude, average absolute, rms, form factor, crest factor, but they hardly throw any light upon possible causes for such motions and forces causing the motion. They are, therefore, a very little relevance to the vibration analyst. Of course, under certain conditions/circumstances, waveform analysis does provide additional/supportive information for diagnosing a problem, but in isolation, they fail to provide the diagnosis. This aspect shall be discussed in the forthcoming sections.

It was shown over one hundred years ago by Baron Jean Baptiste Joseph Fourier that any waveform that one encounters in real life can be generated by adding up sine waves. For example, the waveform shown in the Fig. 7.2(a) can be shown to be primarily consisting of harmonically related sine waves as shown in Fig. 7.3. Conversely, we can break down the real world signal as same sine waves. It may be very clearly remembered/understood that all the three depictions shown in the Figs 7.4(a), (b), and (c) relate to the same vibration phenomenon and no information of the vibration signals is lost when it is shown in depictions 7.4(a), (b), and (c). In fact, the breaking down of the vibration signals into its constituent sine waves provide an yet another dimension of looking at the vibration process.

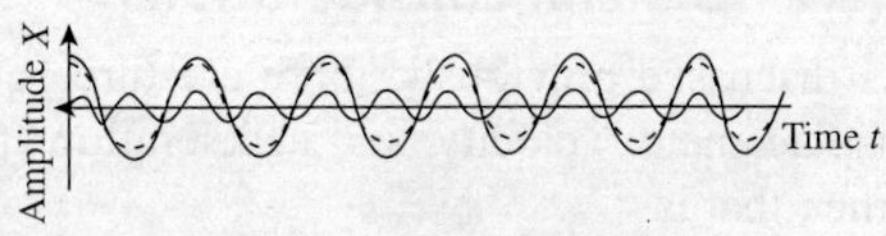

Figure 7.3 Illustration of how the Waveforms Shown in Fig. 7.2(a) can be Generated by Two Harmonally Related Sine Waves

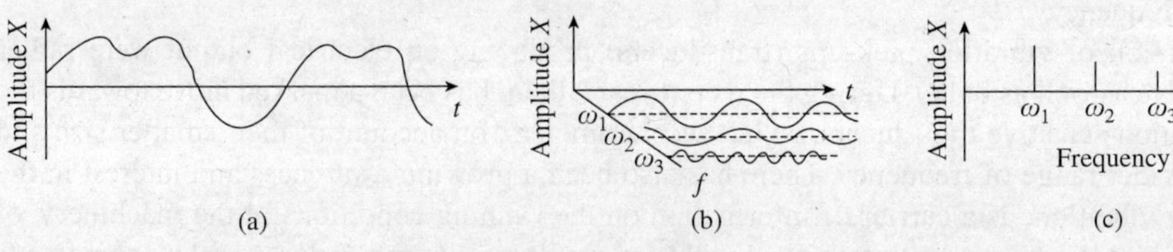

Figure 7.4 Fourier Analysis and Spectrum. Spectrum of Periodic Vibrations, (a) Time-domain Depiction (b) Three-dimensional Coordinates Showing Time, Frequency and Amplitude (c) Frequency-domain Depiction

It can be shown that this combination of sine waves is unique and any real-world signal can be represented by only one combination of specific sine waves. The mathematical theorem formulated by Fourier states that any periodic wave f (t), no matter how much ever complex, might be looked upon as a combination of a number of pure sinusoidal (or cosine) curves with harmonically related frequencies.

$$F(t) = X_0 + X_1 \text{Sin}(\omega t + \Phi_1) + X_2 \text{Sin}(2\omega t + \Phi_2) + \ldots\ldots + X_n \text{Sin}(n\omega t + \Phi_n) \qquad (7.2)$$

Harmonic $X_1\text{Sin}(\omega t + \Phi_1)$ is known as the first or fundamental harmonic whereas the subsequent harmonics having frequency $n\omega$ are known as harmonics of *n*th order, that is, the harmonics will have a frequency equal to 2ω, 3ω, and so on.

In practical cases, the vibrations are caused by a simultaneous action of several periodic disturbing forces, which need not be harmonically related. Consequently, the vibrations can be considered as being made up of harmonics with frequencies equal to ω_1, ω_2, ω_3,.... and so on, which may not be multiples of the overall frequency of vibrations. As mentioned, the breaking down of the vibrations signal into constituent sine waves provide an yet another dimension of looking at the vibration process itself. The frequency-domain representation as shown in the Fig. 7.4(c) is called as the spectrum of the signal. It is important to note that no information of the signal has been lost when viewed in frequency domain. As a matter of fact, it is possible to identify the components (which could ultimately be traced to some forces which could be mechanical / electrical / hydraulic), which constitute the overall signal. This is the reason why spectrum analysis is regarded as the most-powerful diagnostic tool for analysing vibrations / noise problems.

The following sections will deal with various signal sensors, signal processors, data reduction / processing, and the procedure of diagnosis of vibration problems through properly-collated vibration data and spectrum analysis. During these sections, we shall discuss a few case studies to illustrate the applications of the diagnostic procedures for the solution of the vibration problem.

7.2 SENSING AND MEASUREMENTS

7.2.1 General Considerations

As discussed previously, there are three quantities that are of interest in vibration studies, namely, displacement, velocity, and acceleration (peak or rms values). These quantities are related to each other, that is,

$$Displacement\ (Amplitude) = \frac{Acceleration}{(2\pi f)^2} \tag{7.3}$$

$$Velocity\ (Amplitude) = \frac{Acceleration}{(2\pi f)} \tag{7.4}$$

where f is the frequency.

Earlier version of vibration pick-ups (transducers) producing an electrical output were rather velocity-sensitive as well as bulky. During the recent years, there has been a marked move towards the use of acceleration-sensitive transducers, called accelerometers, on account of their smaller size and also a having wider range of frequency. There has also been a growing awareness and interest in the high-frequency vibration, as a carrier of information on the running conditions of the machinery, of the presence of high-frequency pulsating forces that can result into fatigue failures of the components of rotating machinery. The velocity / displacement transducers often miss the detection of such high-frequency components. This is illustrated in Fig. 7.5.

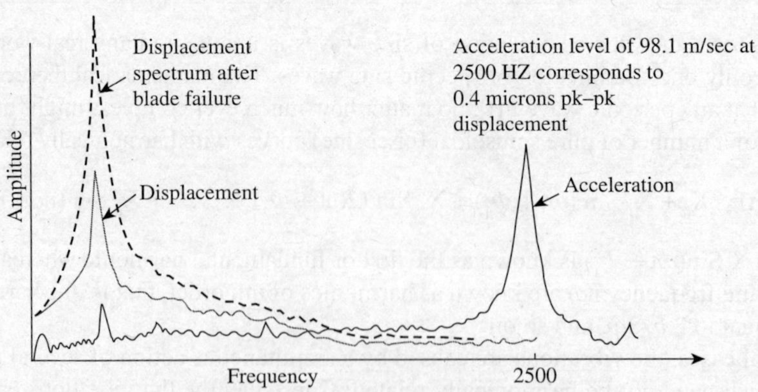

Figure 7.5 Choice of Parameter

10 G force is capable of causing the blade failure by fatigue. When failure takes place, the components in the low-frequency range increase substantially giving rise to a sudden increase in the vibration levels in terms of microns, peak to peak (pk–pk). This is one of the most important reasons for a sudden/unexpected occurrence of the vibration problem. Figure 7.5 shows the spectrum of vibration signals recorded at one of the bearings of a 3000-rpm steam turbine, consisting of multiple stages of stationary and rotating blades. The full line shows the spectrum when displacement is used as the basic measurement parameter. One can notice components of 50 Hz, 100Hz, or 150 Hz, which are harmonics of running speed (50 Hz = 3000 rpm), but we do not see anything in the high-frequency domain. Suppose we have a high-frequency force of 10 G (98.1 metre per second^2) at a frequency of 2500 Hz, which in this case is a diaphragm-passing frequency (number of stator/diaphragm blades × rpm) as indicated in the acceleration spectrum.

The displacement associated with acceleration of 98.1 m/sec^2 at a frequency of 2500 Hz is

$\text{Displacement} = \dfrac{98.1\times10^6}{(2\pi..2500)^2} = 0.4$ micron (pk–pk). This is too small a quantity that can be detected.

The velocity spectrum, of course, does show a peak at 2500 Hz, but if one attempts a diagnosis based upon only pk–pk displacement measurement, the presence of a dynamic force at a high frequency like 2500 Hz is surely missed. If such an enormous force acts on the blades, the blades will certainly fail and may cause a severe unbalance of the rotor. This will show up in the form of a sudden appearance of heavy vibrations in the turbine as shown by the spectrum drawn with heavy lines. The values quoted in this example are somewhat exaggerated with a purpose to drive home the point about selection of vibration parameter for the diagnosis purpose. Sadly, even now many operations and maintenance engineers carry out the vibration-condition monitoring (overall vibration levels) in the critical rotating machinery in the displacement mode and face the problem of sudden and unexpected rise in the vibration levels which forces the equipment out of operations.

The acceleration-based transducers can also yield the information about the velocity and displacement of vibration by carrying single and double integration of acceleration signal, respectively $(\dot{x} = \int \ddot{x}dt, x = \iint \ddot{x}dtdt)$. The electronic integration of the signal is quite easy. On the other hand, if one uses displacement or velocity transducers for getting the acceleration, then they have to carry out single/double differentiation of the signal, which is quite a complex and dubious affair.

In theory, it is irrelevant as to which of the three parameters, acceleration, velocity or displacement, should be chosen to measure vibration. However, if the vibration signal is dominated by a high-frequency component, then the displacement or velocity measurements may not be satisfactory. This condition applies mostly to high-speed turbomachinery such as gas turbines, steam turbines, rotary and high-speed-reciprocating compressors, pumps, and so on. On the other hand, if the vibration process is predominantly in the low-frequency domain as in the case of hydraulic turbines or slow-speed reciprocating pumps, low-rpm diesel engines, and so on, it is appropriate to select the displacement or velocity as a parameter. Displacement is often used as an indicator of unbalance in the rotating machinery because relatively large displacements usually occur at the shaft-rotation frequency, while in some situations, they occur at sub-synchronous frequency. RMS velocity is widely used to determine the vibration severity, as it is indicative of vibrational energy and thus, a destructive potential of the vibration phenomenon. It is customary (not mandatory) to specify pk–pk displacement in microns or mils (1 micron = $10^{(-6)}$ meters, 1 mil = 25 microns).

The frequency range of interest in the vibration measurement has been steadily increasing in the past one or two decades with increasing awareness and understanding of dynamic processes occurring at a higher frequency in the turbomachinery. (We have earlier illustrated this with an example of vibration-signature analysis of a steam turbine.) For example, in hydraulic pumps, one of the severest perturbation forces experienced by the components of the pump occurs at a vane-passing frequency (number of vanes of the impeller × speed of the pump) and its harmonic multiple (2, 3, 4, 5…etc.). As explained earlier, the detection of the presence of these forces by displacement transducers is very difficult as the associated displacements may be too small. The acceleration levels of the order of 10 meter per second^2 (pk-pk) at a vane-passing frequency of 525 Hz of seven-vane pump, running at 75 Hz (4500 rpm), are realistic and are often encountered on high-efficiency pumps. One can verify that an acceleration of 10 meter per second^2 (pk-pk) and a frequency of 525 Hz would produce a vibrational displacement which is less than one μm (micron), pk-pk. Such a low level may possibly be missed or overlooked by the vibration engineer when he/she encounters a relatively higher level of components at a running speed. As mentioned previously, the high-frequency force may cause a fatigue failure of vanes, and the vibration levels in the displacement modes may increase and result in bearing failures too. It is, therefore, preferable and advisable to carry out the vibration measurements in at least two modes, one of which may be acceleration to ensure that no high-frequency components are missed. This is very important to remember. While one can use the displacement and velocity modes of vibration measurement on low/medium speed machinery, for high-speed machinery, the displacement/velocity and acceleration modes for vibration measurement, can be used.

The vibration associated with fluid flow is, quite often, random in nature and must frequently be measured alone or together with periodic-vibration measurements. The measurement technique to be employed in such cases is quite complex.

As mentioned earlier, the diagnosis of vibration problems requires, apart from the measurement of overall vibration, frequency analysis to reveal individual frequency components making up wide-band signals. Traditionally, for this purpose, a filter is included in or attached to the vibration instrument, thus making a frequency analyser. The filter allows frequency components to be measured, which are content in a specific frequency band. The pass band of the filter is moved sequentially over the entire frequency range of interest so that separate vibration-level readings can be obtained at each frequency. The filter section may contain a number of individual, fixed-frequency filters, which are scanned sequentially, or a tunable filter, which can be tuned continuously over the frequency range. With advances in digital signal processing (DSP) technology, it is now possible to compute and display the instantaneous spectrum on the screen of real-time analysers, which also permits a continuous updating of the complete frequency spectrum. There are two types of real-time analysis procedures: digital filtering and fast-Fourier-transform (FFT) methods. While the former is best suited for constant percentage band-width analysis the latter is suitable for constant band-width analysis. The former is used for condition monitoring while the latter is used for diagnosis purposes.

7.2.2 Important Terminologies in Vibration/Noise Measurements and Band-Pass Filter

In vibration analysis, the following terms are often used. Hence, it is necessary that readers may clearly understand that they are (1) decibel (dB) of quantity and (2) octave. A decibel (dB) of a quantity such as power (P) is defined as

$$dB = 10 \log_{10} (P/P_{ref}) \tag{7.5}$$

where P is the power and P_{ref} is a reference value of the power. This quantity is extensively used in the noise analysis where we talk in terms of noise levels in dB.

An octave is the interval between any two frequencies, that is, (f_2/f_1) whose frequency ratio (f_2/f_1) is 2. Two frequencies f_1 and f_2 are said to be separated by a number of octaves N when $f_2/f_1 = 2^N$. Or,

$$N \text{ (in Octaves)} = \log_2(f_2/f_1) \tag{7.6}$$

In this, N can be an integer or a fraction. If N = 1 we have one octave, if N = 1/3, we get one-third octave, and so on. With these definitions and terminologies, we can now discuss about band-pass filter, which is an essential component of a spectrum analyser.

A band-pass filter is a circuit built of resistors, inductors, and capacitors that permits the passage of a frequency component of a signal for a frequency band and reject all other frequency components. Fig. 7.6 shows the response characteristics of a filter whose lower and upper cut-off frequencies are f_1 and f_2. The ideal filter characteristics will be as shown by the dotted rectangle.

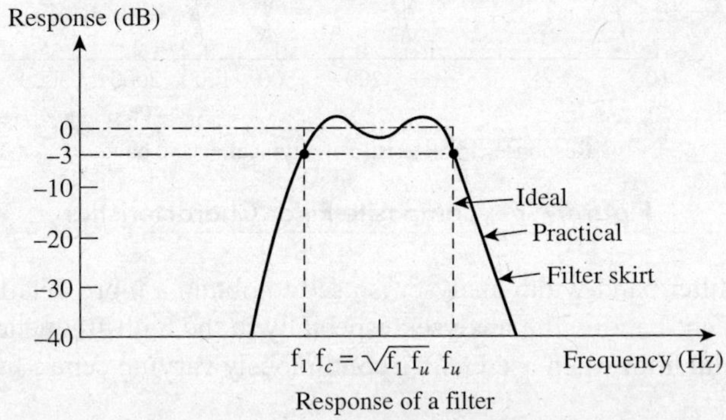

Figure 7.6 Filter Characteristics

A practical filter will have a response characteristic deviating from the ideal rectangle as shown by the full line in Fig. 7.6. A good filter has a small or a minimum ripple content in the band width and steep filter skirt slopes so that the actual band width is close to the ideal band width, $\beta = f_u - f_l$. For a practical filter, the frequencies, f_l and f_u, at which responses 3dB below the mean band-pass response are called the cut-off frequencies. There are two types of band-pass filters used in the signal analysis: constant percent band-width filter and constant band-width filter. For constant percent band-width filter, the ratio of band-width to the centre frequency (tuned frequency), $(f_u - f_l)/f_c$ is the constant. The octave, one-half octaves, one-third octave-band filters are examples of constant percent-band-width filters. The following table shows some of the cut-off limits, the centre frequencies of octave band used in the signal analysis.

Lower cut-off frequency (Hz)	5.63	11.2	22.4	44.7	89.2	178	355	704	1410
Central frequency (Hz)	8.0	16.0	31.5	63.0	125	250	500	1000	2000
Upper cut-off frequency (Hz)	11.2	22.4	44.7	89.2	178	355	789	1910	2820

For a constant band-width filter, band width $f_u - f_l$ is independent of centre (tuned) frequency.

One can notice that in octave-band filters the upper cut-off frequency is twice the lower cut-off frequency. These filters give a less-detailed and too-coarse analysis for practical vibration and noise problems. One-half octave-band (N = ½ in Equation 7.6) filter gives twice the information but requires twice the amount of time to obtain data. A spectrum analyser with a set of octave and one-third octave filters are used for noise analysis. Each filter is tuned to a different centre frequency to cover the entire frequency range. Since the lower cut-off frequency of a filter is equal to the upper cut-off frequency of the previous filter, the composite filter characteristics will appear as shown in Fig. 7.7.

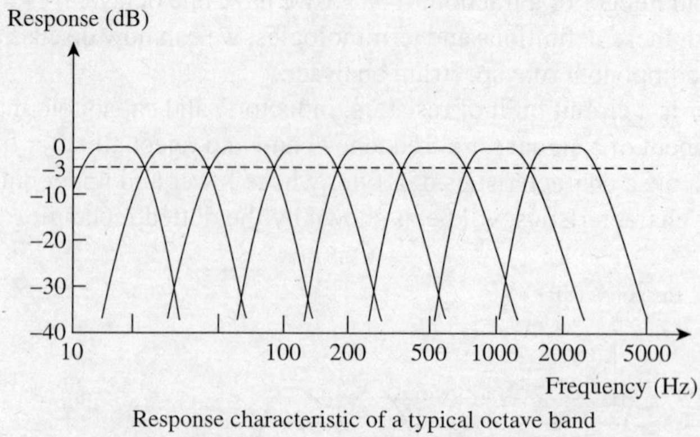

Figure 7.7 Composite Filter Characteristics

A constant band-filter band-width analyser is used to obtain a more detailed analysis than in the case of a constant percent-band-width analyser, especially in the high-frequency range of the signals. The constant band-width filter when used with a continuously varying centre frequency is called wave or heterodyne analyser.

We shall discuss the special features of fast-Fourier-transform analysers (FFTA) in a separate section.

We have briefly discussed some fundamental principles behind spectrum analysers, which may be filter-based or computing analysers (FFT). The signal received from transducers can be due to displacement transducers, velocity transducers, or accelerometers.

7.2.3 Vibration Pick-Ups

(A) Vibrometer/Velometer

The commonly used vibration pick-ups are known as seismic instruments. A seismic instrument consists of a mass–spring–damper system mounted on a vibrating body (Fig. 7.8). The vibratory motion can be measured by finding the displacement of the mass relative to base on which it is mounted. Since the base on which the pick-up is mounted has a vibration, y (t) = Y sinωt, the mass m undergoes deflection x(t), then the displacement of the mass relative to the cage, (which is fixed to the vibrating body),

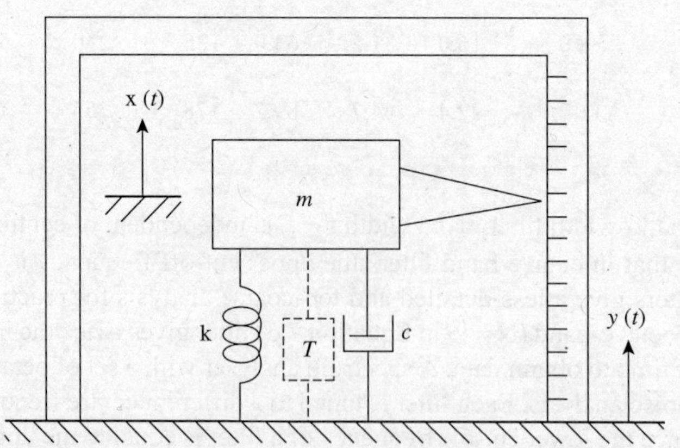

Figure 7.8 Vibration Transducer

is Z = x − y, which can be measured if we attach a pointer to the mass and scale to the cage as shown in the Fig. 7.8. The output Z of the instrument is the relative mechanical motion of the mass which for high-speed operation and accuracy can be converted to electrical signal by a transducer.

The equation of motion of the mass m can be written as

$$m\ddot{x} + \mathrm{f}\,(\dot{x} - \dot{y}) + (x - y) = 0 \tag{7.7}$$

As Z = x − y, that is, the relative motion of mass m, we can write Equation 7.7 as

$$m\ddot{z} + \dot{z} + kz = -m\ddot{y} \tag{7.8}$$

But

$$\ddot{y} = -m\omega^2 y \sin \omega t.$$

This makes the Equation 7.8 as

$$m\ddot{z} + f\dot{z} + kz = m\omega^2 y \sin \omega t \tag{7.9}$$

The solution of Equation 7.9 is given as

$$Z = \frac{r^2 y}{\sqrt{(l-r)^2 (2\xi r)^2}},\ \varphi = \tan^{-1}\left(\frac{2\xi r}{1-r^2}\right) \tag{7.10 \& 7.11}$$

where $r = \omega/\omega_n$ and $\xi = \dfrac{f}{zm\omega_n}$ and $\omega_n = \sqrt{k/m}$ = resonant frequency.

It can be shown that $Z/y \approx 1$ when $\omega/\omega_n = r \geq 3$. Under these conditions, the relative displacement between the mass and the base is essentially the same as the displacement of the base. If we consider the Equation 7.10, we note

$$Z(t) = y\sin(\omega t - \varphi) \tag{7.12}$$

$$\text{If}\ \frac{r^2}{\sqrt{(l-r^2)^2 + (2\xi r)^2}} \approx 1 \tag{7.13}$$

then a comparison of the Equation 7.12 with $y(t) = Y \sin \omega t$ shows that Z(t) gives directly the motion except for phase lag φ. This time lag is not so important as $r = \omega/\omega_n$ has to be large, the natural frequency of the spring–mass–damper system must be very low so that ω/ω_n is ≥ 3. This means that mass m must be large and the stiffness of the spring must be low. These conditions/requirements make the vibrometer transducers very bulky and obsolete in the current use as accelerometers do not possess this deficiency. The same can be said about the velocity transducers called velometers where the velocity of the vibrating body is measured by converting the velocity signal to electrical signal.

(B) Accelerometers

Accelerometers are transducers that measure the acceleration of the vibrating body. From Equations 7.10–7.12, we get

$$-Z(t)\,\omega_n^{\,2} = \frac{1}{\sqrt{(l-r^2)^2 + (2\xi r)^2}}\left[-Y\omega^2 \sin(\omega t - \varphi)\right] \tag{7.14}$$

$$\text{If}\ \frac{1}{\sqrt{(l-r^2)^2 + (2\xi r)^2}} = 1,$$

then Equation 7.14 becomes

$$-Z(t)\,\omega_n^{\,2} \approx -Y\omega^2 \sin(\omega t - \varphi) \tag{7.15}$$

But $-Y\omega^2\sin\omega t = \ddot{y}(t)$, thus the transducers can be made to record directly the value of $\ddot{y}$ i.e., the acceleration of the vibrating body except that phase lags by φ. One way to achieve the condition $\frac{1}{\sqrt{(l-r^2)^2+(2\xi r)^2}} \rightarrow 1$ is r must be very small since $r = \omega/\omega_{n.}$ r becoming very small is possible when ω_n, (the natural frequency of spring mass damper system used in the transducer) is very large. This is possible by using smaller mass and short spring (very stiff). This makes the transducers very small. This characteristic of being miniature in size (compared to vibrometer/velometer) and having a wide range of frequency makes an accelerometer extremely suitable for vibration measurement. The basic parameter measured by the accelerometer is the acceleration of vibration. The acceleration signal can be electronically integrated once or twice to obtain the velocity and displacement of the vibration.

The transducer consists of a small mass and stiff spring. The force created by vibration acceleration creates a voltage or charge (that is proportional to acceleration) in the piezoelectric element. The piezoelectric elements, forming the heart of the accelerometer, are made from artificially polarized ferroelectric ceramic. These elements produce an electrical charge directly proportional to the strain and thus applied force when loaded in tension, compression or shear. In the design of the accelerometers, the piezoelectric elements are arranged such that they are loaded by mass or masses and preloaded by spring or ring.

For frequencies lying well below the resonant frequency of the accelerometer assembly, the acceleration of the masses will be the same as the acceleration of the base and the output signal will be proportional to the acceleration to which the accelerometer is subjected. Fig. 7.9(a) shows two commonly used configurations of the accelerometer, namely, compression and shear type.

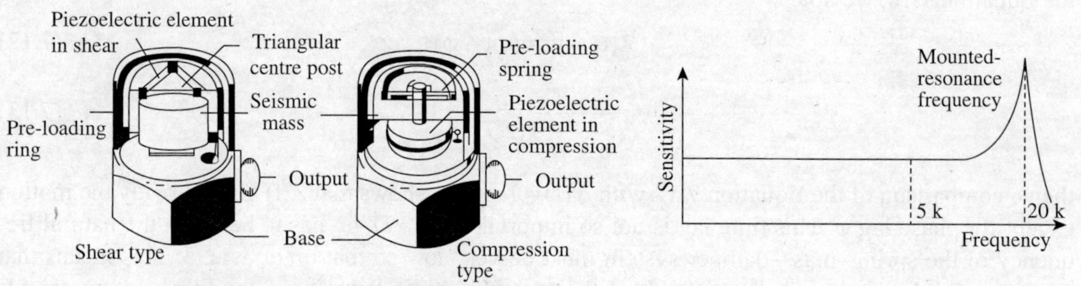

(a) Two-accelerometer configurations (b) Frequency response of a typical accelerometer (Courtesy: Bruet & Kjaer, Denmark)

Figure 7.9 Accelerometers

Figure 7.9(b) shows the frequency characteristics of a typical piezoelectric accelerometer. Measurements are normally confined to using the linear portion of the response curve [Fig. 7.9(b)], which at the high-frequency end is limited by the accelerometers' natural resonance. (As an accelerometer is the assembly of spring, mass, and damper, it has its own natural frequency). As a rule of thumb, the upper-frequency limit for the measurement can be set to one-third of the accelerometer-resonance frequency (for mounted-resonance frequency) to minimize the errors.

As the useful range of accelerometer is limited by the accelerometers' natural frequency (mounted-resonance frequency), it is very important to fix the accelerometer very firmly to the surface where the vibration measurements are being done. The best way to do this is a small stud insertion.

Figure 7.10 shows various methods of mounting the accelerometer. In the arrangement (a), the accelerometer is fixed using a short-length stud or by using a thin film of adhesive (including bee

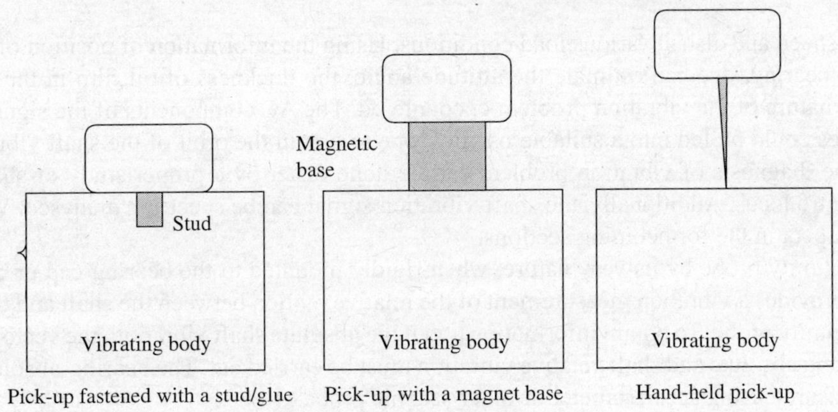

Figure 7.10 Fixing Pick-up

wax), respectively. In these cases, the useful range of accelerometer is 0 to $w_x/3$. In the arrangement shown in 7.10(b), the accelerometer is fixed to the surface using a strong magnetic base (which should be as short as possible). The presence of a magnetic base greatly reduces the natural frequency of the pick-up assembly from w_n to w_n which makes a useful range of accelerometer 0 to $w_n/3$. Thus, for getting accurate results, it is always desirable to use the fixing arrangement as shown in Fig. 7.10(a). In other words, the best mounting arrangement are stud mounting with a mica washer or a thin film of silicone grease or a fixed one using bees' wax or epoxy resins.

(C) Shaft-Vibration Transducers

The traditional method of assessing the vibration behaviour by monitoring the bearing cap/casing vibrations sometimes proves inadequate on account of (variable) transfer impedance between the shaft motion and the casing motion. Many a times, high-shaft vibrations are masked by a very stiff pedestal, because of stiffness and damping provided by the bearing-oil film. A vast majority of machine malfunctions manifest themselves (at lower frequencies, usually less than four times the running speed) in the form of high and abnormal shaft-vibration behaviour. As mentioned earlier, in the event of bearings providing a large damping ratio, these malfunctions may not appear damaging when measurements are made on the bearing cap or casing alone. It is, therefore, prudent to carry out shaft-vibration measurement whenever possible.

Many of the shaft probes operate on the inductive principle and, therefore, do not touch the object to be observed (Fig. 7.11). These probes are completely unaffected by any material in the probe gap that is not electrically conductive; and hence, oil, air, gas, and so on, between the probe and the observed surface has no effect on the probe output. A shaft probe, called the proximity probe, provides two signal outputs. They are (1) voltage proportional to the dynamic motion of the shaft relative to probe mounting (AC signal output) and (2) the voltage proportional to the average gap between the shaft and the probe mounting (the average gap or DC signal output). It is customary to mount two mutually perpendicular shaft probes in the bearing housing as shown in Fig.7.11(a). The DC gap voltages from these probes can be plotted on a polar chart as shown in Fig. 7.11(b). This enables locating the journal (portion of the shaft in the bearing is called the journal) position at all speeds, starting

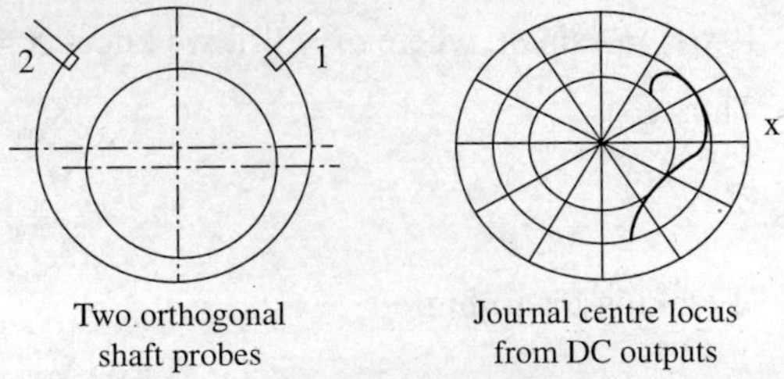

Figure 7.11 Shaft-vibration Probe

from zero speed and also at various load conditions. Using the information of position of shaft centre relative to bearing, one can estimate the attitude angle, the thickness of oil film in the bearing and hence, the nature of the vibration problem encountered. The AC components of the signals from two shaft probes could be fed into a suitable oscilloscope to obtain the orbit of the shaft vibration. Many a times, the diagnosis of vibration problem can be done faster by a proper study of shaft orbit and journal-centre locus. Additionally, the shaft-vibration signal can be spectrum analysed. We shall deal with this aspect in the forthcoming sections.

A proximity probe by its very nature, when rigidly mounted to the bearing cap or casing of the machine, provides a vibration measurement of the relative motion between the shaft and the mounting of the proximity probe. To obtain information about the absolute shaft vibration, the vector summation of the bearing absolute and shaft-relative vibration must be carried out. The bearing-absolute vibration can be measured using accelerometer or other seismic probe.

In some international standards such as API (American Petroleum Institute) standard, the upper limit of permissible shaft-vibration at various operational speeds of shaft rotation are specified and strictly followed. For example, in critical machines such as steam turbines, high-speed compressors, pumps, and so on, used in refinery / petrochemical / fertilizer industries, the machine-health standards are based upon the level of shaft vibrations.

Before we proceed further in signal analysis, let us discuss about the international standards for permissible values of vibration displacement, velocity, or acceleration on low / medium / high-speed machinery. Having understood the basics of vibration transducers including proximity probes, it may be lot easier to understand the basic principles and the logic of these standards.

7.3 VIBRATION NOMOGRAPHS AND VIBRATION CRITERION

Vibrations are experienced on all equipments during the operation and also during natural catastrophes such as earthquakes, to a smaller or a larger extent depending upon various conditions and design features. Nevertheless, the presence of vibration often leads to excessive wear of bearings, formation of cracks, loosening of fasteners, structural and mechanical failures, frequent and expensive maintenance of machines, malfunctioning of electric components through fracture of solder joints, and so on. Additionally, the occupational hazards of exposure of vibration such as pain, very high quality of manufacture, and assembly discomfort become important as the vibration levels go beyond a certain level. To a certain extent, the vibrations can be reduced / eliminated by resorting to a very high quality of manufacture and assembly / installation. However, the cost involved in eliminating the vibration may be too high; so, the designer must compromise between acceptable amount or level of vibration or noise and a reasonable manufacturing, assembly, and installation cost.

The acceptable levels of vibration are often specified in terms of the response of an undamped single degree-of-freedom system undergoing a harmonic vibration. The bounds are shown in a graph, called vibration nomograph, shown in Fig. 7.12.

If $x(t) = X\sin\omega t$, where $\omega = 2\pi f$, we know $X_p = X$, $V_p = 2\pi fX$, and $A_p = -4\pi^2 f^2 X_p$.

Thus,

$$X_{max} = X$$

$$V_{max} = 2\pi fX \tag{7.16}$$

$$A_{max} = -4\pi^2 f^2 X_p = -2\pi f V_{max} \tag{7.17}$$

By taking logarithm,

$$\text{Log } V_{max} = \log 2\pi f + \log X \tag{7.18}$$

$$\text{Log } V_{max} = -\log A_{max} - \log 2\pi f \tag{7.19}$$

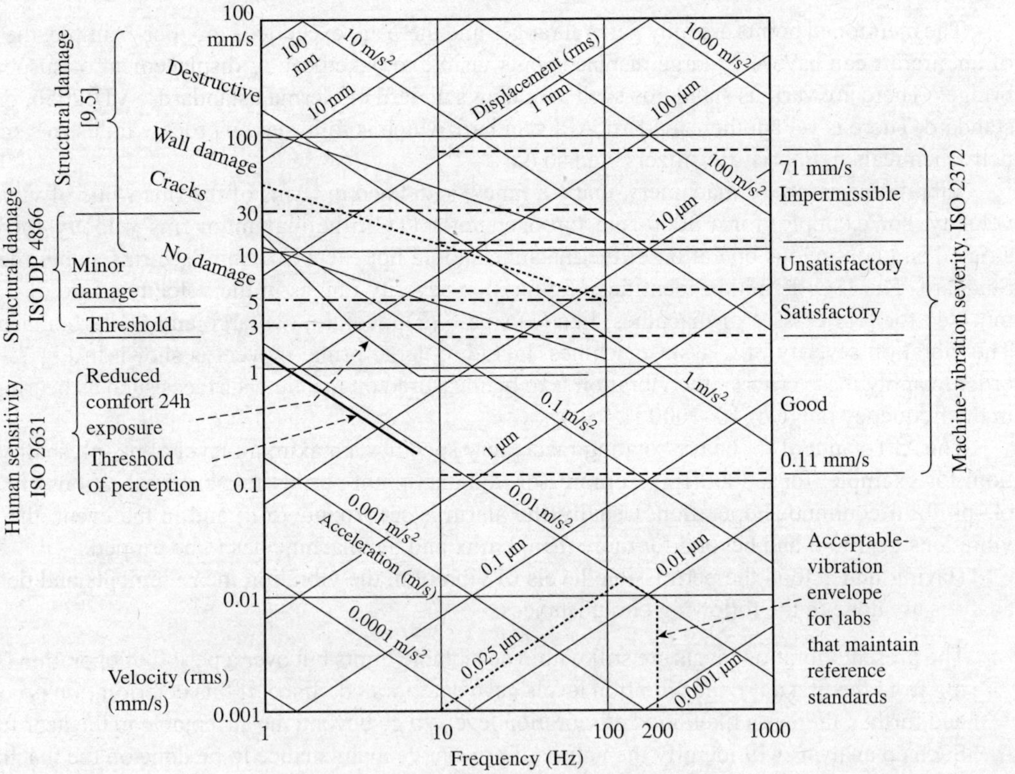

Figure 7.12 Vibration Severity

It can be seen that for a constant value of the displacement X, Equation 7.19 shows that Log V_{max} varies with log $2\pi f$ as a straight line with slope +1. Similarly, for a constant value of acceleration A_{max}, equation 7.20 indicates that Log V_{max} varies with log $2\pi f$ as a straight line with slope −1. These variations are shown in Fig. 7.12. Thus, every point on the nomograph denotes a specified harmonic vibration. Since the vibrations imparted to a human or a machine is composed of many frequencies, the rms values of x(t), v(t), and a(t) are used in specification of vibration levels. Also, it is customary to specify displacement and acceleration in a peak to peak (pk–pk) mode while velocity is specified as peak (pk).

The usual ranges of vibration encountered in different scientific and engineering applications are given as follows:

(1) Atomic vibration: Frequency = 10^{12} Hz and displacement magnitude = 10^{-8} to 10^{-6}mm.

(2) Minor tremors of earth's crust: Frequency 0.1–1 Hz, displacement magnitude = 10^{-5} to 10^{-3}mm. This vibration also denotes the threshold of disturbance of optical, electronic, and computer equipment.

(3) Machinery and building vibration: Frequency = 10–100 Hz and displacement amplitude = 0.01–1 mm. The threshold value of human perception falls in the frequency range of 1–8 Hz. Slow-speed engines and hydraulic turbines are an exception to this as we can have a turbine speed as low as 100 rpm.

(4) Swaying of tall buildings: Frequency range = 0.1–5 Hz and displacement amplitude = 10–1000 mm.

The mentioned points are only general ranges and there are exceptions too. For example, the wing of an aircraft can have very large displacements visible to naked eye or displacement of suspension bridges. There are various standards such as Indian standard or German standard—VDI2056, or ISO standard. There is yet another standard, API standard, which is followed by process industries such as petrochemicals, refineries, fertilizers, and so on.

Vibration severity of machinery, many a times, is defined in terms of the rms value of vibration velocity. For example, for a 3000-rpm turbogenerator (TG), the maximum rms velocity for unrestricted and continuous operation of the machine should not exceed 7.1mm / sec rms as per the ISO Standard. The ISO definition identifies 15 vibration severity ranges in the velocity range of 0.1–71 mm / sec for four classes of machines, like (1) small, (2) medium, (3) large and (4) turbomachines. The vibration severity of Class 3 machines, including large prime movers is shown in Fig 7.12. In order to apply these criteria, the vibration is to be measured on machine surfaces such as bearing cap in the frequency range of 10–1000 Hz.

The API standard for heavy-rotating machinery specifies a maximum level of pk–pk shaft vibration: for example, for a 3000-rpm compressor, the maximum permissible-shaft vibrations are 75 μ pk–pk for a continuous operation. Usually, the alarm is set around 75 μ; and in the event, the shaft vibrations go 100 μ and beyond for more than 2 min, and the machine has to be tripped.

Having understood the permissible levels of vibration, the vibration measurements and detailed analysis are done in the following circumstances:

- The present vibration levels are still within acceptable limits but over a period of operation (starting from the first day), the vibration levels have deteriorated. Since, if this deterioration is continued further, there is a likelihood of vibration levels to go beyond the acceptable in the near future. Such an analysis will identify the nature of preventive maintenance to be done on the machine.
- The levels have reached levels beyond acceptable levels for a continuous operation of the machine.
- In such a case, a total vibration analysis comprising of a measurement of vibration amplitudes, phases, and spectrum analysis is required to be carried out. We already discussed the various aspects of vibration levels, measurement, and spectrum analysis. We shall now discuss the phase measurement.

7.4 VIBRATION ANALYSIS

We have already discussed the principles of vibration measurement and also the basic principles of vibration spectrum analysis. In many cases, it is possible to diagnose the root cause of the vibration problem by measuring the vibration levels and carrying a vibration spectrum analysis. However, there are problems wherein it is not sufficient to use the measured vibration levels and the vibration spectrum details for correct and a complete analysis of the vibration problem. The other important parameter is the phase measurement and also, at times, we need to carry out instantaneous or real-time spectrum analysis. In this section, we shall deal with the phase measurement and real-time spectrum analysis.

7.4.1 Phase Measurement

In the analysis, diagnosis, and correction of vibration problems, phase analysis plays a very important role. Phase in the simplest terms can be defined as that part of the cycle (0–360°) through which one point of a machine has travelled relative to another part or a fixed reference. Phase measurement can also be considered as a means of determining the relative motion of various parts of the machine.

In practice, the measurement of phase is done using (1) stroboscopic light source or (2) phase meter or (3) oscilloscope. We have already explained in the previous chapter the use of stroboscopic light for balancing of rotors. Many of the portable vibration analysers are furnished with a stroboscopic light which when triggered by the vibration measurement, provides a quick and convenient means of obtaining phase readings for analysis and balancing. To obtain phase readings, a reference mark is placed on the rotor at the end of the shaft or some other location, which can be observed. For convenience, an angular reference scaled in degrees (0–360°) can be superimposed on the shaft to note the phase angle, as in Fig.7.14(a).

Using the analyzer, the strobe light can be tuned so that the reference mark appears stationary. Comparative phase readings are obtained by simply shifting/moving the vibration pick-up to another measurement location, but keeping the same reference mark and angular reference that were used for the initial measurement. In some cases, where the end of the shaft is not available for viewing, an angular reference can be placed in the rotating shaft (by placing a graduated tape or marking 0–360° in steps of 5 or 10°). The stationary reference used might be the split line between the upper and lower halves of the bearing or any other convenient stationary component.

The measurement of phase as described earlier pertains to the phase of rotational component ($(1 \times n)$, component, n = shaft rotational speed in Hz) of the machine vibration. However, if the vibration of interest is occurring at $2 \times n$, then the strobe light will flash twice for each revolution of the shaft, and the reference mark will appear at two positions when observed with the light. It would, thus, be virtually impossible to obtain comparative phase readings with two or more identical reference marks visible under the light. When it is desired to obtain comparative phase measurements for submultiple or multiple or non-harmonically-related vibration frequencies, some other techniques such as using a phase meter or an oscilloscope must be used.

A remote read out-of-phase requires a voltage reference pulse at the desired vibration frequency. Where it is desired to observe the phase of the vibration occurring at $2 \times n$, a reference pick-up such as a photocell, an electromagnetic pick-up, or a non-contact pick-up is mounted close to the shaft, which has been properly prepared to trigger the reference pick-up. While using a photocell, it is customary to wrap a black non-reflection tape around the shaft and then paint a white line across the tape or attach a small piece of reflective material to ensure that the photocell responds to the changes in the reflectivity. When a non-contact or an electromagnetic pick-up is used, the target area of the shaft is prepared with an abrupt depression or protrusion. An existing key or a keyway can be used with excellent results.

The $1 \times n$-voltage pulse from the reference pick-up applied to the analyser serves two purposes. First, the reference pulse automatically tunes the filter of the analyser to the rotating frequency so that any speed change during the measurement does not cause errors in the measurement. Second, the reference pulse is electromagnetically filtered when compared with the filtered vibration signal to provide a measure of a relative phase.

It is also possible to measure the phase of vibration using a reference pick-up mounted at a different point of the machine. If the vibration is predominant at the desired frequency, a signal from a reference pick-up can be fed to a vibration meter shown in Fig. 7.13, whose output can be used as a reference signal to the analyser.

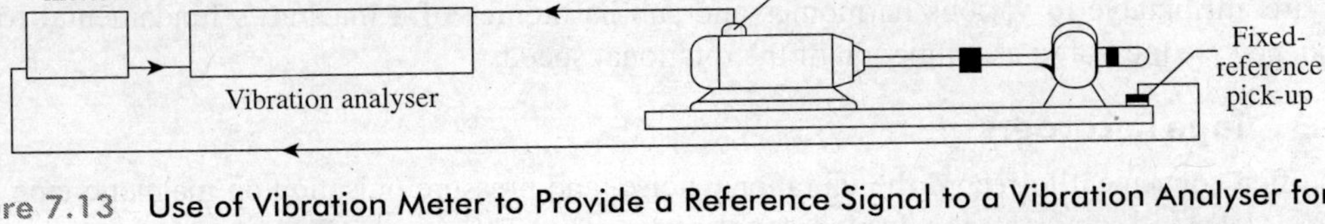

Figure 7.13 Use of Vibration Meter to Provide a Reference Signal to a Vibration Analyser for Phase Measurements

This method, however, cannot be used for balancing since the placement of trial weights for balancing would also alter the reference signal. The signal from reference pick-up can also be fed to a vibration analyser (instead of a simple meter), which can be considered as a reference analyser. The output of the reference analyser is then connected to the input of the main analyser. This method then facilitates the comparative phase measurements for virtually any frequency of interest—synchronous multiples, sub-multiples, non-harmonic frequencies, and so on.

7.4.2 General-Purpose Vibration Analyser

The vibration meter consists of one or more instruments, which amplify the signal received from the transducer. Traditionally, the meter is provided with suitable filters to limit the frequency range at the upper and lower end, so as to avoid the measurement of unwanted signals, noise, and so on. The instrument provides for the integration of the signal to enable the velocity and displacement measurement and facilities for indicating rms or pk–pk of peak values of the signal. Facilities are also provided for connecting the analyser to a tracking or a tunable filter to enable the frequency analysis to be performed, and to a level recorder to plot the spectra. A typical meter may consist of a signal band-pass filter which may be switched to either 30 or 25 per cent ($^1/_3$ octave) band-width and which may be tuned over the frequency range of 0.2–20 KHz in five sub-ranges. Tuning could be done either manually or swept automatically when used in conjunction with a graphic-level recorder. Adding a photoelectric pick-up and trigger, unit / phase meter can further increase the analyser versatility. Thus, a typical vibration analyser consists of a meter, tracking filter, phase-meter, and a photo-optic probe.

Addition of a tracking filter to the general-purpose vibration meter makes a versatile tracking analyzer as shown in Fig 7.14.

(a) (b)

Figure 7.14 (a) Tracking Filter (b) Use of Tracking Filters for Engine

The tracking filter, in addition to being manually tunable, can be turned by virtually any periodic signal form, for example, a tachometer probe on the rotating shaft. This tuning facility enables the vibration signal to be analysed during the machine run-up / run-down. An additional feature of the filter is that it can be tuned to any ratio combination of the tuning (tachometer) signal frequency between 1 / 99 and 99 / 1. This enables the engine-order analysis to be performed, that is, the vibration levels that are attributable to various harmonics and sub-harmonics of a machine's fundamental rotation frequency are measured as a function of the rotational speed.

7.4.3 Tape Recorders

It is often convenient to record the vibrations, noise, and pressure pulsation on magnetic tape for an analysis later and at a greater convenience in the laboratory. This, especially, is the case when one deals with transient signals; for example, when vibration / noise are associated with pressure pulsations in the hydraulic / aerodynamic path of a flow machine such as hydraulic turbine, pump, compressor, path turbine, and so on. In these problems, a simultaneous capture of vibrations / noise / pressure pulsation signal is absolutely essential.

Two recording principles are in common use, direct recording (DR) or frequency modulation (FM). FM recording is normally employed in order to obtain the linearity and low-frequency response that are necessary for many vibration-measurement purposes. A typical DR tape recorder would have a dynamic range (narrow band) of 70 dB and frequency limits from 2.5 Hz to 50 kHz while a typical FM tape recorder would have a dynamic range of 60 dB and a frequency range from a DC to 10 kHz.

The tape recorders, used in vibration measurement are multi-channelled so that several data could be simultaneously recorded. Out of these, four-channel recorders are very popular (Fig 7.15). As the tape recorder is likely to be the most limiting factor in determining the dynamic range of the system, it is wise to choose the parameter for recording (acceleration, velocity, or displacement), which has the flattest spectrum. The conversion between the parameters is quite straightforward once a narrow-band spectrum analysis is carried out.

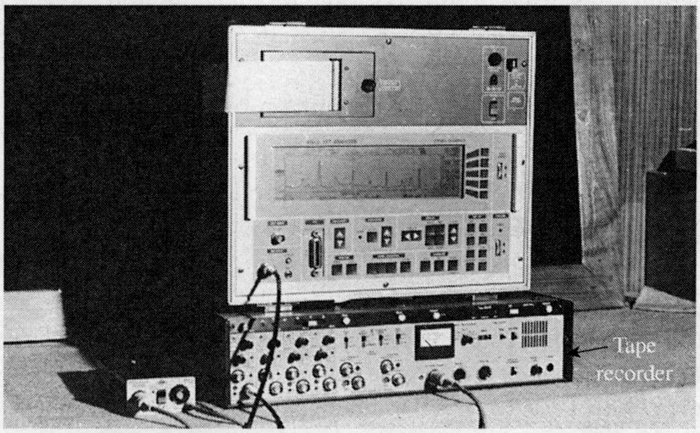

Figure 7.15 Tape Recorder and Analyser

It is convenient, as well as very necessary, to provide pre-amplification as well as conditioning of the signal input to the tape recorder. The pre-amplifier enables the amplification of the signal so as to achieve a full-scale level of the tape recorder even with a small signal. The pre-amplifier usually includes an integrator so that acceleration, velocity, and displacement measurements can be performed. The pre-amplifier also provides a choice of high- and low-pass filters so that the unwanted signals can be prevented from influencing the measurements.

7.4.4 Real-Time Analysers

The most outstanding advantage of real-time frequency analyzers is that they provide an analysis, in all frequency bands over their entire analysis range simultaneously. They give virtually an instantaneous graphical display of the analysed spectrum on a larger built-in screen, which is continuously updated. Real-time analysers (RTAs) are also particularly best suited for the analysis of short-duration signals such as transients, shocks, and so on. The read-out and display of the analysed transient takes place practically at the very instance of capture. This is impossible with serial-frequency analyzing instrumentation prescribed earlier.

Most of the RTAs are based on the FFT procedure while some RTAs are based on recursive digital filtering. The FFT-based RTA produced a constant band-width spectrum on a linear-frequency scale, while the recursive digital-filtering analysers generate a constant percent, $^1/_3$rd or $^1/_1$th octave band-width spectrum. For a vibration-analysis work, FFT analysers are better suited while analysers using digital filters are more suited for condition monitoring and quality control work.

FFT is an algorithm for transferring data from time domain to frequency domain. Because of the computational work involved, transformation must be implemented on a digital computer. With this, algorithm transforms digitized samples from time domain to samples in frequency domain. The governing discrete-Fourier-transform (DFT) equations are as follows:

Forward: $$G(k) = \frac{1}{N}\sum_{n=0}^{N-1} g(x)e^{-2\pi jkn/N} \tag{7.20}$$

Inverse: $$g(x) = \frac{1}{N}\sum_{n=0}^{N-1} G(k)e^{-2\pi jkn/N} \tag{7.21}$$

where G(k) represents the spectrum values at N-discrete frequencies, kΔf and g(x) represents samples of time function at N-discrete time periods, nΔt. The mentioned equations are finite sums when compared to Fourier-transform equations, which involve infinite integrals of continuous functions.

The discrete nature of DFT gives rise to following problems:

- Aliasing caused by the sampling of time signal may result in the appearance of lower frequencies, which actually do not exist in the signal. Using appropriate anti-aliasing eliminates this and the ensuring rate is greater than twice the highest frequency of the input.
- This is in regard to the time-window effect resulting from the finite length of the time record. For example, if the time record contains an integral number of cycles of the input sine waves, the sampled input to FFT may match the actual input. On the other hand, if the input is not periodic in the time record, the FFT algorithm is computed on the basis of the highly distorted waveform as shown in the Fig. 7.16.

A careful examination of Fig. 7.16 shows that most of the distortion is at the edges of the time record; the centre is, however, a good sine wave. If the time record is multiplied by a function that is zero at the ends of the time record and large in the middle, then the concentration of FFT would be on the middle of the time record. Such a function is called "windows functions" as it forces the data to be seen through a narrow window. There are various types of window functions available such as Hanning Window, Rectangular Window, and Flat-top Window. In general, windowing brings a vast improvement in FFT result of input, which is not periodic in time record as shown in Fig. 7.16 earlier.

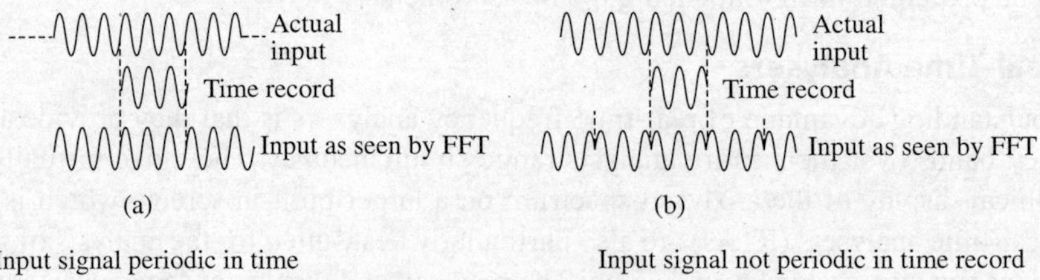

Figure 7.16 Data of FFT

A typical FFT analyser has a transform size N (in Equations 7.20 and 7.21) of 1024 data samples and theoretically gives 1024 frequency values. Moreover, since the data values are real, only 512 positive-frequency values are calculated. Not all of the 512 values are used since low-pass filters are needed for anti-aliasing purposes. For commercial analyser, it is typical to place the filter cut-off so that the first 400 lines are valid and are displayed. Thus, the frequency resolution Δf is 1/400 of the selected full-scale frequency and the automatically-selected sampling frequency is 2.56 times the full-scale frequency. Figs 7.16 and 7.17(a) and (b) show typical FFT-based, real-time analysers.

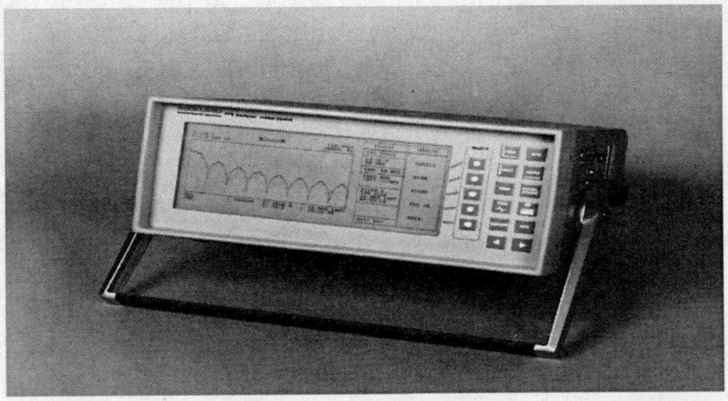

Figure 7.17(a) FFT Analyser

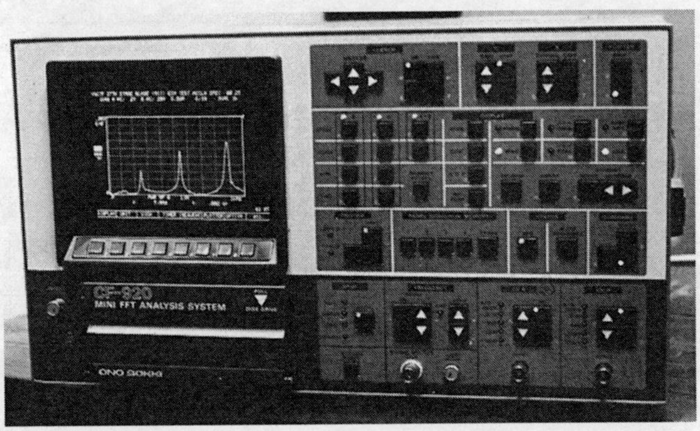

Figure 7.17(b) FFT Analyser

7.4.5 Remote Sensing

In certain situations, it is impossible to have a direct access to the output of the sensor, for example, a censor (accelerometer, strain gauge) mounted on a revolving component. The most widely accepted method of collecting signals from the sensor mounted on a rotating component is the use of slip rings which are mounted at a convenient location on the rotating shaft and using special brushes for collecting the signal brought from the sensors to the slip ring by appropriate wiring. Recently, the trend has shifted towards the use of frequency-modulated (FM) telemetry system on account of the high signal to noise ratio, longer life, and greater simplicity in an experimental set-up caused by the elimination of slip rings.

A typical FM telemetry system shown in the Fig. 7.18 comprises of a FM transmitter, FM receiver, and a lithium-battery (2.9 V) power source.

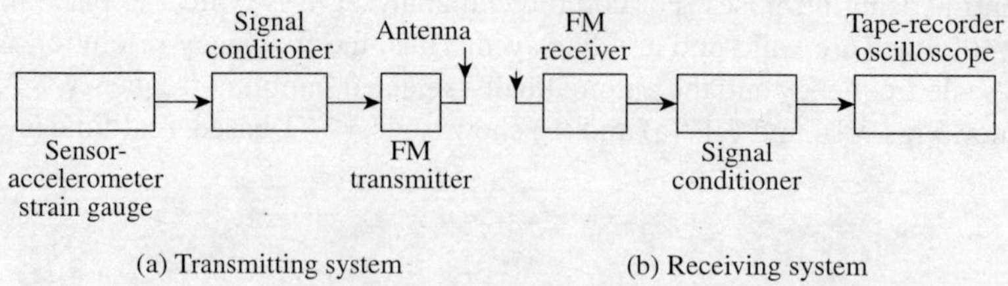

Figure 7.18 FM Telemetry

The transmitter collects the suitably-conditioned signal from the sensor, performs modulation, and transmits the data by radiation through an antenna. The receiver receives the signal from the receiving antenna and then modulates the signal to retrieve the information from it. Signal conditioners are used to adjust the gain, DC offsets, and so on. In a situation where the diameter of the shaft is small and the transmission distance is also small, wire antennae are adequate to radiate and receive radio-frequency power. For larger-shaft diameter and transmission distances, dipole antennae, matched to the transmitter-output impedance, are used.

CASE STUDIES

Case 1: The author of this book has very successfully used a telemetry system for monitoring the vibration displacement and stresses, in the moving blades of a large steam turbine and also on the impellers of a forced-draft (FD) fans, used in large thermal-power plants. These components repeatedly failed in a relatively short time of commercial operation and thus, required detailed investigations. The examination of the fracture surfaces of the steam-turbine blades as well as that of the impellers of the FD-fan under scanning, the electron microscope showed that these failures were caused by fatigue. As explained in the earlier chapters, the operation of the machine elements at resonance or near-resonance condition can result into substantial dynamic stresses (vibratory stresses). Consequently, it was found necessary to measure the natural frequency of the blades and impellers while they rotate at their rated speed and also the total stresses (static plus dynamic) in the components (steam-turbine blades in one case and FD-fan impeller in another case). It may be understood well that the natural frequencies as well as the mode shapes can be determined during the stationary condition of the machines, but the analysis as well as the experience shows that the natural frequencies do change (due to centrifugal-force effects) when the machine rotates. Also, the dynamic-stress measurement during the static test has not got much meaning unless an extensive analysis is carried out to determine the stress-amplification factor at resonance or near-resonance condition.

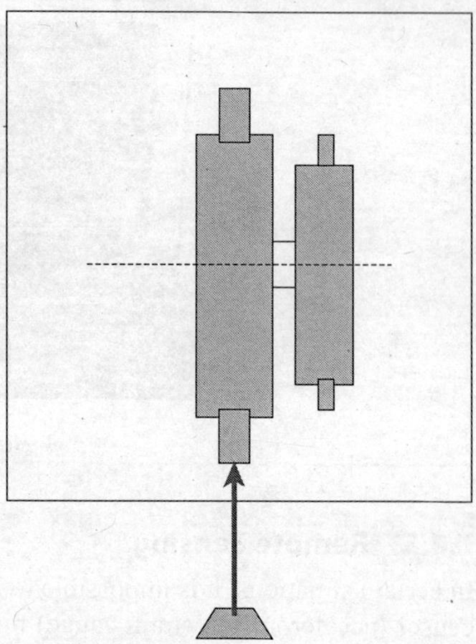

Figure 7.19 Use of Jet

During the running tests, the steam-turbine blades were appropriately strain gauged at several places on the blades starting from the root portion of the blades. Also for monitoring vibrational displacements/accelerations, accelerometers were mounted around the same location. In a similar fashion, FD-fan impeller was instrumented using strain gauges and accelerometers. For exciting the natural frequency modes, while the bladed discs of steam turbine (and also the impeller of the FD-fan) rotate, an air jet impingement was used as shown in Fig. 7.19, 7.20(a), and 7.20(b).

The jet impingement on the rotating discs is equivalent to providing an impulse periodically on the instrumented moving blades. The test results show the case of steam turbine, one of the natural frequencies, which was very close to the diaphragm or guide blade-passing frequency (no. of diaphragm/guide blades × speed (50 Hz in this case)). To eliminate this problem, the number of diaphragm blades was changed from the existing 50 to 62. The modification proved extremely successful in a sense that even after two decades of operations of the machine, no blade failures have been reported. The tests on FD-fan impellers also showed that near-resonance condition existed. The structural modification significantly altered the natural frequencies of the impeller and thus solved the problem of failures on the FD-fan impeller.

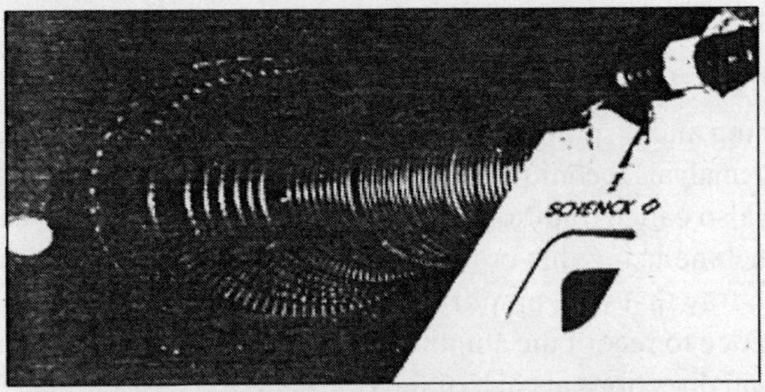

(a)

(b)

Figure 7.20 (a) Turbine Rotor Under Test (b) Turbine Rotor

We shall discuss in more detail the determination of natural frequency and the mode shapes (modal analysis) later in this chapter.

We shall now discuss a very important aspect of diagnosis of the vibration problem. They are data reduction and processing.

7.5 DATA REDUCTION AND PROCESSING

There are many ways of obtaining and processing the vibration data for detecting and identifying problems in turbomachinery including the reciprocating machinery. Important among them are as follows:

(1) Amplitude versus frequency or traditional-vibration spectrum analysis
(2) Amplitude versus frequency versus time—a three-dimensional plot called water-fall diagram
(3) Amplitude versus rpm, also called bode plot
(4) Time waveform
(5) Lissajous patterns, also called orbit plots
(6) Mode-shape analysis (modal analysis)

These techniques are discussed in the following sections with an emphasis on practical applications of diagnosing specific machinery problems / malfunctions.

7.5.1 Vibration Amplitude Versus Frequency Analysis

The procedure of obtaining and displaying the amplitude of vibration for all frequencies present is one of the most useful of all analysis techniques since majority of the practical problems can be identified through this procedure (also called traditional spectrum analysis). However, a successful identification of the problems in a machine using this or any analysis technique requires that the complete data be obtained in a systematic way that will simplify the interpretation.

It is a common practice to record the amplitude versus frequency data measured in the horizontal, vertical, and axial pick-up directions at each of the bearings of the machine being analysed. Whenever it is possible, shaft vibration must also be measured and spectrum analysed. For vertical machines, the measurements are undertaken in radial, tangential, and axial directions. Obtaining measurement in all three directions is extremely important for distinguishing between various mechanical problems. For example, unbalance, misalignment, and bent shaft will generally cause appearance of 1 – n peak (high harmonics will also be present for misaligned or bent shafts). However, unbalance will almost always produce high amplitudes in the radial (vertical / horizontal) direction and relatively lower vibration in axial direction while misalignment of coupling and bearings or a bent shaft will generally show a relatively high amplitude of vibration in the axial direction along with the radial amplitudes. The vibration engineer, as a matter of routine habit, and practice, must scan the bearing from top to bottom, right to left, and also on the joint between the bearing pedestal and the imbedded plate in the concrete on which the bearing pedestal rests—that is, in left / right, inner / outer holding down the bolts and also on the embedded plate grouted in the concrete. It is not very uncommon to see the lose mounting of the pedestal in many practical cases.

In addition to comparing the radial and axial readings, it is also very important to compare the radial-horizontal and radial-vertical readings for the horizontal machines and radial and tangential readings for vertical machines. For vertical hydro machines, the same should be measured and compared along two directions, namely, upstream / downstream side in relation to the penstock / tailrace and across. Much can be learnt about the machine from such comparisons. Horizontal-to-vertical amplitude ratios of 2:1, 3:1, 4:1, and, sometimes, even 5:1 in the horizontal machines can usually be considered normal. As abnormally-high ratio indicates possibilities of resonance, while abnormally-high vertical vibrations may be caused by wiped bearings, excessive-bearing clearance, or other sources of looseness (as pointed earlier).

The data obtained through measurements must be displayed in such a way that it facilitates the interpretation. This is illustrated in the following case study.

CASE STUDIES

Case 2: Figure 7.21 shows the vibration data obtained on a large boiler feed pump operating on a re-circulation mode. The boiler feed pumps, at times, are run in a recirculation mode, when turbine trip takes place and restart of the turbine/boiler is expected to be done shortly.

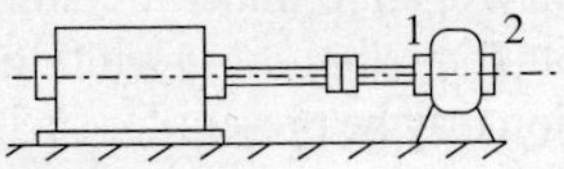

Condition	Probe location		Displacement micron PK–PK	Velocity mm/sec P–P	Acceleration m/sec^2 P–P	Frequency analysis m/sec^2 rms													
						1x	2x	3x	4x	5x	6x	7x	8x	9x	10x	11x	12x	13x	14x
Pump on recirculation discharge pressure 100 kg/cm^2 speed 66 Hz	Drive end (1)	H	45–50	50–55	130–150	-	-	-	-	-	-	43	-	-	-	-	-	-	3.31
		u	10	15	65	-	-	-	-	-	-	22.7	-	-	-	-	-	-	6.18
		^	12	45	55	-	-	-	-	-	-	5.78	-	-	-	-	-	-	7.24
	Non-drive end (2)	H	20–24	28–30	50–55	-	-	-	-	-	-	9.2	-	-	-	-	-	-	19
		u	13–15	10–12	20–22	-	-	-	-	-	-	-	-	-	-	-	-	-	-
		^	12–13	15–17	38–10	-	-	-	-	-	-	12	-	-	-	-	-	-	-

: Vibration amplitude *vs* frequency analysis of a 7-vaned pump

Figure 7.21 Vibration Analysis on Pump

The data shows very high velocity/acceleration levels in horizontal directions, both on DE and NDE bearings. Note that the displacement levels are, although high, are not alarming. In situations where the overall displacement and acceleration levels are not high, the vibration spectrum will have low-frequency peaks dominating. But in situations, as shown in Fig. 7.21, where the acceleration levels are very high when compared with the displacement levels, we must look for the presence of high-frequency components when spectra are seen in the acceleration mode. In fact, in such situation, one should always see the vibration spectrum in NDE bearing, the highest levels are at 7 × running speed.

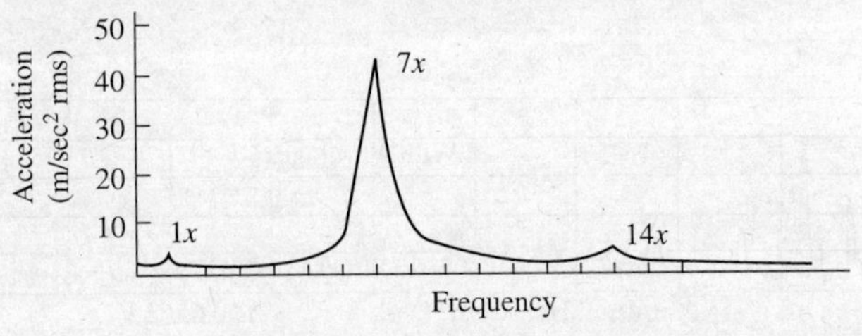

Figure 7.22 Pump-Vibration Spectrum

The pump has seven vanes in the impeller and hence, any defect in radial/axial setting of the pump or flow irregularities, will produce high vibration at blade or vane-passing frequency. Figure 7.22 shows the acceleration spectrum where one can notice hardly any significant peak at 1x, 2x, and so on, and a lone peak at vane-passing frequency. In this particular case, it was noticed that the pump-impeller setting was asymmetrical and also that radial-tip clearances had been kept too tight. In many situations, the pump-impeller stages are staggered so that the excitation at vane-passing frequency minimized (by staggering, it meant that the angular position of vanes for various stages are not the same though each stage has same number of vanes).

The case study shows that while we note down vibration details, we should also include pertinent information of the machine such as number of blades/vanes, guide blades/ vanes, number of stages, machine sketch, the operating parameters, and vibration/noise at various operating conditions. The pertinent information of the machine will vary from machine to machine—for example, it is an electrical motor we need to know information about whether the motor is AC or DC motor, rpm of motor, current drawn, number of poles, the type of bearings used, and so on.

7.5.2 Spectrum Averaging

Standard frequency analysis procedures are carried out with an assumption that the vibration being analysed is reasonably a steady state to permit the frequency range that is to be scanned with the tunable filter. Unfortunately, not all analysis situations meet these ideal requirements. Some machines may operate under a continuously varying speed, load, and temperature while some machines have random vibrations resulting from combustion, flow turbulences, for example. In other cases, the vibration may be present for such a short period of time that the frequency analysis by standard procedure is impossible. In such a situation, one may use FFT-based spectrum analyser, which provides an instantaneous and continually-updated display of the vibration amplitude versus frequency signature so that the analysis is displayed as it occurs (most of the FFTs and RTAs incorporate built-in recorders to provide hard-copy signature). If the vibration spectrum does not fluctuate or change significantly when observed on the analyser-scope display, then a single instantaneous spectrum is adequate to describe the vibration characteristics of the machine. However, when conducting an analysis, it is not unusual to find that amplitudes vary considerably with some average value. In some cases, the spectrum too may vary so much that it is impossible to accomplish the analysis with any confidence on the results. In such cases, spectrum averaging proves extremely beneficial.

Spectrum averaging in its simplest form is the process of taking a number of samples, adding them together, and dividing the sum by the number of samples. The result will approach the value that is expected if an infinitely large number of samples are taken. The modern analysers provide sampling ranging from 1 to 512. A good (and strongly recommended) way to determine the number of samples is to begin with a low number of, say, Δ and increasing them gradually, each time observing the average spectrum on the screen till no such significant difference between the succeeding averages is found as demonstrated in Fig. 7.23.

While performing the vibration amplitude-versus-frequency analysis (traditional spectrum analysis), the following important points must always be remembered:

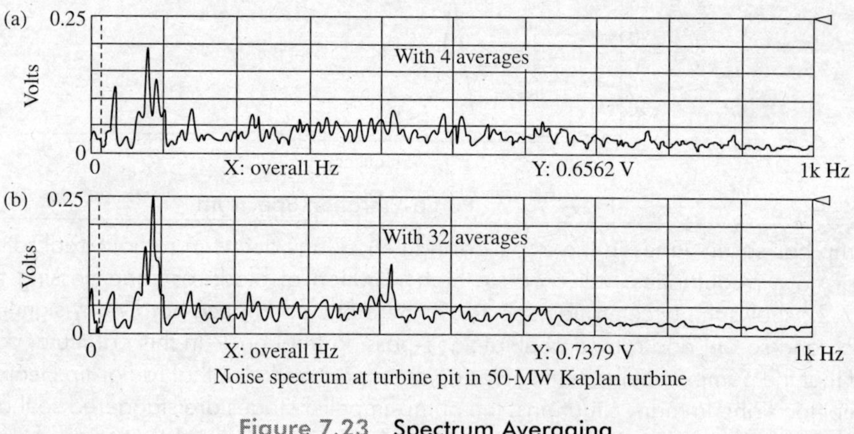

Noise spectrum at turbine pit in 50-MW Kaplan turbine

Figure 7.23 Spectrum Averaging

- See the overall vibration levels in all three modes, namely displacement, velocity, and acceleration. Also choose, approximately, one-third of the pick-up resonance frequency as the upper range. Considering the rotating frequency of the machine, find out the estimated values of velocity and acceleration from the value of overall displacement. If the acceleration values thus estimated are much smaller than the measured acceleration, then the frequency analysis range must be such that it accommodates high frequencies such as blade-passing frequency and harmonics and diaphragm-impulse frequency and harmonics (see Case studies 1 and 2 described previously and also the spectrum shown in Fig. 7.5).

- It is important to note the historical data of vibration, that is, the vibration levels and spectrum, during the run of the machine after initial commissioning, and compare these with the present levels and spectrums of vibrations. Such a comparative analysis throws sufficient light on the factors responsible too.
- In some cases of vibration problems, the vibration severity gradually increases over a period of time, stabilizes for some time, and then decreases. The cycle-vibration behaviour often repeats. The diagnosis of such problems often requires an in-depth analysis of operating parameters, spectrum analysis in a very narrow band, and, at times, in a low-frequency regime. The following two case studies will explain such situations:

CASE STUDIES

Case 3: In a large, 3000-rpm, sea shore-located power plant comprising of HP turbine and LP turbine, the sea water-cooled surface conducers is connected to the LP cylinder through a flexible joint (dog-bone joint) and the condenser is rigidly fixed to the foundation (Fig. 7.24).

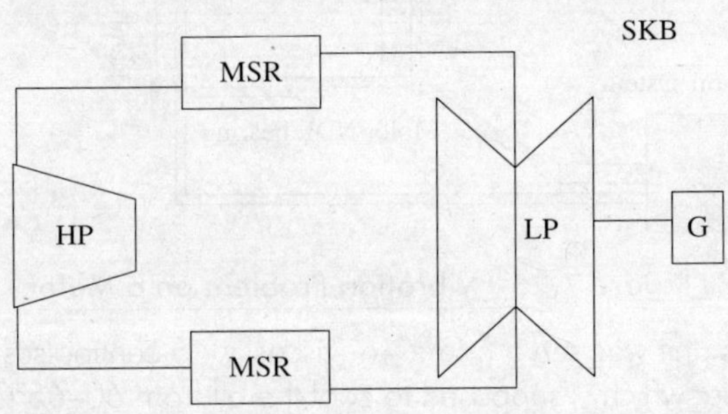

Figure 7.24 **Nuclear Steam-Turbine Generator**

In many of the designs of the steam-power plant, the joint between the LP casing and the condenser is solid and the condenser is spring mounted. However, in this case, the customer had preferred a flexible joint between the condenser and the LP turbine. In this case, any changes in the vacuum that is produced in the condenser, will cause the turbine casing to move up and down. Consequently, the bearings of the LP turbine, which are housed in the turbine casing, will also move up and down depending upon the changes in the vacuum. This movement causes change in the alignments of the HP/LP rotor and generator rotors and would cause an increase in the vibrations, especially that of LP-rear and generator-front bearing. As long as the vacuum in the condenser does not change significantly, the vibration behaviour would be steady. In this particular case, during the high-tide days, the jellyfish would enter the seawater-intake pipe and ultimately get lodged in the water boxes of the condenser causing obstruction to the flow of water in the condenser tubes. This results in the loss of vacuum in the condenser. The condenser develops the rated vacuum once these jellyfish are dislodged and drained out. This phenomenon occurred periodically causing periodic excursions in the vibration levels. The corrective action of preventing the jellyfish from entering the condenser water boxes proved to be the solution for the problem. The case study shows that the vibration engineer, in addition to vibration analysis, must carry out a systematic study of operating parameters to pinpoint the cause of the vibration problem.

Case 4: This case study pertains to the vibration problem appearing in the bearing of the drive motor (2MW, 3000 rpm, AC motor) that drives gas-oil pump in a refinery. As per API standard, the trip limit for the shaft vibration for pump-motor assembly was kept at 80 μ pk–pk. The schematic arrangement is shown in Fig. 7.25.

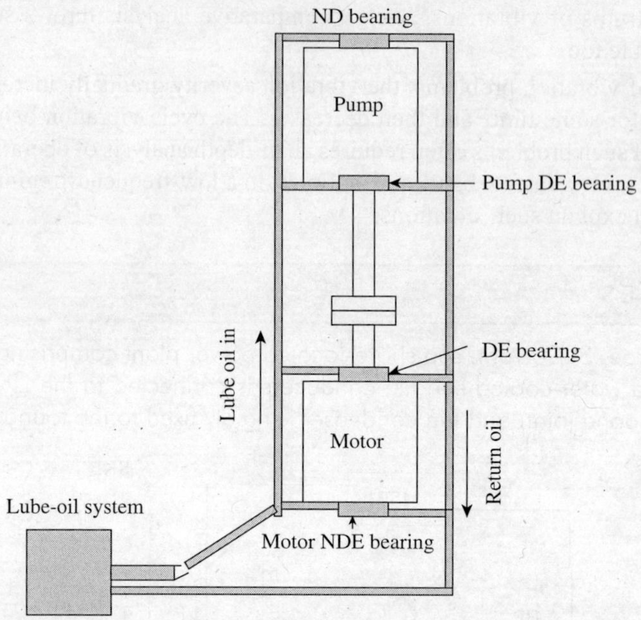

Figure 7.25 Vibration Problem on a Motor

The tube-oil wad for this unit was servo prime 46 (viscosity 46 centipoises at 40°C). The return hot oil is taken to an oil cooler which is supposed to cool the oil from 60–65°C to 40–42°C for feeding in the tube feed line from where the branches would take the tube oil to various bearings.

The unit had a peculiar problem of very high shaft vibration (at times crossing 80 μ pk–pk and even tripping the unit during the starting of the motor). In case the unit does not trip during the starting, after about 45 min to 1 hr, the shaft vibration would slowly come down to 55–60 μ pk–pk. The vibration levels then start increasing after say 4–5 hr of operation and would again start increasing. In case the maximum level reached is lesser than the trip level (80 μ), then the vibrations would gradually come down to 55–60 μ at both drive as well as the non-drive end of motor. However, during the winter months, very often the shaft-vibration levels at late night or early hours in the morning would cross the trip limits and the unit would get tripped. The tripping of the unit at late night or early hours of morning created chaos and panic in the operating staff (being very less in number during the night shift),as the tripping of the running unit called for an immediate starting of the stand by unit which was also prone to tripping during the start.

The O&M engineers with the help of engineers from condition-monitoring cell had carried out a vibration-spectrum analysis based upon which they requested the motor supplier to replace the motor. But even with a new motor, the problem continued.

The problem was subsequently referred to the authors of the book for an in-depth analysis to suggest a permanent solution.

The visit to the site brought out the following important observations:

(1) The construction of lube-oil feed-line feeding lube oil to bearing was defective. The pipe after cooler had a 90° bend very near the branch lines, which fed oil to the motor bearings.

(2) Lube-oil line was not lagged with insulation and at the end where branches for pump bearings were fallen, the lube-oil feed line was cold showing that cold oil was being fed.
(3) The oil temperature leaving the cooler unit was much lower than the recommended 40–45 °C temperature.
(4) Customers' engineers had taken the vibration spectra in the frequency range of 0–5000 Hz due to which the low-frequency components never appeared in the spectra they collected.
(5) The bearings of the motor were designed for oil of the grade servo prime 32 (viscosity 32 centipoises at 40°C).

The vibration-spectrum analysis in the range of 0–250 Hz showed many sub-harmonics such as 1/5, 1/3 and 1x, 2x, of the running speed. It was also observed that pump-motor shaft in the axial direction was not satisfactorily anchored at the thrust bearing located in the pump. This was clearly seen in the axial-shift display.

The analysis had shown that the rotor-bearing system was having a very slow-speed instability especially due to malfunctions of the thrust bearing in the correct positioning of pump-thrust bearing and axial misalignment of the pump rotor. One of the strongest causative factors for slow-speed instability was higher viscosity of the oil reaching the bearing due to

(a) Higher viscosity of servo-prime 46 oil.
(b) Oil temperature at the inlet of bearing being lower due to the malfunctioning of oil cooler.
(c) Lack of thermal insulation on lube-oil feed line.

The other contributing factor would be wrong axial positioning of the pump but that could not be established. Changing the lube oil from servo46 to servo32, providing insulation on the feed line, removal of sharp bends in the lube-feed line did result in the total elimination of the problem of a gradual built-up vibration.

It may be noted that the phenomenon of slow instability of the rotor-bearing system could not come to light unless the frequency range in the spectrum analysis had been done in a smaller frequency range (0–250 Hz).

7.5.3 Amplitude versus Frequency versus Time Analysis

A single signature can reveal only the characteristics of vibration at a single instant in time with the machine operating at a particular speed and under a specific load condition. However, sometimes, it may be very important to note the amplitude versus frequency at various operating conditions (say operating speed for a variable-speed machine), to obtain information about the resonant conditions or critical speeds being excited by various forcing frequencies generated by the machine components, or as explained in the previous case study, the vibration amplitude and frequency characteristics during generation and growth of instability, and so on. The high-speed analysis capability of a real-time analyser is ideally suited for these requirements for the amplitude versus frequency versus time data. For this purpose, the real-time analyser is used in conjunction with a high-speed recorder to provide a chronological signature at a rate, say one every second.

Figure 7.26 illustrates one such data obtained during the start-up of a machine. Plot such as this is called the "water-fall" diagram.

One can easily see the occurrence of an oil whirl during the start-up. The resonant/critical speeds excited by the rotor unbalance can also be seen. The water-fall diagrams are also extremely useful for evaluating the effect of change in the load or other separating parameters of the machine, as the effect of change on each frequency can be detected. The occurrence of new or additional vibration frequencies or the disappearance of existing frequencies can also be detected. For the type of problems described in the previous case study, the water-fall diagram would have proved very useful.

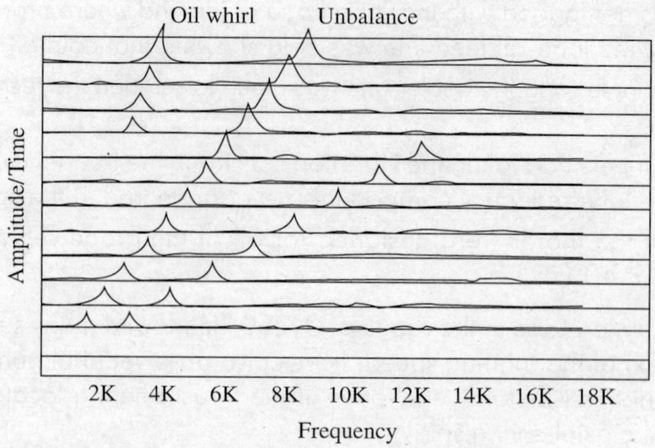

Figure 7.26 Water-fall Diagram

7.5.4 Amplitude / Phase versus rpm Analysis

The machines and their supporting structures have resonant frequencies at which very high amplitudes of vibrations can result even from a relatively small exciting force. Since the machine and their structures are generally complex systems consisting of many elastic (springs) components and masses, a large number of resonant frequencies exist and this resonance is relatively a common problem.

Much information can be had about the response of a machine to the forces that cause vibration by obtaining plots of vibration amplitude and phase as a function of rpm. A typical plot for a utility TG set is shown in Fig. 7.27. Such a plot, sometimes called "bode" plot, clearly identifies the resonant frequencies by the characteristic peak amplitude and (more importantly) corresponding 180° phase shift. From the sample plot in Fig. 7.27 it is apparent that the machine in operation has two resonant frequencies—one approximately at 1300 rpm and the other at 2800 rpm. Should the machine, in its normal running experience an excitation force which corresponds to either of these frequencies (the excitation could be due to unbalance, misalignment, hydraulic / aerodynamic perturbation, torque pulse), severe vibrations would be experienced. The normal exciting frequencies of vibrations inherent to the machine can be found out by the amplitude-versus-frequency analysis of vibration as described in the previous section.

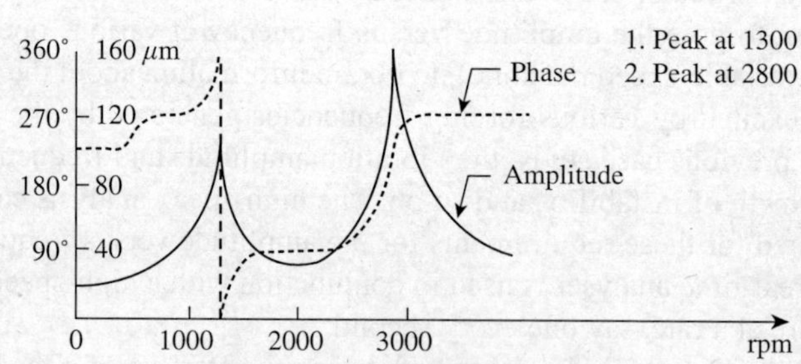

Amplitude/Phase *vs* rpm analysis
(Bode plot) of turbogenerator

Figure 7.27 Bode Plot

The resonant frequencies can also be found out by observing amplitudes of vibrations at various points of time during the run-down of the machine. This, of course, has to be supplemented by the data of speed versus time during the run-down so that amplitude versus speed data can be generated. This method, although correct in principle, is quite liable to many errors. For a truly accurate and complete amplitude/phase-versus-rpm data, such as shown in Fig. 7.27, a vibration analyser with a tracking filter is used. The instrument utilizes a reference pick-up at the shaft of the machine to provide a voltage pulse in each revolution. The 1 × n signal from reference pick-up also controls the reference signal which is actually a DC voltage proportional to shaft rpm. This voltage is utilized to drive the X-axis of the X–Y or XY_1–Y_2 recorder for obtaining the graphic plots of amplitude and/or phase versus rpm. The reference signal provides a fixed reference for comparison with signal from a vibration pick-up, resulting in a DC voltage proportional to the relative phase between the signals. This DC voltage is available to drive the Y-axis of an X–Y or XY_1–Y_2 recorder for plotting the phase versus rpm (phase versus amplitude).

The passage of a machine through resonance during the run-down/run-up (coast-down/coast-up) is accompanied with a peak in the amplitude of vibrations and a characteristic phase shift of approximately 180° (for a very lightly damped system and 90° when the system has damping through the structural damping or damping provided by the oil film in the bearings). However, at times, the signal could be quite confusing as far as amplitudes are concerned. This is illustrated in Fig. 7.28.

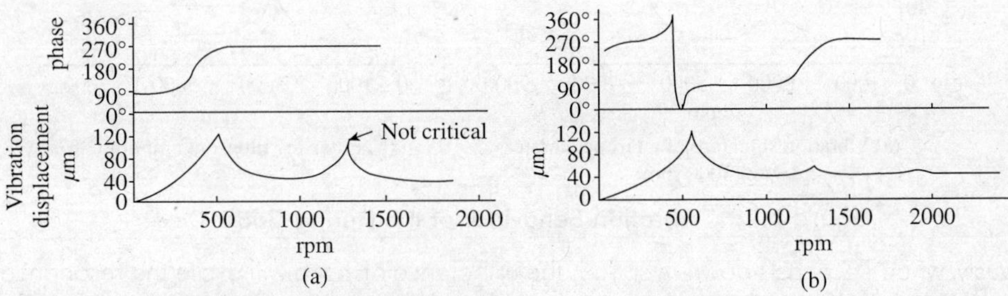

(a) Peak at 500 rpm with + 180° phase shift showing no resonance at 1200 rpm

(b) Resonance at 600 and also at 1300 rpm

Figure 7.28 Run-Down Signature Analysis

Figure 7.28(a) shows two amplitude peaks at 500 rpm as well as at 1200 rpm. However, at 1200 rpm, a phase shift of 180° (approximately) does not exist so that 1200 rpm is not a resonant or a critical speed. Fig. 7.28(b) shows a clear resonant peak at 600 rpm (resonant peak and critical speed peak are the same) and also a 180° phase shift at 1300 rpm but no amplitude peak. The 1300 rpm is also a resonant frequency (critical speed) though amplitude levels are not peaking up at this speed because the pick-up has been, perhaps, placed at a nodal point (in the shaft displacement/deflective curve). It may, therefore, be necessary to alter the pick-up position. The absence of peak at 1300 rpm may also be due to a very low level of excitation force or also due to heavy damping.

At times, the amplitude and phase-versus-rpm data may show an amplitude peak accompanied with a phase shift of 360° instead of 180°. Such a situation arises in the case of two systems in resonance or near the same frequency, for example, a tuned vibration-absorber system would show such behaviour.

The interpretation of amplitude and phase-versus-rpm plot obtained using non-contact (shaft) pick-ups can be extremely confusing, at times, as the non-contact or proximity probe cannot distinguish between actual-shaft vibration, eccentricity (out of roundness) of the journal (journal is the portion of shaft/rotor in the bearing zone) and apparent vibrations caused by the magnetic unevenness of the shaft surface (called run-out or glitch). As a result, the signal from non-contact pick-up would consist of a vector sum of run-out and actual-shaft vibration. It is, therefore, advisable to carefully measure the run-out amplitude (in a slow roll of the rotor) before response measurements are taken or employ instrumented run-out subtraction.

CASE STUDIES

Case 5: It is extremely important to note and remember that in each of the example discussed in the previous paragraphs, the recorded vibration amplitude occurred at the rotating speed frequency (1 × n) of the machine. However, these plots alone may not reveal the system response to other vibration frequencies which can affect the overall machine performance from the vibration point of view. To illustrate this, consider the amplitude-versus-frequency signature of a typical TG set operating at 3000 rpm as shown in Fig. 7.29(a). In this instance, the TG has two significant exciting frequencies, that is, 1 × n due to, perhaps, unbalance and another at 2 × n (6000 rpm), probably due to misalignment. In this typical case study, the TG rotor has critical speeds (resonant frequencies) at 1300 and 2850 rpm.

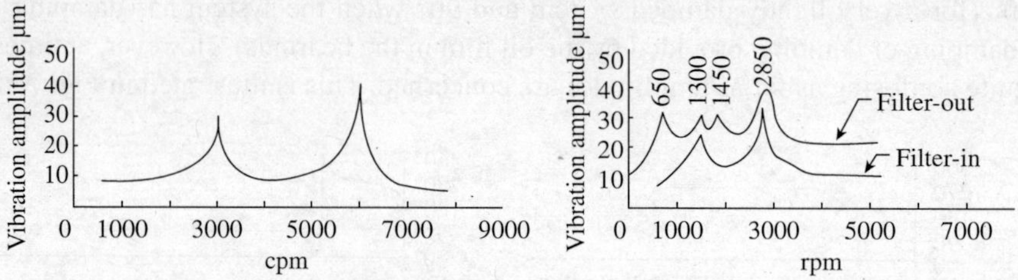

(a) Vibration signature of a turbogenerator set (3000 rpm)

(b) Filter-in / filter-out amplitude-rpm

Figure 7.29 Vibration Behaviour of the During Coast-Down

Obviously, when TG coasts down in speed, the unbalance at 1 × n will excite the resonance when the speed falls down from 3000 rpm (at the instant machine is tripped) to 2850 rpm. In addition, as the TG continues to run down, the resonance will once again get excited when the speed reaches 1425 rpm on account of 2 × n excitation caused by misalignment (of connected rotors/bearings). This excited resonance will not appear in the bode plot at 1425 rpm because the analyser filter is tuned to 1 × n and rejects all other frequencies. Because the filter-in amplitude versus rpm plot (a part of bode plot) does not always give a complete picture of the total system response, it is customary to obtain tow plots of amplitude versus rpm, one filter in, synchronized to 1 × n (conventional bode plot) and second (filter-out) overall amplitude-versus-rpm plot as shown in Fig. 7.29(b). These plots are of great value in ascertaining the nature of distress in the machine.

It is, from this point of view, very important to identify the critical speeds, nature of balance, and alignment condition of the machine at the time of commissioning of the machine through bode plots and also filter-in/filter-out vibration-versus-rpm plots as shown in Fig. 7.29(a) and b. These plots can be used as reference plots for the purpose of condition monitoring of the machine. Any departure from the reference plot indicates a deteriorated condition of the machine and hence, the operator can carry out a preventive maintenance to avoid catastrophic failures. We shall, in the forthcoming sections, discuss a case study where the catastrophic failure of a large steam turbine—generator could be avoided by comparing the bode plot and filter-in/filter-out versus rpm plots of the operating machine with reference plots.

7.5.5 Time Waveform Analysis

Although vibration-spectrum analysis is generally adequate for identifying most of the machinery problems, sometimes an additional information is needed to diagnose a particular defect or to study the dynamic behaviour of a machine under a specific operating condition. One of the additional techniques often used is the observation of the time waveform of the vibration signal of oscilloscope. The oscilloscope observations facilitate an identification of spiky nature (if any) of the signal, beats of the signal, and also enables the confirmation as to whether or not the signal is constituted by harmonically-related frequencies. We shall show an application of this analysis procedure through a few case studies later in this chapter.

7.5.6 Lissajous Pattern (Orbit) Analysis

It has been mentioned in the Section 7.2.3 relating to shaft vibrations that the signal output from two mutually perpendicularly-installed proximity probes around the shaft can be fed to the horizontal and vertical inputs of the oscilloscope to depict the dynamic motion of the journal centre (many modern RTAs do provide this facility). The plots or displays, thus obtained are called Lissajous patterns and are also called as orbits. When observing shaft orbits on an oscilloscope, it is difficult to obtain the frequency information from the display unless some kind of frequency reference such as a synchronous $1 \times n$ pulse is superimposed on the display. The synchronous pulse is obtained by installing an electromagnetic pick-up or a non-contact pick-up observing a protrusion or depression (key or key way) on the shaft. The synchronous reference pulse thus obtained can be applied to the Z-axis (intensity) input of the scope. There are many mechanical problems, which are readily identified by the characteristic (typical) patterns of the shaft orbits. These will be discussed later in this chapter.

7.5.7 Mode-shape Analysis

The mode-shape analysis technique is extremely useful for confirming resonance conditions, identifying nodal (zero displacement) and anti-nodal (maximum displacement) points, and revealing sources of structural weakness. In this, one determines the vibratory mode shape of the vibrating component, by making the measurement of vibration amplitude and phase at various points on the component and by plotting them to obtain the mode shape. For example, Fig. 7.30 shows the mode shape of vibrations of vertical pump structures which had very high vibrations.

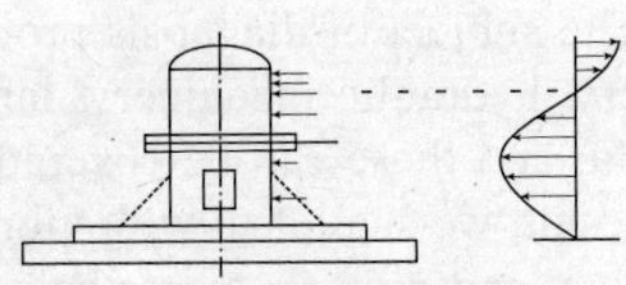

Mode shape of vibrations of a vertical pump

Figure 7.30 Mode-shape Analysis

The mode shape clearly indicates that the structure is vibrating at the second flexural resonance. Not only has the resonance condition been confirmed but the nodal points too have been identified. This is very important especially when a decision to stiffen the structure is reached as the mode-shape analysis will enable an avoiding fixation of stiffening members of the nodal point. If the stiffening members were added the structure at nodal points, little or practically no improvement might be seen. As a matter of fact, the user of this pump had attempted stiffening shown by a thin line of the structure without carrying out an appropriate mode-shape analysis and as expected failed to get any improvement. The stiffener brought down the vibration levels on the pump.

The mode shapes shown in Fig. 7.31 illustrate an identification of sources of weakness of a bearing pedestal. Fig. 7.31(a) shows low levels of pedestal vibrations and high levels of vibration in the bearing block. Any attempt to strengthen the support would be obviously fruitless. The solution lies in

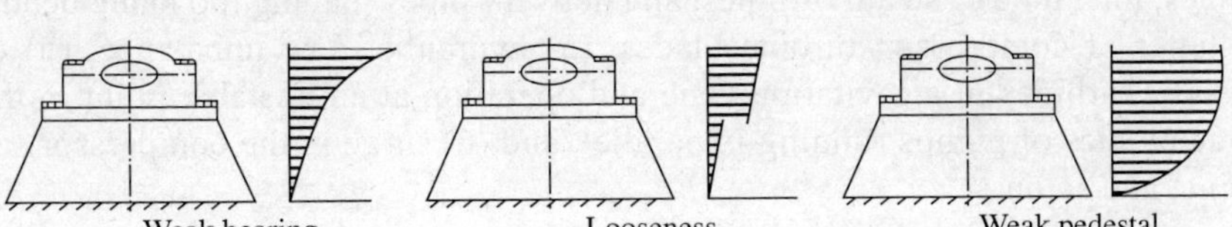

Mode shapes for (a) Weak bearing (b) Looseness (c) Weak pedestal

Figure 7.31 Mode Shapes

strengthening the bearing. Figure 7.31(b) shows that vibrations on the bearing block are high but do not increase significantly with elevation and also that the bearing block is moving back and forth on the top of the pedestal, which is indicative of looseness of the holding-down bolts. Fig. 7.31(c) clearly indicates that the vibrations are due to the weakness of the pedestal. Stiffening of the pedestal in such a case would be highly beneficial.

In short, the mode-shape analysis can be extremely valuable for identifying structural weakness, resonance, the nodal and anti-nodal points before structural modifications are made in an attempt to solve a problem of excessive vibration. This will ensure avoiding of costly and many a time embarrassing trial-and-error approaches for solving the problems.

As mentioned repeatedly in this book, the solution of vibration problem requires an elimination of perturbation forces and/or modifying the response of the system by carrying out the structural modification of the system. Many a time it is impossible to totally eliminate the perturbation forces although they can be brought down by resorting to very accurate assembly, balancing, and so on—these, at times, are extremely expensive.

7.6 DIAGNOSIS AND CORRECTIVE ACTIONS

The subject of diagnosis problems is relevant to a vast variety of machines such as turbomachinery, hydraulic machinery, large electric generators/motors, machine tools, automobiles, and so on; and these can be covered only through a separate book that too to a certain level only. As a result, we shall deal with vibration analysis of turbomachinery, hydraulic machinery, large generators, and motors. Nevertheless, the concepts brought out in this chapter can be applied on other machines/systems.

The perturbation forces responsible for the vibration of turbomachinery, hydraulic machinery, and so on, can be classified into two main categories, which are (1) periodic in nature under steady-state regime and (2) aperiodic during transitions (unsteady) regime of operation. The perturbation forces for each of the regime, that is, steady and unsteady, can further be subdivided into mechanical, hydraulic/aerodynamic, and electrical. The basic perturbation forces experienced in thermal or hydro power plant under each of these heads are listed as follows.

7.6.1 Steady-state Operating Regime

(i) Mechanical perturbation forces are (1) centrifugal forces due to unbalance of rotating parts, (2) elastic forces of shaft which appear when the centre line is disturbed or distorted, (3) frictional forces, and (4) forces caused by incorrect setting of bearings and unfavourable operating conditions of bearings.

(ii) Hydraulic/Aerodynamic perturbation forces appear due to following reasons: (1) presence of vortex in the spiral casing, wicket gates, runner, and draft tube in hydraulic turbines, (2) non-uniform velocity and flow distribution in the flow passages in the pump casing/rotor, steam-turbine guide/moving blades, (3) pressure pulsation in penstocks of hydraulic turbines, inlet pipe of steam turbines, and delivery pipes (having too many bends) in pumps, (4) flutter of compressor/turbine blades, (5) hydraulic/flow unbalance, (5) operation of hydraulic turbines in a cavitation zone and operation at an unstable point in the combined characteristics of pumps running in parallel, and (6) surge in the compressor section of gas turbine, and so on.

(iii) Electrical perturbation forces: (1) unbalanced magnetic pull experienced by generator rotor, (2) non-uniform flux distribution due to wide variation of air gap in motors and generators, (3) shorting of poles, (4) stator-winding failure, (5) rotor-winding failure, and (6) asymmetrical operation.

The mentioned factors are a list of only typical and few of the perturbation forces experienced by turbomachinery / hydraulic machinery / electrical motors and generators. A vibration engineer, investigating a vibration problem in any machine must spend some time in identifying the perturbation force the machine is likely to experience. It must be well appreciated that solving a vibration problem is not mere balancing and alignment of the rotors but involves quite an involved procedure and analysis.

The perturbation forces in the unsteady state-operating regime usually appear during starting/ stopping of the unit, load changes, synchronization of the generator with grid, and so on.

7.6.2 Detection of Perturbation Forces and Corrective Actions

(A) Mechanical Unbalance

The components of rotating machinery which are vulnerable to unbalance are the turbine rotor, generator rotor, pump / compressor rotor, and motor rotor. The unbalance in these components may result from (a) defects / errors in construction / manufacturing, insufficient rigidity of the shaft causing an in-operation bend, unequal weights of rotor blades, improper wedging of poles in hydrogenerator, (b) errors in assembly such as skewed setting of labyrinth seals, and (c) non-uniform wear of blades and impellors due to cavitations / erosion, and so on.

The presence of unbalance of the rotating components / system can primarily be judged by high levels of vibration at rotating-speed frequency in the vibration signature obtained from pick-ups mounted on bearing housing and more preferably obtained from shaft pick-ups. As mentioned previously, the measurement on bearing housing should be done for vertical machines in three directions, namely, radial, transverse (tangential), and axial directions; and for horizontal machines in radial horizontal, radial vertical, and axial directions. Normally, the largest amplitude (both in overall and 1 × n component) would occur in the radial direction. However, in case of overhung rotors, considerably high levels would be experienced in the axial direction also. Similarly, a rotor mounted between bearings and having a substantial couple unbalance would also show a considerable vibration in the axial direction.

Judgement of the nature of unbalance, that is, static, couple, quasi-static, or dynamic unbalance (Fig. 7.32–7.35), can be done by the usual signature analysis coupled with phase measurements as follows:

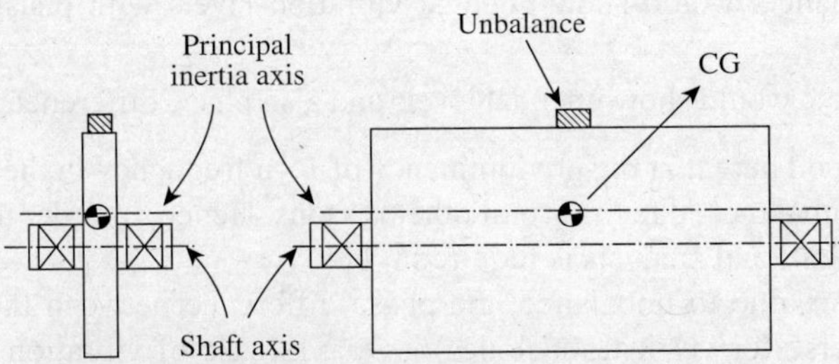

Figure 7.32 Static Unbalance

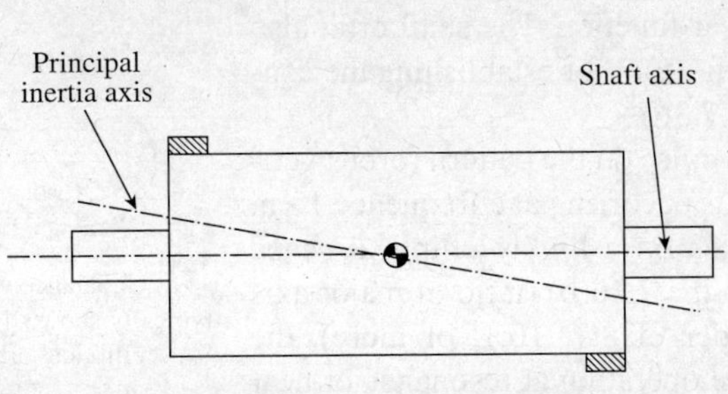

Figure 7.33 Couple Unbalance

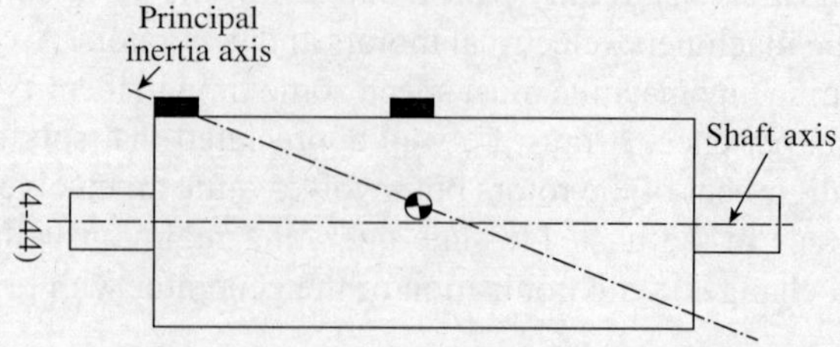

Figure 7.34 Quasi-static Unbalance

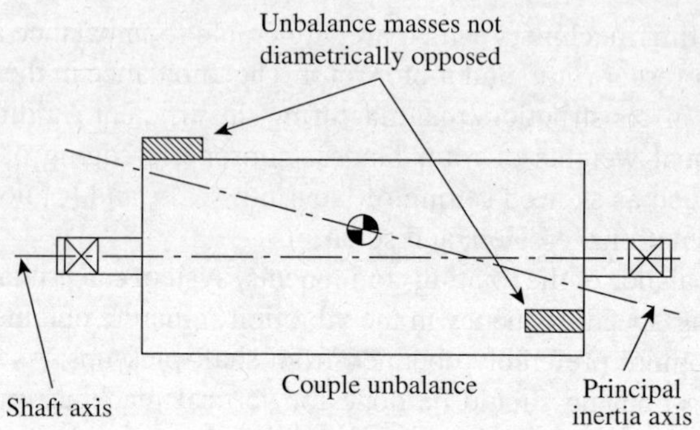

Figure 7.35 Couple Unbalance

- Static unbalance: would indicate almost identical vibration levels at both bearings and also the same phase angle.
- Couple unbalance: would indicate almost identical vibration levels at both bearings but the phase angles would differ by 180°.
- Quasi-static unbalance: would show unequal vibration levels with phases either equal or 180° opposite.
- Dynamic unbalance: would show unequal levels and also phase difference other than zero or 180°.

It may be well understood here that the predominance of $1 \times n$ frequency in the vibration signature can be, in addition to the unbalance, due to several other reasons. Hence, in order to establish that the $1 \times n$ is due to unbalance, additional analysis is required.

In case of vibration due to unbalance, the phase difference between the horizontal radial and vertical radial (transverse for vertical machines) is 90°. In case of vibration problem arising out of resonance, this phase difference would either be zero or 180°. Another way to establish the unbalance is to watch an increase in the $1 \times n$ component as a function of square of rotating speed and confirm whether or not it is a linear function. The shaft orbit also proves to be a very effective tool for establishing the condition of unbalance (Fig. 7.36).

The single reference pulse on the pattern (orbit) verifies that the shaft motion is occurring at a frequency $1 \times n$. If the pattern (orbit) takes on a highly elliptical shape such as the one shown in the 7.36(b) (ratio of major axis to minor axis of the order of 8:1, 10;1, or more), the machine may, perhaps, be operating at resonance or near resonance. As mentioned earlier, this could be easily confirmed by noting the effect of the change of speed.

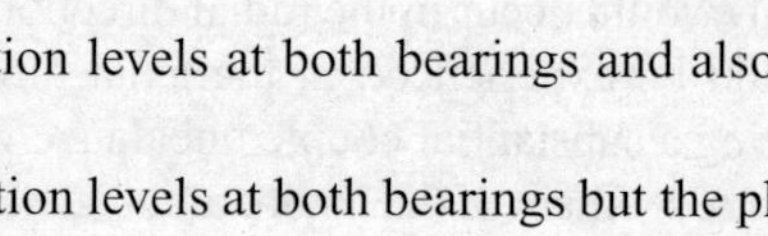

(a) Unbalance will typically show slightly elliptical orbit

(b) Highly elliptical pattern may indicate misalignment, worn-out bearing, or resonance

Figure 7.36 Judging Unbalance

The existence of unbalance (especially dynamic unbalance) in the presence of hydraulic/aerodynamic perturbation forces may lead to situations wherein it may be almost impossible to balance the rotor of the machine and it takes a large number of balancing trials before a satisfactory balance is achieved. Barring such situations, the unbalance can be corrected by resorting to single plane or two (or more) planes balancing depending upon the nature of unbalance.

Rotating electrical machines such as turbogenerator and hydrogenerator motors may give symptoms of electrical/magnetic unbalance in case of large variation in the air gap between the stator/rotor. In many situations, where it is not possible to bring down any variation in the air gap by adjustments at site, the machines are balanced in an excited condition where upon in the de-exited condition they run very rough.

CASE STUDIES

Case 6: Large TG rotors provide ventilating ducts in the rotor construction. The fans attached to the rotor discharge-cooling medium through these ventilation slots so that rotor windings which become hot due to passage of electrical currents, are kept cool up to a certain temperature. In one case of a large 200-MW TGs due to manufacturing defects of loose wedges over the rotor windings these cooling slots got blocked. As a consequence, when the TG was loaded to its full-rated load, the vibration levels (which were well within limits when the TG was rolled, excited, and not fully loaded) became excessive. To establish that the increase in vibration levels was because of a thermal bow, the excitation current was reduced in stages—in each stage of reduction of excitation the current reduction in vibration levels was noticed. The rotor then was sent back to manufacturing works for clearing the ventilation canals in the rotor and putting appropriately tight wedges and shop balancing. The repaired rotor subsequently never showed any signs of mechanical or thermal unbalance.

Similar 1 × n-type vibration problems can be experienced on a gas turbine when some of the combustors (spaced around) fail to fire. The resulting thermal unbalance will appear exactly similar to mechanical unbalance. The mechanical balancing trials in such cases invariably fail to reduce the vibration.

One of the gravest problems encountered on rotating machines (such as hydrogenerators, steam turbines, TG, motors) is 1 × n-frequency vibration problem caused by loose components such as blades, poles, and rotor windings. Mere spectrum analysis would indicate symptoms of unbalance yet the rotor fails to respond to balancing trials—in fact, due to loose components, the phase angle would never remain stationary and hence, positions of balance weights cannot be determined.

Case 7: In a 2-pole, 3000-rpm AC motor meant for driving a pump, the symptoms of looseness never appeared till the motor reached the full speed. It may be understood that the components, which are shrink fit, tend to loosen the grip as the speed increases since the centrifugal force tends to decrease the level of shrink fit. In this case, the looseness was detected by observing a non-stationary phase, the appearance of 1 × n components along with 2nd, 3rd, and 4th harmonic of running speed. Consequently, the motor–rotor assembly had to be redone ensuring an adequate shrink fit.

Case 8: In one case of 120-MW steam TG set, after an annual overhaul, it exhibited severe vibration (1 × n frequency) problem during recommissioning trials. The vibrations started appearing after crossing the speed of 1500 rpm; and afterwards, it was not possible to sustain the machine at 3000 rpm. The m/c prior to overhauling had demonstrated extremely smooth behaviour from the vibration point of view and steam-turbine rotors during the overhauling had not shown any deposits on either diaphragm (guide) blades or on the running blades. Nevertheless, the site people tried *in situ* balancing of the machine. Even 30–40 balancing runs proved to be futile. The investigation which comprised of making the measurement of turbine casing in transverse and axial directions, movements of the connected steam piping, and vibration-spectrum analysis along with phase measurements. It was then discovered that the connected piping (when dismantled from casing connection) was not given a cold pull. As a result, when m/c was rolled, with an increase in the steam parameters (pressure, flow, and temperature), the piping was exerting force on the valve connected to the HP turbine. This force caused the transverse shifting of the casing and disturbed the setting of the turbine. After restoring the cold pull, the machine did not create any 1 × n-rpm vibration problem.

To sum up, it can be said that 1 × n-vibration problem can be caused by many other reasons than only unbalance; and therefore, 1 × n-vibration problem at times, may become very complex to resolve unless one looks at the problem in its totality.

(B) Elastic Forces due to Loss of Centring or Distortion of Shaft Centre Line

This class of problems is popularly described as the misalignment problems. In horizontal machines, they arise due to the offset of coupling (parallel misalignment) and/or angularity between the axes of coupled shafts (angular misalignment) caused by non-perpendicularity of shaft axis with reference to the plane of flange coupling. In heavy-rotating machinery such as steam TG, the static-deflection curve of the connected shaft assembly because of heavy mass of rotors, also known as "shaft catenary", calls for appropriate locations of bearings (called bearing catenary) such that at flange couplings neither parallel nor angular misalignment takes place (Fig. 7.37).

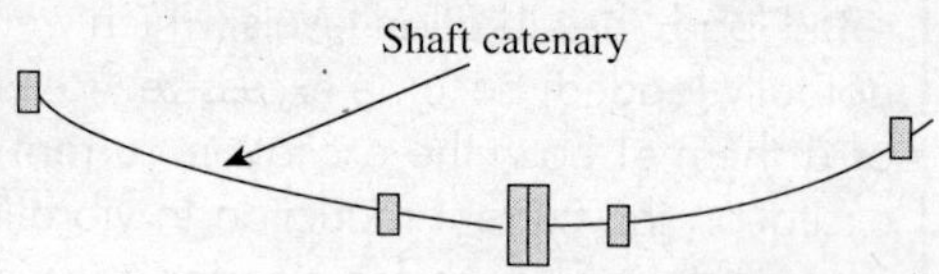

Figure 7.37 Shaft Catenary

In heavy vertical machines such as hydro units, see Fig. 7.38, the distortion of the shaft-centre line and loss of centring could result from (1) loss of perpendicularity between shaft axis and plane of

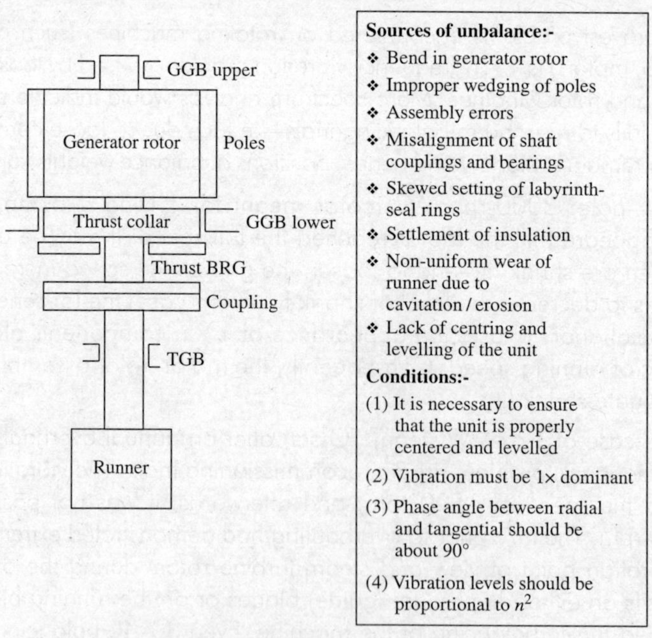

Figure 7.38 Schematic of Hydro Unit

flange coupling giving rise to angular misalignment, (2) plane of the supporting boss of the vertical journal and shaft axis not perpendicular to each other, (3) parallel misalignment between generator and turbine shafts, (4) guide bearings not coaxial with shaft, (5) thrust collar not perpendicular to shaft axis, and (6)thrust disc surface not true.

The misalignment, especially angular misalignment, gives rise to forces and moments; and hence, the vibrations are both in radial and axial directions.

Experience shows that whenever the axial vibration is equal or greater than 50 per cent of the highest radial vibrations, misalignment is likely to be one of the major sources of perturbation forces and vibrations. As indicated in Fig. 7.39(a), the angular misalignment primarily subjects the shaft to axial vibrations at the same frequency as shaft rpm (1 × n).

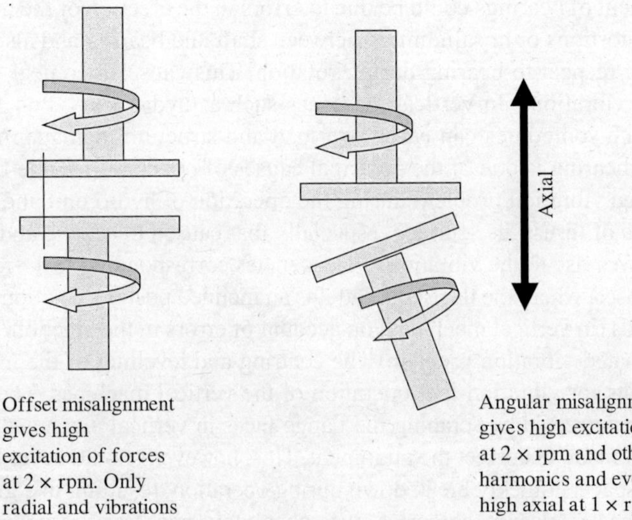

Figure 7.39 Misalignment of Rotors

Parallel or offset misalignment as illustrated in Fig. 7.39(b) produces primarily a radial vibration at twice the rotational frequency of the shaft. It may be appreciated. Fig. 7.39(b) is a highly idealized model using a single-pin connection across the coupling.

In actual practice, multiple-pin connection could produce a highly complex vibration pattern in radial and axial directions and consequently, higher frequencies would be generated in the vibration signature. It is quite usual to see higher harmonies such as 3 × n, 4 × n, and so on, apart from 2 × n components (and also 1 × n component) in the vibration signature associated with the misalignment.

Fig. 7.39 shows that the radial vibration resulting from misalignment will occur predominantly in the direction of misalignment. Thus, if the coupling halves are offset vertically (in horizontal machines), the predominant radial vibration is likely to occur in the vertical direction. As a result, the radial vibrations of misaligned couplings will often be somewhat directional. As a consequence, the shaft orbit may not reveal any circular or slightly elliptical pattern that is characteristic of unbalance but instead may reveal a pattern shaped like elongated ellipse or banana. Sometimes, pattern of figure eight or still more complicated patterns may develop in case of involvement of higher-order frequencies. This is illustrated in Fig. 7.40.

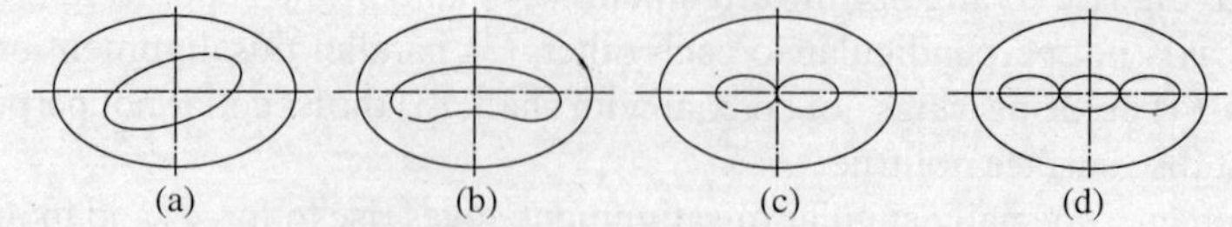

(a) Highly elliptical pattern may indicate heavy misalignment in (b), (c), and (d)

Figure 7.40 Shaft Orbits in Misalignment

In some situations, misalignment does not involve couplings but bearings. In such cases, no vibration problem would be encountered if the state of balance is excellent. With unbalance in the system, the vibrations would be experienced both in radial as well as axial directions.

The misalignment of bearings could be due to errors in the erection of the machine or, sometimes, due to structural distortions or misalignment between shaft and bearing and also gives rise to an inclination of shaft with respect to bearing during rotation. This causes a conical oil film in the bearing giving rise to axial vibrations. In vertical machines such as hydro units, this causes radial thrust on guide bearings which sometimes can cause overload and structural damage. Misalignment of vertical shaft and guide bearing is one of the principal causes of opening of guide-bearing clearances and consequent increased vibration problem during the operation of hydro unit and large vertical pumps.

The unevenness of thrust-disc surface, especially that caused by elastic and thermal deformations during operation, gives rise to the vibration at frequencies corresponding to $1 \times n$, and k xn (k= number of thrust pads). In cases, where the thrust disc attains an inclined positive position due to disturbed levelling of the thrust pads (in vertical machines) on account of errors in the erection/assembly, the machine exhibits $1 \times n$-dominated vibration problem. The centring and levelling of the machines are, therefore, absolute requirements for vibration-free operation of the vertical machines. The misalignment at coupling flanges can be corrected by scraping the flange faces in vertical machines. In some cases, use of spacer/shims is also made to correct the alignment. This, however, should never be considered as a permanent solution as spacers quickly break down during operation, resulting in a greater play of the shaft.

In horizontal, heavy turbomachineries, it is absolutely necessary to ensure parallelity between thrust disc and thrust pads during the setting of the shaft and bearing catenaries.

(C) Frictional Forces and Oil Whirl

Both of these phenomena cause unstable vibrations of rotor/bearing systems. Considering the instabilities associated with these phenomena, it is always necessary to give immediate attention and corrective action. Although both of them relate to rotor instability, their mechanisms are entirely different.

Oil whirl is a problem associated with plain-journal bearing. The shaft operating at an eccentric position from bearing centre (Fig. 7.41) draws oil into a wedge to produce a pressure load carrying film. The oil film is made up of molecules and these molecules adjacent to shaft stick to the shaft rotating at shaft speed. These molecules adjacent to the bearing tend to adhere to the bearing, which is stationary. As a result, the oil film between shaft and bearing will be in shear and tend to rotate at a speed that is average of shaft speed and bearing rpm which is zero. The average rotating speed of the film of lubrication between the shaft and bearing is approximately one-half of the shaft rpm and if friction losses are taken into account, the average speed of the oil film will be slightly less (42–45 per cent) than one-half of shaft rpm. Figure 7.41 shows pressure distributions in the oil film with resultant force R. The force R has two components, R_V—vertical component and R_H—horizontal component. In the normal bearings, R_V balances the load W of the rotor while in cases where whirling takes place, R_V is greater than W. The balance (R_V – W) and R_H tend to push the shaft so that it rides up in the clearance space between the shaft and the bearing with a frequency of rotation of oil film, that is,

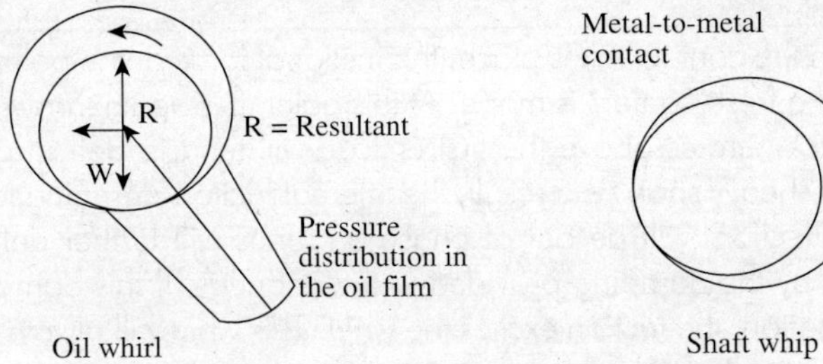

Figure 7.41 Hydrodynamic Bearing

approximately 50 per cent of shaft rpm. Thus, the shaft is simply forced to whirl within the bearing in a manner similar to a boat being pushed by a water wave. The whirling shaft thus wanders around the clearance space between the bearing and shaft and will never have an equilibrium position of the journal and thus will have an instability. Obviously, the phenomenon is due to higher (than required) pressure of oil film in the bearing. This can happen only when a bearing design is defective from a load-carrying point of view or when oil film is too viscous.

CASE STUDIES

Case 9: In a thermal-power plant, it is usual to put a closed circuit for lube-oil system. In these systems, the return oil from bearings through a common pipe (header) is taken to lube-oil system. In order to control the temperature of oil supplied to bearings so that oil of right viscosity is supplied to the bearings, it is customary to provide oil coolers as well as heaters. In one-utility power plant, the oil-intake pipe was exposed to ambient temperature so that during severe winter season, the temperature of oil reaching the bearing was less than the design temperature and also heaters were not put in service. Consequently, the TG system experienced a high vibration at a frequency of 1300 CPM. Lagging the feed line to bearing and putting heaters into service solved the problem.

Case 10: In yet another similar case, the vibration levels despite controlling the oil-inlet temperature and pressure did not come down to safe limits. Obviously, the problem was arising of too-less clearance between the bearing and the shaft. As mentioned, the oil-whirl problem arises when the load to be supported by the oil film is less than the vertical component of the pressure of the hydrodynamic film of lubrication.

Since the problem could not be sorted out by manipulating the oil-inlet temperature and pressure, a deliberate unbalance was put on the shaft so that the bearing load increases. With this, though the vibration levels in the other non-whirling bearings increased to a certain extent, all the bearing-vibration levels came within the acceptable range.

Many a times, a deliberate misalignment in the neighbouring shafts, solves the oil-whirl problem. Nevertheless, it should be borne in mind that the measures such as deliberate unbalance or misalignment and so on, provide only a symptomatic relief. The correct solution is to provide only a correctly designed bearing.

A whirling bearing in a multi-rotor-bearing system can cause damages to other bearings too. Keeping this in mind and also the fact that whirling bearing will be subjected to rapid wear out, it is not advisable to neglect the oil-whirl problem.

Contrary to the phenomenon of the oil whirl, the friction-related whirl, also called the shaft whirl occurs due to the physical contact between the bearing liner and the journal. This can wipe out of the bearing. Here, when the contact takes place, the frictional force at the point of contact drags the rotor up (Fig. 7.41) and a fresh contact is made. At this point, it is again thrown to another position. Normally, a rotor which operates above the first critical, will tend to deflect or bow in a direction opposite the unbalance/heavy spot. As a result, the internal friction-damping factor, which normally works to restrict the deflection, will be out of phase and thus will further deflect the rotor. This is generally kept in check by the damping provided in the bearings. If this damping is low due to an improper bearing lubrication, the friction excites the whirl. This whirl will always occur at a frequency equal to the first critical speed.

Many tests performed on hydro machines and large pumps have shown that the hydrodynamic bearings are lubricated with not just oil alone but a mixture of oil and air, which is present in the form of stable suspension. The air bubbles are usually quite clearly visible and comparable in diameter to the thickness of the oil film. At times, the diameter may exceed the thickness of the oil film. The degree of saturation of the oil with air in thrust bearings depends upon (a) the discharge of oil from the bearing, (b) the turbulence at the surface of the oil tank, (c) foam at the surface due to agitation of the oil caused by the obstruction in its oil path (oil coolers, baffles). The foaming sharply increases in presence of even a small amount of water (which can find its way in the system through several ways). This can cause a direct contact between the bearing and thrust collar giving rise to vibration and bearing-wipe problem.

Figure 7.42 shows the shaft orbits when oil whirl and shaft whirl takes place. In fact, shaft orbits can detect not only the bearing rubs but the rubs between the rotating and stationary components in the rotating machinery. The rubs at bearings are identifiable through the vibration frequency close to the first critical while other rubs are identified through sub-harmonics and super-harmonics depending upon the degree and location of the rub.

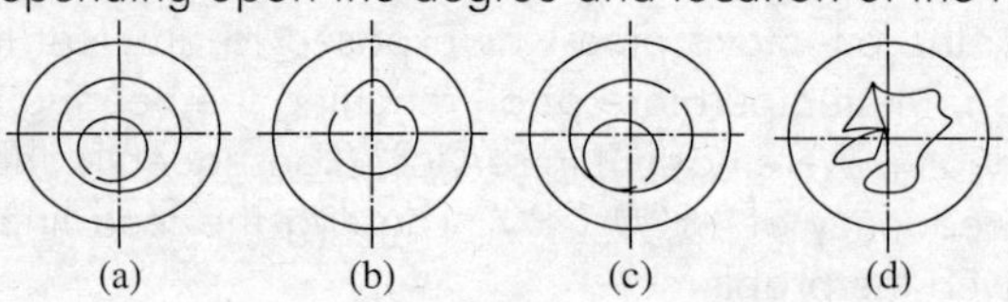

Shaft-orbit patterns (a) with oil which,
(b) mild rub occuring once per revolution,
(c) light rub (d) heavy rub

Figure 7.42 Shaft Orbits with Whirl

(D) Bent Rotor

The rotors of machinery, usually always have a small bend due to the static deflection of the rotor. Such bends cause eccentricity of the rotor during running. Carrying out the balancing of the rotor can compensate the bend caused by the dynamic forces resulting from the unbalance.

The bends larger than the static deflection/eccentricity occur on account of one or more of the following reasons: (1) manufacturing errors causing heavy unbalance, (2) entry of colder stream or water in steam turbines causing thermal bend, (3) thermal bend in generator/motor rotors due to blocked ventilation holes, inter turn shorts and so on, (4) the thermal bend in gas-turbine rotor due to improper distribution of cooling air, and (5) temporary bends caused by rubbing of rotating components with static components giving rise to hot spots.

The symptoms of rotor bend are similar to misaligned shafts (angular misalignment) with $1 \times n$, $2 \times n$, $3 \times n$, and so on, with dominant components in the radial direction and high$1 \times n$ component in the axial direction. The bent condition of rotor is confirmed by taking phase measurements. Figure 7.43 shows the method of confirming the bent condition of the rotor. The rubs, depending upon their severity, also cause local bends. However, the rubs are always identifiable as they will cause excitations of natural

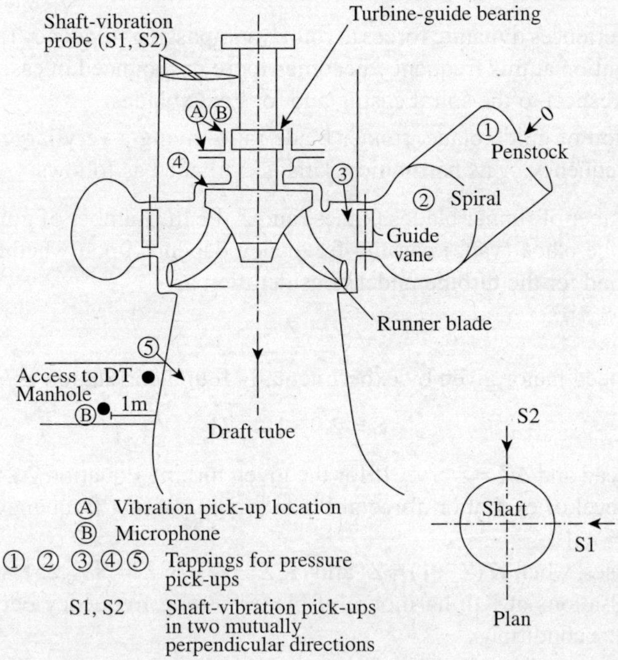

Figure 7.43 Instrumentation for Vibration Analysis of Hydraulic Turbine

frequencies of the rotor in addition to 1×n, 2×n, 3×n, and so on, components in the vibration spectrum. The most effective way to identify rubs by examining the shaft-orbit pattern is shown in Fig. 7.42. The bent rotor always shows multi-lobed orbit pattern similar to heavy-to-very-heavy misalignment.

(E) Hydraulic/Aerodynamic Perturbation forces

The subject of hydraulic/aerodynamic perturbation forces is very vast and can only be discussed thoroughly in a separate book dedicated to the same. We shall, however, deal with preliminary aspects of these perturbation forces.

Primarily, the basic source of hydraulic/aerodynamic perturbation force is the non-uniform flow distribution in the flow path of the machine. The pressure pulsations in the flow-carrying pipes, vortices shed from the bodies in the flow path, flow disturbances caused by the incorrect choice of the number of stationary and rotating blades (vanes), operation at an unstable point in the combined characteristics of pumps running in parallel, and the incorrect design of flow-intake and discharge systems are some of the commonly observed sources of hydraulic/aerodynamic or flow-induced perturbation forces.

We shall now briefly discuss some of these sources of perturbation forces in some typical rotating machinery such as hydraulic turbines, pumps, and steam and gas turbines.

(I) Hydraulic Turbines

Some of the commonly experienced sources of hydraulic perturbation forces are

- Vortices from guide vanes generate forces at vortex-shedding frequency. This frequency, f depends upon Strouhal number, which in turn, depends upon the flow velocity and the characteristic dimension. The Strouhal number is given by

$$f = Sv/t \tag{7.22}$$

where S is the Strouhal number, v = the flow velocity and t = the characteristic dimension. S usually is 0.21.

- Bladed rotor experiences dynamic forces at rotor-vane-passing frequency (number of rotor vanes × RPM). The excitation at this frequency becomes more pronounced in case of improper setting of the runner with respect to the spiral casing and/or guide blades.
- Improper selection of guide blade/runner blade ratio giving a very large excitation of forces at blade-passage frequency or its harmonics. This is explained as follows:

  Let Z_r be the number of runner blades (vanes) and Z_g be the number of guide blades. The severity of excitation at the blade (vane) passing frequency depends upon whether or not the following relation holds good for the turbine under consideration as

$$\Delta Z / Z_r = S \tag{7.23}$$

  where S is the speed factor given by experimentally found correlation as

$$S = 2.9 \times 10^{-3}\sqrt{H} \tag{7.24}$$

  where H is the head and $\Delta Z = Z_g - Z_r$. If for the given turbine Equation 7.24 holds good, there will be a very high level of excitation force at blade (vane) passing frequency. The Equation 7.25 is valid for $Z_g - Z_r = \pm 1$.

  For other cases, when $K(Z_r + 1) = Z_g$ and $(K Z_r - Z_g)/K Z_r = S$ (7.25), severe vibrations, noise, and pressure pulsations at Kth harmonic of blade-passing frequency occur. The above are also called interference conditions.

  The only way to solve this type of vibration/noise problem is to change the existing runner by the runner having a suitable number of blades (vanes).
- Incorrect design of intake and/or discharge system.

CASE STUDIES

Case 11: High vibrations/noise and repeated cracking of runner blades (Fig. 7.45) were experienced on practically all units of 6x60 MW, 167.7 RPM (specific speed 263 metric) hydro station. A large number of cracks were detected in the runner blades of almost all the individual units after about 13000–15000 hours of operation of the machines. The design head for the machines was 66 m; however, due to changes in the hydrological conditions, excessive rainfall in the catchment area of the dam, the machines had operated at a head higher than even 90 m for a considerable amount of time. Additionally, the machines exhibited severe vibration and intense noise behaviour.

The scanning and electron-microscopic examination of the fracture surface of the failed blades revealed that these failures had taken place due to fatigue.

Considering the fact that the machines had been exhibiting high vibrations/noise from the time they were commissioned it was necessary to study the vibration/noise behaviour of the machines at two head conditions, namely 90 m and 60 m at various loads. For this purpose, a detailed investigation comprising of vibration/noise and pressure-pulsation measurements was undertaken. Figure 7.43 shows the location of various transducers used for this purpose. The analysis of vibrations at various bearings of the unit and the analysis of noise at the draft tube and turbine pit showed that at 90 m head, the dynamic behaviour of the machine is extremely rough whereas the same, when the head became 62 m a few months later, became relatively quite smooth. The draft-tube noise level, which at 92 m head was 112 dB, came down to 105.2 dB when the head became 62 m; the noise levels in the turbine pit also came down from 106 dB (at 90 m) to 103.7 dB (at 62 m) when air was injected below the runner. The vibration-spectrum analysis carried out on the turbine-and-generator bearings (head 92 m), showed a prominence of 42 HZ and 245 HZ. Out of these frequencies, the 42 HZ frequency corresponds to runner-blade-passing frequency (15 runner vanes × 2.8 HZ rotational speed). The vibration spectrums at 62 m also showed very low levels of these components (Fig. 7.44). The calculations showed that 245 HZ corresponds to the vortex-shedding frequency.

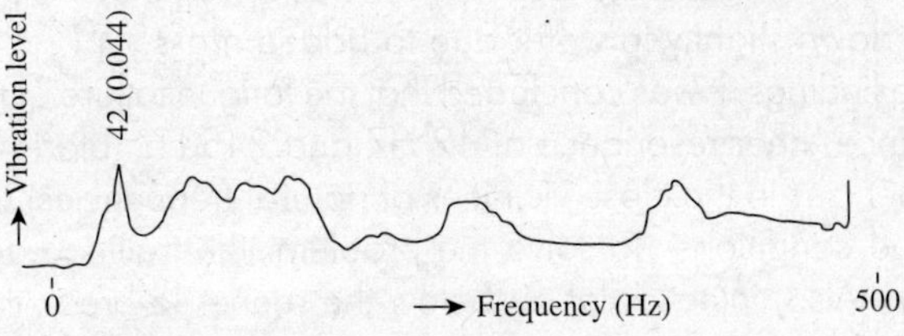

(a)

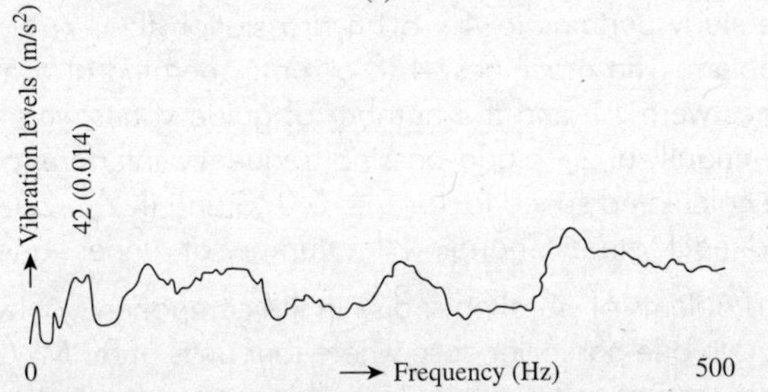

(b)

Figure 7.44 (a) Vibration Spectrum at TGB Radial Direction Full Load 60mw Head 90m. Overall Vibration Level 0.988 m/s^2 (b) Vibration Spectrum at TGB Radial Direction Full Load 60mw Head 62m. Overall Vibration Level 0.624 m/s^2.

Since the failures were found to be caused by fatigue it was necessary to find the natural frequency of the runner vanes. For this purpose, one of the uncracked runners was subjected to modal test to determine the various natural frequencies and the mode shapes.

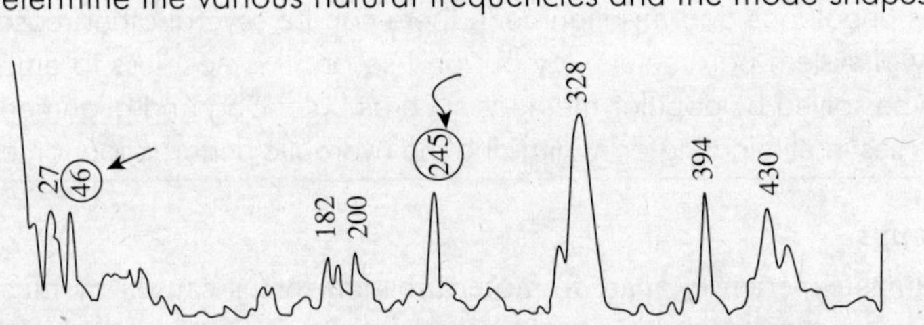

Typical bang-test spectrum

(a) Natural frequency 46Hz very close to blade passing frequency

Table

Head (m)	65	70	75	80	85	90
Gate opening (%)	100	80	69	62	56	52
Approach velocity (ms^{-1})	26.3	26.3	25.9	25.75	25.76	25.76
Vortex-shedding frequency (Hz)	248	248	238	235	235	235

(b) Vortex-shedding frequency very close to natural frequency (245 Hz) of the blades (a) and (b) show near resonance condition of blades and hence failure.

Figure 7.45 Natural Frequencies of Blades and Vortex-Shedding Frequency at Various Heads

The test (Fig. 7.45) revealed that the runner has two frequencies, 46HZ and 245HZ, in the domain of our interest. Out of these, 46 HZ and 245 HZ correspond to the bending and torsional modes. It may be noted that these frequencies were detected on the runner in air; and thus, they may come down slightly in water due to added mass.

Based upon these findings, it was concluded that the fatigue failure of the blades had occurred on account of resonance/near resonance at 42 HZ and 245 HZ (blade-passage frequency and vortex-shedding frequencies in the close vicinities of natural frequencies) under prolonged operation at off-design head conditions. To solve the problem, the trailing edges were thinned down and chamfered to 45°. Also, injection of air below the runner to break the trailing-edge vortices was suggested. The problem of runner cracking ceased after the modifications.

Case 12: The case study pertains to 4 x 80 hydro station (Fig. 7.49) having high vibrations and intense noise problem. The machines, 428.6 rpm, operated at a net head of 251 m,. the number of runner vanes were 11 and the number of guide vanes were 24. The vibrations and noise occurred predominantly at 2 x blade-passing frequency. Interference calculations (Equation 7.26) showed that the calculated speed factor S = 0.070 and $(K Z_r - Z_g)/K Z_r = 0.09$ were in the near neighbourhood of each other. Changing the number of runner vanes to 13 was suggested.

There are certain problems of vibration/noise in the components of hydro unit which arise out of poor tail-race design. In one particular case where four units of 80 MW were operating, the tail-race boundary in front of the first unit was too close which seriously affected the evacuation of water from that unit. This caused a serious vibration and noise and thrust/guide-bearing failure problem on that unit. The other units had not experienced this problem since the tail-race boundary was at a sufficiently long distance from these units. Based upon the detailed model studies, the tail-pool boundary in front of Unit 1 was shifted by about 5 metres resulting into smooth operation of Unit 1.

In hydro units, the penstock vibrations, at times result into vibration/noise of the machine. In one case, the flutter of butterfly between penstock and the spiral casing had caused both penstock as well as machine vibrations. These case studies do show that apart from the normal causes of vibrations such as unbalance and misalignment, there can be several other reasons which require an in-depth study of system parameters, lay out and so on. It is needless to emphasize that these problems cannot be solved by popular methods such as balancing and alignment. For such problems, the solution lies in eliminating or minimizing the hydraulic perturbation forces.

(II) Hydraulic Pumps

The centrifugal pumps experience, apart from perturbation forces caused by mechanical conditions such as unbalance, misalignment, and so on, several other perturbation forces which mainly are hydraulic in nature. The vibrations caused by the hydraulic perturbation forces can never be corrected by measures such as balancing of rotor, improving the accuracy/quality of coupling between the pump and prime mover such as electric motor or steam turbine. Eliminating or minimizing the hydraulic perturbation forces can only solve them. Some of the most important hydraulic perturbation forces are

- Perturbation forces at blade-passage frequency and its harmonics due to disturbed centring of the vaned impeller with respect to stator. The high-efficiency pumps, which on account of requirement of high-hydraulic efficiency, employ very small radial clearances between the rotor and the stator. Any error in the setting of the pump can cause intense noise and high vibrations at the blade-passing frequency and its harmonics. The flow disturbances in the gap can be severe enough to dislodge a small amount of material from the stator and/or the rotor. These small pieces can come in the flow path and cause seizure of the pump impellers.
- The axial float of the pump impeller is also a very important factor. Incorrect setting/location of the pump rotor in the axial direction may cause unsatisfactory film of lubrication at the thrust

bearing and cause a wide fluctuation in the axial shift of the pump impeller and trigger an unstable vibration behaviour.

- Mismatch of rotor/stator locations may cause a separated flow from the pump impeller. At times, this causes cavitation and high vibrations. The vibrations signature does reveal the ongoing separated flows and also the cavitations.
- The boiler-feed pumps of large power plants handle high-temperature boiler-feed water. As a result, the connected piping must be given a cold pull, so that in hot condition the piping does not exert forces on pump casing. Quite a large number of vibration problems investigated by the author were due to piping forces on the pump casing.
- The impellers in a multi-stage pump must be staggered which means that the angular positions of the identical impellers discs should not match. At times, since the impellers are shrunk on the shaft, this aspect may be missed at the manufacturing stage. The non-staggered arrangement may give rise to the hydraulic unbalance and consequent vibration problem.
- Incompatible stator/rotor-vane ratio (as in the case of hydraulic turbines).

(III) Steam Turbines

Apart from unbalance, misalignment, and so on, the vibration excitation can come through one or more of the following:

- If the number of guide blades is Z_g, then there are Z_g streams of flow before the rotating blades. Thus, each blade experiences Z_g impulses in one revolution. So, the moving blades are subjected to an excitation force of frequency Z_g x rps, called Diaphragm Impulse Frequency (DIF). If there is a considerable deviation in guide-blade pitch, the excitation at this frequency would be very significant. This is the principal reason for the high cost of diaphragms in steam turbine as it involves a very high level of precision in manufacturing. This is also the principal reason for the requirement of a very high precision required in the dressing up of diaphragms during the maintenance of the steam turbines.
- Obstructions in the flow path, wakes from the struts or braces ahead of a stage usually excite forces at frequencies ranging from $2 \times n$ to $6 \times n$ (n = rotational speed).
- Mismatch of guide blades at the parting plane resulting in a disturbing force of frequency $2 \times n$.
- An extremely bad condition results if the guide-blade pitch is equal to the rotating-blade pitch or if there is a common divisor between the number of guide-blades and the number of rotating blades.
- Water induction. This may cause a temporary or permanent bend of the rotor depending upon the severity of the thermal shock.

The steam turbines driving electrical generators in large/mega thermal and nuclear power plants run at a speed of 3000 rpm (when grid frequency is 50 HZ, as in India) or 3600 rpm (60 HZ, as in USA). Since the power to be generated is very high, the turbine blades handle a large quantity of steam due to which they are usually very long. Blade lengths of the order of 40–44 inches are thus not unusual. These large blades are quite susceptible to fatigue failures due to their exposure to the perturbation forces as mentioned earlier. It is, therefore, necessary to design the blades from the point of view of natural-frequency characteristics. The natural frequencies and the associated mode shapes can be found by either experimental or by theoretical methods such as Finite Element Method (FEM). The susceptibility of the blades to failure is judged by studying the closeness of the natural frequencies with various excitations such as $1 \times n$, $2 \times n$, $3 \times n$ …. DIF, 2XDIF, and so on, at various speeds up to an operating speed of power plant turbine or various operating speeds of variable speed-drive steam turbines (used in refineries, fertilizers, petrochemical plants). Such studies are possible through diagrams called Campbell diagrams as shown in Fig. 7.46.

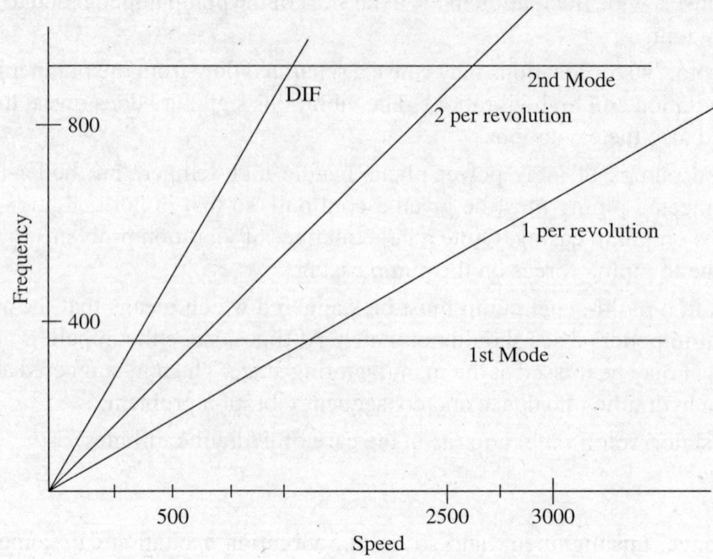

Figure 7.46 Campbell Diagram

The abscissa in the Campbell diagram is the speed of the machine in revolutions per minute (rpm) and the ordinate is frequency in HZ. The diagram shows f_1, f_2, f_3… f_n (various natural frequencies) lines found either by experimental techniques or analytical techniques. The increase in natural frequencies with rpm is because of centrifugal force-induced stiffening at the blade roots of the joint between the wheel and the blade root).

Lines 1X, 2X, 3X … and so on, are called the engine orders. Thus, 1X line will have an ordinate of 50HZ at a speed of 3000 rpm, 2X line will have an ordinate of 100HZ at 3000 rpm, and so on,. the point where the engine order crosses the natural frequency line is the resonance point. For example, at point A 3X line crosses f_1 natural frequency at 950 rpm, which means when the speed of the machine is 950 rpm, f1 natural frequency mode will be excited, and keeping machine at this speed would be undesirable. Similarly, at point B, 4X line crosses f_2 line at a speed of 1800 rpm. Running the machine at this speed for a longer time would be very undesirable. Hence, if the machine happens to operate continuously at these speeds, the resonance condition may cause fatigue failures. The safe-operating speed range of a typical utility (power plant) turbine is shown in Fig. 7.40. One can see that there are no resonance conditions in the continuous-operating speed range 2900–3100 rpm and hence, no blade failures due to fatigue will take place in the turbine in this speed range.

Campbell diagrams are extremely useful for variable-speed steam turbines driving large compressors/pumps. One can always avoid resonance-speed range once the Campbell diagram is available.

CASE STUDIES

Case 13: A large utility turbine comprises of HP turbine, IP turbine, and LP turbine on a common shaft driving an AC generator, experienced repeated failures of one of the stages of IP turbine. The stage consists of blades provided with a lacing wire to increase the natural frequency from 147 HZ to 153HZ of the free-standing blade to 160 HZ to 165 HZ so that the third-harmonic excitation (around 150 HZ) does not cause high vibrations of the blades and consequent failures. At a particular site, a few of these blades failed during the service causing damages to the diaphragm blades. The failed blades were replaced with new blades and the lacing wire was rebraced. The damaged diaphragms were repaired at site by grinding (dressing) and the machine was put back into service. The particular stage failed once again after about 4–6 weeks and was again replaced. The phenomenon of blade failure once again occurred after another 6 weeks of operation.

The author of this book had an opportunity to investigate the failures, which occurred for the third time. The visit to site showed the diaphragm blades which failed in the first instance of failure were hand-dressed using a portable grinder and during this activity the guide blades were dressed by removing the material of guide blades such that guide blade surface becomes smooth. Apparently, the maintenance staff were not aware of the fact that such dressing would result into a variable pitch of diaphragm blades and would give rise to a substantial excitation of forces at 1X, 2X, 3X...DIF, 2 DIF, and so on. Also, it was found that the workmanship of brazing the lacing wire to the blades was sub-standard in a sense that the natural frequency of the unbrazed blade (147Hz) went up to 152 Hz to153 Hz after brazing the lacing wire. Since the set, at times, had run at a grid frequency of 50.5 Hz–51Hz in the night hours, the moving blades had seen the resonant condition due to (1) higher-grid frequency, (2) closeness of natural frequency to 3X frequency perturbation force, and (3) high excitation at 3X frequency due to large variation in the pitch of diaphragm blades.

It must always be remembered that a high-speed machinery such as steam turbine/gas turbine requires a very high degree of accuracy in manufacture—especially in the blade pitches, blade roots, and fixing of devices such as lacing wire and shroud bands; and that is the reason they are so expensive.

In this particular case, the customer was suggested to replace the repaired diaphragm by a new diaphragm and improve upon the quality of brazing of the lacing wire. No failure at this stage has been reported after implementing the suggested corrective action plan.

A vibration engineer must keep himself aware of various perturbation forces experienced by the machine under investigation. We have discussed a few typical perturbation forces such as mechanical, hydraulic/aerodynamic forces experienced by machinery such as pumps, turbines, compressors, fans, generators, and so on. We shall now discuss the electrical perturbation forces.

(E) Electrical Perturbation Forces

The electrical perturbation forces normally encountered on motors and generators are (a) a periodic component of magnetic pull, (b) forces due to non-uniform air gap between the stator and the rotor, and (c) forces due to the short circuiting of rotor windings.

The frequency of electrical perturbation forces is fairly high (50Hz and above) under steady-operating conditions. There have been cases where the state of balance gets significantly changed when the generator is excited and loaded. This, sometimes, leads to the necessity of carrying out compromise balancing so that the machine exhibits somewhat higher vibrations in the mechanical run but smooth vibration behaviour in the excited and loaded condition. The induction motors show a typical vibration behaviour because of the slip of motor. Because of slip, the rotating speed of the motor (multiplied by a number of pole pairs) is slightly small than the impressed frequency of input power (say, 50Hz). Thus, for a two pole, 3000-rpm motor having 2 per cent slip, experiences a mechanical unbalance at ($3000 - 0.02 \times 3000$) 2940 rpm and electrical frequency of 3000 rpm (50Hz).

Simultaneous action of these forces will show a typical beat type of vibration phenomenon. As long as the unbalance in the rotor is within the specified limits, high/low values (also called envelope strength) will not be significant. Nevertheless, the vibration levels as monitored on a healthy AC motor by a vibration meter, will show changing amplitudes at beat frequency (1Hz in the case cited). However, if the rotor has a high level of unbalance or a cracked rotor bar or any other such defect, the envelope will become highly pronounced or large. In case the vibration levels drop immediately upon switching off the motor, we can conclude the vibration problem is because of an electrical defect requiring correction.

7.7 MODAL ANALYSIS

As has been discussed previously, one of the reasons for high vibrations and consequent failures of rotating as well as static equipments is operation of the equipment at resonance or near-resonance condition. One can reduce to a certain extent the levels of perturbation forces by measures such as balancing, alignment, and so on; and in some cases, putting flow straighteners in the flow path of turbomachinery. Unfortunately, there is a limit for reducing these forces and, many a time, it may not be possible at all to effect such reduction. Also it is impossible (almost in all cases) to alter the frequency of the perturbation forces. The only way out, in such cases, is to alter the natural frequency by some means stiffening the components or making them flexible. Also the location at which this can be done can be known only after determining the deflected shape (mode shape) of the component whose natural frequency is equal to or in the near neighbourhood of the frequency (ies) of perturbation force (s). Determination of natural frequencies and the associated modes is called "modal analysis".

Modal analysis can be performed either by theoretical methods or by experimental methods. Barring a few simple geometries of the component / structure, most of the engineering components / structures prove to be highly complex for carrying out modal analysis by analytical methods. However, at the design stage the computational methods such as FEM can be used to estimate the natural frequencies and the mode shapes—provided exact boundary conditions can be specified, which at times is quite complex. For example, accurate FEM analysis of blade discs require a correct specification of the root fixity as well as the fixity condition of shroud bands, lacing wire, and so on. Blade-to-blade or packet-to-packet (blades usually are grouped so that the bladed disc comprises of several packets) variation, which depends upon the manufacturing and assembly process, and cannot be easily accommodated in the FEM analysis. Nevertheless, FEM is used at the design stage but once the components are manufactured, it is prudent to experimentally find out the natural frequencies and the mode shapes, that is carry out the experimental modal analysis so that corrections, if any, can be incorporated in the component / assembly / structure.

Over a period of last one-and-half decade, the experimental modal-analysis techniques have become very sophisticated, user-friendly due to vast improvements in the transducer technology, analyzing equipments, and computer softwares.

The simplest method for carrying out the modal analysis of component / structure consists of exciting the vibrations of them by means of variable-frequency shaker and identifying those frequencies where responses are the highest and the points where these occur. The natural frequency, the point of maximum response (antinodal point) and minimum (or zero) response (nodal point) can be identified by such a theoretically simple shaker test. There are, however, some difficulties in carrying out such test. Some of them are (a) the excitation force changes with excitation frequency, (b) the shaker point adds to the mass especially when we test components such as blades, (c) difficulties in mounting the shaker. Hand-held shakers do not give trustworthy results, and (d) if the point of the application of shaker happens to be a nodal point, no trustworthy results can be obtained.

The other popular method of extracting the information of natural frequencies and the mode shapes is the impact-hammer test (also called bang test or bump test). The impact hammer is a hammer with built-in-force transducer in its head. The impact hammer is used to hit or impact the structure or machine being tested to excite a wide range of frequencies without causing the problem of mass-loading experienced in the shaker test. The impact force caused by the hammer, which is nearly proportional to the mass of the hammerhead and the impact velocity, can be found from the force transducer embedded in the head of the hammer. As shown in the previous chapters, the response of the structure or the component of the machine to an impulse is composed of excitations at each of the natural frequencies of the component / structure.

A very common application of impact test is the railway bogie/wagon wheels being tested by wheel trappers when the train is at rest at railway stations. The wheel trapper can judge from the sound whether or not the wheel is cracked. An uncracked wheel will emit a ringing sound at a frequency, which the wheel trapper judges from the quality of the sound. A cracked or defective wheel will emit a thud like sound. Although the wheel trapper uses no sophisticated instruments, the sound emitted by the wheel enables him to judge the integrity of the wheels.

Although the impact hammer is simple, portable, inexpensive, and much faster to use than shake, it is, in some cases, not capable of imparting a sufficient energy to obtain an adequate response signal in the frequency range of interest. Usually, the useful range of frequency excitation is limited by a cut-off frequency, ω_c, which means that the structure/component/machine did not receive sufficient energy to excite modes beyond ω_c. The value of ω_c is often taken at a frequency where the amplitude reduces by 10 to 20 dB from its maximum value. To overcome this problem, at least partially, one can use hammers with variety of hammer tip, for example, rubber, plastic, steel, and so on.

The most difficult and involved task in the impact-hammer test is to identify the real signal from the structure/component, as the input and output data measured by the transducers usually contain some random or (electronic) noise. There are several procedures such as finding auto-correlation function, power-spectral density, and coherence function which makes identification of natural frequencies and mode shapes possible despite the electronic noise. For further details, readers may refer to "Vibration Testing" brought out by BRÜEL & KJAER, Naerum, Denmark.

In order to find the natural frequencies alone, one can use a single-channel FFT analyser. The peaks in the response spectrum identify the natural frequencies. However, if the mode shape is also required to be determined, one has to use a two-channel FFT analyser. One channel receives the signal X from the vibration pick-up mounted on the structure/component and the other channel records the signal Y from the force transducer mounted on the hammer. The complex (force and response, in general, have a phase lag) ratio X/Y is the frequency response. The two-channel FFT analyser has a capability to separate out the real and imaginary parts. The imaginary part then determines the mode shape (recall the discussions we had on force/response characteristics in the earlier chapters 2, 3, and 4). For a detailed information on two-channel FFT analysers, refer literature from standard manufacturers like BRÜEL & KJAER, ONNO SOKKI.

7.8 VIBRATION CONTROL

We have dealt with this topic at different places in this book. The principal methods of vibration control are

- Reduction at source. Improving the balancing and improved quality of alignment are typical examples.
- Modification of natural frequency by structural alteration such as stiffening, in case vibrations are due to resonance or near-resonance condition.
- Use of dampers or energy-dissipation devices such as damping pins, lacing wire and turbine blades.
- By reducing transmission from one part of the machine to another part by vibration isolators
- By vibration absorbers or tuned absorbers
- Magnetic bearings

Considerable work is being done in the area of active control of vibrations using magnetic bearings. This subject is beyond the present scope of this book.

CONCLUSION

We, in this chapter, have dealt with the methodology of diagnosing and solving vibration problems. The importance of the proper choice of vibration parameter (displacement, velocity, and acceleration) that are to be used has been discussed. We also focused on the fundamentals of frequency analysis. We also briefed about the necessity of looking into high-frequency components in the vibration spectrum considering their damage potential.

In this chapter, we briefly discussed the theory behind the working of various vibration transducers, phase measurement, and so on. We dealt in detail the methodology of processing and reduction of vibration data and also the traditional spectrum analysis, water-fall diagram, Bode plot and identification of critical speeds of rotating machinery, shaft-orbit analysis, and so on, with a number of case studies pertaining to various perturbation forces causing vibrations in various machinery.

We dealt with the experimental methods for modal analysis and finally corrective steps to be taken for solving the vibration problems too.

References

1. *Vibration Testing,* Brüel and Kjaer, Denmark, 1983.
2. D.J. Ewins, "Modal Analysis as a Tool for Studying Structural Vibrations" in *Mechanical Signature Analysis: Theory and Applications,* S. Braun (ed.), Academic Press, London, 1986.
3. B. Nevada, *Rotating Machinery System and Service*, Bentley Nevada, Minden.
4. S.K. Bhave, Ch.B. Murthy and S.K. Goyal, "Investigation into Blade Failures of Francis Turbines," Water Power and Dam Construction, January 1986.
5. *Mater Catalogue–Electronic Instruments*, Brüel and Kjaer, 1989.
6. J.P. Den Hartog, *Mechanical Vibrations*, McGraw-Hill, New York, 1956.
7. Hewlett-Packard, "Fundamentals of Signal Analysis–Application Note 2.43," Hewlett-Packard.
8. "Product Catalogue of Endevco Corporation," San Juan, Capistrand, The United States.
9. H.C. Radhakrishna, S.K. Bhave, and Ch.B. Murthy, "Vibration and Noise Measurements in Francis Turbine Power Plants," International Symposium on Large Hydraulic Machinery and Associate Equipment, May 1989, Beijing, China.
10. S.K. Bhave, "Case Studies in Failure Analysis of Electrical Equipments," All India Symposium on Reliability of Heavy Electrical Equipments, Institution of Engineers (India), Bangalore, 1980.
11. S.K. Goyal, and S.K. Bhave, "Analysis of High Vibrations / Eccentricity Problems in Steam Turbines–Case Studies," IInd Indo-German Power Plant Symposium, New Delhi, 1982.
12. Ch. Bhavnarayana Murthy and S.K. Bhave, "Diagnosis of Hydraulic Perturbation Forces in Hydro Turbines," Eighth Congress of Asia-Pacific Division, IAHR, CWPRS (Pune), India, 1992.
13. Ch.B. Murthy, "Vibration and Noise in Francis Turbine Power Plants," PhD Thesis, IIT (Madras), 1990.
14. B.L. Jaiswal, S.K. Goyal and S.K. Bhave, "Structural Analysis of Large Size Bladed Impellers," International Modal Analysis Conference, Florida, The United States, 1993.
15. S.K. Bhave, "Diagnosis," in *Vibration and Oscillation of Hydraulic Machinery,* H. Ohashi (ed.), Hydraulic Machinery Book Series, Averbury Technical, Gower Publishing Company.
16. B.L. Jaiswal and S.K. Bhave, "Experimental Evaluation of Damping in Bladed Disc Model," *Journal of Sound and Vibration*, 1994, 177(1), 111–120.

EXERCISES

7.1 Explain through appropriate mathematical analysis why vibrometer is very bulky and not useful in many practical applications.

7.2 Explain through appropriate mathematical analysis, why accelerometer is a preferred transducer for measurements of vibrations.

7.3 What is mounted-resonance frequency of an accelerometer? Derive the equation for the mounted-resonance frequency of an accelerometer. Explain why the transducer should be stud mounted or fixed with a suitable adhesive?

Hint: Treat accelerometer as a single degree-of-freedom system.

7.4 A vibration pick-up has a natural frequency of 5 Hz and damping ratio of 0.5. Find the lowest frequency that can be measured with error not exceeding 1%.

Answer: 35.25 Hz.

7.5 A spring–massdamper, having an undamped natural frequency of 100 Hz and a damping constant of 20 N-s/m, is used as an accelerometer to measure the vibration of a machine operating at a speed of 3000 rpm. If the actual acceleration is 10 m/sec^2 and the recorded acceleration is 9 m/s^2, find the mass and spring constant of the accelerometer?

Answer: $m = 19.4$ g, $k = 7623$ N/m

7.6 A critical high-speed compressor in a petrochemical plant is included in the extensive condition-monitoring program wherein the vibration levels are monitored on hourly basis. An unexpected failure of the compressor occurred due to failure of a blade. The shaft vibration levels prior to failure were maximum 25 microns pk–pk and the spectrum of bearing vibrations had shown displacement level of 10 microns at the frequency of rotational speed of 3000 rpm. Explain what must have gone wrong in the condition-monitoring program and give your recommendations.

7.7 Explain the reasons for using the information of phase angle shift of 90° for identifying the critical speeds of a rotating machine. You may idealize the system as a single degree-of-freedom system subjected to a harmonic excitation force.

7.8 The following are typical vibration spectra obtained on two bearings A and B of a rotating machine operating at site.

Bearing A: horizontal 150 microns, 1 × 145 microns/phase angle 0°, 2 × 10 microns.

Bearing B: vertical 90 microns, 1 × 85 microns/phase angle 90° 2 × negligible.

Identify the causes for the above vibration problem and suggest a suitable remedial measure. The bearings are end-shield mounted. Give the reasons for your recommendation.

Answer: Heavy unbalance. Balance the rotor.

7.9 Explain the phenomenon of oil whirl. Explain the role played by the grade of lubricating oil used for the bearings.

7.10 A motor-driven pump assembly shows a high vibration behaviour at motor drive end-bearing as well as the front-bearing of the pump. Figure below shows the schematic of the system. The salient features of the vibration data are as follows.

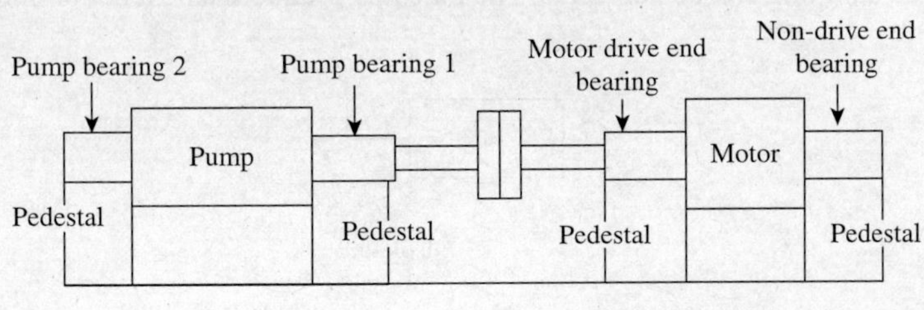

Motor drive end-bearing horizontal vibration 95 μ pk–pk, vertical 20 μ pk–pk, axial 10 μ pk–pk.

Pump bearing 1 horizontal 85 μ pk–pk, vertical 15 μ pk–pk, and axial μ 20 pk–pk.

What inference can you draw from the above data? Suggest suitable further investigations. The pump is driven by a 3000 rpm AC induction motor.

Another important aspect of vibration analysis is vibration isolation. In these exercises, we shall include some problems on vibration isolation although the basics of the same have been covered elsewhere in this book.

Concepts of Vibration Isolation

The emphasis in this chapter is on diagnosis and control of practical vibration problems. We concluded that for resolving the problem, we must identify the perturbation forces and eliminate them or at least reduce their intensity to the extent possible. We also discussed that in some practical situations, we may not be able to handle the issue of perturbation forces and thus, we need to alter the structural behavior of the component/structure and/or introduce additional damping devices. The other important aspect of this problem is vibration isolation which is procedure by which the undesirable effects of vibration are reduced. The isolation of vibrating system is also extremely important as in absence of it; the vibrations can harm the personnel operating the machine as well as they may harm the surroundings in which the machine is operating.

The vibration-isolation system could be passive in which we utilize a resilient member (stiffness) such as metal springs, pneumatic springs and an energy dissipater (damping) such as cork, felt and elastomers. We dealt with such systems in earlier chapters of this book (chapters 2, 3 and 4). The vibration-isolation system can also be active and is comprised of a servomechanism with a sensor, signal processor and an actuator. Active-vibration isolation is a very complex subject and hence, it is beyond the scope of the present book.

The effectiveness of an isolator is stated in terms of its transmissibility. We had defined this as the ratio of the amplitude of the force transmitted to that of excitation force. We, in this exercise shall deal with some complex problems of passive-vibration isolation.

7.11 A heavy machine of mass m is mounted through a resilient system on a foundation. The resilient system comprises of a spring of stiffness k and a viscous damper with damping constant c. the machine produces an excitation force $F(t) = F_0 \sin \omega t$. Derive the formula for the total force transmitted to the foundation. Prove that the forcing frequency has to be greater than $\sqrt{2}$ times the natural frequency of the system in order to achieve isolation of vibration.

7.12 A sensitive instrument in the cockpit of the aircraft is required to be isolated from the vibrations experienced by the aircraft. Idealize the instrument as a mass m and assume that the base on which it is mounted undergoes a harmonic motion. The isolator comprises of a spring and a damping device.

7.13 The figure below shows a machine with isolator on a flexible foundation. Derive the formula for the force transmitted to the supporting structure and transmissibility of the isolator. A represents the machine having a mass m_1 and B represents a supporting structure having a mass m_2. The isolator has stiffness k and the damping can be neglected. The machine produces an excitation force $F = F_0 \sin \omega t$.

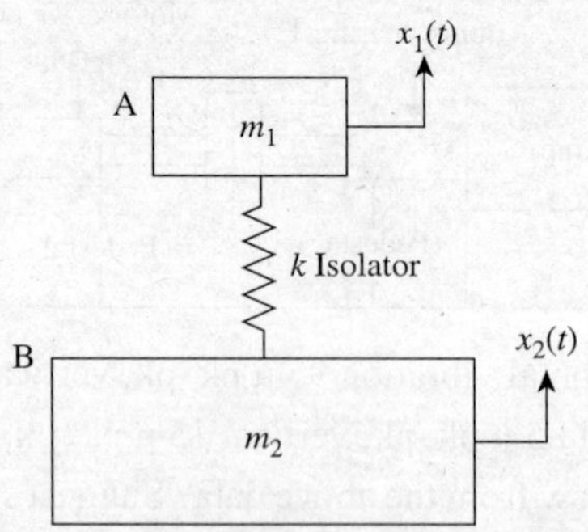

Problem 7.13

Hint: Treat the system as two degrees-of-freedom system. Arrive at the equations of motion. Get the frequency equation and the natural frequencies. The force transmitted to the supporting structure is $m_2\ddot{x}_2$.

7.14 Assuming that supporting structure as shown in the figure above rests on a column structure fixed to the ground which can be modelled as a massless spring, derive the equations of motion.

7.15 An air compressor of mass 1000 kg has an eccentricity of 100 kg-cm and operates at a speed of 300 rpm. The compressor is to be mounted on of the following mountings: (a) an isolator consisting of a spring with negligible damping, (b) a shock absorber having a damping ratio of 0.15 and (c) both of (a) and (b). Select a suitable mounting and specify the design details by considering static deflection of the compressor not exceeding 5 mm, the transmission ratio and amplitude of vibration of the compressor.

Hint: Maximum value of the spring stiffness = 1000 × 9.81/0.005 kg/m. The excitation force is 100 × ω^2. The speed of the compressor is 300 rpm.

7.16 The figure below shows schematic of a vibration absorber.

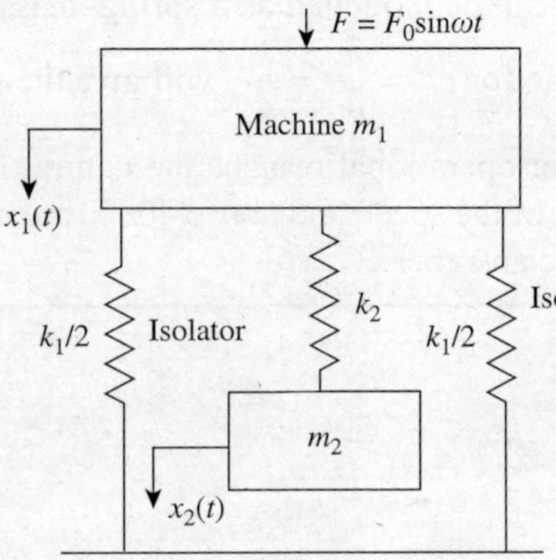

Problem 7.16

Derive the equations of motion and show how the assembly comprising of spring k_2 and mass m_2 can be tuned such that vibration of the machine disappears. Also show that the dynamic absorber, while eliminating vibration at known impressed frequency ω, introduces two resonant frequencies and Ω_1 and Ω_2 at which the amplitude of vibration of the machine is infinite. Evaluate these frequencies. Thus define the safe-speed zone. What will happen if we introduce damping element in the isolator?

Answer:
$$\left.\begin{matrix}(\frac{\Omega_1}{\omega_2})^2\\ (\frac{\Omega_2}{\omega_2})^2\end{matrix}\right\rangle = \frac{\{[1+(1+\frac{m_2}{m_1})(\frac{\omega_2}{\omega_1})^2]\mp\{[1+(1+\frac{m_2}{m_1})(\frac{\omega_2}{\omega_1})^2-4(\frac{\omega_2}{\omega_1})^2\}^{1/2}\}}{2(\frac{\omega_2}{\omega_1})^2}$$

7.17 A turbine-driven compressor (shown in figure below) operates in the speed range of 2000–4000 rpm and usually it runs around 3000 rpm. The vibration behaviour at and around 3000 rpm is extremely rough despite achieving a very high degree of balancing of both the turbine and compressor rotors done in the balancing tunnels of the manufacturer. The vibration-spectrum analysis done on the unit ruled out the possibilities of vibrations arising due to misalignment and other possible defects in the assembly of the unit. It was therefore decided to use a tuned absorber to reduce the severity of vibration on the turbine-compressor unit. The proposed tuned absorber is a cantilever beam carrying a mass of 20 kg. The absorber natural frequency is adjusted to 3000 rpm. Find the natural frequencies of the assembly. Assume the mass of turbine-compressor assembly as 1000 kg.

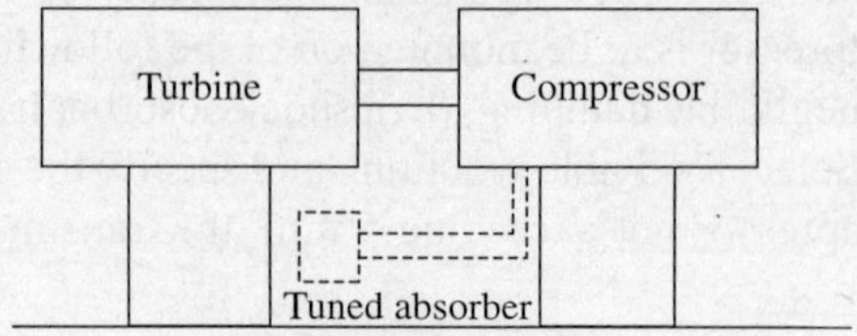

Problem 7.17

Hint: Natural frequency of tuned absorber is 50 Hz. Find k_2 (stiffness of the cantilever). Assume that the turbine-compressor unit can be modelled as a spring–mass system as shown in Problem 7.16 and write the equations of motion $\omega^2 = \frac{k_2}{m_2} = \frac{k_1}{m_1}$, will give the value of k_1.

7.18 For the above problem for operational reasons the natural frequencies are required to be less than 2000 rpm at lower limit of the speed and above 4000 rpm at the higher limit of the speed of turbine compressor. Design the absorber.

8 Finite Element Method

8.1 INTRODUCTION

In the previous chapters (Chapters 3, 4, and 6), we explained the method of idealizing/simulating a continuous system as an assemblage of multiple masses, springs, and damper elements to obtain (at least) a first-level evaluation of natural frequencies, mode shapes, and the response of the system to a given system of excitation forces. However, at times, such approximation, which uses lumped masses, springs, and dampers does not yield a reasonably-acceptable/satisfactory result.

While dealing with the analysis of vibrations of beams for various conditions of fixity, we have seen that the solutions are in the form of infinite series, and larger is the number of terms we consider in the series, higher is the accuracy of the result. Alternatively, we have some approximate methods such as Dunkerley's method, Raleigh–Ritz method etc., to evaluate the natural frequencies and the corresponding mode shapes of transversely-loaded beams/rotors. Rigorous analysis as well as experience shows that the level of accuracy obtained using these methods is quite acceptable in practical engineering situations. However, for complex structures involving plates, shells, beams, columns, etc., it is not possible to make assumptions as we made in simple-beam structures; also it is extremely difficult to idealize/simulate them by the lumped spring, mass systems and obtain any sensible solution. Finite Element Method (FEM) provides a reasonably correct and accurate solution for such complex problems.

The FEM essentially is a numerical method that can be used for a variety of engineering analysis problems such as stress analysis, fluid dynamics, vibration analysis, heat transfer, electromagnetics, etc. In this chapter, we deal with the application of FEM for vibration analysis. In this method, the actual structure/component of the machine or the entire assembly of the machine (which we will now refer to as structure) is considered as assemblage of several pieces of elements joined at points called joints/nodes. Each of the elements is assumed to behave as a continuous structural member called finite element. As mentioned earlier, the elements are assumed to be interconnected at certain points known as joints or nodes.

It may be clearly understood that it is extremely difficult to find an exact and closed-form solution for the vibration displacements of the original structure under specified time-dependent loads/forces. In order to overcome this problem, a convenient (for computations) approximate solution is assumed at each finite element. If the solutions of various elements are selected properly, these elements can be made to converge to the exact solution of the entire structure. This requires that we divide the structure into large number of small size elements. This process is termed as discretization of the total structure.

8.2 IMPORTANT CONDITIONS TO BE SATISFIED

There are two principal conditions that need to be satisfied for each element as well as for the entire assemblage of elements so that the solution obtained for the entire structure is accurate. These conditions are:

- Equilibrium of forces at nodal points giving due consideration to the boundary forces and displacements.
- Compatibility of displacements between elements. This means that if a node say J is shared by say ith, $i - 1$, and $i + 1$th element etc., then displacements at J are same regardless of the element we choose as long as the node belongs to the element. In simpler words, the deflected shape of the structure is such that there are no gaps/voids/discontinuities and the structure in dynamic environment maintains the condition for continuum.

In this chapter, we discuss the basic procedure of the FEM in vibration analysis and deal with its application to simple vibration problems. For advanced applications, please refer to any one of large number of books on vibration analysis using FEM.

In this chapter the shall derive element-stiffness matrix, mass matrix, and force vectors for simple elements such as bar element, torsion element, and beam element. In addition, at this stage it is necessary to understand that each of the elements in FEM analysis has its own local coordinate axes (called local coordinate axes) and the entire structure would have its own coordinate axes (called global axes). The transformation of element matrices and vectors from local-coordinate system to global-coordinate system (of the entire structure) is a important issue that needs clear understanding.

The ultimate objective of vibration analysis by FEM is to arrive at equations of motion of the complete system of assembled finite elements with due consideration to the boundary conditions of forces/displacements, and solve them. It may not be out of place to mention that there are several commercial grade FEM softwares that can be used for vibration analysis. However, they will yield correct results only when the user employs/selects proper elements, suitable meshing strategies, and appropriate boundary conditions. In absence of these, the software will not deliver appropriate results. In this chapter, we shall only deal with one-dimensional elements; however, the technique presented in this chapter can be applied to more complex problems involving 2-D and 3-D finite elements. However, before we discuss the 1-D problems, it will be appropriate to describe how modelling is done and how appropriate elements are chosen. We shall illustrate this by considering application of FEM concepts to a large milling machine.

8.3 MODELLING

Figure 8.1 shows sketch of a milling machine. The important components of the milling machine are 1) columns, 2) overarm, 3) cross-slide, and 4) tool holder. The components of the machine are subjected to dynamic loads caused by the cutting forces having components $F_X(t)$, $F_Y(t)$, and $F_Z(t)$ where t is time.

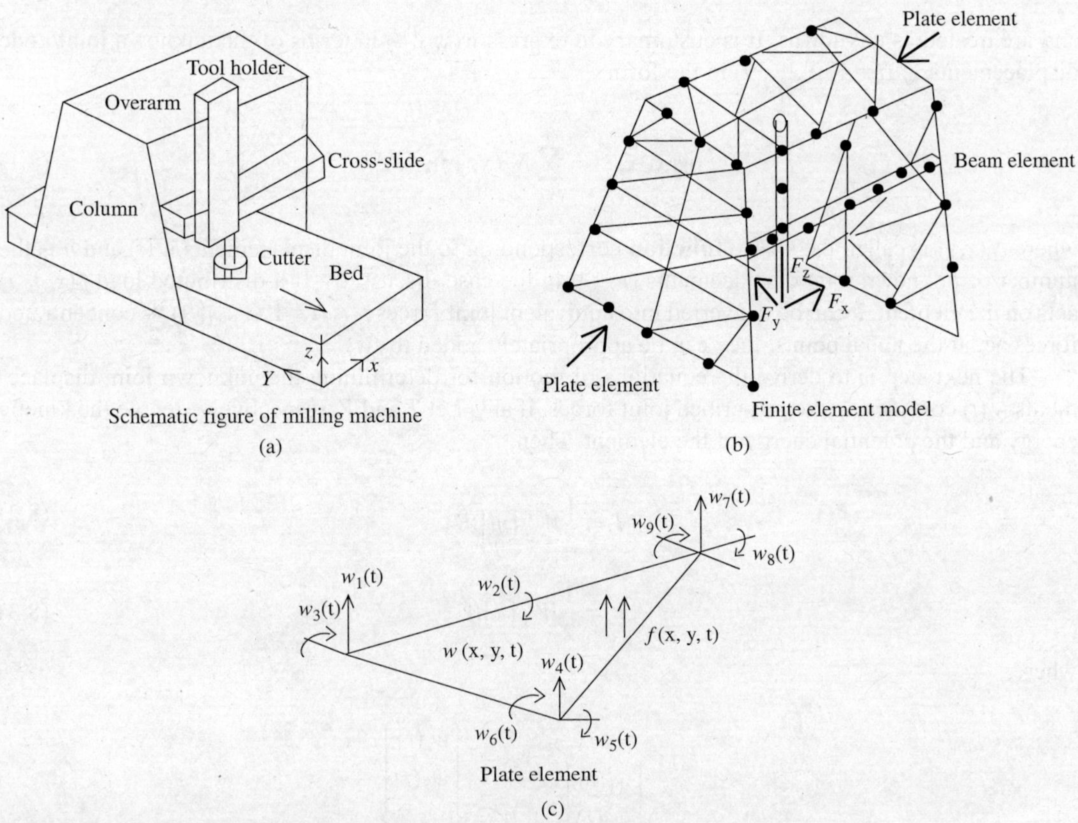

Figure 8.1 (a) Machine Structure (b) Finite Element Model (c) Plate Element

Figure 8.1 shows the global-coordinate system *X*, *Y*, and *Z* for the entire assembly. It is necessary to understand why we have chosen the particular element for the components of the machine. For example, the entity called column will hardly behave as a structural column since its lateral dimension (depth) is quite substantial as compared to true column whose lateral dimensions are much smaller than its length. This member is more likely to behave as a plate. Similarly, the overarm has large lateral dimension and therefore will behave as a plate. Thus, both column and the overarm are modelled as plates and plate elements as shown in Fig. 8.1(c). The cross slide as well as tool holders are modelled as beams because of the lateral loading they experience. Thus, it is apparent that the accuracy and the correctness of the FEM solution strongly depend upon how correctly we choose the elements for the sub-components of the assembly. Figure 8.1(a) also shows the time-dependent cutting forces acting on the system.

Figure 8.1(c) shows various displacements w_i ($i = 1, 2, \ldots, 9$) at three nodes (1, 2, and 3) whereas displacement at any point within the element are given by $w(x, y, t)$, where (x, y) is the local-coordinate system. Since the forces acting on the system are time-dependent, the displacements are also functions of time *t*. The boundary forces acting on the element (if any) are denoted by $f(x, y, t)$.
The displacement and the slopes of the deflection curves at nodes 1, 2, and 3 are denoted as

$$w_1(x_1, y_1, t), \frac{\partial w}{\partial x}(x_1 y_1, t)\ldots\ldots\ldots, \frac{\partial w}{\partial y}(x_3, y_3, t)$$

and are treated as unknowns. It is customary to express $w(x, y, t)$ in terms of the unknown joint/node displacements w_i ($i = 1, 2, \ldots, 9$) in the form

$$w(x, y, t) = \sum_{i=1}^{n} N_i(x, y) w_i(t) \tag{8.1}$$

where $N_i(x,y)$ is called the **shape function** corresponding to the joint displacements $w_i(t)$ and n is the number of unknown joint displacements ($n = 9$ in the case discussed). If a distributed load $f(x, y, t)$ acts on the element, it can be converted into equivalent joint forces $f_i(t)$ ($i = 1, 2, \ldots, 9$). If concentrated forces act at the nodal points, they can be appropriately added to $f_i(t)$.

The next step is to derive the equations of motion for determining the unknown joint displacements $w_i(t)$ considering the prescribed joint forces, if any. Let T and V, respectively, denote the kinetic energy and the potential energy of the element. Then,

$$T = \frac{1}{2}\dot{\vec{W}}^T [m] \dot{\vec{W}} \tag{8.2}$$

$$V = \frac{1}{2}\vec{W}^T [k] \vec{W} \tag{8.3}$$

where,

$$\vec{W} = \begin{Bmatrix} w_1(t) \\ w_2(t) \\ .. \\ w_n(t) \end{Bmatrix} \qquad \dot{\vec{W}} = \begin{Bmatrix} \dot{w}_1(t) \\ \dot{w}_2(t) \\ . \\ . \\ \dot{w}_n(t) \end{Bmatrix}$$

Also, $[m]$ and $[k]$ are the mass and stiffness matrices of the element. Using Equations 8.2 and 8.3 in Lagrange's equations, we obtain the equation of motion of the finite element as

$$[m]\ddot{\vec{W}} + [k]\vec{W} = \vec{f}$$

$$\vec{f} = joint\ \ force\ vector \tag{8.4}$$

The acceleration vector is given by

$$\ddot{\vec{W}} = \begin{Bmatrix} \ddot{w}_1 \\ \ddot{w}_2 \\ . \\ . \\ \ddot{w}_n \end{Bmatrix} = \begin{Bmatrix} \dfrac{d^2 w_1}{dt^2} \\ \dfrac{d^2 w_2}{dt^2} \\ etc \end{Bmatrix}$$

It may now be appreciated that equation of motion has been expressed in terms of nodal displacements and accelerations, and thus for the entire system we can find the equations of motions in terms of nodal displacements/acceleration and we can form as many equations of motion as the number of unknown nodal displacements are. Of course, the procedure is not as simple as it may appear.

It must be remembered that the shape of the finite elements and the number of unknown joint displacements may differ for different applications. Also, Equation 8.4 is not useful directly to the assemblage of the elements. What we need to do is to assemble these elements, transformed from the local-coordinate system to the global-coordinate system of the entire system.

At this stage, it is useful to get some conceptual ideas about the shape functions used in the FEM analysis.

8.4 SHAPE FUNCTIONS

Let us consider a three-nodded triangular element as shown in Fig. 8.2.

For a three-nodded triangular element, the displacement at any point in the element is assumed as

$$u(x, y) = a + bx + cy \tag{8.5}$$

(For brevity we have dropped the time-variable in the above equation)
The nodal displacements then are given by

$$\begin{aligned} u_1(x_1, y_1) &= a + bx_1 + cy_1 \\ u_2(x_2, y_2) &= a + bx_2 + cy_2 \\ u_3(x_3, y_3) &= a + bx_3 + cy_3 \end{aligned} \tag{8.6}$$

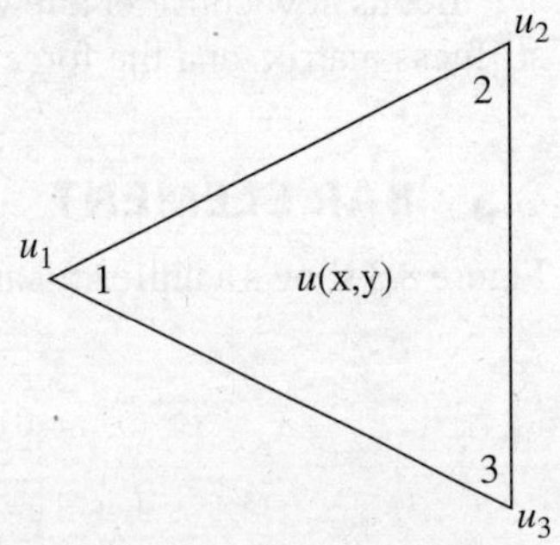

Figure 8.2 Triangular Three-Node Element

We can solve the above equations for the constants a, b, and c and substituting them in Equation 8.5, we get (details omitted)

$$u = N_1(x, y)u_1 + N_2(x, y)u_2 + N_3(x, y)u_3$$

where

$$\begin{aligned} N_1(x, y) &= \frac{1}{2A}\left[(x_2y_3 - x_3y_2) + (y_2 - y_3)x + (x_3 - x_2)y\right] \\ N_2(x, y) &= \frac{1}{2A}\left[(x_3y_1 - x_1y_3) + (y_3 - y_1)x + (x_1 - x_3)y\right] \\ N_3(x, y) &= \frac{1}{2A}\left[(x_1y_2 - y_1x_2) + (y_1 - y_2)x + (x_2 - x_1)y\right] \end{aligned} \tag{8.7}$$

where A = Area of the triangle

The following are the important points to be noted:

- Each of the shape function $N_i(x, y)$ has a value 1 at the node whose number it bears and zero at all other nodes. This means that

$$N_1(x_1, y_1) = 1$$
$$N_2(x_1, y_1) = N_3(x_1, y_1) = 0$$

  and so on
- The sum of the shape functions $\Sigma N_i(x, y) = 1$

- It is most important to note that the interpolation scheme for a given element will match that of adjacent element when they share two nodes in common and the function varies linearly between them. Thus, the displacement of a node that is common with adjacent elements is unique and single-valued. Thus there will not be any gaps/voids/discontinuities. This is extremely important to satisfy the condition of compatibility/continuum.
- The shape functions are therefore based upon a firm logic and are not arbitrary. For example, assuming a second order or cubic polynomial for triangular element will not give the shape continuity when adjacent elements are considered.

Let us now consider a few 1-D vibration-analysis problems. First, we shall deal with mass matrix, stiffness matrix, and the force vector for a bar element.

8.5 BAR ELEMENT

Figure 8.3 shows a uniform-bar element. For this 1-D element, the two end-points form the nodes/joints.

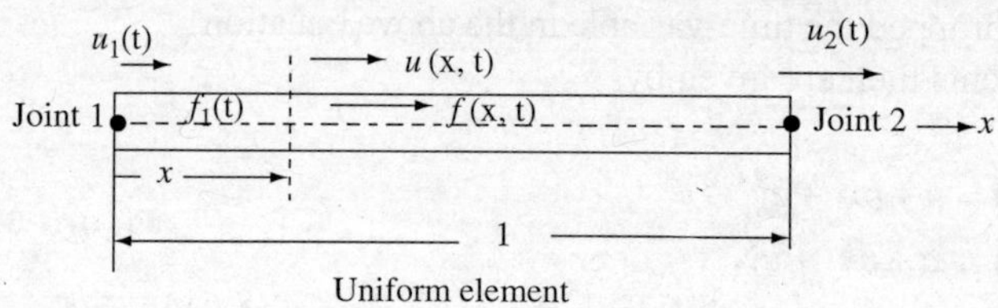

Figure 8.3 Bar Element

The axial displacement within the element, which is loaded by the external force $f_1(t)$ or more, is assumed to be linear in x as

$$u(x, t) = a(t) + b(t)x \tag{8.8}$$

The joint displacements $u_1(t)$ and $u_2(t)$ are the unknowns. Considering Joint 1 as the origin, we have,

$$u(0, t) = u_1(t) \text{ and } u(1, t) = u_2(t) \tag{8.9}$$

Using Equations 8.8 and 8.9, we get,

$$a(t) = u_1(t)$$

and

$$b(t) = [u_2(t) - u_1(t)]/l \tag{8.10}$$

Readers may note that we have exactly followed the procedure described in the previous Section 8.4. With the constants in equation thus determined, Equation 8.8 becomes

$$u(x, t) = [1 - (x/l)]u_1(t) + (x/l)\, u_2(t) \tag{8.11}$$

Equation 8.11 can also be written as

$$u(x, t) = N_1(x)\, u_1(t) + N_2(x)\, u_2(t) \tag{8.12}$$

In Equation 8.11,

$N_1(x) = [1 - (x/l)]$ and $N_2(x) = (x/l)$ are the shape functions. Thus, we can verify $N_1(0) = 1$ $N_1(1) = 0$ and $N_2(0) = 0$, $N_2(1) = 1$. Also, we can verify that $N_1(x) + N_2(x) = 1$. Thus, the conditions mentioned in the previous section are totally satisfied.

Our next step is to derive the equations of motion for this element. For this we require to evaluate the kinetic and the strain energy in the element. First, we derive equation for the kinetic energy. Let ρ be the density of the material of the bar, with cross-sectional area A. The kinetic energy $T(t)$ is given by

$$T(t) = \frac{1}{2}\int_0^l \rho A \left\{ \frac{\partial u(x,t)}{\partial t} \right\}^2 dx = \frac{1}{2}\int_0^l \rho A \left[(1-x/l)\frac{du_1(t)}{dt} + (x/l)\frac{du_2(t)}{dt} \right]^2 dx$$

$$= \frac{1}{2}\int_0^l \rho A \left[(1-x/l)\frac{du_1(t)}{dt} + (x/l)\frac{du_2(t)}{dt} \right]^2 dx$$

$$= \frac{1}{2}\frac{\rho A l}{3}(\dot{u}_1^{\,2} + \dot{u}_1\dot{u}_2 + \dot{u}_2^{\,2})$$

$$\dot{u}_1 = \frac{du_1(t)}{dt}, \dot{u}_2 = \frac{du_2(t)}{dt} \tag{8.13}$$

Equation 8.13 can also be written in the matrix form as

$$T(t) = \frac{1}{2}\vec{\dot{u}}(t)^T [m] \vec{\dot{u}}(t)$$

(Superscript T denotes the transpose.)

$$\vec{\dot{u}}(t) = \begin{Bmatrix} \dot{u}_1(t) \\ \dot{u}_2(t) \end{Bmatrix} \tag{8.14}$$

Examination of Equation 8.14 shows that the mass matrix is given by

$$[m] = \frac{\rho AL}{6}\begin{bmatrix} 2 & 1 \\ 1 & 2 \end{bmatrix} \tag{8.15}$$

The next step is to calculate the strain energy of the bar element. This is done as follows:

$$V(t) = \frac{1}{2}\int_0^l EA\left(\frac{\partial u(x,t)}{\partial x} \right)^2 dx$$

$$= \frac{1}{2}\int_0^l EA\left(-\frac{1}{l}u_1(t) + \frac{1}{l}u_2(t) \right)^2 dx = \frac{1}{2}\frac{EA}{l}(u_1^2 - 2u_1u_2 + u_2^2)$$

$$u_1 = u_1(t), u_2 = u_2(t) \tag{8.16}$$

Where E is the Young's modulus. We can express Equation 8.16 in the matrix form as

$$V(t) = \frac{1}{2}\vec{u}(t)^T [k] \vec{u}(t)$$

$$\vec{u}(t) = \begin{pmatrix} u_1(t) \\ u_2(t) \end{pmatrix}, \ \vec{u}^T = \{u_1(t) u_2(t)\} \tag{8.17}$$

Examination of Equation 8.17 shows that the stiffness matrix is given by

$$[k] = \frac{EA}{l}\begin{bmatrix} 1 & -1 \\ -1 & 1 \end{bmatrix} \tag{8.18}$$

We now evaluate the force vector. The element has been shown to a distributed axial load $f(x, t)$. We must find out the equivalent loads at the nodes due to distributed axial load $f(x, t)$. This can be evaluated by using the principle of virtual work. The virtual work δW can be expressed as

$$\delta W(t) = \int_0^l f(x,t)\delta u(x,t)dx = \int_0^l f(x,t)\left\{\left(1-\frac{x}{l}\right)\delta u_1(t) + \frac{x}{l}\delta u_2(t)\right\}dx \tag{8.19}$$
$$= \left(\int_0^l f(x,t)\left(1-\frac{x}{l}\right)dx\right)\delta u_1(t) + \left(\int_0^l f(x,t)\left(\frac{x}{l}\right)dx\right)\delta u_2(t)$$

Equation 8.19 can be expressed in matrix form as

$$\delta W(t) = \vec{\delta u}(t)^T \vec{f}(t) = f_1(t)\delta u_1(t) + f_2(t)\delta u_2(t) \tag{8.20}$$

Equation 8.20 shows that we can express the equivalent joint/node forces as

$$f_1(t) = \int_0^l f(x,t)\left(1-\frac{x}{l}\right)dx$$
$$f_2(t) = \int_0^l f(x,t)\left(\frac{x}{l}\right)dx \tag{8.21}$$

Till now we have considered a single-bar element. However, the structure might contain several bar elements that need not be collinear. Also, we need to find the equations of motion for the entire structure. We therefore need to understand how the equations of motion for the entire assemblage are derived from the equations of motion of individual elements. For example, consider an assemblage of bar elements as shown in Fig. 8.4.

Figure 8.4 shows assembly of four-bar elements. At Joint 1 we have $u_1(t)$ in the local x-axis. At Joint 2, there are three displacements, namely, $u_1(t)$ of the bar 2–3 in its X direction as shown, $u_1(t)$ of the bar 2–4 in its local X direction as shown, and $u_2(t)$ of bar 1–2 in its local x-axis as shown in Fig. 8.4. Joint 3 has two displacements and the Joint 4 has two displacements. Thus, unless we evaluate the joint displacements referred to global coordinates X–Y, we cannot derive the equations of motion for the assemblage of Fig. 8.4. The joint displacements referred to global-coordinate system are shown as $U_1(t)$, $U_2(t)$, $U_3(t)$, $U_4(t)$, $U_5(t)$, $U_6(t)$, $U_7(t)$ and $U_8(t)$, that is, ($U_i(t)$, $i = 1, 2, .., 8$).

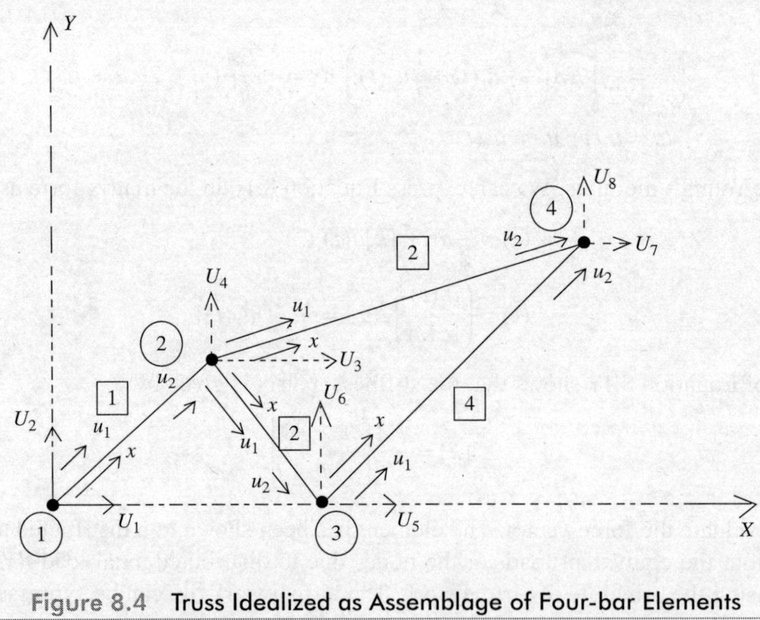

Figure 8.4 Truss Idealized as Assemblage of Four-bar Elements

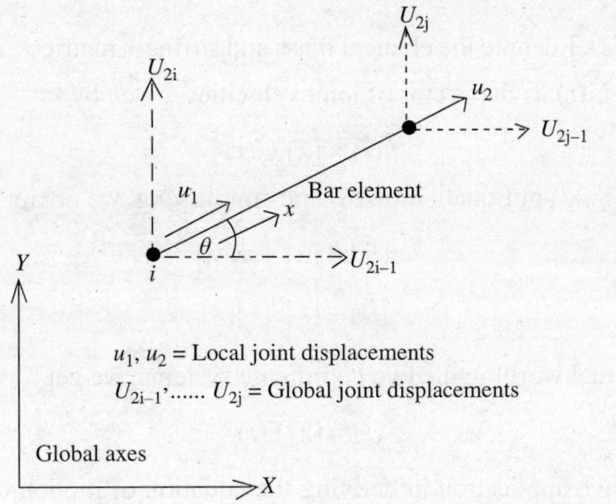

Figure 8.5 Local- and Global-Coordinates

The joint displacement in the local- and the global-coordinate system for a typical bar element is shown in Fig. 8.5. The two sets of joint displacements are related as shown in Fig. 8.5.

$$\begin{aligned} u_1(t) &= U_{2i-1}(t)\cos\theta + U_{2i}(t)\sin\theta \\ u_2(t) &= U_{2j-1}(t)\cos\theta + U_{2j}(t)\sin\theta \end{aligned} \tag{8.22}$$

Equation 8.22 can also be written as

$$\vec{u}(t) = [\lambda]\vec{U}(t) \tag{8.23}$$

where, λ the transformation matrix is given by

$$[\lambda] = \begin{bmatrix} \cos\theta & \sin\theta & 0 & 0 \\ 0 & 0 & \cos\theta & \sin\theta \end{bmatrix} \tag{8.24}$$

Thus, the joint vectors are

$$\vec{u}(t) = \begin{Bmatrix} u_1(t) \\ u_2(t) \end{Bmatrix}, \quad \vec{U}(t) = \begin{Bmatrix} U_{2i-1}(t) \\ U_{2i}(t) \\ U_{2j-1}(t) \\ U_{2j}(t) \end{Bmatrix} \tag{8.25}$$

Having found the transformation relations, it will be very useful to find mass matrix, stiffness matrix, and joint-force vector of an element in terms of global-coordinate system for finding the dynamic response of the complete system. It is well known that the energies of the elements are independent of coordinate system. This means that

$$\begin{aligned} T(t) &= \frac{1}{2}\vec{\dot{u}}(t)^T[m]\vec{\dot{u}}(t) = \frac{1}{2}\vec{\dot{U}}(t)^T[\bar{m}]\vec{\dot{U}} \\ V(t) &= \frac{1}{2}\vec{u}(t)^T[k]\vec{u}(t) = \frac{1}{2}\vec{U}(t)^T[\bar{k}]\vec{U} \end{aligned} \tag{8.26}$$

In these equations, $[\bar{m}],[\bar{k}]$ denote the element mass and stiffness matrices, respectively, in the global-coordinate system and $\dot{\vec{U}}(t)$ is the vector of joint velocities, given by

$$\dot{\vec{u}}(t) = [\lambda]\,\dot{\vec{U}}(t) \tag{8.27}$$

Substituting Equation (8.27) in Equation (8.26) and simplifying, we obtain

$$[\bar{m}] = [\lambda]^T [m][\lambda] \tag{8.28}$$

$$[\bar{k}] = [\lambda]^T [k][\dot{\lambda}] \tag{8.29}$$

and by equaling the virtual worth in the two coordinate systems, we get

$$\vec{\bar{f}}(t) = [\lambda]\vec{f}(t) \tag{8.30}$$

These equations have application in deriving the equation of motion of the complete system of finite elements. The procedure used is as follows.

Let the vector $\vec{U}*$ represent the joint displacements of the structure in the global-coordinate system as

$$\vec{U}*(t) = \begin{Bmatrix} U_1(t) \\ U_2(T) \\ .. \\ .. \\ U_M(t) \end{Bmatrix} \tag{8.31}$$

Let us now consider the joint displacements at a particular element e in the assemblage and denote its displacement vector as $\vec{U}^e(t)$. Since the joint vector of the element e can be identified in the vector given by Equation 8.31, the vectors $\vec{U}*(t)$ and $\vec{U}^e(t)$ are related by

$$\vec{U}^e(t) = [A^{(e)}]\,\vec{U}*(t) \tag{8.32}$$

$[A^{(e)}]$ is a rectangular matrix composed of ones and zeros. To understand this clearly, consider the element 1 assemblage shown in Fig. 8.4. The nodal/joint displacements for this element are

$$\vec{U}^{(1)} = \begin{Bmatrix} U_1(t) \\ U_2(t) \\ U_3(t) \\ U_4(t) \end{Bmatrix} = \begin{bmatrix} 1 & 0 & 0 & 0 & 0 & 0 & 0 & 0 \\ 0 & 1 & 0 & 0 & 0 & 0 & 0 & 0 \\ 0 & 0 & 1 & 0 & 0 & 0 & 0 & 0 \\ 0 & 0 & 0 & 1 & 0 & 0 & 0 & 0 \end{bmatrix} \begin{Bmatrix} U_1(t) \\ U_2(t) \\ U_3(t) \\ U_4(t) \\ . \\ . \\ U_8(t) \end{Bmatrix} \tag{8.33}$$

The *KE* of the complete structure can be obtained by adding the *KE* of individual elements. This means

$$T = \sum_{e=1}^{E} \frac{1}{2}\, \dot{U}^{(e)T}\, [\bar{m}]\, \dot{U}^{(e)} \tag{8.34(a)}$$

E is the number of elements in the assemblage. One can also note Equation 8.32 can be differentiated to obtain

$$\dot{\vec{U}}^e(t) = [A^{(e)}]\,\dot{\vec{U}}^*(t) \tag{8.34(b)}$$

Substituting the Equation [8.34 (b)] in Equation [8.34 (a)], we obtain,

$$T = \frac{1}{2}\sum_{e=1}^{E} \dot{\vec{U}}^T [A^{(e)}]^T [\bar{m}^{(e)}][A^{(e)}]\dot{U}^* \tag{8.35}$$

A close examination of Equation 8.35 reveals that the mass matrix for the entire assemblage is given by

$$[M] = \sum_{e=1}^{E} [A^{(e)}]^T [\bar{m}^{(e)}][A^{(e)}] \tag{8.36}$$

Similarly, we can show that stiffness matrix for the entire assemblage is given by

$$[K] = \sum_{e=1}^{E} [A^{(e)}]^T [\bar{k}^{(e)}][A^{(e)}] \tag{8.37}$$

$$force\ vector = \vec{F}^* = \sum_{e=1}^{E} [A^{(e)}]^T \vec{f}^e \tag{8.38}$$

We can now write the equation of motion for the entire assemblage as

$$[M]\,\ddot{\vec{U}}^* + [K]\,U^* = \vec{F}^* \tag{8.39}$$

8.6 BOUNDARY CONDITIONS

In the earlier analysis, we did not consider the boundary conditions. Thus, the complete structure is capable of undergoing rigid-body motions and the stiffness matrix will become singular. This is irrelevant to us as the structure will have some points which do not move. Thus, it is necessary to incorporate appropriate fixity conditions. A simple method of incorporating the zero-displacement conditions is to eliminate the corresponding rows and columns from matrices [M] and [K] and the vector $\vec{F}^*$. The final equations of motion of the restrained structure can be expressed as $[M]\ddot{\vec{U}}^* + [K]U^* = \vec{F}^*$, where [M] is $N \times N$ matrix, [K] is $N \times N$ matrix, $\ddot{\vec{U}}$, U^*, and $\vec{F}^*$ are $N \times 1$ vectors and B denotes the number of free joints of the structure (total number of joints minus number of fixed joints).

The methodology described in this section can be applied to various other elements provided we know the shape function for that element. We already described the triangular element where we showed that shape function linear both in x and y satisfies the basic requirement of analysing continuum by approximating the continuum as an assemblage of elements. Assuming shape function in FEM is similar to assuming sum of assumed functions, where each function denotes a deflection shape of the entire structure in the Rayleigh–Ritz method that we discussed earlier. Also, the levels of accuracies increase when we employ large number of elements and correct discretization strategies. Several textbooks are available to learn the finer tricks in FEM analysis. The aim of this chapter is only to provide a sort of familiarization to FEM. We now consider torsion element followed by beam element.

8.7 TORSION ELEMENT

Figure 8.6 shows a typical uniform-torsion element with the x-axis taken along the centroid axis.

Let I_P denote the polar moment of inertia about the centroid axis and GJ represents the torsional stiffness, A is the area of cross-section and ρ is the density of the material. The length of the bar element is l. The torsional displacement within the element is assumed to be linear in x-axis as

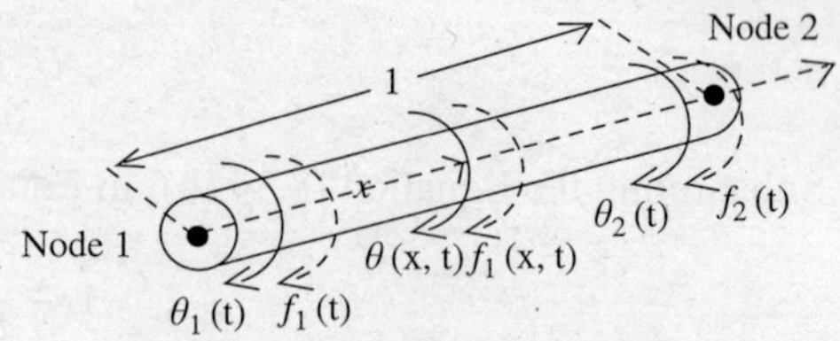

Figure 8.6 Uniform Torsion Element

$$\theta(x,t) = a(t) + b(t)x \tag{8.40}$$

The joint (node) displacements $\theta_1(t), \theta_2(t)$ are the unknown. As we did in bar element we express the torsional displacements within the element in terms of nodal displacements.

$$\theta(x,t) = N_1(x)\theta_1(t) + N_2(x)\theta_2(t) \tag{8.41}$$

We can see the similarity between bar element and torsion element. Proceeding in the manner with which we found the mass and stiffness matrices for bar element, we obtain the mass and stiffness matrices for torsion element as

$$[m] = \frac{\rho l I_P}{6}\begin{bmatrix} 2 & 1 \\ 1 & 2 \end{bmatrix}, \tag{8.42}$$

$$[k] = \frac{GJ}{l}\begin{bmatrix} 1 & -1 \\ -1 & 1 \end{bmatrix}, \tag{8.43}$$

$$\vec{f} = \begin{Bmatrix} f_1(t) \\ f_2(t) \end{Bmatrix} = \begin{Bmatrix} \int_0^l f(x,t)\left(1 - \frac{x}{l}\right)dx \\ \int_0^l f(x,t)(x/l)dx \end{Bmatrix} \tag{8.44}$$

Rest of the procedure is obvious.

8.8 BEAM ELEMENT

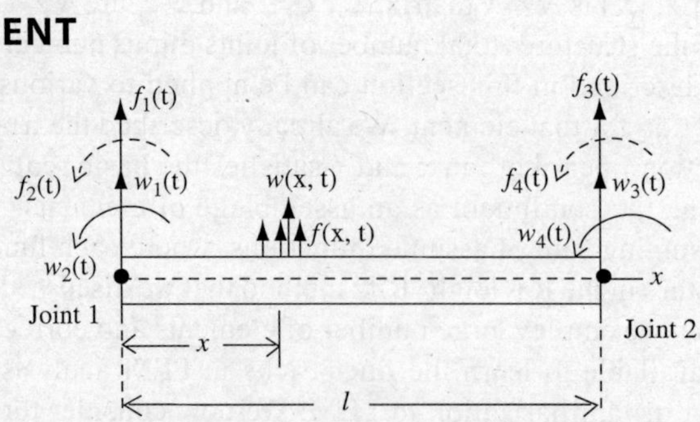

Figure 8.7 Beam Element

We now consider a uniform beam element subjected to transverse-force distribution, $f(x, t)$. In the beam element the nodes/joints experience both translational and rotational displacements. The translational displacements are $w_1(t)$ at node1 and $w_3(t)$ at node 2. The corresponding linear joint forces are designated as $f_1(t)$ and $f_3(t)$. The rotational displacements are $w_2(t)$ at node 1 and $w_4(t)$ at node 2. The corresponding rotational (bending) moments are $f_2(t)$ and $f_4(t)$. The transverse displacement within the element is $w(x, t)$ and is expressed by a cubic polynomial in x (as in the case of static deflection of a beam) as

$$w(x,t) = a(t) + b(t)x + c(t)x^2 + d(t)x^3 \tag{8.45}$$

As one can see that there are four unknowns and therefore we must have four boundary conditions. The unknown joint displacements must satisfy the conditions

$$\begin{aligned} w(0,t) &= w_1(t), \frac{\partial w}{\partial x}(0,t) = w_2(t) \\ w(l,t) &= w_3(t), \frac{\partial w}{\partial x}(l,t) = w_4(t) \end{aligned} \tag{8.46}$$

Using Equations (8.45) and (8.46), we obtain

$$\begin{aligned} a(t) &= w_1(t), b(t) = w_2(t) \\ c(t) &= \frac{1}{l^2}\left[-3w_1(t) - 2w_2(t)l + 3w_3(t) - w_4(t)l\right] \\ d(t) &= \frac{1}{l^3}\left[2w_1(t) + w_2(t)l - 2w_3(t) + w_4(t)l\right] \end{aligned} \tag{8.47}$$

Substituting Equation 8.47 in Equation 8.45, we obtain

$$\begin{aligned} w(x,t) = {} & \left(1 - 3\frac{x^2}{l^2} + 2\frac{x^3}{l^3}\right)w_1(t) + \left(\frac{x}{l} - 2\frac{x^2}{l^2} + \frac{x^3}{l^3}\right)lw_2(t) \\ & + \left(3\frac{x^2}{l^2} - 2\frac{x^3}{l^3}\right)w_3(t) + \left(-\frac{x^2}{l^2} + \frac{x^3}{l^3}\right)lw_4(t) \end{aligned} \tag{8.48}$$

We can write the Equation 8.48 as

$$w(x,t) = \sum_{i=1}^{4} N_i(x)w_i(t) \tag{8.49}$$

Thus we can identify the shape functions as

$$N_1(x) = 1 - 3\left(\frac{x}{l}\right)^2 + 2\left(\frac{x}{l}\right)^3 \tag{8.50}$$

$$N_2(x) = x - 2l\left(\frac{x}{l}\right)^2 + l\left(\frac{x}{l}\right)^3 \tag{8.51}$$

$$N_3(x) = 3\left(\frac{x}{l}\right)^2 - 2\left(\frac{x}{l}\right)^3 \tag{8.52}$$

$$N_4(x) = -l(x/l)^2 + l\left(\frac{x}{l}\right)^3 \tag{8.53}$$

One can verify that

$$\sum_{1}^{4} N_i(x) = 1$$
$$N_1(0) = 1$$
$$N_2(0) = N_3(0) = N_4(0) = 0\, etc.$$

The kinetic energy, the strain energy, and the virtual work of the element can be expressed as

$$T(t) = \frac{1}{2}\int_0^l \rho A \left\{\left[\frac{\partial W(x,t)}{\partial t}\right]^2\right\} dx \equiv \frac{1}{2}\dot{\vec{w}}(t)^T [m]\dot{\vec{w}} \tag{8.54}$$

$$V(t) = \frac{1}{2}\int_0^l EI \left\{\left[\frac{\partial^2 w(x,t)}{\partial x^2}\right]^2\right\} dx \equiv \frac{1}{2}\vec{w}(t)^T [k]\vec{w}(t) \tag{8.55}$$

$$\delta W(t) = \int_0^l f(x,t)\delta w(x,t)dx \equiv \delta\vec{w}(t)^T \vec{f}(t) \tag{8.56}$$

$$\rho = \text{density}, E = YM, I = MI, A = \text{Area}$$

$$\vec{w}(t) = \begin{Bmatrix} w_1(t) \\ w_2(t) \\ w_3(t) \\ w_4(t) \end{Bmatrix}, \dot{\vec{w}}(t) = \begin{Bmatrix} \frac{dw_1}{dt} \\ \frac{dw_2}{dt} \\ \frac{dw_3}{dt} \\ \frac{dw_4}{dt} \end{Bmatrix}, \delta\vec{w}(t) = \begin{Bmatrix} \delta w_1(t) \\ \delta w_2(t) \\ \delta w_3(t) \\ \delta w_4(t) \end{Bmatrix}, \vec{f}(t) = \begin{Bmatrix} f_1(t) \\ f_2(t) \\ f_3(t) \\ f_4(t) \end{Bmatrix}$$

By substituting Equation 8.48 into Equations 8.54−8.56 and carrying out the necessary integrations, we obtain the mass and stiffness matrices as,

$$[m] = \frac{\rho A l}{420}\begin{bmatrix} 156 & 22l & 54 & -13l \\ 22l & 4l^2 & 13l & -3l^2 \\ 54 & 13l & 156 & -22l \\ -13l & -3l^2 & -22l & 4l^2 \end{bmatrix} \tag{8.57}$$

$$[k] = \frac{EI}{l^3}\begin{bmatrix} 12 & 6l & -12 & 6l \\ 6l & 4l^2 & -6l & 2l^2 \\ -12 & -6l & 12 & -6l \\ 6l & 2l^3 & -6l & 4l^2 \end{bmatrix}, \tag{8.58}$$

$$f_i(t) = \int_0^l f(x,t) N_i(x) dx, \; i = 1 \text{ to } 4 \tag{8.59}$$

One can easily see the level of computations required when the structure is divided into several finite elements. Also, it is necessary to remember that the type of element used for various components of the structure need not be the same. We have discussed this while we discussed FEM analysis of the milling machine. We understood that the basic mode in which a component is likely to behave (beam/plate, 2-D or 3-D) and accordingly we choose the element. In absence of understanding these basic aspects, the FEM programs will yield a result that cannot be trusted. The purpose of this chapter is to provide the basic concepts used in FEM. Once these are understood well, FEM analysis through commercial grade software such as ANSYS, NISA, etc. will be highly useful.

MATLAB—TOOL FOR COMPUTATION

INTRODUCTION

MATLAB (an abbreviation of MATrix LABoratory) is computer software developed by Math Works Inc. This software is widely used in the fields of science and engineering. MATLAB is an interactive program for numerical computations and data visualization and is supported on UNIX, Macintosh, and windows environments.

MATLAB integrates mathematical computing and visualization. It is easy to use MATLAB to explore data, create algorithms etc., and is useful for engineering students as well as professionals.

For more information on MATLAB, one can contact **The Math Works.com**. Typical applications include: (1) numeric computation and development of algorithms, (2) modelling, simulation, (3) data analysis and signal processing, and (4) engineering graphics and scientific visualization. We in this section will discuss salient features of MATLAB. The matter presented here is not a package for total familiarization of MATLAB but aims at giving a broader ideas about what this package is capable of.

(I) Display Windows

MATLAB has three types of windows. They include the following.

- **A command window,** which is used to enter commands and the data to display plots and graphs.
- **A graphics window** used for displaying plots and graphs.
- **An edit window** used for creating and modifying M-files that contain a program or script of MATLAB commands. These commands are case-sensitive and lower-case letters are used. To execute an M-file, one can simply enter the name of the file without its extension.

MATLAB capabilities can be learnt by entering *demo* command. One can learn the method of operating MATLAB through demo versions. The intent of this section is to explain various capabilities of the program.

(II) Arithmetic Operations

They are as follows.

1) Addition. Symbol is + as in 7 + 1 = 8
2) Subtraction. Symbol is − as in 7 − 1 = 6
3) Multiplication. Symbol is * as in 7 * 2 = 14
4) Right division. Symbol is / as in 8/2 = 4
5) Left division. Symbol is \ as 8\2 = 2/8 = 1/4
6) Exponentiations. Symbol is ^ as 6^3 = (6^3) = 216

MATLAB has several different screen-output formats. These commands are (a) **format-short**—fixed point with four decimal places (like 50.1429), **(b) format-long**—fixed point with 14 decimal digits and (c) **format-short e**—scientific notation with four decimal digits (like 5.0143e+001). There are other formats also such as format-long e, format-short g, format-long g, format-bank etc. One can choose one from them depending upon the requirement of accuracy.

(III) Built-in Functions

(a) Common Math Functions

- **abs(x)** computes absolute value of **x**
- **sqrt(x)** square root of **x**
- **rounds(x)** rounds **x** to nearest integer
- **fix(x)** truncates/rounds **x** to the nearest integer towards zero.
- **floor(x)** rounds **x** to the nearest integer towards $-\infty$
- **ceil(x)** rounds **x** to the nearest integer towards ∞
- **sign(x)** returns a value of -1 if **x** is less than zero, a value of 0 if **x = 0,** and a value of it if otherwise
- **rem(x, y)** returns the remainder of x/y. **rem(21, 5)** is 1 and **rem(100, 21)** is 16. This function is also called **modulus**
- **exp(x)** computes e^x. e is the base of natural logarithm
- **log(x)** computes natural logarithm of **x** (to the base e)
- **log 10(x)** computes logarithm to the base 10
- **sin(x), cos(x), tan(x)** compute sine, cosine, and tangent of **x**, respectively (trigonometric functions)
- **asin(x)** computes inverse sine of **x** (**x** must be between -1 and 1. The function returns an angle in radians between $-\pi/2$ and $\pi/2$
- **acos(x)** computes inverse cosine of **x (x** must be -1 and 1. The function returns the angle in radians between 0 and π
- **atan(x)** computes inverse tangent of **x**. The angle in radians between $-\pi/2$ and $\pi/2$
- **atan2(y, x)** computes inverse tangent of the value of *y*/*x*. The function returns an angle in radians between $-\pi$ and π, depending upon the signs of **x** and **y**
- **sinh(x)** computes hyperbolic sine ($= \{e^x - e^{-x}\}/2$)
- **cosh(x)** computes hyperbolic cosine
- **tanh(x)** computes hyperbolic tangent
- **asinh(x), acosh(x), atanh(x)** compute inverse hyperbolic sine, cosine, and tangent, respectively. These are natural logarithms of $\left(x+\sqrt{x^2+1}\right), x+\sqrt{x^2-1}$, and $\sqrt{\dfrac{1+x}{1-x}}$, respectively.
- **conjugate(x)** computes complex conjugate of complex number *x*
- **real(x), imag(x), abs(x)** computes real part, imaginary part, and the absolute value of the complex number *x*. For complex number **x, angle(x)** computes the angle using the value of **atan2(image(x),real(x))**. Arithmetic operations such as addition, subtraction, multiplication, division, mod value are done when complex numbers are used in computations. In MATLAB, variables can be named using letter(s) and digits. These names are case-sensitive. The names of built-in functions for a variable are not permitted since once a function is used to define a variable, the function cannot be used. MATLAB has predefined variables such as **ans, pi, eps, inf, I, j, NaN, clock, date,** etc. One can know these through demo packages.

- There are commands for managing variables. Some of these are: **clear** (which removes all variables from memory), **clear x,y,** (clears only variables *x, y*, and *z* from the memory **z** (lists the variables currently in the workspace) etc.
- There are general commands for online-help workspace information, directory information, and termination. It is necessary that the user is familiar with all these.

Let us now understand how we can input various quantities such as row vector, column vector, and matrix. Also at times we need to address arrays. We shall show this through some examples.

Row vector In this the elements are entered with a space or comma between the elements inside the square brackets. For example,

$$x = [7 \quad -1 \quad 2 \quad -5 \quad 8]$$

Column vector In a column vector, the elements are entered with a semicolon between the elements inside the square brackets. For example,

$$x = [7; \ -1 \quad 3 \quad -5 \quad 7]$$

(IV) Matrix

The MATLAB input command is a row followed by a semicolon by another. For example $\begin{bmatrix} 1 & 2 & 3 \\ 4 & 5 & 6 \\ 7 & 8 & 9 \end{bmatrix}$ is written as [1 2 3 ; 4 5 6 ; 4 5 6 ; 7 8 9] or for complex matrix

$$\begin{bmatrix} -5x & \ln 3x + 5y \\ 4i & 5-3i \end{bmatrix}$$ is written as [−5*x log(3*x+5*y) ; 4i 5−3i]

For addressing arrays, MATLAB uses a colon to address a range of elements in a vector or a matrix. For example,

- **Va**(:) refers to all the elements of the vector Va (which could be a row vector or a column vector
- **Va**(*m*:*n*) to elements *m* through *n* of the vector Va
- **A**(:, *n*) refers to the elements in all the rows of a column *n* of the matrix A
- **A**(*n*, :) refers to the elements in all the columns of row *n* of the matrix A
- **A**(:, *m*:*n*) refers to the elements in all the rows between column *m* and *n* of the matrix A
- **A**(*m*:*n*, :) refers to the elements in all the columns between rows *m* and *n* of matrix A
- **A**(*m*:*n, p*:*q*) refers to the elements in rows *m* through *n* and columns *p* through *q* of the matrix A

These commands give extreme versatility in handling matrices for various operations. Also, we can add elements to a vector or matrix or delete some elements.

For matrix operations, MATLAB has some built-in functions such as **length(A),** which shows the number of elements in the matrix A, **size(A)**, which means a row vector (*m n*). For example, for a matrix [2 3 0 8 11; 6 17 5 7 1) the command is >>size(A) and the answer is 2 5.

Another important command is **reshape(A, m, n)**. For example, A is [3 3 4; 0 0 2], the **reshape(A, 3, 2)** would mean a matrix B = [3 0; 0 4; 3 2] that is, >>b=reshape(A, 3, 2)

- **diag(v)** when v is a vector creates a square matrix with the elements of v in the diagonal. Suppose v = [3 2 1] then
- >>**A**= diag(v) will create a square matrix A = [3 0 0; 0 2 0; 0 0 1]. On the other hand suppose we give the same command for a matrix A, MATLAB creates a vector from the diagonal elements of A.

Adding or subtracting their corresponding elements obtains the matrix addition and subtractions of the two arrays. These operations are performed with arrays of identical size (same number or rows and columns)

- **Dot product** is a scalar computed from two vectors of the same size. Thus A.B = $\sum_{i=1}^{n} a_i b_i$. If A and B are matrices, the dot product is a row vector containing the dot products for the corresponding columns of A and B.
- **The array multiplication** the value in position c_{ij} of the product C of the matrices A and B is the dot product of row i of the first matrix and column j of the second matrix.
- **The array division** operation can be explained by means of the identity matrix and the matrix inverse operation.
- **The transpose** of a matrix **A** is denoted by **A′**
- **Determinant A** MATLAB computes the determinant of a square matrix by **det(A)**
- **Array division** MATLAB provides two types of array division, namely left division and the right division.

  Left division is used to solve matrix equation Ax = B where x and B are **column vectors**; thus x = A^{-1}B. In MATLAB it is written as **x = A\B**

  Right division is used to solve matrix equation xA = B where x and B are **row vectors**. In MATLAB this is written as **x = B/A**
- **eigenvalues and eigenvectors**

Command **eigen(A)** computed a column vector containing the eigenvectors of A. There is one important command **[Q, d]**, which computes a square matrix **Q** containing eigenvectors of A as columns and a square matrix d containing the eigenvalues (λ) of A on the diagonals. The values of Q and d are such that Q*Q is the identity matrix and A*X equals λ times X. There are other commands for triangular factorization or lower–upper factorization. Singular value decomposition **[svd]** decomposes a matrix A ($m \times n$) into a product of three matrix factors **[U S V] = svd(A)**. In this U and V are orthogonal matrices and S is a diagonal matrix.

- **Arithmetic operators for matrices are: (1) + addition, (2) − subtraction, (3) * multiplication, (4) Λ exponentiation. (5) / left division** and (6) **\ right division.**
- **Arithmetic operators for array operators are: (1) + addition, (2) − subtraction, (3) .* array multiplication, (4) .^ array exponentiation, (5) ./ array left division and (6) .\ array right division**
- There are built-in functions like **mean(A),** which denotes mean value of elements in vector A; **C = max(A),** which denotes largest element in vector A; **min(A); sum(A); sort(A); median (A); dot**(a, b), which calculates scalar (dot) product of vectors a and b; **cross** (a, b), which does the cross product of vectors a and b; **inv(A) gives inverse of the matrix A.**
- MATLAB has commands for random-number generation, which has many applications in physical sciences and engineering.

(V) Polynomials:

- Suppose we have a polynomial $f(x) = a_0x^n + a_1x^{n-1} + a_2x^{n-2} + ... + a_{n-1}x^1 + a_n$. A vector represents a polynomial in MATLAB. This is entered as data in MATLAB by entering each coefficient of the polynomial into vector in descending order. The above polynomial is entered as

```
>>x = [a0    a1    a2    an-1    an]
x =  a0    a1    a2    an-1    an
```

- MATLAB contains functions that perform **polynomial multiplications and divisions**. For example, **conv(a, b)** computes a coefficient vector that contains the coefficients of the product of polynomials represented by coefficients in **a** and **b**. These vectors need not have the same size. **[q, r]** = **deconv(n, d)** returns two vectors. The first vector contains the coefficients of the quotient and the second vector contains the coefficients of the remainder polynomial.
- MATLAB function for determining the roots of a polynomial is **root(a)**, where **a** represents vector consisting of coefficient as mentioned. Conversely, if roots are known, the coefficients of the polynomial are determined using **poly(r)**. Let us consider a polynomial $5s^5 + 7s^4 + 2s^2 - 6s + 10$. Let us find roots of this polynomial. We use **root(a)** for this purpose. We write the following.

```
>>roots([5  7  0  2  −6  10])

ans =

    −1.8652
    0.4641 +1.0832 i
    0.4641 +1.0832 i
    0.6967 + 0.5355 i
    0.6967 − 0.5355 i
```

Note that the coefficient of cubic term is placed zero in the vector. Also note that solution contains complex numbers and there are five roots of this fifth-order polynomial.

Let us now see how polynomial multiplication and division is done. Let the two polynomials be $x = 2x + 5$ and $y = x^2 + 3x + 7$. For this we write

```
>>x = [2  5]
>>y = [1  3  7]
>>z = conv(x, y)
z = 34  2  11  29  35
```

This is a cubic polynomial with four terms. Let us now see the division. For this we use the command **deconv** for dividing z as found above by $2x + 5$ as follows.

```
z = [2  11  29  35]; x = [2  5]
>> [g, t] = deconv(z, x)
g = 1  3  7
t = 0  0  0  0
```

Note that since z is a cubic polynomial, it has four terms and since $2x + 5$ is one of the roots of z, the remainder is zero for all four terms.

(VI) System of Linear Equations

The system of (non-singular) equations **Ax** = **B** can be solved either by (a) matrix-division method or (b) matrix-inversion method. In the former, we use **X** = **A/B**. The vector **X** contains the values of **x**. In the matrix-inversion method, we give a command **x** = **inv (A)*B**. We can also use the command **x = B*inv (A)**

It should always be remembered that matrix expression is enclosed in square brackets [ ], blanks or commas separate the column elements and semicolons separate rows. Also the basic computational unit in MATLAB is the matrix. Let us show the use of MATLAB with illustrative examples

(1) Let us consider a simple matrix $\begin{bmatrix} 1 & 2 & 3 & 4 \\ 5 & 6 & 7 & 8 \\ 9 & 10 & 11 & 12 \end{bmatrix}$. Let us write transpose of this matrix

We now write the matrix as >> A = [1 2 3 4 ; 5 6 7 8 ; 9 10 11 12]
>> B = A'

$$B = [1 \quad 5 \quad 9\,;2 \quad 6 \quad 10\,;3 \quad 7 \quad 11\,;4 \quad 8 \quad 12] \quad \text{or} \quad B = \begin{bmatrix} 1 & 5 & 9 \\ 2 & 6 & 10 \\ 3 & 7 & 11 \\ 4 & 8 & 12 \end{bmatrix}.$$

Let us now carry out A*B. For this we give command

>> C = A*B.

The result is $C = \begin{bmatrix} 107 & 122 & 137 & 152; \\ 122 & 140 & 158 & 176; \\ 137 & 158 & 179 & 200; \\ 152 & 176 & 200 & 224; \end{bmatrix}$, which is a square matrix

(2) Let us find inverse of a matrix $D = \begin{bmatrix} 1 & 2 \\ 3 & 4 \end{bmatrix}$. For this we write

>> D = [1 2 ; 3 4]
>> E = inv(D)

$$E = [-2.0000 \quad 1.000\,;1.5000 \quad -0.5000] \quad \text{or} \quad E = \begin{bmatrix} -2.0000 & 1.0000 \\ 1.50000 & -0.50000 \end{bmatrix}$$

(3) Let us find the eigenvalues of matrix D as given in Example 2 above. For this we write >> eig(D). (Remember that D is written as [1 2 ; 3 4]).

ans =

[−0.3723 ; 5.3723].

Note that this is a row vector.

(4) We have two matrices A and B as given here

$$A = \begin{bmatrix} 1 & 0 & 1 \\ 2 & 3 & 4 \\ -1 & 6 & 7 \end{bmatrix} \quad \text{and} \quad B = \begin{bmatrix} 7 & 4 & 2 \\ 3 & 5 & 6 \\ -1 & 2 & 1 \end{bmatrix}.$$

Let us carry out some matrix operations

$$\mathbf{A + B, AB, A^2, B^{-1}, B^TA^T}$$

The first step in this is to write the matrices as per MATLAB practice.

```
>> A = [1  0  1 ; 2  3  4 ; −1  6  7]
>> B = [7  4  2 ; 3  5  6 ; −1  2  1]
```

>> C = A + B = [8 4 3 ; 5 8 10 ; −2 8 8]

>> D = A*B = [6 6 3 ; 19 31 26 ; 4 40 41]

$$D = \begin{bmatrix} 6 & 6 & 3 \\ 19 & 31 & 26 \\ 4 & 40 & 41 \end{bmatrix}$$

>> E = $A^{\wedge}2$ **= [0 6 8 ; 4 33 42 ; 4 60 72]**

$$E = \begin{bmatrix} 0 & 6 & 8 \\ 4 & 33 & 42 \\ 4 & 60 & 72 \end{bmatrix}$$

>> H = inv(B) = [0.1111 0.0000 −0.2222 N ; 0.1429 −0.1429 0.5714 ; −0.1746 0.2857 −0.3651]

$$H = \begin{bmatrix} 0.1111 & 0.0000 & -0.2222 \\ 0.1429 & -0.1429 & 0.5714 \\ -0.1746 & 0.2857 & -0.3651 \end{bmatrix}$$

We can find the transposes of matrices B and A using commands described earlier and then evaluate their product B^TA^T using appropriate commands. We can also find the det(A), det(A*B) and also the eigenvalues and eigenvector.

(5) Solve the equations

$$y - 3z = -5$$

$$2x + 3y - z = 7$$

$$4x + 5y - 2z = 10$$

These are linear-algebraic equations in *x*, *y*, and z. For use in MATLAB, we rename these as x_1, x_3, x_3. Thus, we can identify the matrix and the vectors, which generate the above equations. They are

$$\mathbf{A} = \begin{bmatrix} 0 & 1 & -3 \\ 2 & 3 & -1 \\ 4 & 5 & 2 \end{bmatrix}, \mathbf{B} = \begin{bmatrix} 5 \\ 7 \\ 10 \end{bmatrix} \text{ and } \mathbf{X} = \begin{bmatrix} x_1 \\ x_2 \\ x_3 \end{bmatrix}.$$

We thus have

$$\mathbf{AX} = \mathbf{B}$$

$$\mathbf{A}^{-1}\mathbf{AX} = \mathbf{A}^{-1}\mathbf{B}$$

$$\mathbf{IX} = \mathbf{A}^{-1}\mathbf{B}$$

$$\mathbf{X} = \mathbf{A}^{-1}\mathbf{B}$$

We can now use MATLAB commands as given here.

```
>> A = [0   1   −3 ; 2   3   −1 ; 4   5   −2] ;
>> B = [−5; 7; 10]
>> X = inv(A)*B
```

We can alternatively use command

```
>> X = A\B
```

One of the most significant features of MATLAB is that in this we can write programs, which are known as the M-files. M-files are ordinary ASCII text files written in MATLAB language. A function file is a sub-program in such programs. In MATLAB, we have relational/logical operators, built-in logical functions, conditional statements, and also powerful graphics commands that generate 2-D plots, overlay plots, 3-D plots, mesh, and surface plots. Due to its capability to carry out complex matrix calculations such as matrix inversion, determination of eigen values and eigen vectors, it is of great value in vibration analysis.[1, 2] Also, the mesh generation capabilities along with matrix calculations also make it useful in FEM analysis for stress analysis as well as computational fluid mechanics. With elementary orientation given in this book, the readers should attempt computations pertaining to polynomials, system of equations, eigen values, and eigen vectors. The other aspects such as using plot commands, writing programs will become clear using demo package. This approach, is better than providing solutions to numerical problems using C++ or FORTRAN programming.

With MATLAB, it is possible to solve the differential equations, differentiation, integrations LAPLACE and other transforms, etc. The demo program of MATLAB describes these capabilities and one can self-learn the MATLAB in due course.

CONCLUSION

MATLAB is a very powerful tool for carrying out computational analysis for stream, vibration, fluid mechanics etc., as it enables solution of the governing equalities in quick time. The versatility of matrix operations is the greatest advantage. In vibration analysis, before attempting solution using commercial softwares, MATLAB can be very useful to know the nature of correct solution by working approximate simulation of the existing system. (See Fig. 6.3 where building is simulated as lumped system).

References

1. O.C. Zienkiewez, *The Finite Element Method* (4th ed.) McGraw−Hill, London, 1987
2. S.S. Rao, *The Finite Element Method in Engineering* (3rd ed.), Butterworth−Heinemann, Boston, 1999

Exercises

8.1 Derive the expression for the shape function for a uniform-bar element.

8.2 For a bar element, we choose the relationship for the axial displacement as $u(x, t) = a(t) + b(t)\,x$. This shows a linear relationship. Explain the logic behind this. What would happen if we choose a polynomial expression for the axial displacement?

Hint: See the analysis of structure comprising of four-bar elements shown in Fig. 8.4. At point 1, we have $u_1(t)$ in the local x-axis. At Joint 2, there are three displacements, namely $u_1(t)$ of the bar 2–3 in its x direction as shown, $u_1(t)$ of the bar 2–4 in its local x direction as shown and $u_2(t)$ of bar 1–2 in its local x-axis as shown in Fig. 4. Joint 3 has two displacements and the Joint 4 has two displacements. Show that unless we use linear-shape function the condition of shape compatibility cannot be satisfied.

8.3 Find the natural frequencies of simply-supported beam. You may use a single element.

Answer: $\omega_1 = \left(\dfrac{120EI}{\rho A l^4}\right)^{1/2}, \quad \omega_2 = \left(\dfrac{2520EI}{\rho A l^4}\right)^{1/2}$

8.4 Find the natural frequencies of rod fixed at both ends. You may idealize the rod as an assemblage of two-bar elements.

Answer: The exact natural frequency $\omega_{exact} = 3.142\sqrt{\dfrac{E}{\rho L^2}}$

$$\omega = 3.464\sqrt{\frac{E}{\rho L^2}}$$

8.5 A stepped bar as shown in figure below is subjected to axial vibrations. Derive the equation of eigen value problem using FEM approach.

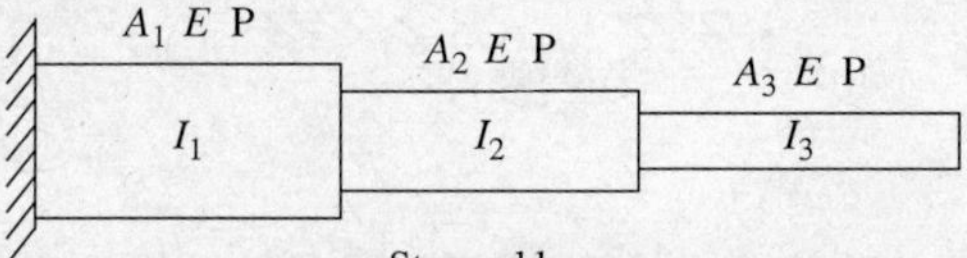

Stepped bar

Hint: Consider the rod to be assemblage of bar elements. The stiffness and mass matrices for the assembly are:

$$[K] = \begin{bmatrix} A_1E/l_1 & -A_1E/l_1 & 0 & 0 \\ -A_1E/l_1 & (A_1E/l_1) + (A_2E/l_2) & -A_2E/l_2 & 0 \\ 0 & -A_2E/l_2 & (A_2E/l_2) + (A_3E/l_3) & -A_3E/l_3 \\ 0 & 0 & -A_3E/l_3 & A_3E/l_3 \end{bmatrix}$$

$$[M]=\frac{1}{6}\begin{bmatrix} 2\rho A_1 l_1 & \rho A_1 l_1 & 0 & 0 \\ \rho A_1 l_1 & 2\rho A_1 l_1 + 2\rho A_2 l_2 & \rho A_2 l_2 & 0 \\ 0 & \rho A_2 l_2 & 2\rho A_2 l_2 + 2\rho A_3 l_3 & \rho A_3 l_3 \\ 0 & 0 & \rho A_3 l_3 & 2\rho A_3 l_3 \end{bmatrix}$$

The frequency equation is given by

$$\left[[K]-\omega^2[M]\right]\begin{Bmatrix} u_1 \\ u_2 \\ u_3 \end{Bmatrix}=\begin{Bmatrix} 0 \\ 0 \\ 0 \end{Bmatrix}$$

In the above, the nodal points are at the intersections of the areas and u_i are the nodal displacements.

8.6 Find the frequency equation for a beam of length fixed at both ends. The beam has uniform cross-sections A, E, and I. Consider the beam to be assemblage of three-beam elements each of length $l/3$.

Hint: Use MATLAB program called **Program17.m**

8.7 Figure below shows a two bar truss. The length of bars is 25 cm. The truss is subjected to vibratory force F in the horizontal direction. Show how you will evaluate the natural frequencies of the structure.

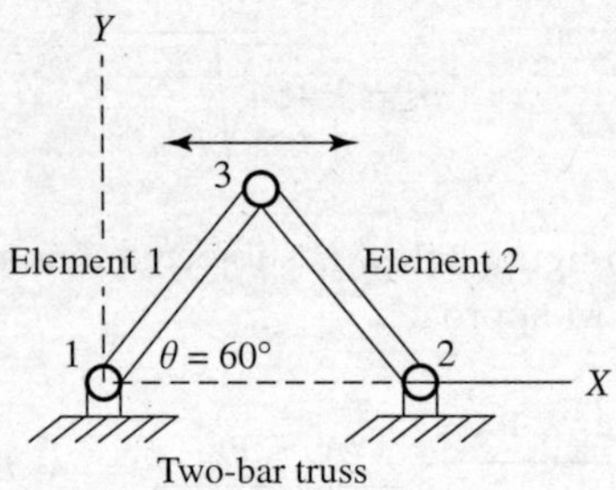

Two-bar truss

Hint: (1) Make the finite element model as shown below. Follow the procedure outlined in paragraph 8.4 for four-bar truss.

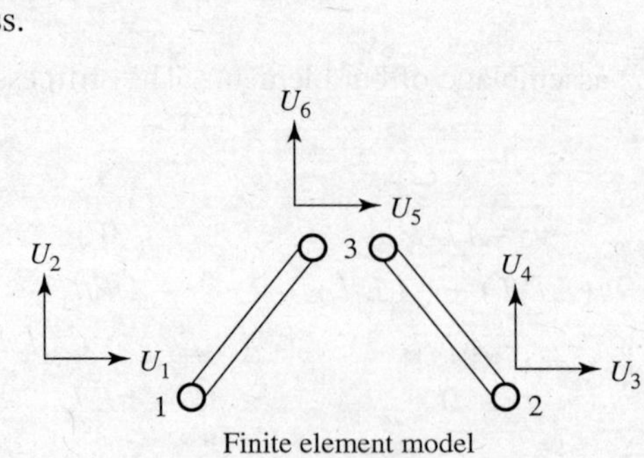

Finite element model

9

Fundamentals of Experimental Modal Analysis

9.1 INTRODUCTION

Modal analysis is a technique to find the intrinsic properties of components of a machine/structure in terms of their natural frequencies and the mode shapes in each natural frequency mode of vibrations. We know that these largely depend upon the mass distribution, deformations, as well as the damping in the system. For simplicity, we generally ignore the internal damping in the system. We shall explain the basic concepts of modal analysis by analysing the modes of vibration of a simple plate, which is freely suspended as shown in Fig. 9.1.

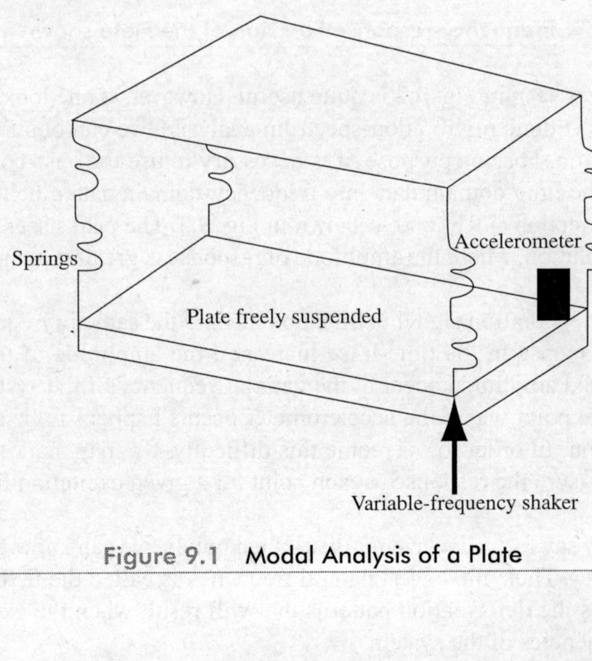

Figure 9.1 Modal Analysis of a Plate

This configuration is called free–free configuration. A constant force will be applied to one corner of the plate as shown in the Fig. 9.1, but the frequency of the excitation will be changed in a sinusoidal fashion. The accelerometer placed at a corner of the plate (Fig. 9.1) measures the response of the plate excited by a sinusoidal force of constant amplitude but with varying frequency.

The time-response of the accelerometer in terms of the level of amplitude of vibration, which changes as the frequency of vibratory force changes, is shown in Fig. 9.2.

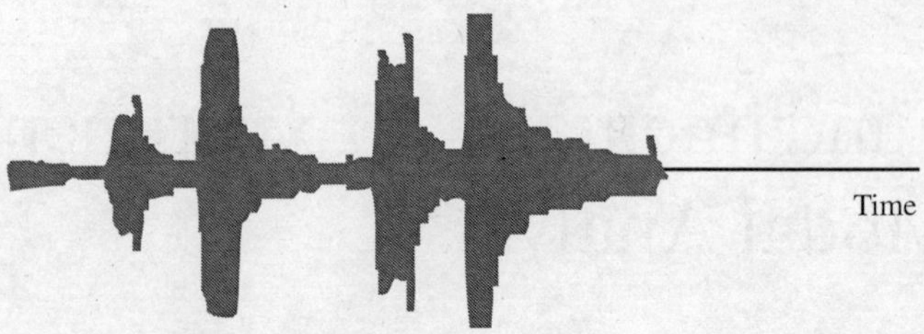

Figure 9.2 Vibration Response as a Function of Time for the Plate in Fig. 9.1

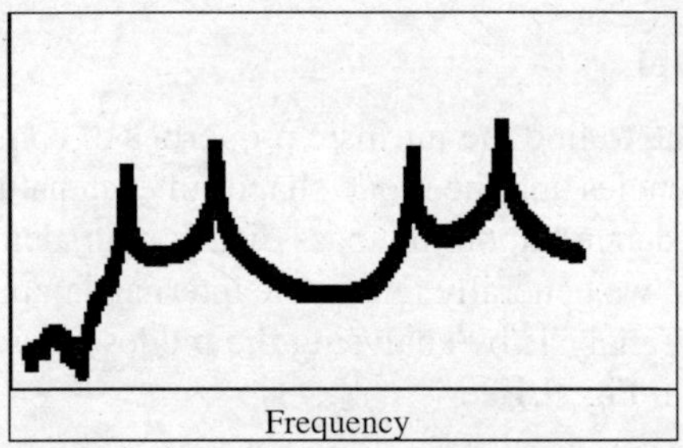

Figure 9.3 Frequency-response Function of the Plate shown in Fig. 9.1

The time-response as seen in Fig. 9.2 is quite useful. However, if one looks to this information in a frequency domain (as is done in vibration-spectrum analysis), we can obtain some very interesting and important information. For this purpose, it is necessary to use the Fast-Fourier-Transform (FFT) analyser that converts the time-domain data into frequency-domain data. Such a plot is loosely called Frequency-Response Function (FRF) and is shown in Fig. 9.3. The peaks seen in Fig. 9.3 correspond to the frequency of oscillation, where the amplitude of response is greatest, which in other words mean the natural frequencies.

Figure 9.4 shows the vibration signal in time domain and the same in frequency domain together. As the amplitude of response in the time-trace increases, the amplitude of the FRF also increases. These points of increased amplitude occur at the natural frequency of the system. However this may not always happen if the point where the accelerometer occurs happens to be a nodal point for a particular mode of vibration. In order to overcome this difficulty, we may mark several points (through grid structures) and measure the response at each point for a given excitation frequency that gives the highest response.

Thus, if we identify say forty-five evenly-distributed points, we can know the deformation pattern at each natural frequency. Thus, the lowest natural frequency is called the first or fundamental mode. The figure below shows the deformation patterns that will result when the excitation coincides with one of the natural frequencies of the system.

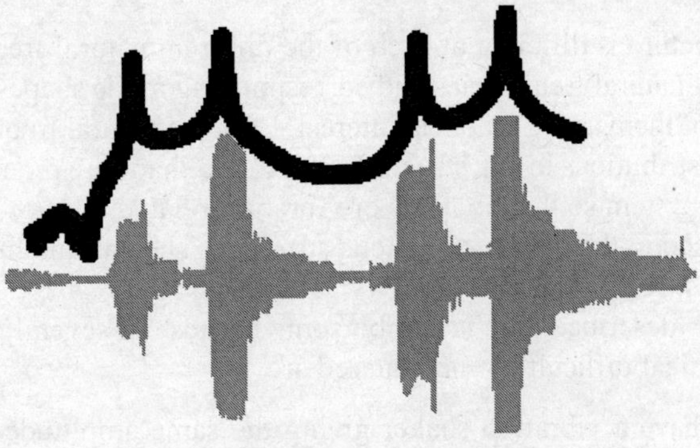

Figure 9.4 Frequency-response and Time Response of the Vibration of Plate in Fig. 9.1

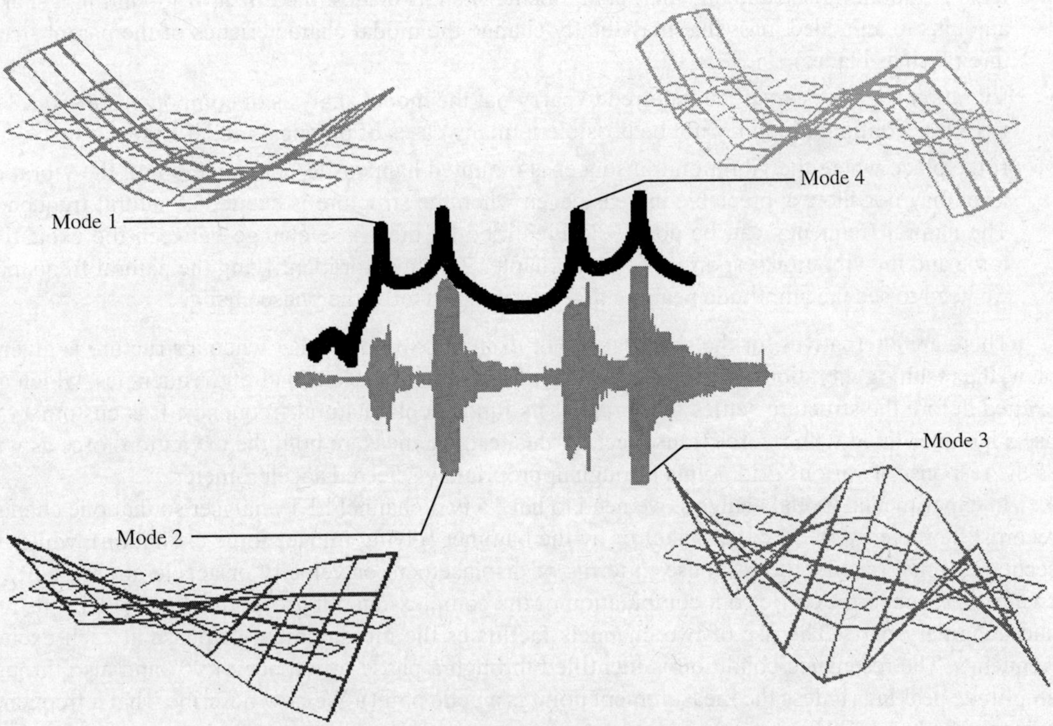

Figure 9.5 Various Mode Shapes of the Plate

The first frequency is called the fundamental natural frequency. When excited at this frequency, the plate vibrates in the bending mode. The shape of the plate is termed as the fundamental mode or mode 1 as shown in Fig. 9.5. Once we cross this frequency, the amplitude of vibration decreases until such time that the next natural frequency is approached. The next frequency at which the vibration response reaches maximum is called the second natural frequency and the corresponding mode is called the second mode. The shape of deformation of the plate in Fig. 9.5 shows a twisting motion. This is followed by the third natural frequency, where the mode of vibration is again a bending mode, called the second bending mode. Figure 9.5 shows the fourth mode, which is a second twisting mode.

The response of the structure is different at each of the different natural frequencies. In theory, there are an infinite number of natural frequencies and corresponding mode shapes. However, as the experience shows, only a few of them are of practical interest. Both the natural frequency (which depends on the mass and stiffness distributions in any structure) and mode shape are required to be used during the design of any structural system so that no noise or vibration-related failures occur. Modal analysis is, therefore, extensively used in the design of various structures such as automobiles, aircrafts, rockets, spacecraft, tennis rackets, etc.

The methodology as described here is cumbersome because of several practical problems and is not used often. The practical difficulties encountered are

- It is impossible to have a vibration shaker giving the same amplitude of force for a large frequency range.
- It is impractical to use shaker for structures such as turbine blades. Besides the complex problems in mounting shakers on small components such as blades, the effect of mounting a shaker amounts to an added mass that may totally change the modal characteristics of the parent structure (such as blades)
- Vibration engineers are often required to carry out the modal analysis of components *in-situ*. The free–free configuration may not be possible in many cases of practical modal analysis.
- If the place where the vibration transducer is mounted happens to be a node point, the vibration level may not show appreciable increase even when the structure is excited at natural frequency. The natural frequency can be positively identified by the phase change between the excitation force and the vibration response (Refer to chapter 2). Thus, for identifying the natural frequency, we need to see the amplitude peaking and more importantly, the phase change.

There are alternatives for the shaker test. For example, we know that when a structure is given a blow, the resulting vibration waveform contains information of various natural frequencies, which get excited before the structure settles to vibrate at its fundamental natural frequency. It is customary to use a hammer fitted with a force transducer. In this test, we measure both the excitation force as well as the response at various grid points through appropriately selected accelerometer.

In experimental modal analysis, we need to have a two-channel FFT analyser so that one channel records the force exerted by the shaker or by the hammer (giving impact-force excitation), while the second channel records the response in terms of displacement or velocity or acceleration. The two-channel FFT analyzer carries out computation of the complex-quantity response/force (FRF), its real and imaginary parts. The use of two channels facilitates the measurement of phase at each excited frequency. The resonance condition is identified through a phase difference of 90° and also through amplitude peaking (unless the measurement point is a node point). We now describe what a frequency response function (FRF) means.

9.2 FREQUENCY-RESPONSE FUNCTION

Systems respond differently to inputs (in the form of disturbing force, which could be from a shaker or impact) of different frequencies. At certain frequency, we may see considerable amplification of the vibration response, while at some frequency we may see attenuation of the vibration response. The way the system output (vibration) is related to the system input (disturbing/excitation force) for different frequencies is called the frequency response of the system.

The frequency response, popularly called the Frequency-Response Function' (FRF), is the relationship between the system input and output in the Fourier Domain.

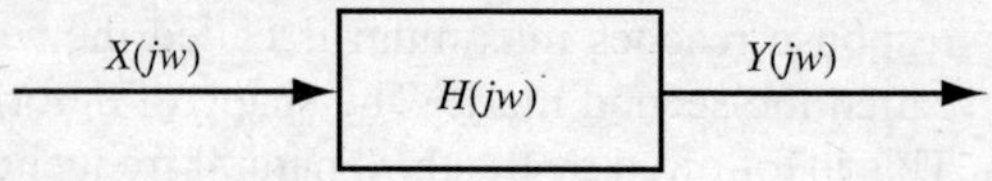

Figure 9.6 Frequency-response Function

In the system, $X(j\omega)$ is the system input (excitation force), $Y(j\omega)$ is the system output (vibration), and $H(j\omega)$ is the FRF. We can define the relationship between these functions as

$$Y(j\omega) = H(j\omega)\, X(j\omega) \tag{9.1}$$

$$\frac{Y(j\omega)}{X(j\omega)} = H(j\omega) \tag{9.2}$$

As mentioned earlier, $Y(j\omega)$, $H(j\omega)$, and $X(j\omega)$ are complex quantities. The reason that we need a two-channel FFT analyser is that we feed $X(j\omega)$ in one channel and $Y(j\omega)$ in the other channel. The computing FFT analyser evaluates $H(j\omega)$. $H(j\omega)$, a complex quantity, has two parts, namely, real part (which shows the amplitude of vibration response) and the imaginary part that gives the information of the phase.

We know from previous chapters that at resonance, the phase difference between the excitation force and the response is 90° and the vibration levels, depending upon the location of the transducer, may or may not show amplification of vibration levels. If the transducer happens to at or near a node point (zero displacement), we may not get vibration-level amplification at resonance condition, but phase shift of 90° will certainly be there to conform the resonance condition.

There are various ways the quantity (output/input) $H(j\omega)$ is considered depending upon the selection of vibration parameter (displacement/velocity/acceleration) used in the analysis. These are described in Tables 9.1 and 9.2.

Table 9.1 Frequency-response function names

Dimension	Displacement/Force	Velocity/Force	Acceleration/Force
Name	Admittance, Compliance, Receptance	Mobility	Accelerance, Inertance

Table 9.2 Frequency-response function names

Dimension	Force/Displacement	Force/Velocity	Force/Acceleration
	Dynamic stiffness	Mechanical impedance	Apparent mass

Let us see the application of modal analysis of a cantilever beam as shown in Fig. 9.7. Figure 9.7 shows four modes of vibrations. For the purpose of modal testing, we identify four locations for mounting the vibration transducer. One may also note that we can apply the hammer force at the four locations. We can thus have $4 \times 4 = 16$ FRFs. In a typical test, we can keep the point of application of the excitation force at fixed location and obtain the vibration response at four points as shown in Fig. 9.8. We can thus obtain four FRFs. These are sufficient to provide the information of natural frequencies as well as the mode shapes.

We shall now see the basic logic behind the concept of FRF. For the sake of simplicity, we shall discuss the case of single degree-of-freedom system. The concepts brought out in this case are readily extendible to multiple mass–spring–damper including complex multi-component continuous systems.

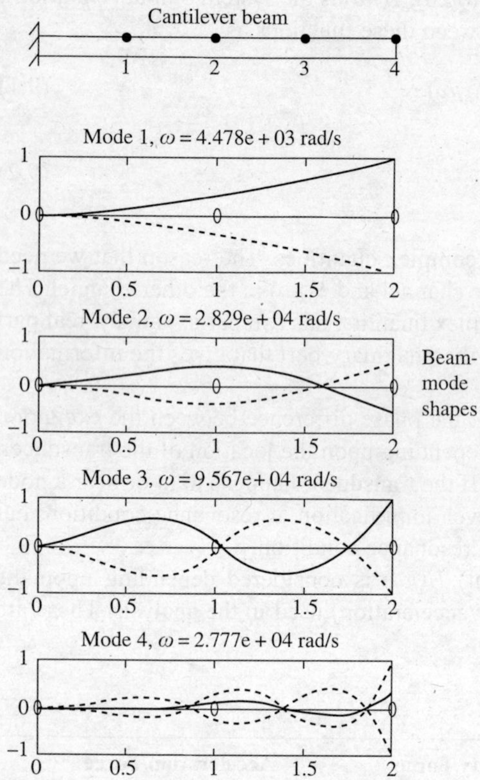

Figure 9.7 Mode Shapes of Cantilever Beam

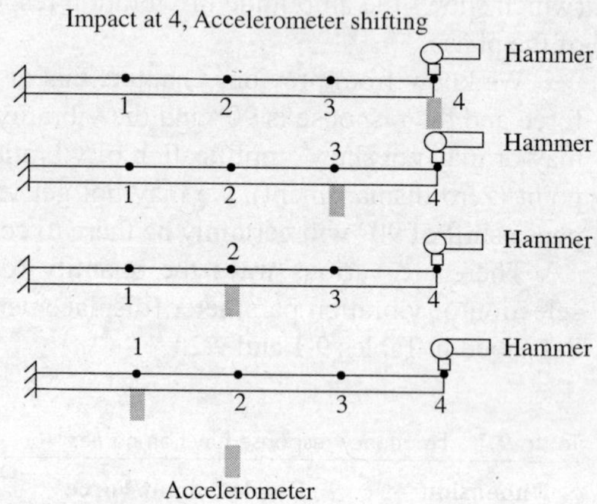

Figure 9.8 Impact Test for Modal Analysis

9.2.1 Frequency-Response Function—Basic Principles

(a) Sinusoidal Force Excitation

Let us first consider the case where we evaluate the FRF using sinusoidal-force excitation. Let the force vector be $F = F_0\, e^{i\omega t}$. We know that the steady-state response lags behind the excitation force. The velocity vector is $\dot{x} = i\omega.x$ and the acceleration vector is $\ddot{x} = -\omega^2 x$. Therefore, the mechanical impedances of spring, mass, and damper systems are

$$\text{Spring} = k$$
$$\text{Mass} = -m\omega^2$$
$$\text{Damper} = ic\omega$$

Let the harmonic-forcing function be expressed as $F(t) = F_0\, e^{i\omega t}$ so that the equation of motion becomes

$$m\ddot{x} + f\dot{x} + kx = F_0 e^{i\omega t} \tag{9.3}$$

One should remember that the actual excitation is given only by the real part of $x(t)$, where $x(t)$ in Equation 9.3 is a complex quantity. By assuming the particular solution $x_p(t)$

$$x_p(t) = X\, e^{i\omega t} \tag{9.4}$$

and by substituting this in Equation 9.3, we get,

$$X = \frac{F_0}{(k - m\omega^2) + if\omega} \tag{9.5}$$

Equation 9.5 can also be written as

$$Z(i\omega)X = F_0, \text{ where } Z(i\omega) = -m\omega^2 + if\omega + k$$

$Z(i\omega)$ is the impedance of the system as defined above

Multiplying the numerator and denominator on RHS of Equation 9.5 by $[(k - m\omega^2) + if\omega]$ and separating real and imaginary parts, we get,

$$X = F_0\left[\frac{k - m\omega^2}{(k - m\omega^2)^2 + f^2\omega^2} - i\frac{f\omega}{(k - m\omega^2)^2 + f^2\omega^2}\right] \tag{9.6}$$

We know that $x + iy = Ae^{i\Phi}$ where $A = \sqrt{(x^2 + y^2)}$, $\tan \Phi = y/x$; Equation 9.6 can be expressed as

$$X = \frac{F_0}{[(k - m\omega^2)^2 + f^2\omega^2]^{1/2}} e^{-i\Phi} \tag{9.7}$$

where

$$\Phi = \tan^{-1}\left(\frac{f\omega}{k - m\omega^2}\right) \tag{9.8}$$

The steady-state solution is given by

$$x_p(t) = \frac{F_0}{\sqrt{[(k - m\omega^2)^2 + (f\omega)^2}} e^{i(\omega t - \Phi)} \tag{9.9}$$

These manipulations are done to bring out a very important concept in vibration measurements and analysis, known as frequency-response function (FRF).

Equation 9.9 can be rewritten as

$$\frac{kX}{F_0} = \frac{1}{1 - r^2 + i2r\xi} \equiv H(i\omega) \tag{9.10}$$

where $H(i\omega)$ is known as complex-frequency response of the system.

$$|H(i\omega)| = \left|\frac{kX}{F_0}\right| = \frac{1}{\sqrt{(1 - r^2)^2 + (2r\xi)^2}} \tag{9.11}$$

Equation 9.11 denotes the magnification factor described earlier.

Recalling that $e^{i\Phi} = \cos \Phi + i \sin\Phi$, we can show that

$$H(i\omega) = |H(i\omega)|\, e^{-i\Phi} \tag{9.12}$$

where

$$\Phi = \tan^{-1}(2r\xi/(1 - r^2)) \tag{9.13}$$

Thus, we can obtain the following equation

$$x_p(t) = \frac{F_0}{K}|H(i\omega)|e^{i(\omega t - \Phi)} \tag{9.14}$$

It is important to note that the complex-response function, $|H(i\omega)|$ contains both magnitude and phase information of the steady-state response. The use of this function, is made in the experimental determination of system parameters (m, f, and k), natural frequencies, and the mode shapes.

(b) Impulse-force Excitation

Let us once again consider a single degree-of-freedom system. We write the equation of motion as follows:

$$\ddot{y} + 2\zeta\omega_n\dot{y} + \omega_n^2 y = x(t),$$

where

$$x(t) = \frac{F(t)}{m}, \quad \omega_n = \sqrt{\frac{k}{m}}, \zeta = \frac{c}{c_c} \tag{9.15}$$

$$c_c = 2km$$

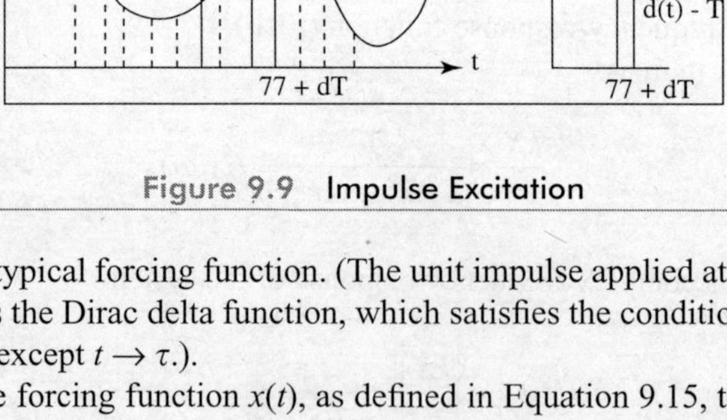

Figure 9.9 Impulse Excitation

Figure 9.9 shows a typical forcing function. (The unit impulse applied at $t = \tau$ is denoted as $x(t) = \delta(t - \tau)$, where $\delta(t - \tau)$ is the Dirac delta function, which satisfies the condition $\delta(t - \tau) \to \infty$ as $t \to \infty$ and $\delta(t - \tau) = 0$ for all t except $t \to \tau$.).

We can consider the forcing function $x(t)$, as defined in Equation 9.15, to be made up of a series of impulses of varying magnitudes as shown in Fig. 9.9 At time $t = \tau$, the impulse is $x(\tau)d\tau$. If $y(t) = h(t - \tau)$ denotes the response to the unit impulse excitation $\delta(t - \tau)$, it is called the impulse response function. The total response of the system at time t can be found out by superposing the responses to impulses of magnitude $x(\tau)d\tau$ applied at different values of $t = \tau$. The response to the excitation $x(\tau)d\tau$ will be $[x(\tau)d\tau]$ and the response to the total excitation will be given by the integral

$$y(t) = \int_{-\infty}^{t} x(\tau)h(t-\tau)d\tau \tag{9.16}$$

Proof of

$$\frac{Y(j\omega)}{X(j\omega)} = H(j\omega)$$

The transient function $x(t)$ can be expressed in terms of its Fourier transform $X(\omega)$ as

$$x(t) = \frac{1}{2\Pi} \int_{\omega=-\infty}^{\infty} X(\omega)e^{i\omega t} d\omega \tag{9.17}$$

Thus $x(t)$ can be considered as the superposition of components of different frequencies ω. If we consider the forcing function of unit modulus as $\tilde{x}(t) = e^{i\omega t}$, its response can be denoted as $\tilde{y}(t) = H(\omega)e^{i\omega t}$. $H(\omega)$ is called the complex-frequency response function. Since the actual excitation is given by the superposition of components of different frequencies (Equation 9.17), the total response of the system is obtained from

$$y(t) = H(\omega)x(t) = \int_{-\infty}^{\infty} H(\omega)\frac{1}{2\Pi}X(\omega)e^{i\omega t} d\omega = \frac{1}{2\Pi}H(\omega\int_{-\infty}^{\infty})X(\omega)d\omega \tag{9.18}$$

Let us denote the Fourier transform of the response function $y(t)$ as $Y(\omega)$. We, then obtain $y(t)$ in terms of $Y(\omega)$ as

$$y(t) = \frac{1}{2\Pi}\int_{-\infty}^{\infty} Y(\omega)e^{i\omega t} d\omega \tag{9.19}$$

Comparing Equations 9.18 and 9.19, we obtain

$$Y(\omega) = H(\omega)X(\omega) \tag{9.20}$$

The quantities in the above equation are complex quantizes. The FRF is therefore expressed as the ratio-vibration response/input force.

The author of this book has used air jets impinging upon rotating steam-turbine blades to excite their natural frequencies. The excitation force is described in Figure below.

The analysis of the waveform shown in Fig. 9.10 can be shown to be expressed as

$$p(t) = \frac{a_0}{2} + \sum_{m=1}^{\infty}[a_m \cos m\omega t + b_m \sin m\omega t]$$

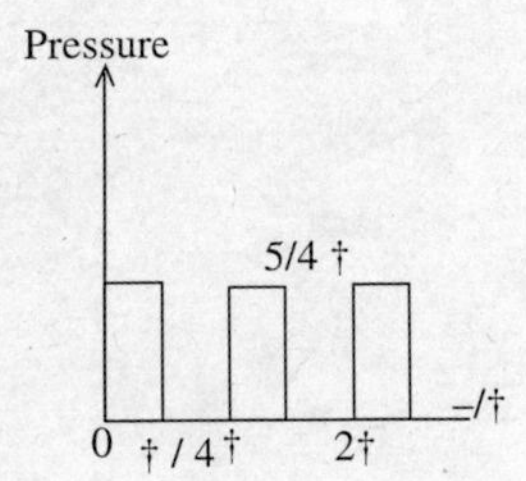

Figure 9.10 Use of Air Jets for Modal Analysis

We can show that in this waveform, only odd harmonics exist. This means that the spectral components of the force exerted by air jets on the rotating blades will be $1 \times \omega$, $3 \times \omega$, $5 \times \omega$, etc. If the number of moving blades is Z and the rotor rpm is N, then ω is given by

$$\omega = \frac{2\Pi ZN}{60} \text{ rad/sec}$$

The quantity ZN is called moving blade-passing frequency (cycles per minute). Several accelerometer transducers mounted at pre-determined places provide the response data over the entire blade, while the impinging jets provide the excitation force. The vibration data is transmitted through telemetry technique. We can obtain data of all possible modes of vibration as the rotor speed can be varied. We have discussed this as a case study in Chapter 7.

For detailed reading on Modal Analysis, readers may refer to 'Modal Analysis of Large Structures—Multiple Exciter Systems' by K Zaveri, Bruel & Kjaer, Denmark, 1984 and 'Modal Analysis as a Tool for Studying Structural Vibrations' in Mechanical Signature Analysis: Theory and Applications, S. Braun (ed.), Academic Press, London, 1986.

Exercises

9.1 Explain the importance of modal analysis in vibration analysis. Why is experimental modal analysis considered as one of the most important steps required in resolving the vibration and vibration-related problems?

9.2 What is frequency-response function? Derive the expression for the frequency-response function for a spring–mass, damper system subjected to sinusoidal excitation from a shaker. Why shaker test is not preferred in practice?

9.3 Derive the expression for the frequency-response function for a single degree-of-freedom system subjected to impulse excitation. Why the impulse-excitation method is a preferred method for modal analysis.

9.4 What are sources of errors in impulse-excitation method of modal analysis?

How can one be sure about truthfulness of the result of the modal-analysis test?

Hint: See Appendix D

10

Miscellaneous Topics in Vibration Analysis and Introduction to Noise Analysis

In this chapter, we shall deal with some topics, which might be useful in dealing with some complex problems encountered in practical life. The aim of this chapter is to introduce some important aspects of these topics. The topics that are being deal there include

- Flow-induced vibrations
- Noise analysis
- Unsteady (non-stationary) vibrations
- Rotor dynamics

Of these, flow-induced vibrations and the noise are somewhat interrelated. Therefore, we shall deal with them in some detail.

The subject of unsteady (non-stationary) vibrations is quite complex from the point of view of mathematical analysis and the solution of the governing equations. However, we shall deal with the experimental aspects of measurement and the analysis of non-stationary vibrations, since in practice, we come across such problems quite frequently.

We have seen that the solution of the governing equations of spring–mass–damper system comprises of a complimentary function (representing the transient response which with passage of time, disappears) and the particular integral, which represents the steady-state response of the vibrating system. However, there are some vibrating systems, which show occurrence of some perturbation forces at certain times. The conventional spectrum analysis, where spectrums are time-averaged, fails to identify these perturbation forces. For analysing such type of unsteady vibrations, one can resort to analysis of the vibration spectrums in real time, but this can be highly cumbersome and at times, quite misleading. For the analysis of such unsteady/non-stationary vibration problems, another technique, called wavelet analysis is found to be extremely useful. Although the present book does not deal with the theory of unsteady vibrations, it would be beneficial to know how the unsteady non-stationary vibrations can be analysed in practice and the fundamental principles behind these wavelet analyses.

Certain issues related to the spring/damping elements and the vibrations of vehicles have been dealt through some solved numerical the problems in the previous chapters. We shall also discuss some fundamentals of rotor dynamics.

10.1 FLOW-INDUCED VIBRATIONS

Flow-induced vibrations occur in all types of turbomachinery such as steam/water/gas turbines and pumps/compressors. They also occur in reciprocating engines, pumps, and compressors. Flow-induced vibrations is a serious consideration in the design of heat exchangers, nuclear reactors, chemical reactors, and many important components in refinery, petrochemical plants, aircraft wings and fuselage, rockets/missiles, space crafts, etc.

The basic underlying reasons behind the flow-induced vibrations (and the accompanied noise) are the pressure pulsations in the flow path. These pulsations take place because of several reasons such as accelerating/decelerating flows, obstructions in the flow path, vortex formation, etc. We first deal with vortex-induced vibration (VIV).

Vortex-induced vibration (VIV) are motions induced by a body immersed in an external flow by periodical irregularities in the flow. The classical example is the vortex-induced vibration of an underwater cylinder. We can see how this happens by putting a cylinder into the water and moving through the direction perpendicular to its axis. Since the real fluids always possess viscosity, the flow around the cylinder will be slowed down while in contact with its surface, forming a boundary layer (zero-fluid velocity at the boundary). Depending upon the velocity distribution, the boundary layer flow could be laminar or turbulent. However, at some point along the direction of the flow this boundary layer can separate from the body because of its excessive curvature. Vortices are then formed changing the pressure distribution along the surface. When the vortices are not formed symmetrically (which usually happens in practical cases) around the body (with respect to its mid-plane), different lift forces develop on each side of the body, thus leading to motion transverse to the flow. Figure 10.1 shows the vortex phenomenon.

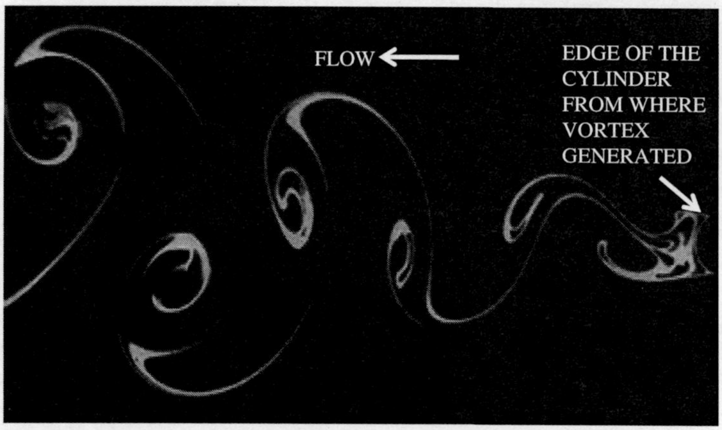

Figure 10.1 Vortex Shedding—Vortex Street

Vortex-induced vibration manifests itself on many different branches of engineering, from cables (galloping transmission lines) to blades of rotating machinery (turbines, pumps, compressors, etc), heat-exchanger tube arrays, offshore structures for oil/gas exploration, etc.

Vortex-induced vibration of turbomachinery such as pumps, turbines, compressors is one of the major factors that influence their vibration behaviour and vibration-related failures. In case, the vortex-shedding frequency matches with or is in close proximity of the natural frequency of the components in flow path (resonance or near-resonance condition), the amplified dynamic stresses caused by the resonance vibration may lead to fatigue failures. The vortex phenomenon may also lead to the cavitation phenomenon, which when combined with vibratory stresses can result in much earlier failures.

Vortex-induced vibrations (VIV) of ocean structures is a major factor affecting all stages of development of offshore structures (conceptualization, design, analysis, construction, and monitoring) and governs the arrangement of risers, details during fabrication, method of installation, and instrumentation. Advances to deeper waters in search for crude oil have resulted in multi-billion-dollar offshore projects world over. In these projects, prediction of VIV response is of tremendous importance. A recent estimate by British Petroleum puts the estimated cost of countering VIV to approximately 10% of the project cost itself.

Thus, study of VIV is a part of a number of disciplines, incorporating fluid mechanics, structural mechanics, vibrations, computational fluid dynamics (CFD), etc. Other important applications of vortex-induced vibrations are design of bridges, transmission lines, aircraft-control surfaces, etc.

Analysis of flow around a circular cylinder or aerofoil or any arbitrarily shaped body is one of the classical problems analyzed in fluid mechanics. These bodies are called the bluff bodies. The flow past a bluff body creates alternating low-pressure vortices on the downstream side of the object. As a result the object tends to move towards the low-pressure zone. At very low Reynolds' numbers, the streamlines of the resulting flow are symmetric. However, as the Reynolds' number is increased, the flow becomes asymmetric and the so-called **Von Karman Vortex Street** occurs. These vortices are shed at a particular frequency related to the velocity of flow and the characteristic dimension of the body (diameter in case of cylinder). These are related by the equation.

$$S = fD/U \tag{10.1}$$

In the above equation, f is the vortex-shedding frequency, U is the velocity of ambient flow, D is the characteristic dimension, and S is called the **Strouhal** number. The Strouhal number for a cylinder is around 0.2 over a wide range of flow velocity. The "lock-in" phenomenon happens when the vortex-shedding frequency is in the near neighbourhood of the natural frequency of the body/structure in the flow path. At this condition, the structure experiences very large vibrations because of which it may fail also.

The vortex-induced vibrations are not a small perturbation superimposed upon a mean steady motion. It is an inherently non-linear, self-governed/regulated and multiple degrees-of-freedom phenomena. Thus its a very complex subject requiring special knowledge of multi-disciplinary subjects and requires extremely-complex computational work. It is quite beyond the scope of this book to discuss these aspects. However, we shall consider some important concepts behind identifying experimentally the causes of flow-induced vibration and noise of fluid machinery.

Figure 10.2 shows the schematic arrangement of a typical turbomachinery. The guide blades are supposed to direct the flow to the moving blades such that the flow passes over the moving blades without shock. Even if this is ensured by very high-quality manufacturing, the very presence of guide blades will cause as many streams of flow as the number of guide blades are, and these will be passed on to the moving blades. Thus, a given moving blade will be subjected to as many impingements of streams as those coming from the guide blades in one revolution. Thus, the moving blade experiences a vibration-causing perturbation force occurring at a frequency equal to the number of guide blades multiplied by the rotational speed of the runner blades. In hydraulic turbines, this frequency is called the guide-vane-passage frequency (number of guide blades × rotational speed of the turbine), while in steam turbines, it is called diaphragm impulse frequency. It can be seen from Fig. 10.2 that both guide blades as well as runner blades behave as bluff bodies in the flow path of the fluid and hence they experience pressure pulsations at frequencies corresponding to the guide-blade-passage frequency as well as runner-blade-passage frequency and their harmonic multiples. If any of the frequencies happen to match (or in close proximity) with natural frequency of the blades, the blades may experience very high vibrations and consequential failures.

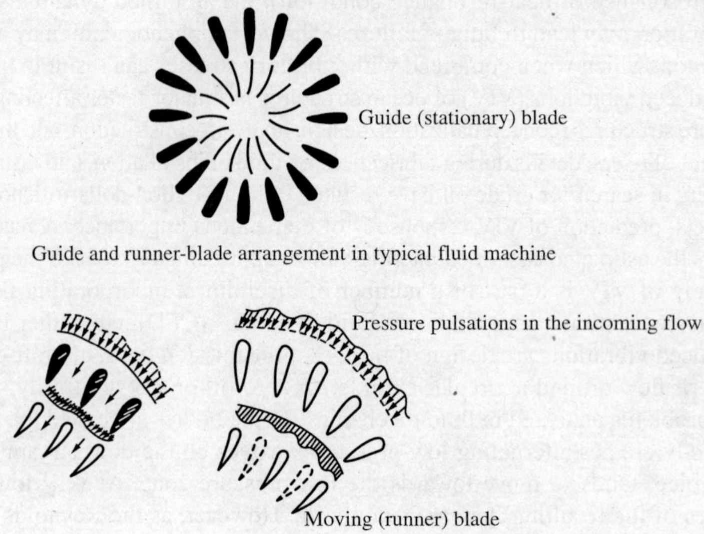

Figure 10.2 Guide/Runner-Blade Combination

Smooth entry of fluid on moving blades is very important, which means that the angle of stream exiting the guide blade must match with the angle at which it meets the moving blade. These angles depend upon not only the profile angles but also the fluid velocity. Any non-matching will result in separated flows that can generate vortices which could be damaging. This is shown in Fig. 10.3 and Fig. 10.4.

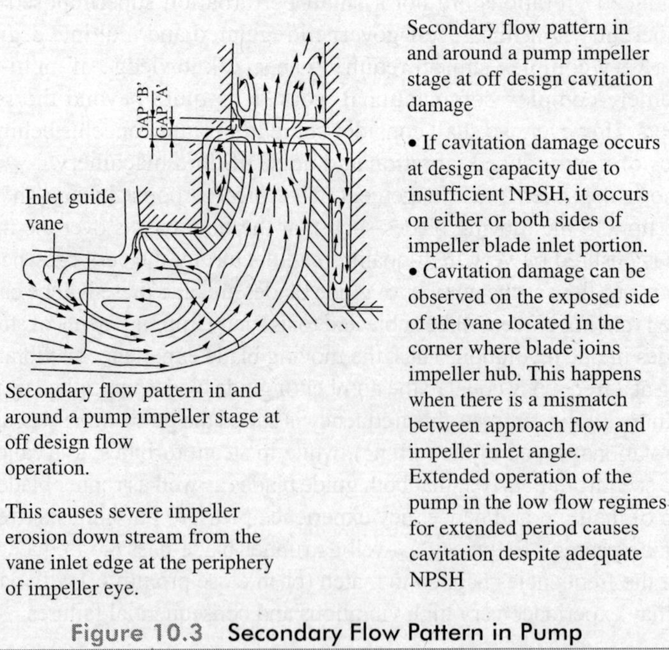

Figure 10.3 Secondary Flow Pattern in Pump

Pumps/turbines/compressors behave extremely rough from vibration/noise point of view when they operate in off-design load regimes. Figure 10.3 shows the flow pattern when the pump operates at off-design conditions. One can see prominently recirculating flow pattern that promote flow separation and thus vortex formation. These conditions favor cavitation pitting, in addition to high vibration/noise of the equipment.

Figure 10.4 shows how the mismatch between blade-inlet angle and flow incidence angle causes recirculating flows in impeller vane passages. At times, the flow gets separated causing favourable conditions for cavitation-induced pits. The figure also explains how hydraulic imbalance occurs in rotating turbomachinery. It should be remembered that hydraulic unbalance cannot be overcome by carrying out mechanical-balancing activities. The only way to solve this problem is to remove the mismatch of blade-inlet angle and the flow-incidence angle. This is possible only with very high level of accuracy in the manufacturing process.

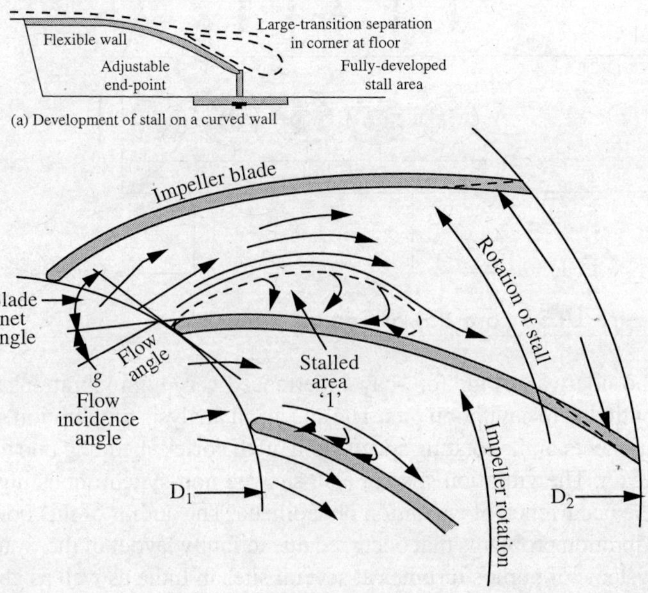

- Mis-match between blade inlet angle and the flow incidence angle causes recirculating flows at the impeller eye. This is called rotating stall-cavitation.
- One can notice underfilled and overfilled regions. This causes very heavy hydraulic unbalance which can never be balanced by mechanical balancing activities.

Figure 10.4 Recirculating Flow in Pump/Turbine

In hydraulic turbines, under certain operating conditions, severe vortex formation takes place beneath the runner and cause very large pressure pulsations, vibrations, as well as cavitation damage. The frequency of vortex shedding, called Rheingan's frequency, is approximately 0.2–0.3 times the speed of the runner. Specific corrective actions such as supplying pressurized air beneath the runner are required to overcome this problem.

The vibration-signature analysis does pinpoint the fluid-flow-induced vibration phenomena. For details, see the chapter on diagnosis in this book.

There are several flow-induced vibration problems that are caused by the design of the layout of water-conductor system. For example, in a thermal-power plant, severe vibration- and cavitation-related problems occurred on a pump located near the wall of the fore-bay (see the Fig. 10.5). The visual analysis showed a very high level of disturbance in flow near the wall. This was also confirmed by observing the vibration pattern on pump—motor assembly at this location.

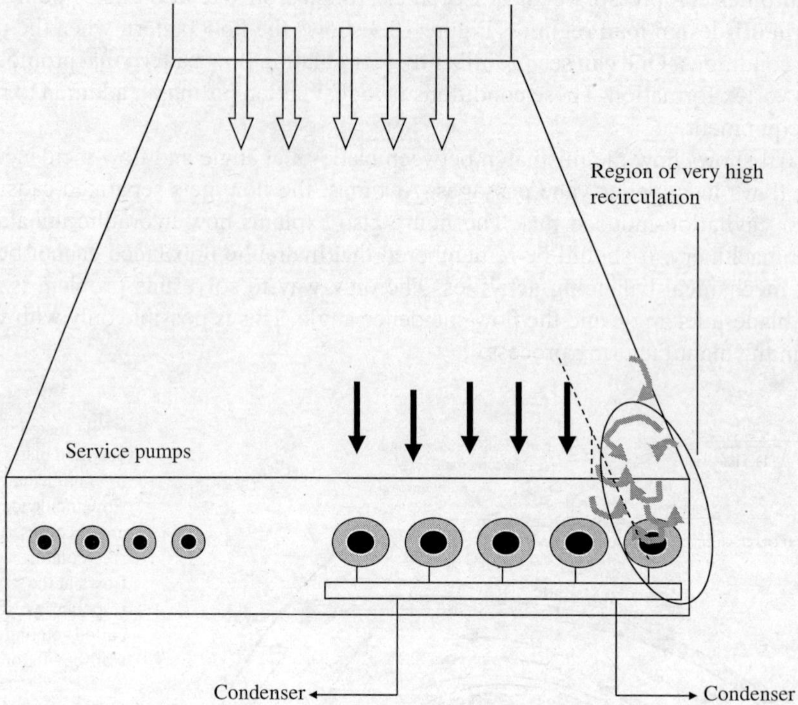

Figure 10.5 Fore-Bay of Cooling Water Plant

The pump located near the side-wall of the fore-bay experienced very high vibrations and frequent damages of the pump bell-mouth due to cavitation pits. The spectrum analysis of vibration showed presence of sub-synchronous components of the rotating speed along with vortex shedding due to flow disturbances, which occurred upstream. The vibration spectra also showed non-synchronous high-frequency components, which indicate the occurrence of cavitation phenomena. The author of this book has investigated several flow-induced vibration problems that occurred due to faulty layout of the water-conductor system in various hydraulic systems of pumps/turbines at several sites in India as well as abroad.

Singing of tubes in boilers as well as heat exchangers is a phenomenon that is caused by the flow-induced vibrations and the associated noise.

10.2 ACOUSTICS AND ANALYSIS OF NOISE

We shall discuss in this section, some fundamental aspects of noise analysis. The subject of noise analysis has become important since noise, apart from giving discomfort to the people in the environment can also result in very serious health issues such as total impairment of human hearing capability, psychological distress, and variety of stress-related complex diseases. Noise analysis can also lead to the identification of the reasons for the distress in the machinery/systems. This is because vibrations and noise go together. In this section, we shall discuss how noise is analysed.

10.2.1 Basics of Sound

The basic definitions and other aspects related to the basics of sound and noise are presented in this section. Most definitions have been internationally standardized and are listed in standards publications such as **IEC 60050-801 (1994).**

Noise can be defined as "disagreeable or undesired sound" or other disturbance. From the point of view of acoustics, sound and noise constitute the same phenomenon of pressure fluctuations about the mean atmospheric pressure. The criterion for specifying as well as recognizing it differentially is greatly subjective. What is sound to one person can very well be (intolerable) noise to somebody else. The recognition of noise as a serious health hazard is a development, which has happened only a few decades ago.

The rapid industrialization, which has happened in the last 4 to 5 decades has given rise to multitude of sources of sound and noise. This has caused accelerated noise-induced hearing loss to such an extent that industry-specific tolerable (safe) have now been specified in the international standards such as **OSHA**. High noise levels with amplified music also take its toll. While amplified music may be considered as sound (not noise) and giving pleasure to many youngsters, in the long term, it does cause hearing impairment. The excessive noise of much of modern industry probably gives pleasure to very few and none to almost all people.

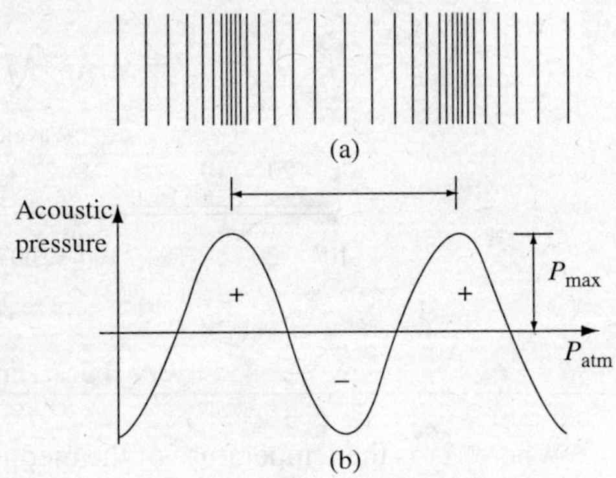

Figure 10.6 Representation of a Mono Tone Sound Wave. (a) Compressions and Rarefactions caused in Air by the Sound Wave. (b) Graphic Representation of Pressure Variations Above and Below the Atmospheric Pressure.

Sound (or noise) is the result of pressure variations, or oscillations, in an elastic medium. This medium could be air, water, solids. These are usually generated by a vibrating surface, or turbulent fluid flow or flow associated with vortices and separated flows.

Sound propagates in the form of longitudinal (as opposed to transverse) waves, involving a succession of compressions and rarefactions in the elastic medium. This is illustrated in Fig. 10.6(a) When a sound wave propagates in air (which is the medium considered in this section), the oscillations in pressure are above and below the ambient atmospheric pressure.

10.2.2 Amplitude, Frequency, Wavelength, and Velocity

Sound waves that consist of a pure tone/mono-tone (only one frequency) are characterized by the amplitude of pressure changes, which can be described by the maximum pressure amplitude, P_M, or the root-mean-square (RMS) amplitude, P_{RMS}, and is expressed in Pascal (Pa).

For finding the RMS values, the instantaneous sound pressures (which can be positive or negative) are squared, averaged, and the square root of the average is taken. This quantity is related to maximum amplitude by

$$P_M = 0.707\ P_{RMS}$$

The wavelength (λ) is the distance travelled by the pressure wave during one cycle;

The speed of sound propagation, c, the **frequency**, f, and the wavelength, λ, are related by the following equation:

$$c = f\lambda \tag{10.2}$$

The speed of propagation, c, of sound in air is 343 m/s at 20°C and 1 atmosphere pressure.
At other temperatures (not too different from 20°C), it may be calculated using the equation

$$c = 332 + 0.6T_C \tag{10.3}$$

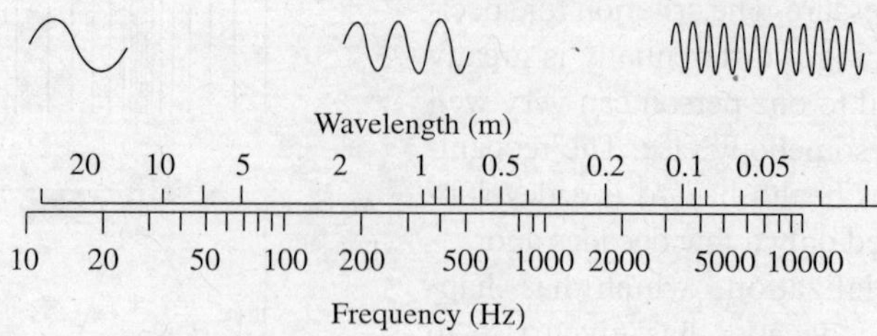

Figure 10.7 Frequency and Wavelength

Where, T_C is the temperature of the medium. Alternatively, making use of the equation of state for gases, the speed of sound may be written as

$$c = \sqrt{\gamma RT / M} \tag{10.4}$$

where R is the universal gas constant (8.314 J per mole), M is the molecular weight (0.029 kg/mole for air) and $\gamma = 1.402$.

All of the properties (except the speed of sound) apply only to a pure tone (single frequency) which is described by the oscillations in pressure shown in Fig. 10.6. However, sounds usually encountered are not pure tones. In general, sounds are complex mixtures of pressure variations that vary with respect to phase, frequency, and amplitude. For such complex sounds, there is no simple mathematical relation between the different characteristics. However, any signal may be considered as a combination of a certain number (possibly infinite) of sinusoidal waves, each of which may be described as outlined above. These sinusoidal components constitute the frequency spectrum of the signal.

To illustrate longitudinal wave generation, as well as to provide a model for the discussion of sound spectra, the example of a vibrating piston at the end of a very long tube filled with air will be used, as illustrated in Fig. 10.8.

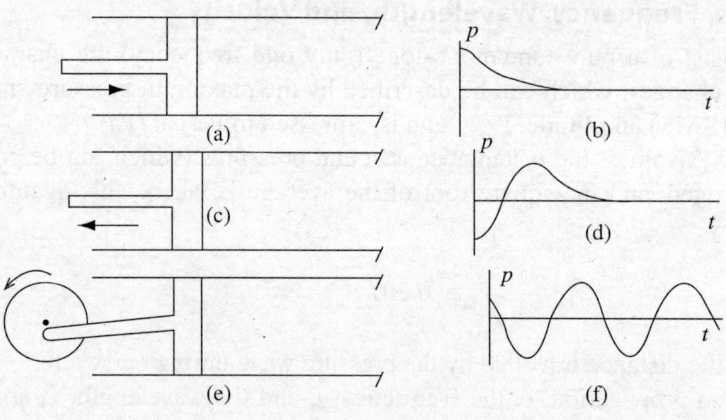

Figure 10.8 Sound generation. (a) The Piston Moves Right, Compressing Air as in (b). (c) The Piston Stops and Reverses Direction, Moving Left and Decompressing Air in Front of the Piston, as in (d). (e) The Piston Moves Cyclically Back and Forth, Producing Alternating Compressions and Rarefactions, as in (f). In all Cases, the Disturbances Move to the Right with the Speed of Sound.

Let the piston in Fig. 10.8 move forward. Since the air has inertia, only the air immediately next to the face of the piston moves at first; the pressure in the element of air next to the piston increases. The element of air under compression next to the piston will expand forward, displacing the next layer of air and compressing the next elemental volume. A pressure pulse is formed, which travels down the tube with the speed of sound, c. Let the piston stop and subsequently move backward; a rarefaction is formed next to the surface of the piston, which follows the previously formed compression down the tube. If the piston again moves forward, the process is repeated with the net result being a "wave" of positive and negative pressure transmitted along the tube.

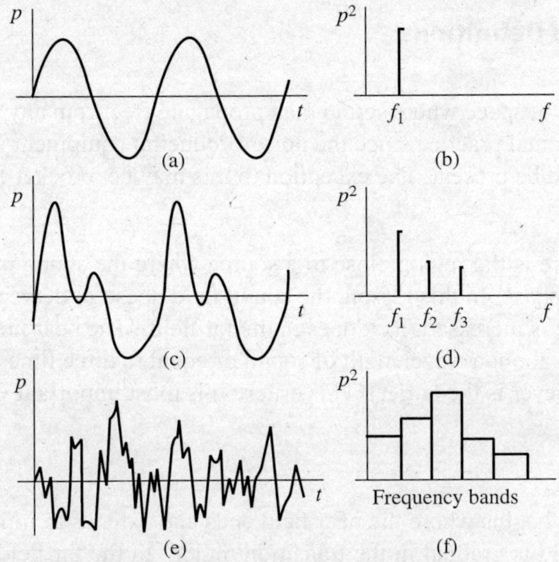

Figure 10.9 Noise-spectrum Analysis. (a) Monotone Noise. Disturbance p Varies sinusoidally with Time t at a Single Frequency f_1, as in (b). (c) Noise From Three Sources of Noise. Disturbance p Varies Cyclically with Time t as a Combination of Three Sinusoidal Disturbances of Fixed Relative Amplitudes and Phases; the Associated Spectrum has Three Single-Frequency Components f_1, f_2, and f_3, as in (d). (e) Disturbance p Varies Erratically with Time t, with a Frequency-Band Spectrum as in (f).

If the piston moves with simple harmonic motion, a sine wave is produced; that is, at any instant the pressure distribution along the tube will have the form of a sine wave, or at any fixed point in the tube the pressure disturbance, displayed as a function of time, will have a sine wave appearance. Such a disturbance is characterized by a single frequency. The motion and corresponding spectrum are illustrated in Fig. 10.9(a) and (b).

If the piston moves irregularly but cyclically, for example, so that it produces the waveform shown in Fig. 10.9(c), the resulting sound field will consist of a combination of sinusoids of several frequencies. The spectral (or frequency) distribution of the energy in this particular sound wave is represented by the frequency spectrum of Fig. 10.9(d). As the motion is cyclic, the spectrum consists of a set of discrete frequencies.

Although some sound sources have single-frequency components, most sound sources produce a very disordered and random waveform of pressure versus time, as illustrated in Fig. 10.9(e). Such a wave has no periodic component, but by Fourier analysis, it may be shown that the resulting waveform may be represented as a collection of waves of all frequencies. For a random type of wave, the sound pressure squared in a band of frequencies is plotted as shown; for example, in the frequency spectrum of Fig. 10.9(f).

It is customary to refer to spectral density level when the measurement band is 1 Hz wide, to one-third octave or octave-band level when the measurement band is one-third octave or one-octave wide and to spectrum level for measurement bands of other widths.

Two special kinds of spectra are commonly referred to as white-random noise and pink-random noise. White-random noise contains equal energy per Hertz and thus has a constant spectral-density level. Pink-random noise contains equal energy per measurement band and thus has an octave or one-third octave-band level, which is constant with frequency.

10.2.3 Sound Field Definitions

Free Field

The free field is a region in space where sound may propagate free from any form of obstruction. This usually is irrelevant in actual practice since the noise-producing equipment is placed in a plant where many obstructions would be present. The exception to this may be a rocket-launching station.

Near Field

The near field of a source is the region close to a source where the sound pressure and acoustic particle velocity are not in phase. In this region, the sound field does not decrease by 6 dB each time the distance from the source is increased (as it does in the far field). The near field is limited to a distance from the source equal to about a wavelength of sound or equal to three times the largest dimension of the sound source (whichever is the larger). This criterion is most important while making sound-level measurements.

Far Field

The far field of a source begins where the near field ends and extends to infinity. Note that the transition from near to far field is gradual in the transition region. In the far field, the direct field radiated by most machinery sources will decay at the rate of 6 dB each time the distance from the source is doubled. For line sources, such as traffic noise, the decay rate varies between 3 and 4 dB. It is interesting to note that the noise of power plants (especially the gas-turbine-based) located at sea shore may travel to habitation depending upon the direction of the wind. At certain times, these plants may appear extremely noisy. The plantation created near the power plant usually dampens this effect significantly and is thus one of the methods to control-transmitted noise.

Direct Field

The direct field of a sound source is defined as that part of the sound field, which has not suffered any reflection from any room surfaces or obstacles. This is rarely experienced as the industrial sources of noise do see obstacles.

Reverberant Field

The reverberant field of a source is defined as that part of the sound field radiated by a source, which has experienced at least one reflection from a boundary of the room or enclosure containing the source. Power and industrial plants are the examples of reverberant field in addition to auditoriums, cinema halls, which need to control these reverberating sounds.

Frequency Analysis

Frequency analysis may be thought of as a process by which a time-varying signal in the time domain is transformed to its frequency components in the frequency domain. It can be used for quantification of a noise problem, as both criteria and proposed controls are frequency dependent.

In particular, tonal components that are identified by the analysis may be treated somewhat differently than broadband noise. Sometimes frequency analysis is used for noise source identification and in all cases, frequency analysis will allow determination of the effectiveness of controls.

There are a number of instruments available to carry out a frequency analysis of arbitrarily time-varying signals. To facilitate comparison of measurements between instruments, frequency-analysis bands have been standardized. Thus, the International Standards Organization has agreed upon "preferred" frequency bands for sound measurement and analysis.

The widest band used for frequency analysis is the octave band; that is, the upper-frequency limit of the band is approximately twice the lower limit. Each octave band is described by its 'centre frequency', which is the geometric mean of the upper- and lower-frequency limits. The preferred octave bands are shown in Table 10.1, in terms of their centre frequencies.

Occasionally, a little more information about the detailed structure of the noise may be required than the octave band will provide. This can be obtained by selecting narrower bands; for example, one-third octave bands. As the name suggests, these are bands of frequency approximately one-third of the width of an octave band. Preferred one-third octave bands of frequency have been agreed upon and are also shown in Table 10.1.

Instruments are available for other forms of band analysis. However, they do not enjoy the advantages of standardization so that the inter-comparison of readings taken on such instruments may be difficult. One way to ameliorate the problem is to present such readings as mean levels per unit frequency. Data presented in this way are referred to as spectral density levels as opposed to band levels. In this case, the measured level is reduced by ten times the logarithm to the base ten of the bandwidth. For example, referring to Table 10.1, if the 500 Hz octave band that has a bandwidth of 354 Hz were presented in this way, the measured octave band level would be reduced by 10 log10 (354) = 25.5 dB to give an estimate of the spectral-density level at 500 Hz.

The problem is not entirely alleviated, as the effective bandwidth will depend upon the sharpness of the filter cut-off, which is also not standardized. Generally, the bandwidth is taken as lying between the frequencies, on either side of the pass band, at which the signal is down 3 dB from the signal at the centre of the band.

There are two ways of transforming a signal from the time domain to the frequency domain. The first involves the use of band-limited digital or analogue filters. The second involves the use of Fourier analysis where the time-domain signals is transformed using a Fourier series. This is implemented in practice digitally (referred to as the DFT—Digital Fourier Transform) using a very efficient algorithm known as the FFT (Fast Fourier Transform).

One-Third Octave-Band-Centre Frequencies

The one-third octave-band-centre frequency numbers have been chosen so that their logarithms are one-tenth decade numbers. The corresponding frequency pass bands are a compromise; rather than follow a strictly octave sequence which would not repeat, they are adjusted slightly so that they repeat on a logarithmic scale. For example, the sequence 31.5, 40, 50, and 63 has the logarithms 1.5, 1.6, 1.7, and 1.8. The corresponding frequency bands are sometimes referred to as the 15th, 16th, etc., frequency bands.

Note that in IEC 61260, index numbers are used instead of band numbers.

Usually logarithmic scales are used in noise-spectrum plots. In such situation, it is desirable to remember the one-third octave band centre frequencies. For example, the centre frequencies given above will lie respectively at 0.5, 0.6, 0.7, and 0.8 of the distance on the scale between 10 and 100. The latter two numbers in turn will lie at 1.0 and 2.0 on the same logarithmic scale.

Quantification of Sound

Sound Power (W) and Intensity (I) Sound intensity is a vector quantity determined as the product of sound pressure and the component of particle velocity in the direction of the intensity vector. It is a

Table 10.1 Preferred octave and one-third octave bands

Band Number	Octave Band Centre Frequency	One-Third Octave Band Centre Frequency	Band Limit Lower	Band Limit Upper
14		25	22	28
15	31.5	31.5	28	35
16		40	35	44
17		50	44	57
18	63	63	57	71
19		80	71	88
20		100	88	113
21	125	125	113	141
22		160	141	176
23		200	176	225
24	250	250	225	283
25		315	283	353
26		400	353	440
27	500	500	440	565
28		630	565	703
29		800	707	880
30	1000	1000	880	1130
31		1250	1130	1414
32		1600	1414	1760
33	2000	2000	1760	2250
34		2500	2250	2825
35		3150	2825	3530
36	4000	4000	3530	4400
37		5000	4400	5650
38		6300	5650	7070
39	8000	8000	7070	8800
40		10000	8800	11300
41		12500	11300	14140
42	16000	16000	14140	17600
43		20000	17600	22500

measure of the rate at which work is done on a conducting medium by an advancing sound wave and thus the rate of power transmission through a surface normal to the intensity vector. It is expressed as Watts per square meter (W/m^2).

In a free-field environment, that is, no reflected sound waves and well away from any sound sources, the sound intensity is related to the RMS acoustic pressure as follows:

$$I = \frac{P_{\text{RMS}}^2}{\rho c} \tag{10.5}$$

where, ρ is the density of air (kg/m^3), and c is the speed of sound (m/s). The quantity, ρc is called the 'acoustic impedance' and is equal to 414 Ns/m^3 at 20°C and one atmosphere. At higher altitudes, it is considerably smaller.

The total sound energy emitted by a source per unit time is the sound power, W, which is measured in watts. It is defined as the total sound energy radiated by the source in the specified frequency band over a certain time interval divided by the interval. It is obtained by integrating sound intensity over an imaginary surface surrounding a source. Thus, in general the power, W, radiated by any acoustic source is

$$W = \int_A I \cdot ndA \tag{10.6}$$

where, the dot multiplication of I with the unit vector, n, indicates that it is the intensity component normal to the enclosing surface that is used. Most often, a convenient surface is an encompassing sphere or spherical section, but sometimes other surfaces are chosen, as dictated by the circumstances of the particular case considered. For a sound-source producing uniformly spherical waves (or radiating equally in all directions), a spherical surface is most convenient, and in this case, the above equation leads to the following expression:

$$W = 4\Pi r^2 I \tag{10.7}$$

Where, the magnitude of the acoustic intensity, I, is measured at a distance r from the source. In this case, the source has been treated as though it radiates uniformly in all directions.

Sound-pressure Level The range of sound pressures that can be heard by the human ear is very large. The minimum acoustic pressure audible to the young human ear judged to be in good health, and not much exposed to excessively loud music, is approximately 20×10^{-6} Pa, or 2×10^{-6} atmospheres. The minimum audible level occurs at about 4000 Hz and is a physical limit imposed by molecular motion. Lower than this, sound-pressure levels would be swamped by thermal noise due to molecular motion in air.

For the normal human ear, pain is experienced at sound pressures of the order of 60 Pa or 6×10^{-6} atmospheres. Evidently, acoustic pressures ordinarily are quite small fluctuations about the mean.

A linear scale based on the square of the sound pressure would require 10^{13} unit divisions to cover the range of human experience; however, the human brain is not organized to encompass such a range. The remarkable dynamic range of the ear suggests that some kind of compressed scale should be used. A scale suitable for expressing the square of the sound pressure in units is best matched to subjective response is logarithmic rather than linear. Thus, the unit Bel was introduced which is the logarithm of the ratio of two quantities, one of which is a reference quantity.

A factor of 10 is introduced to avoid a scale that is too compressed over the sensitivity range of the ear. This has given rise to the decibel scale frequently used in noise analysis. The level of sound pressure p is then said to be L_p decibels (dB) greater or less than a reference sound-pressure (p_{ref}), according to the following equation.

$$L_P = 10\log_{10}\frac{p_{rms}^2}{p_{re}^2} = 20\log_{10}\frac{p_{rms}}{p_{ref}} = 20\log_{10} p_{rms} - 20\log_{10} p_{ref}\ \text{(dB)} \tag{10.8}$$

For the purpose of absolute-level determination, the sound pressure is expressed in terms of a datum pressure corresponding to the lowest sound pressure, which the young normal ear can detect.

The result is called the sound-pressure level, L_p (or SPL), which has the units of decibels (dB). This is the quantity, which is measured with a sound level meter.

The sound pressure is a measured RMS value and the internationally-agreed reference pressure $pref = 2 \times 10^{-5}\ \mathrm{Nm^{-2}}$ or 20 μPa . When this value for the reference pressure is substituted into the previous equation, the following convenient alternative form is obtained.

$$L_P = 20\log_{10} p_{rms} + 94 \text{ (dB)} \tag{10.9}$$

where, the pressure p is measured in Pascals. Some feeling for the relation between subjective loudness and sound-pressure level may be gained by referring to Fig. 10.10, which illustrates sound-pressure levels produced by some noise sources.

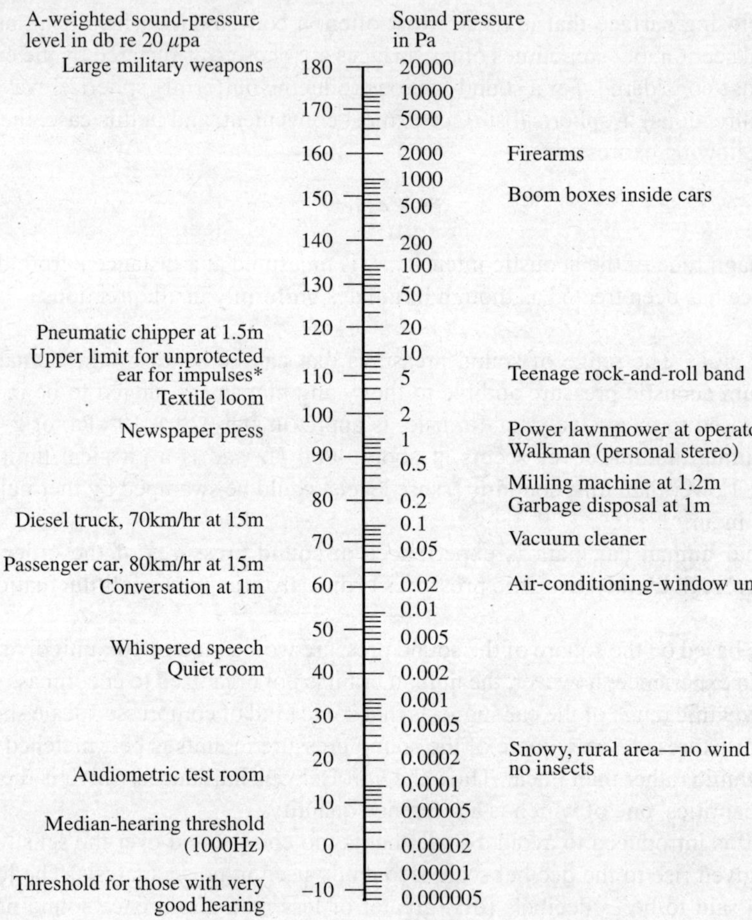

Figure 10.10 Typical Noise Levels

Sound-intensity Level A sound-intensity level, L_I, may be defined as follows:

$$L_I = 10\log_{10}\frac{\text{Sound intensity}}{\text{Reference sound intensity}} \text{ (dB)} \tag{10.10}$$

Internationally-agreed reference intensity is 10^{-12} Wm^2. Using this in the above equation, we get

$$L_W = \log_{10} I + 120 \text{ (dB)} \tag{10.11}$$

Use of the relationship between acoustic intensity and pressure in the far-field of a source gives the following useful relationships:

$$L_I = L_P + 10\log_{10}\frac{400}{\rho c}$$

$$L_I = L_P + 26 - 10\log_{10}(\rho c) \qquad [10.12 \text{ (a)\&(b)}]$$

At sea level and 20°C, the characteristic impedance, ρc, is 414 kg m^{-2} s^{-1} so that for both plane and spherical waves,

$$L_I = L_P - 0.2 \text{(dB)} \tag{10.13}$$

Sound-power Level The sound power level, L_w (or PWL), may be defined as follows:

$$L_w = 10\log_{10}\frac{\text{Sound power}}{\text{Reference sound power}} \tag{10.14}$$

Internationally agreed reference power is 10^{-12} W. Using the same approach as sound intensity we get,

$$L_W = \log_{10} W + 120 \text{ (dB)} \tag{10.15}$$

where, the power, *W*, is measured in Watts.

Combining Sound Pressures In a typical industrial plant, several machines operate simultaneously and they create a sound, which is the result of additions (not scalar) of the levels of various sources of noise. These form a system of incoherent sources of noise. However, for two sound sources having almost equal frequencies, the resultant noise will be of the type of beats.

Sources of Noise and Noise Reduction As mentioned each, the noise is associated with pressure pulsations to which our ears respond. The pressure pulsations occur due to (1) mechanical vibrations, (2) magnetic noise, and (3) flow-induced vibrations.

Mechanical noise is usually associated with rattling sound from components improperly assembled or due to their typical mode shapes. For example, the stampings of electrical motor, when not properly fixed, will emit rattling sound. Similarly, the heat-exchanger tubes may vibrate such that they rub against the baffle plates or the mode of vibration may be such that vibrating tubes touch each other. For eliminating/reducing the noise levels in electrical machines caused by improper construction, the construction errors are required to be corrected. For eliminating/reducing the noise in heat exchange equipments, such as boiler, heat exchanger, it may be necessary to alter the tube-support arrangement or change (if possible) flow parameters or provide flow straighteners. One of the reasons for the rattling sound from turbomachineries, such as steam turbine/gas turbine is the friction between the blades and damping wire/pins.

The magnetic noise usually is due to non-uniform magnetic fields and stator/rotor-bar bouncing. It is important to keep the air gap between the stator and rotor as uniform as possible as it also results into high vibrations if the air gap is not uniform. The frequency of noise is usually 100 Hz (in India) and its harmonic multiples.

In hydraulic turbines, the extreme noise can occur due to incorrect choice of number of guide blades and the number of runner vanes. The frequency of the noise in such cases will be the runner blade passing frequency (number of runner blades × speed of the runner or its harmonic multiples). Since it is often impossible to change the number of guide blades as they form the part of structure supporting the weight of the entire machine, the only way left is to change the blades of the runner.

In hydraulic pumps, the problem of high noise comes if the radial gap between the impeller tip and the volute casing is reduced. Undoubtedly, reduced radial gaps would, in theory, reduce hydraulic losses; in reality they make the flow in the gap extremely disturbed thereby creating a very high pitch noise at the impeller vane passing frequency (and/its harmonic multiples). At times, it may cause mechanical failures of the tips of the impeller and the debris so generated would come in the flow path to cause the seizure of the pump. The investigation carried out in the pump development laboratory shows a three-fold increase in pressure pulsations in the radial gap when the radial gap is reduced from 5% to 2% for increasing the hydraulic efficiency of the pump. However, the price to be paid for such an improvement is tremendously high if we consider the reliability and availability of the pump. It is therefore prudent to go for higher gap so that although the hydraulic efficiency of the pump would be sacrificed/reduced, the problem of extremely high-pitch noise as well as the problem of seizure of pump can be avoided. The pump in such case will have high reliability as well as high availability. Flow-induced noise generally shows up at blade-passing frequency or its harmonics. The pressure pulsation signal also shows these components. The vibration/pressure pulsation signal would also reveal if there is vortex shedding phenomenon. The selection of number of guide blades and runner blades is very crucial. At times, it becomes necessary to change the number of guide blades in steam turbine and runner blades in case of hydraulic turbine/pumps.

Noise occurring due to cavitation of the components in the flow path can be identified through spectrum analysis where the spectrum would show harmonically unrelated high-frequency components.

The control of noise can be achieved by removing the root cause of noise. If this is not possible, acoustic enclosures are used. These enclosures provide additional resistance for the noise to propagate. The design of acoustic enclosures is a subject by itself. Another way to control the noise is active noise control but it is too complex and still, a hot topic for research. The best noise-control strategy is to control the noise at source, which can be done by certain design modifications only.

The subject of vibration and noise analysis in rotating machinery is a very wide subject. What has been presented in this section is only for making the readers familiar with the subject at the elementary level.

10.3 NON-STATIONARY (UNSTEADY) VIBRATIONS

We, in this section, shall deal with some elementary aspects of measurement and analysis of unsteady (non-stationary) vibrations. Unsteady/non-stationary vibrations are those vibrations, which vary with time. These are different from the initial transient vibrations experienced by the spring–mass–damper system when set into vibration mode since these transient vibration will die down and disappear after certain time. Our primary interest is in the analysis of vibration signal, which varies with time and remains so for considerably longer time as compared with time of transient vibrations.

Since the vibrations vary with time, the frequency response (vibration spectrum) also varies with time.

Typically, unsteady vibrations comprise of steady part, which does not vary with time and unsteady part, which appears for certain durations. For example, in an hydraulic turbine, we have steady vibrations caused by perturbation forces such as those resulting from mechanical unbalance, misalignment, etc.

and unsteady vibrations due to perturbation forces caused by load modulation, grid-frequency variation, periodic vortices, experienced at certain load regimes, etc. In gas turbines, we may get unsteady perturbation forces if the firing in the combustors located in the annular space becomes disturbed due to some aerodynamic disturbances from the compressor section of the gas turbine. It is important to note that these non-stationary forces remain in the system until such time till the disturbances continue to exist. As per the experience of the author of this book, such non-stationary perturbation forces, at times, do cause substantial deterioration of the vibration behaviour of the machine.

When we resort to conventional spectrum analysis where samples of vibration–time signals are time averaged, it is quite possible to miss identification of the non-stationary components in the vibration spectrum and thus very valuable information is missed out. Luckily, we now have computing Fast Fourier Transform-based real time vibration analyzers, which do provide almost instantaneous spectrum of the signal at the instant signal is fed. A rudimentary procedure is to see the vibration spectrum in real time (no averaging done) and distinguish between those frequency components in spectrum, which do not change with time and the components which appear at certain times or whose levels change with time. This procedure may not be of much use unless the process of unsteadiness is relatively slow and more importantly it is deterministic.

The technique of water-fall diagram described in the chapter on diagnosis may provide certain information such as when the oil whirl started or when unbalance component in the vibration signal started increasing. Fig. 10.11 shown below is reproduced from chapter on diagnosis.

For instance, the spectrum marked A (at time $t = T_A$) does not show any signs of oil whirl. However, at time $t = T_B$, we do see manifestation of oil whirl and as time progresses, we have significant oil whirl at $t = T_D$ and significant increase in the rotating-speed component (marked unbalance).

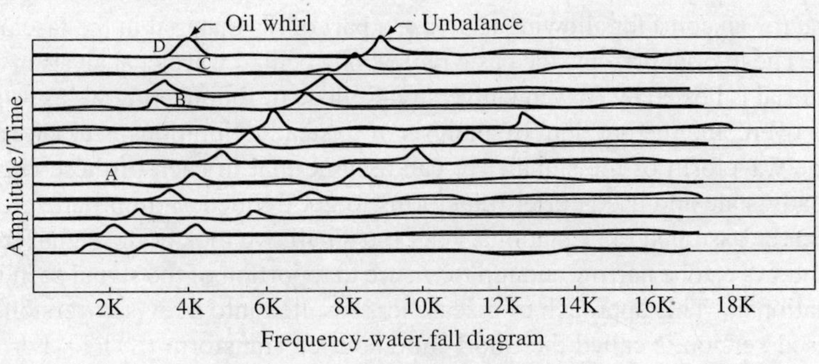

Figure 10.11 Water-fall Diagram

Logical thing to do is to investigate what happened during the period from $t = T_A$ to $t = T_B$ when symptoms of oil whirl started showing up. Usually it is quite possible to identify a change/disturbance in some operating parameter, which may not be easily correctible. Thus, use can see effectiveness of using waterfall diagram for the first-level assessment of a time-dependent vibration process.

It is quite easy to understand that what we need for analysing the non-stationary vibration signal is the information of time–frequency variation, which does not come from FFT analysis. We may have two different signals having the same spectral components with one major difference. Let us assume that one of the signals has four spectral components, which occur at all times while the other signal has the same components but which appear at different times. It is quite evident that the vibration spectra of these signals will be entirely different from each other. More importantly, the conventional spectrum analysis will not be able to pinpoint the severity of perturbation forces, which vary with time.

Let us recall that the Fourier transform decomposes a signal in complex exponential functions of different frequencies. This is defined as

$$X(f)=\int_{-\infty}^{\infty} x(t).e^{-2i\pi ft}\,dt \tag{10.16}$$

$$X(t)=\int_{-\infty}^{\infty} x(f)e^{2\pi ift}\,dt \tag{10.17}$$

In the Equations t is time, f is the frequency, x denotes the signal in time domain.

Equation 10.16 is the Fourier transform of $x(t)$, while Equation 10.17 is the inverse Fourier transform of $X(f)$. If the frequency content of particular frequency $f=f_1$ is negligible in the vibration signal $x(t)$, then Equation 10.16 for that particular frequency will give zero or negligible value. Note that integration limits are $-\infty$ to $+\infty$. Thus if the result of infinite summation done in the integration process is a large or significant, then we say the signal has a dominant spectral component. Equation 10.16 also shows that the frequency component is always present. This, in other words, means that the Fourier transform and consequently the conventional spectrum analysis will give information of perturbation forces causing vibration only when the vibration process is a steady-state or stationary process.

The problem in the conventional analysis is the fact that the vibration spectrum does not show at all at what point of time the frequency components occur. In other words, it does not give the information regarding frequency—time data as it just cannot do it. Researchers all over the world looked into this problem and came out with a new approach called wavelet analysi**s**. We do not intend to give the rigorous mathematical theory behind wavelet analysis as it is rather quite complex. What we present in this section is the basic philosophy behind the wavelet analysis. The author of this book is grateful to Dr. Robi Polikar (polikar@rowan.com) for allowing us to use a part of his material in the fascinating tutorials on wavelet analysis. The tutorial presents the basic philosophy behind wavelet analysis in a very lucid manner. (Wavelet tutorial is hosted by Rowan University, College of Engineering web services.)

One way to overcome the problem of analysis of unsteady vibrations is to take a few regions (of time) in the time-waveform of the signal. We can assume that in each of these regions of time, the signals were steady-state and use Fourier transforms to get frequency–amplitude plots. If the regions where signals can be assumed to be stationary are too small, we look at the signal from **narrow** windows. These windows can be narrow enough to ensure that portion of the signal seen in these windows is reasonably stationary. This approach of researchers resulted into a revised version of Fourier transform. This revised version is called the Short-time Fourier Transform (STFT). For STFT, a window function 'w' is chosen. It is quite obvious that the width of this window must be equal to the segment of the signal where its condition of being stationary is fulfilled. This window function is first located at the beginning ($t=0$) of the signal of duration T. This duration T is the width of the window. At time $t=0$, the window function w will overlap with first $T/2$. The window function and the signal (in the window) are then multiplied. This process implies that by doing this, only the first $T/2$ s of the signal is being chosen with appropriate weighting of the window (rectangular, Hanning, etc). The result of this is a product, which is nothing but yet another signal, whose Fourier transform can be taken. The result of this transformation is the Fourier transform of first T/2 s signal. It is quite reasonable to assume that this portion is stationary and the frequency obtained is the true frequency present in the chosen small duration. We, then, can shift to next time slot and the process can be done on the time waveform of the signal.

The STFT can be expressed as

$$STFT_X^{(\omega)}(t,f)=\int_t [x(t)\omega^*(t-t')]e^{2\pi ift}\,dt \tag{10.18}$$

In the Equation 10.18 * denotes complex conjugate, $x(t)$ is the vibration–time domain signal and $\omega(t)$ is the window function. It is quite evident that what we obtain using Equation 10.18 is the time—frequency information.

Readers are advised to refer to Dr. Polikar's wavelet tutorial for clear understanding of the process of arriving at Discrete Wavelet Transform (DWT).

The wavelet tutorial of Dr. Polikar has presented some excellent examples of the procedure of wavelet analysis. The following is an illustrative example of unsteady vibration waveform and the result of wavelet analysis. The total time spread of the signal is 1 s and has the following frequency components, which appear at different times. They include the following.

1. 0–250-ms sinusoidal wave of frequency 300 Hz
2. Next 250-ms sinusoidal wave of frequency 200 Hz
3. 500–750-ms sinusoidal wave of frequency 100 Hz
4. 75–1000-ms sinusoidal wave of frequency 50 Hz
   Obviously, the signal is non-stationary.
   See Fig. 10.12.

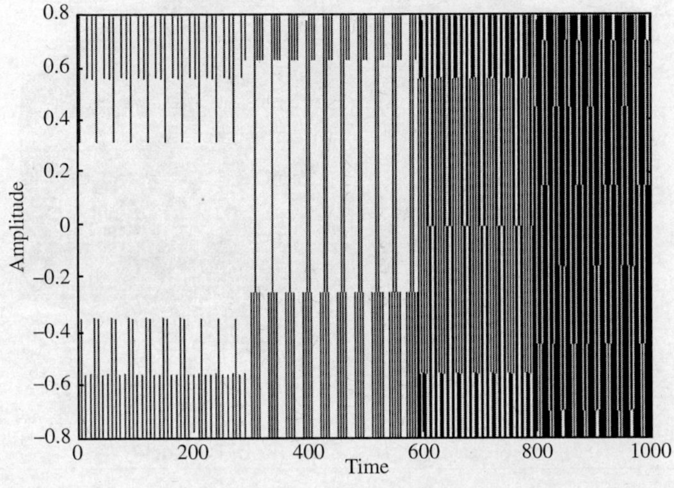

Figure 10.12 Unsteady Vibration-time Waveform (Robi Polikar)

The following figure shows the three-dimensional plot comprising of information of time–frequency–amplitude. For these the window functions used are shown in Fig. 10.13.

Figure 10.14 shows the STFT of the time signal shown in Fig. 10.12.

STFT undoubtedly gives time–frequency representation. In this, we do know the time intervals in which a certain band of frequencies exist but one cannot say exactly what spectral components exist at what instances of time (in line with Heisenberg's uncertainty principle which is applicable here also). The problem with resolution can be easily appreciated if we see Equations 10.16, 10.17, and 10.18 from the point of limits of integration. Equations 10.16 and 10.17 show limits of integration from $-\infty$ to $+\infty$ and the function and the kernel of window function is $\exp\{i\omega t\}$, where $\omega = 2\Pi f$ and hence there is no problem of the resolution. However, the resolution in STFT depends upon the width of the window function chosen in Equation 10.18. A narrow window gives good resolution of time but poor resolution of frequency while wider window would give poor resolution of time but good resolution of frequency.

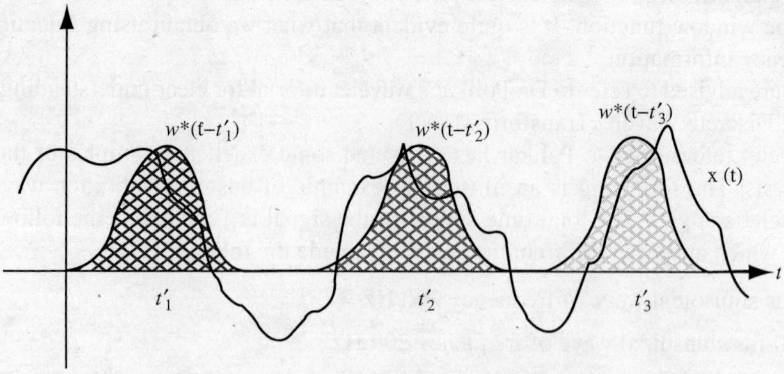

Figure 10.13 Window Functions Used

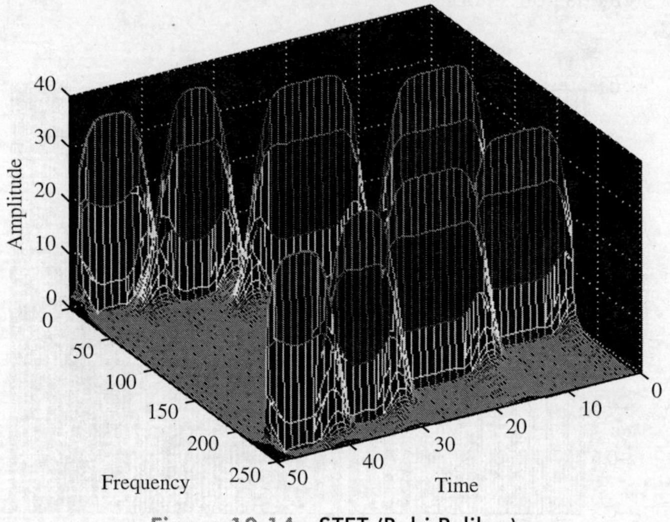

Figure 10.14 STFT (Robi Polikar)

It is not the intention here to discuss what window functions should be used to get the best possible resolution of both time and frequency. Careful reading of all the four parts of the tutorials followed by advanced-level reading of wavelet-transform literature is recommended for further insight in the subject.

10.4 ROTOR DYNAMICS AND HYDRODYNAMIC BEARINGS

Fluid-film bearings are commonly used in heavy-rotating machines because they possess not only high load-carrying capacity but also the inherent damping provided by them. As against rolling-element bearings, the fluid-film bearings, also called the hydrodynamic bearings, offer, in addition to stiffness element provided by the oil film, substantial damping element. These additional springs and damping element provided by the oil film enable the rotor-bearing system to become dynamically stable. This happens because the stiffness and the damping provided by the oil film significantly alter the critical speeds and the unbalance response of the rotor. Reduction of the order of 20–25% in critical speed as well as the unbalance response is possible in practice.

A rotor comprising of a disc and supported on two hydrodynamic bearings is a classical model studied in the subject of rotor dynamics. Such a rotor is popularly known as the Jeffcott rotor and is the first level of analytical model in rotor dynamics. Figure 10.15 shows the Jeffcott rotor supported on hydrodynamic bearings.

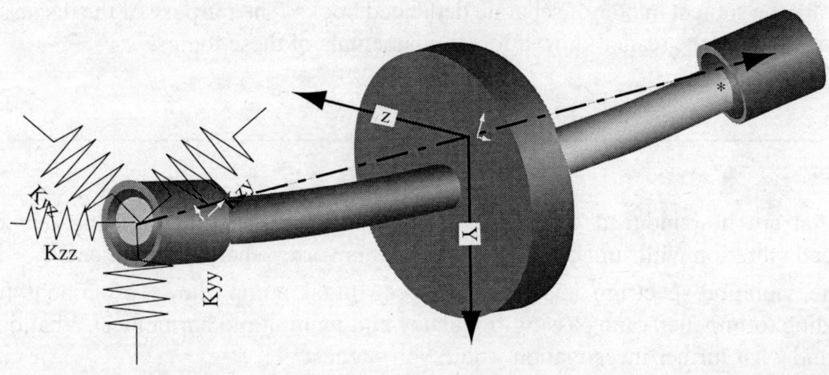

Figure 10.15 Rotor Supported on Hydrodynamic Bearings—Jeffcott Rotor

In this model, the bearings are simulated (modelled) as an assemblage of springs and dampers as shown in Fig. 10.16.

Because of the vibratory motion of the shaft/rotor both in the Y and Z directions, we designate the springs and the dashpots (dampers) of the bearing oil film as K_{ZZ}, K_{YY}, K_{ZY}, K_{YZ} (springs) and C_{YY}, C_{ZZ}, C_{YZ} and C_{ZY} (dampers). The stiffness and the damping of the shaft are given by K and C, respectively.

When the shaft rotates with an eccentricity '*e*' of the centre of mass, dynamic motions occur in two directions (Y and Z). The governing equations of the motion are given by:

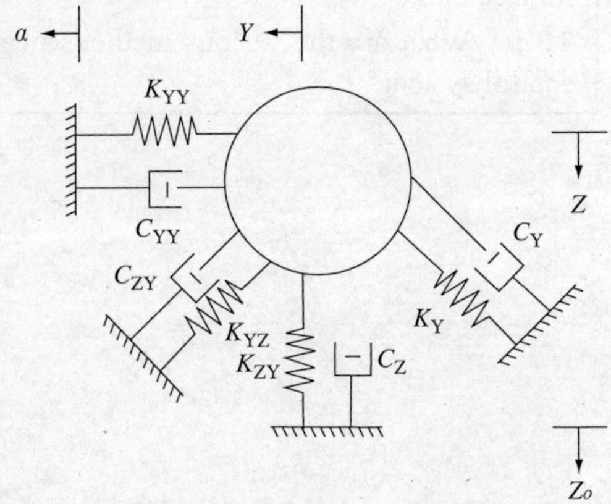

Figure 10.16 Simulation of Bearing as Spring Damper Model

$$M\frac{d^2}{dt^2}(Z+e\cos\omega t)+K(Z-Z_O)+C(\dot{Z}-\dot{Z}_0)=0$$

$$M\frac{d^2}{dt^2}(Z+e\sin\omega t)+K(Y-Y_0)+C(\dot{Y}-\dot{Y}_0)=0$$

$$K(Z-Z_0)+C(\dot{Z}-\dot{Z}_0)=2K_{ZZ}Z_0+2C_{ZZ}\dot{Z}_O+2K_{ZY}Y_0+2C_{ZY}\dot{Y}_0$$

In the equations '.' represent time-derivative.

The stiffness and the damping coefficients for the bearings are obtained from the solution of Reynold's equation. For details, one can refer to books on the subject of Tribology and go through the part on hydrodynamic bearings.

In this chapter, we discussed some additional aspects of vibration analysis such as flow-induced vibrations, noise analysis, unsteady vibrations and the rotor dynamics. It should be noted that each of these topics are the subject matter of separate dedicated books. The purpose of the discussion given in this chapter is to make the reader know a few fundamentals of these topics.

Exercises

10.1 What are flow-induced vibrations? Explain the necessity of carrying out analysis of flow-induced vibration with suitable examples. What is vortex-shedding frequency?

10.2 The vibration-spectrum analysis of a centrifugal pump shows prominent frequencies corresponding to impeller vane passing frequency and its multiple harmonics. What do you infer from this and what further investigation would you suggest?

10.3 Explain the terms: (1) sound pressure, (2) sound power, and (3) decibel.

10.4 What are the sources of noise in boiler furnace? How can we control the noise in the furnace of a boiler?

10.5 What are the various methods of noise control? Explain the concept of passive noise-control system.

APPENDIX - A

Mathematical Relationships

Vibration analyses frequently require referring to some of the relationships from trigonometry, algebra, and calculus. A few of them are given as follows:

$$\sin(\alpha \pm \beta) = \sin\alpha\cos\beta \pm \cos\alpha\sin\beta$$

$$\cos(\alpha \pm \beta) = \cos\alpha\cos\beta \mp \sin\alpha\sin\beta$$

$$\sin(\alpha+\beta)\sin(\alpha-\beta) = \sin^2\alpha - \sin^2\beta = \cos^2\beta - \cos^2\alpha$$

$$\cos(\alpha+\beta)\cos(\alpha-\beta) = \cos^2\alpha - \sin^2\beta = \cos^2\beta - \sin^2\alpha$$

$$\sin\alpha\sin\beta = \frac{1}{2}[\cos(\alpha-\beta) - \cos(\alpha+\beta)]$$

$$\cos\alpha\cos\beta = \frac{1}{2}[\cos(\alpha-\beta) + \cos(\alpha+\beta)]$$

$$\sin\alpha\cos\beta = \frac{1}{2}[\sin(\alpha+\beta) + \sin(\alpha-\beta)]$$

$$\sin\alpha + \sin\beta = 2\sin\left(\frac{\alpha+\beta}{2}\right)\cos\left(\frac{\alpha-\beta}{2}\right)$$

$$\sin\alpha - \sin\beta = 2\cos\left(\frac{\alpha+\beta}{2}\right)\sin\left(\frac{\alpha-\beta}{2}\right)$$

$$\cos\alpha + \cos\beta = 2\cos\left(\frac{\alpha+\beta}{2}\right)\cos\left(\frac{\alpha-\beta}{2}\right)$$

$$\cos\alpha - \cos\beta = -2\sin\left(\frac{\alpha+\beta}{2}\right)\sin\left(\frac{\alpha-\beta}{2}\right)$$

$$A\sin\alpha + B\cos\alpha = \sqrt{A^2+B^2}\cos(\alpha-\phi_1) = -\sqrt{A^2+B^2}\sin(\alpha+\phi_2)$$

where

$$\phi_1 = \tan^{-1}\frac{A}{B}, \quad \phi_2 = \tan^{-1}\frac{B}{A}$$

$$\sin^2\alpha + \sin^2\alpha = 1$$

$$\cos 2\alpha = 1 - 2\sin^2\alpha = 2\cos^2\alpha - 1 = \cos^2\alpha - \sin^2\alpha$$

Law of cosines for triangles: $c^2 = a^2 + b^2 - 2ab\cos C$

$\Pi = 3.14159$ rad, 1 rad = 57.295°, 1° = 0.017453 rad, $e = 2.71828$

$$\log a^b = b \log a,\ \log_{10} x = 0.4343 \log_e x,\ \log_e x = 2{,}3026 \log_{10} x$$

$$e^{ix} = \cos x + i \sin x$$

$$\sin x = \frac{e^{ix} - e^{-ix}}{2i}, \quad \cos x = \frac{e^{ix} + e^{-ix}}{2}$$

$$\sinh x = \frac{1}{2}(e^{x} - e^{-x}), \quad \cosh x = \frac{1}{2}(e^{x} + e^{-x}), \quad \cosh^2 x - \sinh^2 x = 1$$

$$\frac{d}{dx}(uv) = u\frac{dv}{dx} + v\frac{du}{dx}, \quad \frac{d}{dx}\left(\frac{u}{v}\right) = \frac{1}{v}\frac{du}{dx} - \frac{u}{v^2}\frac{dv}{dx} = \frac{v\frac{du}{dx} - u\frac{dv}{dx}}{v^2}$$

$$\int e^{ax}\,dx = \frac{1}{a}e^{ax}$$

$$\int uv\,dx = u\int v\,dx - \int\left(\frac{du}{dx}\int v\,dx\right)dx$$

$$z = x + iy \equiv Ae^{i\theta}, \quad A = \sqrt{x^2 + y^2}, \quad \theta = \tan^{-1}\left(\frac{y}{x}\right)$$

$$z_1 = x_1 + iy_1, \quad z_2 = x_2 + iy_2, \quad z_1 \pm z_2 = (x_1 \pm x_2) + i(y_1 \pm y_2)$$

$$z_1 z_2 = (x_1 x_2 - y_1 y_2) + i(x_1 y_1 + x_2 y_2)$$

$$\frac{z_1}{z_2} = \frac{(x_1 x_2 + y_1 y_2) + i(x_2 y_1 - x_1 y_2)}{\sqrt{x_2^2 + y_2^2}}$$

If $z_1 = e^{i\theta_1}$, $z_2 = e^{i\theta_2}$, then $z_1 + z_2 = Ae^{i\theta}$ and

$$A = [A_1^2 + A_2^2 - 2A_1A_2\cos(\theta_1 - \theta_2)]^{1/2} \text{ and}$$

$$\theta = \tan^{-1}\left[\frac{A_1 \sin\theta_1 + A_2 \sin\theta_2}{A_1 \cos\theta_1 + A_2 \cos\theta_2}\right]$$

$$z_1 z_2 = A_1 A_2 e^{i(\theta_1 + \theta_2)}, \quad \frac{z_1}{z_2} = \frac{A_1}{A_2} e^{i(\theta_1 - \theta_2)}$$

APPENDIX - B

Deflection of Beams and Plates

Beam/Plate	Deflection
Cantilever 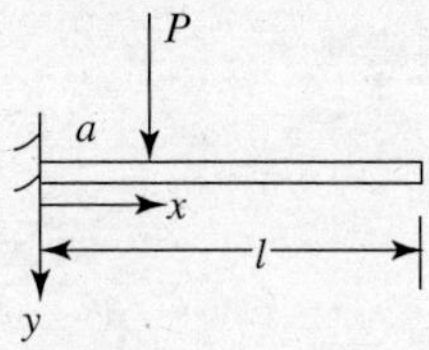	$$y(x)=\frac{Px^2}{6EI}(3a-x);\quad 0\le x\le a$$ $$y(x)=\frac{Pa^2}{6EI}(3x-a);\quad a\le x\le l$$
Simply supported 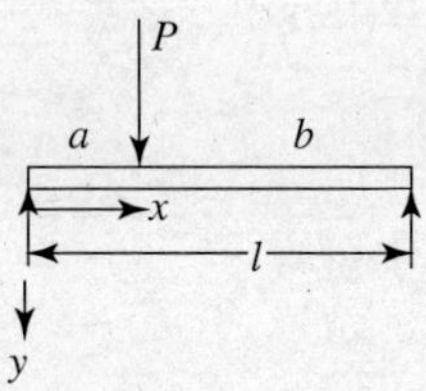	$$y(x)=\frac{Pbx}{6EIl}(l^2-x^2-b^2);\quad 0\le x\le a$$ $$y(x)=\frac{Pa(l-x)}{6EIl}(2lx-x^2-a^2);\quad a\le x\le l$$
Fixed-fixed 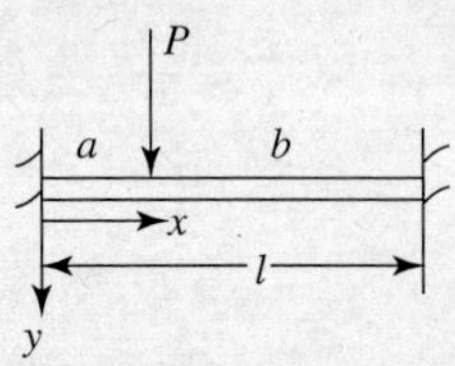	$$y(x)=\frac{Pb^2x^2}{6EIl^3}[3al-x(3a+b)];\quad 0\le x\le a$$ $$y(x)=\frac{Pa^2(l-x)^2}{6EIl^3}[3bl-(l-x)93b+a)];\quad a\le x\le l$$
Simply supported beam with overhang 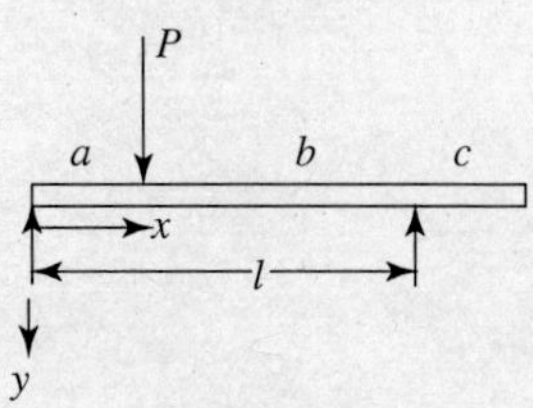	$$y(x)=\frac{Pbx}{6EIl}(l^2-x^2-b^2);\quad 0\le x\le a$$ $$y(x)=\frac{Pa}{6EIl}(l^2-a^2)(x-l);\quad l\le x\le l+c$$
Simply supported with overhanging load 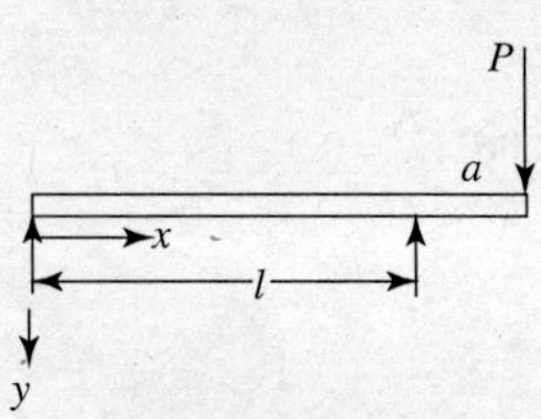	$$y(x)=\frac{Pax}{6EIl}(x^2-l^2);\quad 0\le x\le l$$ $$y(x)=\frac{P(x-l)}{6EIl}\left[a(3x-l)-(x-l)^2\right];\quad l\le x\le l+a$$

(Continued)

Beam/Plate	**Deflection**

Fixed-fixed beam with end-displacement

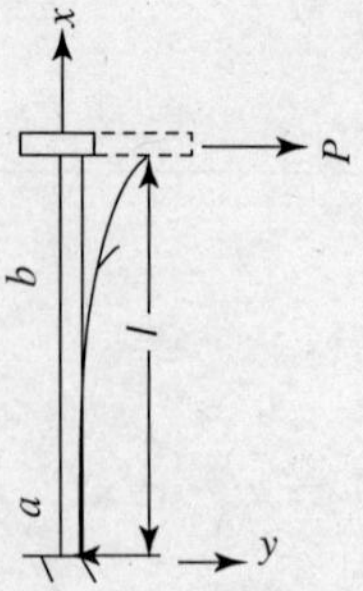

$$y(x) = \frac{P}{12EI}(3lx^2 - 2x^3)$$

Simply supported circular plate

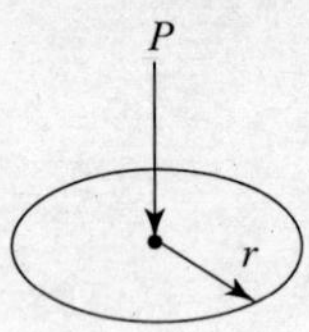

$$y_{centre} = \frac{\Pr^2(3+v)}{16\pi D(1+v)}; \quad D = \frac{Et^3}{12(1-v^2)},$$

t = thickness
v is Poisson's ratio.

Fixed circular plate

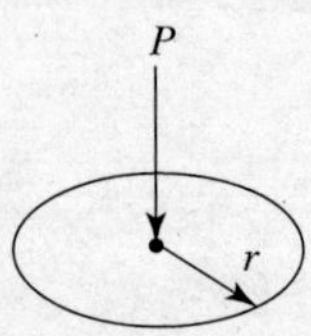

$$y_{centre} = \frac{\Pr^2}{16\pi D}; \quad D = \frac{Et^3}{12(1-v^2)}$$

Square plate simply supported at all sides

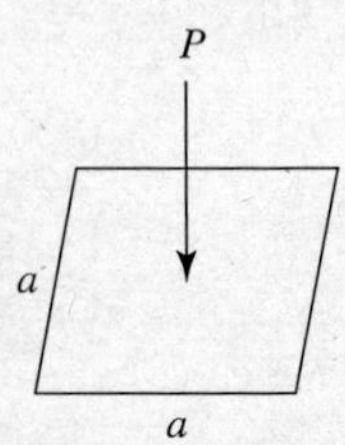

$$y_{centre} = \frac{\alpha Pa^2}{Et^3}; \quad \alpha = 0.1267, v = 0.3$$

Square plate fixed on all sides

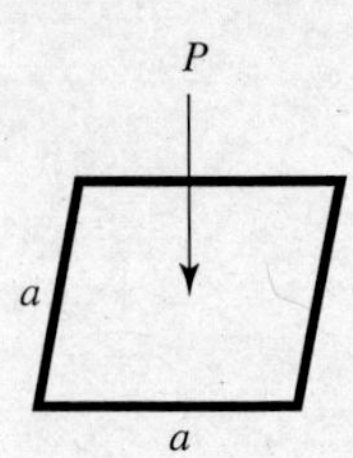

$$y_{centre} = \frac{\alpha Pa^2}{Et^3}; \quad \alpha = 0.0611, v = 0.3$$

APPENDIX - C

Matrices

(I) BASICS

Matrix is a rectangular array of numbers. The array may consist of *m* rows and *n* columns enclosed in brackets. Such an array is called *m* by *n* matrix. A matrix [A] consisting of *m* rows and *n* columns is written as

$$[A] = [a_{ij}] = \begin{bmatrix} a_{11} & a_{12} & . & . & a_{1n} \\ a_{21} & a_{22} & . & . & a_{2n} \\ . & . & . & . & . \\ . & . & . & . & . \\ a_{n1} & a_{n2} & . & . & a_{nn} \end{bmatrix} \tag{C1}$$

In the above form, the numbers a_{ij} are called the **elements** of the matrix A. The first subscript *i* denotes the row and the second subscript *j* denotes the column in which the element a_{ij} appears. The matrix is called **square** matrix when the number of rows *m* is equal to the number of columns *n* ($m = n$). Such a matrix is called a **square matrix of order *n*.**

Column Matrix is a matrix comprising of only one column. This means that we have a $m \times 1$ matrix. Such a matrix is commonly called a **column vector.** Thus, if $\vec{a}$ is a column vector having *m* elements, it can be represented as

$$\vec{a} = \begin{Bmatrix} a_1 \\ a_2 \\ . \\ . \\ a_n \end{Bmatrix} \tag{C2}$$

Row Matrix is a matrix consisting of only one row. This means that we have a $1 \times n$ matrix. Such a matrix is commonly called the **row vector.** Thus, it can be denoted as

$$[b] = \begin{bmatrix} b_1 & b_2 & . & . & b_n \end{bmatrix} \tag{C3}$$

Diagonal Matrix is a matrix in which all the elements are zero except those on the principal diagonal. Thus, if [A] is a diagonal matrix of order *n*, it is written as

$$[A] = \begin{bmatrix} a_{11} & 0 & . & . & 0 \\ 0 & a_{22} & 0 & . & 0 \\ 0 & 0 & a_{33} & . & 0 \\ & & & a_{44} & \\ & & & & a_{nn} \end{bmatrix} \tag{C4}$$

Identity Matrix is a diagonal matrix with all diagonal elements having a value 1. It is usually denoted as **[I]**.

Zero Matrix (Null Matrix) is a matrix having all elements as zero. Thus, a 2×2 zero (null matrix) is given by

$$[0] = \begin{bmatrix} 0 & 0 \\ 0 & 0 \end{bmatrix} \tag{C5}$$

Symmetric Matrix is a square matrix in which the elements in the *i*th row and the *j*th column are the same as the one in *j*th row and *i*th column. The symmetricity of matrix is possible only when it is square, that is, having $a_{ij} = a_{ji}$. For example, a 3 × 3 symmetric matrix can be written as

$$[A] = \begin{bmatrix} 1 & 3 & 5 \\ 3 & 2 & 4 \\ 5 & 4 & 6 \end{bmatrix} \tag{C6}$$

The above matrix is a symmetric matrix of Order 3.

Transpose of Matrix. The transpose of an $m \times n$ matrix [A] is the n × m matrix obtained by interchanging the rows and columns of [A] and is denoted as $[A]^T$. Thus, if

$$[A] = \begin{bmatrix} 2 & 4 & 5 \\ 3 & 1 & 8 \end{bmatrix}$$

Then, $[A]^T$ is given by

$$A^T = \begin{bmatrix} 2 & 3 \\ 4 & 1 \\ 5 & 8 \end{bmatrix}$$

It is obvious that the transpose of a column matrix (vector) will be a row matrix (vector), and vice versa.

Trace. The sum of the main diagonal elements of a **square matrix** $[A] = [a_{ij}]$ is called the trace of [A] and is given by

$$trace\,[A] = a_{11} + a_{22} + \ldots\ldots + a_{nn} \tag{C7}$$

Determinant. If [A] denotes a square matrix of order *n*, then the determinant of [A] is denoted as $\|[A]\|$. Thus,

$$\|[A]\| = \begin{vmatrix} a_{11} & a_{12} & . & . & a_{1n} \\ a_{21} & a_{22} & . & . & a_{2n} \\ . & . & & & \\ & & & & \\ a_{n1} & . & & & a_{nn} \end{vmatrix} \tag{C8}$$

The value of the determinant can be found by obtaining the minors and cofactors of the determinant. It is, therefore, necessary to understand these terms. The *minor* of the element a_{ij} of the determinant $\|[A]\|$ of order *n* is a determinant of the order $(n - 1)$ obtained by deleting the row '*i*' and the column '*j*' of the original determinant. The minor of a_{ij} is denoted as M_{ij}.

The *'cofactor'* of element a_{ij} of the determinant $\|[A]\|$ of order *n* is the minor of the element a_{ij} with either a plus or a minus sign attached. It is defined as follows:

$$\text{Cofactor of } a_{ij} = (-1)^{i+j} M_{ij} \tag{C9}$$

where M_{ij} is the minor of a_{ij}. As an example, consider a determinant det[A].

$$\det[A] = \begin{vmatrix} a_{11} & a_{12} & a_{13} \\ a_{21} & a_{22} & a_{23} \\ a_{31} & a_{32} & a_{33} \end{vmatrix}. \tag{C10}$$

The cofactor of the element a_{32} of det[A]

$$\beta_{32} = (-1)^5 M_{32} = -\begin{vmatrix} a_{11} & a_{13} \\ a_{21} & a_{23} \end{vmatrix} \tag{C11}$$

The second-order determinant ||[A]|| I is evaluated as

$$\det[A] = \begin{vmatrix} a_{11} & a_{12} \\ a_{21} & a_{22} \end{vmatrix} = a_{11}a_{22} - a_{12}a_{21} \tag{C12}$$

The nth-order determinant ||[A]|| is given by

$$\det[A] = \sum_{j=1}^{n} a_{ij}\beta_{ij} \quad \text{for any specific row '}i\text{'}$$

or

$$\det[A] = \sum_{i=1}^{n} a_{ij}\beta_{ij} \quad \text{for any specific column '}j\text{'} \tag{C13}$$

Let us illustrate the procedure by an example:

$$\det[A] = \begin{vmatrix} 2 & 2 & 3 \\ 4 & 5 & 6 \\ 7 & 8 & 9 \end{vmatrix}$$

Let us select the first column for expansion. We obtain

$$\det[A] = 2\begin{vmatrix} 5 & 6 \\ 8 & 9 \end{vmatrix} - 4\begin{vmatrix} 2 & 3 \\ 8 & 9 \end{vmatrix} + 7\begin{vmatrix} 2 & 3 \\ 5 & 6 \end{vmatrix} = -3$$

The evaluation of the determinant of a matrix is of great importance while solving a set of linear algebraic equations, which we get while analyzing a multi degrees-of-freedom vibration systems. It is worthwhile understanding some of the important properties of determinants. They are

- The value of the determinant is not affected if rows (or columns) are written as columns (or rows) in the same order.
- If all the elements of a row (or a column) are zero, then the value of the determinant is zero.
- If any two rows (or columns) are interchanged, the value of the determinant is multiplied by −1.
- If all the elements of one row (or one column) are multiplied by the same constant a, then the value of the determinant is a times the value of the original determinant.
- If the corresponding elements of two rows (or two columns) of a determinant are proportional, then the value of the determinant is zero. As an example, consider a determinant $\det[A] = \begin{vmatrix} 4 & 7 & -8 \\ 2 & 5 & -4 \\ -1 & 3 & 2 \end{vmatrix} = 0$. We can see that the first and the third columns are proportional (by a factor −2).

Adjoint Matrix. The adjoint matrix of a square matrix $[A] = [a_{ij}]$ is defined as the matrix obtained by replacing each element a_{ij} by its cofactor β_{ij} and then transposing. Thus,

$$\text{adjoint}[A] = \begin{bmatrix} \beta_{11} & \beta_{12} & . & . & \beta_{1n} \\ \beta_{21} & \beta_{22} & & & \beta_{2n} \\ . & & & & \\ . & & & & \\ \beta_{n1} & \beta_{n2} & . & . & \beta_{nn} \end{bmatrix}^T = \begin{bmatrix} \beta_{11} & \beta_{21} & . & . & \beta_{n1} \\ \beta_{12} & \beta_{22} & . & . & \beta_{n2} \\ . & & & & \\ . & & & & \\ \beta_{1n} & \beta_{2n} & . & . & \beta_{nn} \end{bmatrix} \tag{C14}$$

We now deal with a very important aspect of the matrix analysis. This is an inverse of the matrix which is extremely important in arriving at a frequency equation, modal shapes, and, in general, modal analysis.

Inverse Matrix. The inverse of matrix has meaning only when it is a square matrix. The inverse of a square matrix [A] is written as $[A]^{-1}$ and is defined by the following relationship:

$$[A]^{-1}[A] = [A][A]^{-1} = [I] \tag{C15}$$

where $[A]^{-1}[A]$ denotes the product of the matrix $[A]^{-1}$ and [A]. In practice, the inverse matrix of [A] is evaluated by

$$[A]^{-1} = \frac{\text{Adjoint}[A]}{\det[A]} \tag{C16}$$

The inverse exists only when the determinant det[A] is not equal to zero. See, for example, the determinant

$$\det[A] = \begin{vmatrix} 4 & 7 & -8 \\ 2 & 5 & -4 \\ -1 & 3 & 2 \end{vmatrix} = 0$$

Let us now illustrate how the inverse of a matrix is evaluated. We consider a matrix

$$[A] = \begin{bmatrix} 2 & 2 & 3 \\ 4 & 5 & 6 \\ 7 & 8 & 9 \end{bmatrix}.$$

The det[A] = −3. The cofactor of a_{11} is

$$\beta_{11} = (-1)^{2|\,|}\begin{vmatrix} 5 & 6 \\ 8 & 9 \end{vmatrix} = -3.$$

In a similar fashion, we can find the other cofactors.

We shall obtain the adjoint of matrix [A] as

$$[\text{adjoint } A] = \begin{bmatrix} -3 & 6 & -3 \\ 6 & -3 & 0 \\ -3 & -2 & 2 \end{bmatrix}.$$

Thus,
$$[A]^{-1} = [\text{Adjoint } A/\det[A]] = \begin{bmatrix} 1 & -2 & 1 \\ -2 & 1 & 0 \\ 1 & 2/3 & -2/3 \end{bmatrix}$$

Singular Matrix. A square matrix is considered as singular when its determinant is zero.

(II) BASIC MATRIX OPERATIONS

Equality of Matrices. Two matrices [A] and [B], having the same order, are equal if only if $a_{ij} = b_{ij}$ for every '*i*' and '*j*'.

Addition and Subtraction of Matrices. The sum of two matrices [A] and [B] having the same order, is given by the sum of corresponding elements. This means that if [C] = [A] +[B], then $c_{ij} = a_{ij} + b_{ij}$. Similarly, the difference between two matrices [C] = [A] − [B], means that $c_{ij} = a_{ij} - b_{ij}$.

Multiplication of Matrices. The product of two matrices [A] and [B] is defined only if they are conformable, which means that the number of columns of [A] is equal to number of rows of [B]. If [A] is of order $m \times n$ and [B] is of order of $n \times p$, then the product $[C] = [c_{ij}] = [A]\,[B]$ is of the order of $m \times p$ with

$$c_{ij} = \sum_{k=1}^{n} a_{ik} b_{kj} \tag{C17}$$

This means that c_{ij} is the quantity obtained by multiplying the ith row of [A] and jth column of [B] and summing these products. For example, if

$$[A] = \begin{bmatrix} 2 & 3 & 4 \\ 1 & -5 & 6 \end{bmatrix}; \quad [B] = \begin{bmatrix} 8 & 0 \\ 2 & 7 \\ -1 & 4 \end{bmatrix},$$

then we observe that they are multipliable as the number of columns in [A] is equal to number of rows in [B]. Applying Equation (C17), we get

$$[C] = [A][B] = \begin{bmatrix} 18 & 37 \\ -8 & -11 \end{bmatrix}.$$

It is obvious that

$$[B][A] \neq [A][B] \tag{C18}$$

If the matrices are conformable, then the matrix-multiplication process is associative

$$([A][B])[C] = [A]([B][C]) \tag{C19}$$

and is distributive

$$([A] + [B])[C] = [A][C] + [B][C] \tag{C20}$$

The product [A][B] denotes the pre-multiplication of [B] by [A] or the post-multiplication of [A] by [B].

The transpose of a matrix product can be found by the product of the transposes of the separate matrices in the reverse order. Thus, if [C] = [A][B], then the transpose of [C] is given by

$$[C]^T = ([A][B])^T = [B]^T[A]^T \tag{C21}$$

The inverse of a matrix product can be determined from the product of the inverse of the separate matrices in the reverse order. Thus, if [C] = [A][B], then

$$[C]^{-1} = ([A][B])^{-1} = [B]^{-1}[A]^{-1} \tag{C22}$$

Note that [A]/[B] is not definable. But we can always find $[C] = [A][B]^{-1}$ as $[B]^{-1}$ is definable and can always be evaluated.

APPENDIX - D

Some Important Issues in Signal Processing

The vibration-response signal, obtained from the vibration transducer, after suitably conditioning, is sent to an analyser. A commonly used analyser is called Fast Fourier Transform (FFT) analyser. Such an analyser receives analogue-voltage signals, which may represent displacement, velocity, or acceleration/strain from a signal-conditioning unit. It computes the discrete frequencies of the individual signals. The most important question is about the truthfulness of the output data and the assurance that the analysis does not contain analysis of any no-rogue signals. The data received by the analyser essentially has several samples of the data from the transducer and the signal-conditioning units.

If n samples of the output $x(t)$ are taken at discrete values of time t_i, the data ($[x_1(t_i), x_2(t_i) \dots x_n(t_i)]$ can be used to obtain the discrete form of Fourier transforms as

$$x_j = x_j(t) = \frac{a_0}{2} + \sum_{i=1}^{N/2}\left(a_i \cos\frac{2\pi i t_j}{T} + b_i \sin\frac{2\pi i t_j}{T} \right); \quad j = 1, 2, \dots N. \tag{D1}$$

The digital spectral coefficients are given by

$$a_0 = \frac{1}{N}\sum_{i=1}^{N} x_j, \tag{D2}$$

$$a_i = \frac{1}{N}\sum_{j=1}^{N} x_j \cos\frac{2\pi i t_j}{N}, \tag{D3}$$

$$b_j = \frac{1}{N}\sum_{j=1}^{N} x_j \sin\frac{2\pi i t_j}{n} \tag{D4}$$

Figure D1 shows the representation of signals in different forms.

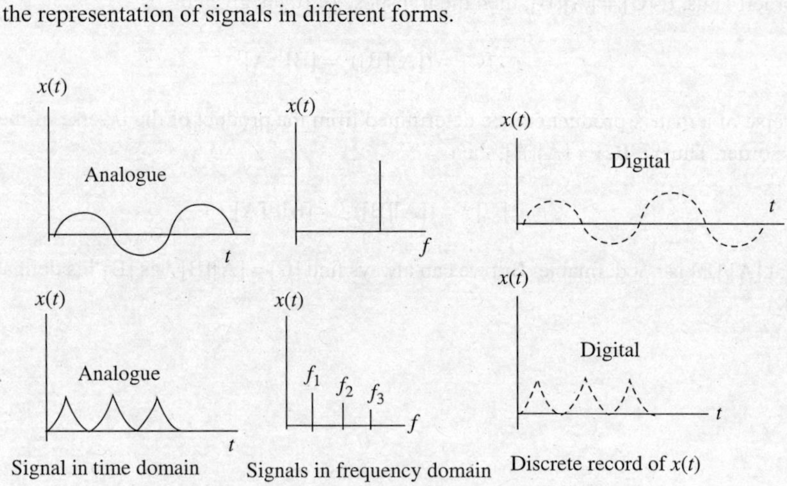

Figure D1 Time, Discrete Time, and Frequency Spectra

The number of samples could be 256, 512, or 1024 depending upon the analyser. The Equations D2, D3, and D4 denote N algebraic equations for each of the N samples. These equations can be expressed in matrix form as

$$\vec{X} = [A]\vec{d} \tag{D5}$$

$\vec{X} = \{x_1 x_2 ... x_n\}$ is the vector of the samples, $\vec{d} = \{a_0 a_1 a_2 ... a_{N/2} b_1 b_2 ... b_{N/2}\}^T$ is the vector of the spectral coefficients, and [A] is the matrix composed of the coefficients $\cos\frac{2\pi i t_j}{T}$ and $\sin\frac{2\pi i t_j}{T}$ of Equation D1. The frequency content of the signal or response of the system can be obtained from the solution of the equation

$$\vec{d} = [A]^{-1}\vec{X} \tag{D6}$$

where $[A]^{-1}$ is computed by the analyser using FFT.

One of the major problems in the input and the output data seen by the transducer is the presence of some random component or noise. Due to this, it is rather somewhat difficult to analyse the data in a deterministic manner. Also, the use of random excitation force is common in vibration testing. If *x*(*t*) is a random signal as shown in Figure D2, its average or mean is denoted as

$$\bar{x} = \lim_{T\to\infty} \frac{1}{N}\int_0^T x(t)\,dt \tag{D7}$$

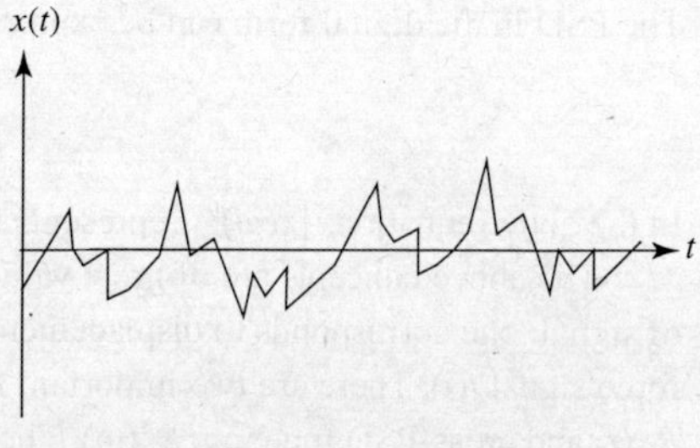

Figure D2 Random Signal x(t)

The above equation for a digital signal can be expressed as

$$\bar{x} = \lim_{N\to\infty} \frac{1}{N}\sum_{j=1}^{N} x(t_j). \tag{D8}$$

We can assume that the random signal *x*(*t*) has a zero mean. Hence, we define the mean-square value or variance of *x*(*t*) as

$$\bar{x}^2(t) = \lim_{N\to\infty} \frac{1}{T}\int_0^T x^2(t)\,dt \tag{D9}$$

We, for digital signal, can write the above equation as

$$\bar{x}^2 = \lim_{N\to\infty} \frac{1}{N}\sum_{j=1}^{N} x^2(t_j) \tag{D10}$$

The root-mean square (RMS) value of *x*(*t*) is given by

$$X_{\text{RMS}} = \sqrt{\bar{x}^2} \tag{D11}$$

The autocorrelation function *R*(*t*) of a random signal *x*(*t*) gives a measure of the speed with which the signal changes in time domain and is defined as

$$R(t) = x^2 = \lim_{T\to\infty} \frac{1}{T}\int_0^T x(\tau)x(t+\tau)\,d\tau \tag{D12}$$

For a digital signal, the above equation can be written as

$$R(n.\Delta t) = \frac{1}{N-n}\sum_{j=0}^{N-n} x_j x_{j+n} \tag{D13}$$

In the above equation, N is the number of samples, Δt is the sampling interval, and n is the adjustable parameter which can be used to control the number of points used in the computation. $R(0)$ represents the mean-square value ($\overline{x}^2$) of $x(t)$.

The autocorrelation function can be used for identifying the presence of periodic components buried in a random signal. Thus, if $x(t)$ is purely random, then $R(t) \rightarrow 0$ as $T \rightarrow \infty$. However, if $x(t)$ contains periodic components (as they exist in the vibration response), $R(t)$ will also be periodic.

The **Power Spectral Density (PSD)** of a random signal $x(t)$, denoted by $S(\omega)$, gives a measure of the speed with which the signal changes in the frequency domain and is defined as the Fourier transform of $R(t)$.

$$S(\omega) = \frac{1}{2\pi} \int_{-\infty}^{\infty} R(\tau) e^{-i\omega\tau} \, d\tau \tag{D14}$$

The PSD in the digital form can be expressed as

$$S(\Delta\omega) = \frac{|x(\omega)|^2}{N \Delta t} \tag{D15}$$

In the above equation, $|x(\omega)|^2$ represents the magnitude of the Fourier transform of the sampled data of $x(t)$.

The above concepts are of great value in vibration testing. In vibration testing, we have two different types of signal: one corresponds to displacement/acceleration response $x(t)/\ddot{x}(t)$ and the other corresponds to applied-force signal $f(t)$. There are two important functions relating them. They are called (a) cross-correlation function, $R_{xf}(t)$ and cross-PSD function, $S_{xf}(\omega)$. They are expressed as

$$R_{Xf}(t) = \lim_{T \to \infty} \int_{0}^{T} x(\tau) f(\tau + t) \, d\tau \tag{D16}$$

and

$$S_{xf}(\omega) = \frac{1}{2\pi} \int_{-\infty}^{\infty} R_{xf}(\tau) e^{-i\omega\tau} \, d\tau \tag{D17}$$

Equations D16 and D17 enable us to determine the transfer function of the structure or a component being tested. If we replace $f(\tau + t)$ by $x(\tau + t)$, we obtain $R_{xx}(t)$. We now substitute $R_{xx}(t)$ in Equation D17. This will give us $S_{xx}(\omega)$. The frequency-response function, $H(i\omega)$, which we discussed in the chapter of experimental-modal analysis, is related to the PSD function as

$$S_{xx}(\omega) = |H(i\omega)|^2 S_{ff}(\omega) \tag{D18}$$

$$S_{fx}(\omega) = H(i\omega) S_{ff}(\omega) \tag{D19}$$

$$S_{xx}(\omega) = H(i\omega) S_{xf}(\omega) \tag{D20}$$

In Equations (D18–D20), $f(t)$ and $x(t)$ denote the random force input and the resulting output, respectively. $S_{xx}(\omega)$ given by Equation D18 contains information about the magnitude of the transfer function of the system (structure or a component), while $S_{fx}(\omega)$ and $S_{xx}(\omega)$ given by Equations D19 and D20 contain information about both magnitude and phase. In vibration testing, the spectrum analyser first computes different spectral-density functions from the transducer output and then computes the frequency-response function $H(i\omega)$ of the system using Equations D19 and D20.

Coherence Function (β) is a function which measures the extent of noise in the signal. It is defined as

$$\beta(\omega) = \left(\frac{S_{fx}(\omega)}{S_{ff}(\omega)} \right) \left(\frac{S_{xf}(\omega)}{S_{xx}(\omega)} \right) = \frac{|S_{xf}(\omega)|^2}{S_{xx}(\omega) S_{ff}(\omega)} \tag{D21}$$

If the measurements of x and f are only noises, then $\beta = 0$ and if the measurements of x and f are totally uncontaminated, $\beta = 1$. Experience shows that at natural frequencies β is ~ 1 because signals are large and are less influenced by the noise.

INDEX

W